Kazuo Yamazaki

Partial Differential Equations in Fluid Mechanics

De Gruyter Series in Nonlinear Analysis and Applications

Volume 48

Kazuo Yamazaki

Partial Differential Equations in Fluid Mechanics

Deterministic and Stochastic Models

DE GRUYTER

Mathematics Subject Classification 2020
Primary: 35A02, 35R60, 35Q35; Secondary: 76D05, 76F30

Author
Ass. Prof. Kazuo Yamazaki
University of Nebraska-Lincoln
Department of Mathematics
Avery Hall 203
68588-0130 Lincoln, NE
USA
kyamazaki2@nebraska.edu

ISSN 0941-813X
ISBN 978-3-11-134975-6
ISBN 978-3-11-134998-5 (PDF)
ISBN 978-3-11-135018-9 (E-PUB)
DOI https://doi.org/10.1515/9783111349985

Library of Congress Control Number: 2026943012

Bibliographic information published by the Deutsche Nationalbibliothek
The Deutsche Nationalbibliothek lists this publication in the Deutsche Nationalbibliografie; detailed bibliographic data are available on the Internet at https://dnb.dnb.de.

De Gruyter and Walter de Gruyter GmbH are part of De Gruyter Brill.
www.degruyterbrill.com

Questions about General Product Safety Regulation: productsafety@degruyterbrill.com

Printing: CPI books GmbH, Leck

MIX
Papier | Fördert gute Waldnutzung
FSC® C083411

Contents

Preface

There are many excellent textbooks and lecture notes on partial differential equations (PDEs) of fluid mechanics in the literature (e. g., [97, 259, 318, 319]). Some greatly emphasize applications of such PDEs in harmonic analysis (e. g. [12, 79, 236]) and stochastic analysis (e. g., [33, 117, 118, 125, 140]).

The motivation of this monograph is three-fold.

1. It uniquely combines the tools of harmonic and stochastic analysis applied to PDEs.
2. It is timely in discussing recent developments of the convex integration technique with focus on stochastic partial differential equations (SPDEs), PDEs of fluid mechanics with random external force (see Chapter 7). Although excellent surveys on the convex integration technique applied to deterministic PDEs have appeared in the literature (e. g., [48, 49]), the author is not aware of such in the stochastic case except short surveys in [129, 365].
3. It touches singular SPDEs, which are, informally stated, PDEs forced by rough random force such that the solution becomes rough and consequently the product within its nonlinear term becomes ill-defined by Bony's paraproducts (see Corollary 8.1.5). Specifically, we discuss their recent developments toward global-in-time solution theory in both unique and nonunique cases, which is currently an active research direction of importance and relies heavily on both harmonic and stochastic analysis tools (see Chapters 8–9).

Additionally, from the author's viewpoint, there are at least two remarkable connections and potentials by combining the harmonic and stochastic analysis techniques and applying them to singular SPDEs.

1. The solutions constructed via convex integration are generally irregular, whereas the solutions to singular SPDEs, for many of which even their existence locally in time is an open problem, are always irregular. Most theories of singular SPDE solutions, if any, start by mollifying the noise by its convolution with a smooth cut-off function (see Chapter 8). The author believes that convex integration, which starts instead by approximation via adding an external force that vanishes in the limit, may shed light and even be able to construct solutions in the locally critical/supercritical cases (see Definition 8.5.2). In fact, solutions constructed via convex integration can have higher regularity than the solutions constructed otherwise (see Section 9.2.2).
2. Tao [314] initiated the global regularity phenomena of logarithmically supercritical PDEs in the deterministic case (see Section 5.2). Heuristically, analogously to the Brezis–Wainger inequality (see Section 3.4), logarithmic divergence can be overcome. Completely independently, Hairer and Rosati [176] realized that for singular SPDEs with Lions' critical threshold (see (5.5)), renormalization constant grows logarithmically, which can be overcome using techniques from Anderson Hamiltonian

https://doi.org/10.1515/9783111349985-203

(see Chapter 9). In both [314] and [176], to handle the logarithmic divergence, splitting the Littlewood–Paley dyadic partition (see (3.43)) with time-dependent threshold has played a key role.

Besides reviewing the recent developments on such important research areas, the monograph is intended to be as accessible as possible to a wide audience, including graduate students with interests and basic background in graduate-level PDEs (e. g., [132]) and measure theory (e. g., [32, 147, 294]). For this reason, Chapters 2, 3, 4, and 8 are devoted to preliminaries respectively on measure theory and functional analysis, harmonic analysis, stochastic analysis, and singular SPDEs; the author intentionally separated Chapter 8 because the studies on singular SPDEs require more technical background.

Trying to be relatively complete, we mention many preliminary results without proofs, only referring interested readers to appropriate resources, so that the monograph focuses on key results that are especially effective and useful. We hope that some of our tricks and simple computational techniques, even if not sharp, will be useful for some readers. Moreover, although there are many interesting research directions in the analysis of deterministic PDEs, such as finite-time blowups and inviscid damping, and SPDEs, such as ergodicity and large deviation principle, this monograph is intended to focus mostly on the global-in-time existence, uniqueness, and nonuniqueness of relatively weak solutions, especially because even the local-in-time existence of solutions is not known for many singular SPDEs.

Specifically, we outline the structure of the monograph as follows.

(a) For readers' convenience, we start with a table of symbols and notation used in the book.
(b) In Chapter 0, we collect the basic abbreviations, notations, and conventions, divided into three categories: Analysis, PDE, and Stochastic.
(c) Chapter 1 starts with introduction of many PDEs of relevance split into two types; PDEs in fluid mechanics in Section 1.1 and SPDEs in mathematical physics in Section 1.2. The first group consists of the following equations: Navier–Stokes equations, Euler equations, Burgers' equation, surface quasi-geostrophic (SQG) equations, Boussinesq system, magnetohydrodynamics (MHD) system, Hall-MHD system, electron-MHD system, and hyperbolic Navier–Stokes equations. The second group consists of the following equations: Kardar–Parisi–Zhang (KPZ) equation, parabolic Anderson model, stochastic Yang–Mills equation, Φ^4 model, and Cahn–Hilliard equation. This monograph, as the title suggests, mostly focuses on the PDEs of fluid mechanics; however, because the SPDEs in mathematical physics arise naturally in the discussion of singular SPDEs in Chapters 8–9, we also introduce them. Higher-dimensional Burgers, Kuramoto–Sivashinsky, transport, nonlinear Schrödiner (NLS), and wave equations are also briefly introduced subsequently in (6.53), (6.54), (7.1), (6.8), and (5.15), respectively.

(d) Chapters 2, 3, and 4 are devoted to sufficient, sometimes more than enough, preliminaries on measure theory, functional analysis, harmonic analysis, and stochastic analysis. Chapter 3 on harmonic analysis will be noticeably more in depth because advanced techniques and fluency in this area are crucial for subsequent discussions. For example, within the research on singular SPDEs, the theory of paracontrolled distributions (see Chapters 8–9) heavily relies on harmonic analysis tools such as Besov spaces, Bony's paraproduct resonances, and commutator estimates. On the other hand, preliminaries on advanced stochastic analysis needed for singular SPDEs will be deferred to Chapter 8, so that we can jump into research discussions in Chapters 5–7 and avoid delaying research discussions too long.

    To make the monograph accessible to a wide range of readers, Section 3.10.1 includes a complete proof of the global-in-time existence of weak solutions to deterministic Cahn–Hilliard equation using the Galerkin approximation. Additionally, Section 4.5.1 includes a complete proof of the global-in-time existence of a martingale solution to stochastic momentum SQG equations using the probabilistic Galerkin approximation. Motivation for such additions of classical techniques is to clarify the sharp contrast to the solutions to be constructed by convex integration technique in Chapters 6–7.

(e) Chapter 5 discusses the global regularity issue of the deterministic MHD and Hall-MHD systems, as well as logarithmically supercritical PDEs. Four open problems are discussed on the MHD and Hall-MHD systems (see Problems 1–4). We discuss the logarithmically supercritical PDEs due to their interesting connection with our discussions in Chapter 10 on the global-in-time solution theory of singular SPDEs for which the renormalization constant grows at most logarithmically.

(f) Chapter 6 starts with a brief history of Onsager's conjecture, Taylor's conjecture, anomalous dissipation, and Nash–Kuiper theorem. We describe the pioneering works of Scheffer, Shnirelman, and Christ, especially those of Christ, because they interestingly consist of deterministic analogues of renormalization and Wick products (see Section 6.1.3). Then we review the developments in the past two decades concerning the adaptation of the convex integration to deterministic PDEs in fluid mechanics. It includes some preliminaries for the convex integration, an example of a theorem and its proof for hyperbolic Navier–Stokes equations, and concludes with two open problems (see Problems 5–6).

(g) Chapter 7 discusses recent developments concerning the application of the convex integration technique to SPDEs in case the noise is white only in time. We describe three types of noise: additive, linear multiplicative, and transport. We briefly describe a few difficulties that arise upon attempts to apply convex integration to SPDEs. This research area is, at the time of writing this monograph, flooded with new results, and we conclude by listing various new recent works, all of which we cannot completely explain.

(h) Chapter 8 discusses preliminaries on singular SPDEs. We discuss the notion of locally subcritical SPDEs introduced by Hairer (see Definition 8.5.2), critical dimensions, precise regularity of noise such as STWN, Young's integral, rough path theory, Wick products, renormalization constants, and Da Prato–Debussche trick applied on singular SPDEs. We describe the breakthrough theories of regularity structures (RS) by Hairer [167] and paracontrolled distributions by Gubinelli, Imkeller, and Perkowski [159] with an example on the 3D MHD system in Section 8.5.3. We suggest one open problem (see Problem 7).
(i) Chapter 9 discusses recent developments in the global-in-time solution theory of singular SPDEs, from both uniqueness and nonuniqueness perspectives. The uniqueness direction focuses on the recent approach of Hairer and Rosati [176] with an example on the 2D MHD system in Section 9.1.2. The nonuniqueness direction describes [189, 190, 183] via convex integration technique. We conclude with two open problems (see Problems 8–9).
(j) Chapter 10 concludes with a review of Chapters 1–8 and lists nine open problems discussed previously for readers' convenience. Again, we acknowledge that there are many important interesting open problems in various research areas of PDEs in fluid mechanics. Those listed in Chapter 10 represent only a handful of most relevance to discussions in this monograph.
(k) Appendices A and B include detailed computations and additional preliminaries, which we chose to defer for better organization.
(l) For readers' convenience, the appendices are followed by an index of terminology used in this monograph.

Although the monograph reviews many previous results, the following seem to be new: a commutator estimate in Proposition 3.8.3, the Galerkin approximation of the deterministic Cahn–Hilliard equation with minimum regularity for existence in Section 3.10.1, and the probabilistic Galerkin approximation of the generalized momentum SQG equations in Section 4.5.1 (see Remark 7.3.6). We also often include details that were omitted in previous publications, e. g., Section A.9.

Acknowledgments

The author thanks Prof. Jie Xiao for his generous invitation to write this monograph. The author also thanks Prof. Jiahong Wu for his valuable teaching on PDEs in fluid mechanics, Prof. David Ullrich for teaching the author countless number of tricks in analysis, especially harmonic analysis, and Prof. Carl Mueller for always willing to teach the author stochastic analysis and SPDEs. The author thanks Prof. Anna Mazzucato, Prof. Hongjie Dong, and Prof. Adam Larios for consistent support throughout the author's career, and Prof. Matthew Novack for valuable discussions related to some of the contents within this monograph. Upon consulting with Prof. Jiahong Wu and Prof. Anna Mazzucato regarding writing this monograph, the author received kind encouragement, without which this monograph would not have been written. The author extends his thanks to the PDE group at the University of Nebraska-Lincoln, namely Prof. George Avalos, Prof. Mikil Foss, and Prof. Petronela Radu, his past Ph.D. student Dr. Rahman Mohammad Mahabubur, and Prof. Ujjwal Koley, Prof. Manil Mohan, and Elliott Walker for fruitful collaborations. The author also thanks a countless number of other colleagues from whom he learned so much through discussions and students from whom he also learned so much through teaching and discussions. Finally, the author thanks his wife for her constant support.

The author gratefully acknowledges support from the National Science Foundation (NSF Award No. 2531744) and Simons Foundation (MPS-TSM-00962572). The author also benefited from valuable discussions with the participants of "Conference: EPSCoR States Workshop for Research and Workforce Development in Fluid Mechanics," supported by NSF Grant No. 2436937.

https://doi.org/10.1515/9783111349985-204

Table of symbols and notations

For convenience of readers, we include a table of symbols and notations used in the monograph.

Tab. 1: Symbols.

Symbols	Description/purpose	Location
$\bullet, \vert, \wr$	Certain tree diagrams involving space-time white noise	p. 171
$\prec, \succ, \circ$	Bony's paraproducts	(3.83a)
$A \lesssim_{\alpha,\beta} B$	$A \leq C(\alpha,\beta)B$	p. 2
$A \overset{(\cdot)}{\lesssim} B$	$A \lesssim B$ due to $(\cdot)$	p. 2
$A \lesssim_{\alpha,\beta} B$	A is approximately B	p. 2
$\alpha(d)$	Volume of a d-dimensional ball	p. 2
$[B,C] = BC - CB$	Lie bracket or commutator of operators	p. 53 and p. 10
$\mathcal{B}$	Inverse divergence operator	(6.14)
$\mathcal{B}_X$	Borel σ-algebra	p. 1
$B_b(H)$	Set of all bounded Borel functions from H to $\mathbb{R}$	Definition 4.6.6(a)
$B_r(x) = B(x,r)$	Ball of radius r centered at x	p. 1
$B_{\mathfrak{s}}(0,1)$	Ball of radius 1 in the distance $d_{\mathfrak{s}}$	p. 158
$\dot{B}^s_{p,r}(\mathbb{R}^d)$, $B^s_{p,r}(\mathbb{R}^d)$	Homogeneous and inhomogeneous Besov spaces	(3.46)
$(\dot{B}^{s_1}_{p,q_1})_h(\dot{B}^{s_2}_{p,q_2})_v$	Anisotropic Besov spaces	p. 52
$C(0,r_1,r_2)$	Annulus of inner and outer radii r_1 and r_2	p. 1
$\mathcal{C}$	Annulus	(3.36)
$C_b(H)$	Set of all continuous bounded functions from H to $\mathbb{R}$	Definition 4.6.6(b)
$C^s(U)$	Hölder space over U	p. 15
$C_0^\infty(\mathbb{R}^d)$	Set of smooth functions with decay	(2.9)
$C^s_c(U), C_b(U), C^\infty(U)$	Spaces related to continuous functions over U	(2.8)
$\mathcal{C}^\alpha$	Besov–Hölder space	(3.50)
$\mathcal{C}^\alpha_{\mathfrak{s}}$	Besov–Hölder space w. r. t. $\mathfrak{s}$	Definition 8.1.1
$\mathcal{C}(f,g,h)$	Special commutator	(3.103)
$\mathrm{Cov}(X), \mathrm{Cor}(X,Y)$	Covariance and correlation operators	(4.2a), (4.2b)
$\lceil x \rceil$	Smallest integer equal to or larger than x	p. 1
Δ	Laplacian	p. 2
δ_λ	Dilation by $\lambda > 0$	(3.2)
δ_0	Dirac delta distribution at the origin	Example 1
$\Delta_j, \dot{\Delta}_j, S_j, \dot{S}_j$	Littlewood–Paley operators and lower frequency cut-off	(3.41)
$d_{\mathfrak{s}}((t,x),(s,y)), \vert\mathfrak{s}\vert$	Distance w. r. t. $\mathfrak{s}$, scaling dimension w. r. t. $\mathfrak{s}$	(8.2)
$\mathbb{E}, \mathbb{E}^{\mathbb{P}}$	Mathematical expectation w. r. t. $\mathbb{P}$	Definition 4.1.1
$\mathbb{E}[X\vert\mathcal{H}]$	Conditional expectation	Definition 4.1.4
$\mathcal{E}(t)$	Energy of MHD	(5.3a)
$f(t+), f(t-)$	Limits from right and left	p. 2
$\fint_X f(x)\,dx$	Average value of f over X	p. 2
$\mathcal{F}f = \widehat{f}$	Fourier transform of f	(3.1)
$\mathcal{F}^{-1}f$	Fourier inverse of f	(3.6)

https://doi.org/10.1515/9783111349985-205

Tab. 1 (continued)

Symbols	Description/purpose	Location
$\{\mathcal{F}_t\}_{t\geq 0}$	Filtration	Definition 4.1.2
$\mathcal{F}(\{v^m(\tau)\}_{\tau\in[0,T]})$	Filtration generated by v^m	(4.5.1)
$\nabla\times$	Curl operator	p. 2
$(\nabla_{\text{sym}}\phi)_{i,j}, (\nabla_{\text{anti}}\phi)_{i,j}$	Symmetric and antisymmetric gradients	(9.17)–(9.18)
$H_n(x)$	Hermite polynomials	(4.72)
$\mathcal{H}_n$	Special subspace of $L^2(\Omega,\mathcal{F},\mathbb{P})$	p. 97
$\mathcal{H}_b(t), \mathcal{H}_u(t)$	Magnetic and cross helicities	(6.5)
$\dot{H}^s(\mathbb{R}^d)$	Homogeneous Sobolev space	(3.13)
$H^s(\mathbb{R}^d)$	Inhomogeneous Sobolev space	(3.15)
$\dot{H}^{s,s'}$	Anisotropic Sobolev space	p. 52
H^s_σ	H^s with divergence-free property	(3.105)
$\mathcal{H}$	Hilbert transform	Definition 3.4.3
$\mathcal{H}^1(\mathbb{R}^d)$	Hardy space	p. 47
$\check{\mathfrak{h}}_\lambda, \check{\mathfrak{l}}_\lambda, \mathcal{H}_\lambda, \mathcal{L}_\lambda$	Lower and higher frequency projections	(8.43)–(8.44)
$\mathcal{H}_0$	Set of all linear combinations of $\mathcal{F}_0$	p. 187
$\mathcal{I}$	Abstract integration map in regularity structures	(8.5.10)
Id	Identity matrix	p. 1
$I_m(f)$	Multiple stochastic integrals	p. 166
$\int_0^t f(s)\circ dW(s)$	Stratonovich integral	Remark 4.2.10
$\mathcal{J}=(\mathcal{A},\mathcal{T},\mathcal{G})$	Regularity structures	Definition 8.5.7
$L(X,Y)$	Set of all bounded linear maps from X to Y	p. 17
$L_1(H)$	Space of trace class operators	(2.21)
$L_2(H,H), L_2(H)$	Space of Hilbert–Schmidt operators	(2.22)
$L^p(U;\mathbb{R})$	Lebesgue space over U	p. 14
$L^p(\Omega)$	Lebesgue space over Ω	(4.1)
$l^p(\mathbb{Z})$	Lebesgue space with counting measure	p. 14
$\mathcal{L}(X)$	Law of a random variable X	p. 68
$\|f\|_{L^p(U)}$	L^p-norm	(2.5)
$\|f\|_{C^s}$	C^s-norm	(2.7)
$[f]_{\dot{C}^s}, [f]_{\dot{C}^{s,\gamma}(\overline{U})}$	Hölder seminorm	(2.10)
$\|f\|_{C^{s,\gamma}(\overline{U})}$	Hölder norm	(2.11)
$\|u\|_{H^s(\mathbb{T}^d)}, \|u\|_{\dot{H}^s(\mathbb{T}^d)}$	H^s-norm on torus	(3.16)
$\|u\|_{\dot{B}^s_{p,r}}, \|u\|_{B^s_{p,r}}$	Norms of $\dot{B}^s_{p,r}(\mathbb{R}^d)$ and $B^s_{p,r}(\mathbb{R}^d)$	(3.46)
$\|(t,x)-(s,y)\|_{\mathfrak{s}}$	Almost-norm w. r. t. $\mathfrak{s}$	(8.3)
$\|a\|_a$	Norm of components of a in T_a	Definition 8.5.7
$\mathcal{M}(H)$	Space of all bounded measures on $(H,\mathcal{B}_H)$	p. 98
$\mathbb{M}^d$	Set of all $d\times d$ matrices	p. 1
$\mathcal{M}^2_T(E)$	Space of E-valued continuous square-integrable martingales M such that $M(0,\cdot)=0$ $\mathbb{P}$-a. s.	p. 71
$\mathbb{N}$	Set of natural numbers	p. 1
$\mathbb{N}_0$	Set of natural numbers and zero	p. 1
$\mathcal{N}$	Normal distributions	p. 72
$(\Omega,\mathcal{F},\mathbb{P})$	Probability space	p. 68
$\otimes$	Tensor matrix	p. 1

Tab. 1 (continued)

Symbols	Description/purpose	Location
$X^{\otimes 2}$	$X \otimes X$	p. 171
$\mathring{\otimes}$	Tensor matrix minus its trace	(6.12)
$f \otimes_r g$	Contraction of r indices of f and g	Definition 8.3.1
$u \otimes_s v, u \otimes_a v$	Symmetric and antisymmetric tensors	(9.15)–(9.16)
p'	Hölder's conjugate of p	(2.3)
∂_i	Partial derivative w. r. t. x_i	p. 2
∂_t	Partial derivative w. r. t. time	p. 2
$\phi_x^\lambda(y)$	Certain rescaling in regularity structures	(8.62)
P_t	Heat kernel	(8.9)
$\mathcal{P}_L$	Leray projection operator	p. 46
$\mathbb{P}_{\neq 0}$	Projection onto mean-zero	(6.46)
$\mathcal{P}_\infty$	Special σ-algebra	Definition 4.1.2
$A \propto B$	A is proportional to B	3
$\mathcal{Q}_\alpha a$	Component of a in $\mathcal{T}_\alpha$	p. 182
$\langle\langle X \rangle\rangle_t^{(p)}$	pth variation process	(4.5)
$\langle\langle M \rangle\rangle_t, \langle\langle M_1, M_2 \rangle\rangle_t$	Quadratic variation of $\{M\}_{t\geq 0}$ and cross-variation of $\{M_1\}_{t\geq 0}$ and $\{M_2\}_{t\geq 0}$	Definition 4.1.7
$\mathcal{R}$	Riesz transform vector	Definition 3.4.3
$\mathfrak{s}$	Parabolic scaling	(8.1)
$\lvert\cdot\rvert_{\mathfrak{s}}$	Scaled degree	(8.4)
$\mathbb{S}^{d-1}$	d-dimensional sphere	p. 1
$\mathcal{S}(\mathbb{R}^d), \lVert\cdot\rVert_{k,\mathcal{S}}, \mathcal{S}'(\mathbb{R}^d)$	Schwartz space, its norm, and space of tempered distributions	Definition 3.1.2
$\mathcal{S}_h'(\mathbb{R}^d)$	Certain subset of $\mathcal{S}'(\mathbb{R}^d)$	(3.23)
$(\mathscr{S}_{\mathfrak{s},\bar{z}}^\delta \varphi)(z), \mathscr{S}_{\mathfrak{s}}^\delta((t,x))$	Scaling maps in regularity structures	(8.5)
$\sigma(\{X(t)\} : t \in [0,T])$	Sigma-algebra generated by $\{X(t)\}_{t\in[0,T]}$	p. 3
$\Subset$	Compact embedding	p. 1
τ_a	Translation by a	(3.2)
$\mathbb{T}^d$	d-dimensional torus	p. 1
$\triangleq$	Definition	p. 1
$\mathcal{T}_\epsilon f = f_\epsilon$	f mollified	(2.25)
$T(f,g), R(f,g)$	Bony's paraproducts	(3.84)
$\mathcal{T}_{poly}$	Space of polynomials in regularity structures	Example 18
$u_k \overset{w}{\rightharpoonup} u$	Weak convergence	(2.18)
$u_k \overset{w^*}{\rightharpoonup} u$	Weak star convergence	(2.19)
$a \wedge b$	Minimum of numbers a and b	p. 3
$\lVert f\rVert_{\dot{W}^{s,p}}, \lVert f\rVert_{W^{s,p}}$	General Sobolev space	(3.17)
$x^\perp$	Vector perpendicular to x	p. 2

0 Basic abbreviations, notations, and conventions

In this chapter, we clarify a minimum of basic abbreviations, notations, and conventions that will be used in this book, which, given its breadth, may be considerable. For clarity, we split them among Sections 0.1, 0.2, and 0.3, respectively, on analysis, PDEs, and stochastic areas, while acknowledging that many certainly overlap.

0.1 Analysis

0.1.1 Abbreviations

(a) "dD" = "d-dimensional,"
(b) "a. e." = "almost every,"
(c) "w. r. t." = "with respect to,"
(d) "w. l. o. g." = without loss of generality,"
(e) "o. n. b." = "orthonormal basis,"
(f) "a. k. a." = "also known as."

0.1.2 Notations

The notation $\triangleq$ is used for definitions. With that, we define $\mathbb{N} \triangleq \{1, 2, \ldots\}$ and $\mathbb{N}_0 \triangleq \mathbb{N} \cup \{0\}$. $\lceil x \rceil$ denotes the smallest integer greater than or equal to x, whereas $\lfloor x \rfloor$ denotes the greatest integer less than or equal to x. We define $\mathbb{R} \triangleq (-\infty, \infty)$, $\mathbb{R}_{>0} \triangleq \{r \in \mathbb{R} : r > 0\}$, and $\mathbb{R}_{\geq 0} \triangleq \{r \in \mathbb{R} : r \geq 0\}$. We denote the $d \times d$ identity matrix by Id, the tensor product by $\otimes$, and $\mathbb{M}^d \triangleq \{A = (a_{i,j})_{1 \leq i,j \leq d}\}$.

For two sets A and B, we write $A \subset B$ if for all $a \in A$, we have $a \in B$. Hence $A \subset B$ allows the possibility that $A = B$; in case we want to emphasize that $A \neq B$, we will write $A \subsetneq B$. By $\overline{A}$ we denote the closure of a set A, e. g., $\overline{\mathbb{R}} \triangleq [-\infty, \infty]$. Additionally, $A \Subset B$ means compact inclusion of a set A in a set B (see Definition 2.4.2(d) of compact operator). The Borel σ-algebra on a set X will be denoted as $\mathcal{B}_X$ or $\mathcal{B}(X)$. We denote by $\mathbb{T}^d \triangleq (\mathbb{R}/\mathbb{Z})^d$ the dD torus for $d \in \mathbb{N}$; sometimes, a torus may be $[-\pi, \pi]$, but such difference will not be crucial in this monograph. We denote by $\mathbb{S}^{d-1} \triangleq \{x \in \mathbb{R}^d : |x| = d\}$ the $(d-1)$D sphere. We denote the open ball centered at x with radius $r > 0$ by $B_r(x) = B(x, r)$ and the annulus (ring) of inner and outer radiuses r_1 and r_2 by $C(0, r_1, r_2)$. Finally, when dealing with complex number, we write $z = x + iy$, so that its complex conjugate is $\overline{z} \triangleq x - iy$.

0.1.3 Conventions

We fix a convention that $\frac{1}{\infty} \triangleq 0$, $0^0 \triangleq 1$, and $\sum_{1 \leq r \leq 0} c_r = 0$ for any $c_r \in \mathbb{R}$.

https://doi.org/10.1515/9783111349985-001

0.2 PDEs

0.2.1 Notations

Unless explicitly written otherwise, we always assume that a measure space is equipped with Lebesgue measure. The spatial domain in most of our discussions will be $\mathbb{R}^d$ or $\mathbb{T}^d$. We denote the standard basis of $\mathbb{R}^d$ as $\{e_j\}_{j=1}^d$. The differentiation operators are denoted by $\partial_t \triangleq \frac{\partial}{\partial t}$, $\partial_t^2 \triangleq \frac{\partial^2}{(\partial t)^2}$, $\partial_i \triangleq \partial_{x_i} \triangleq \frac{\partial}{\partial x_i}$, $\partial_{ij} \triangleq \partial_{x_i}\partial_{x_j}$ for $i,j \in \{1,\dots,d\}$, etc. For $f : \mathbb{R} \to \mathbb{R}$, we define $f(t+) \triangleq \lim_{h\searrow 0} f(t+h)$ and $f(t-) \triangleq \lim_{h\searrow 0} f(t-h)$. We write L_x^p instead of $L^p(U)$ when the domain is clear; similarly, we shortly write H_x^s, C_t^1, etc.

In the 2D case, we denote $x^\perp \triangleq (-x_2, x_1)$ for $x = (x_1, x_2)$. We recall (e. g., from [132, Appendix A]) that a multiindex is a vector $\alpha = \langle \alpha_1, \dots, \alpha_d \rangle$ with $\alpha_j \in \mathbb{N}_0$. Its order is defined by $|\alpha| \triangleq \sum_{j=1}^d \alpha_j$, and for $k \in \mathbb{N}$,

$$D^\alpha u(x) \triangleq \frac{\partial^{|\alpha|} u(x)}{\partial x_1^{\alpha_1} \dots \partial x_d^{\alpha_d}}(x) \triangleq \partial_{x_1}^{\alpha_1} \dots \partial_{x_d}^{\alpha_d} u(x), \tag{1a}$$

$$D^k u(x) \triangleq \{D^\alpha u(x) : |\alpha| = k\}. \tag{1b}$$

Additionally, we define the Laplacian operator by $\Delta \triangleq \sum_{i=1}^d \partial_{x_i}^2$ and $\operatorname{div} F = \sum_{j=1}^d \partial_{x_j} F_j$. The curl operator is defined as $\nabla \times F$ in case $d \in \{2,3\}$, and it is possible to extend the curl operator to higher dimensions via

$$\Omega(F) \triangleq (\Omega_j^i(F))_{1\le i,j\le N} \quad \text{where } \Omega_j^i(\Gamma) \triangleq \partial_j F^i - \partial_i F^j \tag{2}$$

(e. g., [79, p. 8]). Especially, when discussing convex integration, it is useful to know the following definition of the divergence of a matrix:

$$\operatorname{div}(a_{ij})_{1\le i,j\le d} = \begin{pmatrix} \sum_{j=1}^d \partial_{x_j} a_{1j} \\ \vdots \\ \sum_{j=1}^d \partial_{x_j} a_{dj} \end{pmatrix}. \tag{3}$$

Concerning inequalities, we write $A \lesssim_{\alpha,\beta} B$ whenever there exists a constant $C = C(\alpha,\beta) \ge 0$ such that $A \le CB$ and $A \approx_{\alpha,\beta} B$ in case $A \lesssim_{\alpha,\beta} B$ and $A \gtrsim_{\alpha,\beta} B$. We write $A \overset{(\cdot)}{\lesssim} B$ whenever $A \lesssim B$ due to $(\cdot)$ and similarly with $A \overset{(\cdot)}{=} B$.

We denote the gamma function by Γ and by $\alpha(d) \triangleq \frac{\pi^{\frac{d}{2}}}{\Gamma(\frac{d}{2}+1)}$ the volume of a d-dimensional unit ball; due to the translation-invariance of the Lebesgue measure, we may assume that it is $B_1(0)$. With the same reasoning, we denote by $d\alpha(d)$ the surface area of any unit sphere, by $r^d\alpha(d)$ the volume of a dD ball with radius r, and, finally, by $dr^{d-1}\alpha(d)$ the surface area of dD ball with radius r. When no confusion arises, we denote by $|A|$ the volume of a set $A \subset \mathbb{R}^d$. We denote $\fint_X f(x)dx \triangleq \frac{1}{|X|}\int_X f(x)dx$.

0.2.2 Conventions

We follow the typical convention that a universal constant in a series of inequalities is denoted by C.

0.3 Stochastics

0.3.1 Abbreviations

(a) "a. s." = "almost surely,"
(b) "i. i. d." = "independent and identically distributed,"
(c) "SDE" = "stochastic differential equation."

0.3.2 Notations

We define $a \wedge b \triangleq \min\{a, b\}$ and write $A \propto B$ if A is proportional to B. We also denote the σ-algebra generated by a process $\{X(t)\}_{t\in[0,T]}$ by $\sigma(\{X(t)\} : t \in [0, T])$.

1 Introduction

In this chapter, we recall the basic types of nonlinearities and thereafter introduce all the PDEs of our interest. The definitions of semilinear, quasilinear, and fully nonlinear PDEs seem to be inconsistent across some disciplines of applied mathematics, and hence we start with Definition 1.0.1 for clarity. Then we introduce PDEs in fluid mechanics and SPDEs in mathematical physics in Sections 1.1 and 1.2, respectively. We emphasize that, as the title suggests, in this monograph, we focus on the PDEs of fluid mechanics; however, in mathematical physics, singular SPDEs naturally arise (see Chapters 8–9), so we decided to include them here. Some notations such as Riesz transform $\mathcal{R}$, fractional Laplacian Λ, and random force ξ will be formally introduced in Chapters 3–4.

Definition 1.0.1 (e. g., [132, Chapter 1.1]). Given a_α and f, a PDE of the form

$$\sum_{|\alpha|\leq k} a_\alpha(x) D^\alpha u(x) = f(x)$$

is called linear. If a PDE is not linear, then it is called nonlinear. There are three types of nonlinearity.

(a) A PDE of the form

$$\sum_{|\alpha|=k} a_\alpha(x) D^\alpha u(x) + a_0(D^{k-1}u, \ldots, Du, u, x) = 0,$$

where $D^k u \triangleq \{D^\alpha u : |\alpha| = k\}$, is called semilinear.

(b) A PDE of the form

$$\sum_{|\alpha|=k} a_\alpha(D^{k-1}u, \ldots, Du, u, x) D^\alpha u(x) + a_0(D^{k-1}u, \ldots, Du, u, x) = 0$$

is called quasilinear.

(c) A PDE is called fully nonlinear if it depends nonlinearly on highest order derivatives, for example, it contains $\sum_{|\alpha|=k} |D^\alpha u|^2$, where k is the highest degree of derivatives in the PDE.

1.1 PDEs in fluid mechanics

In this section, we introduce the Navier–Stokes and Euler equations, Burgers' equation, surface quasigeostrophic (SQG) equations, Boussinesq system, MHD system, Hall-MHD system, electron MHD system, and hyperbolic Navier–Stokes equations. The Kuramoto–Sivashinsky and transport equations will also be briefly introduced in (6.54) and (7.1), respectively. Although there are many other PDEs of interest, such as the micropolar fluid system or compressible case, we will not discuss them for brevity.

https://doi.org/10.1515/9783111349985-002

For simplicity of introduction, we focus on the case $x \in U = \mathbb{R}^d$ or $\mathbb{T}^d$ and present in this section all the PDEs without external force.

1.1.1 Navier–Stokes and Euler equations

The Navier–Stokes and Euler equations have a wide variety of applications in the real world, such as fluid mechanics, aerodynamics, medicine, and finance. We denote by $u : \mathbb{R}_+ \times U \to \mathbb{R}^d$ the velocity field, by $\pi : \mathbb{R}_+ \times U \to \mathbb{R}$ the pressure field, and by $\upsilon \geq 0$ the kinematic viscosity (reciprocal of the Reynolds number). Then the Navier–Stokes equations can be written as

$$\partial_t u + (u \cdot \nabla)u + \nabla\pi = \upsilon\Delta u, \quad \nabla \cdot u = 0. \tag{1.1}$$

The case $\upsilon = 0$ reduces to the Euler equations. We notice that (1.1) is semilinear if $\upsilon > 0$ and quasilinear if $\upsilon = 0$.

In the 2D and 3D cases, denoting the vorticity by $\omega \triangleq \nabla \times u$ and applying the curl operator to (1.1), we deduce

$$\partial_t \omega + (u \cdot \nabla)\omega = \upsilon\Delta\omega + (\omega \cdot \nabla)u, \quad \nabla \cdot \omega = 0, \tag{1.2}$$

where the vorticity-stretching term $(\omega \cdot \nabla)u$ vanishes in the 2D case. We will discuss this advantage in the 2D case in Section 5.1. In contrast, due to the presence of the vorticity-stretching term in the 3D case, finite-time singularities may occur, and we refer the interested readers to, e. g., [108, 130] and references therein.

1.1.2 Burgers' equation

The Burgers' equation was initially introduced by Bateman [16] in 1915 and later studied by Burgers [51]. It is widely applied in fluid mechanics, gas dynamics, and traffic flows. Mathematically, it can be seen as a pressureless Navier–Stokes equation without the divergence-free condition solved by $\theta : \mathbb{R}_+ \times U \to \mathbb{R}$:

$$\partial_t \theta + \frac{1}{2}\partial_x \theta^2 = \upsilon\partial_x^2\theta. \tag{1.3}$$

It has been investigated for many decades as a prototype 1D toy model of (1.1). The viscous Burgers' equation is semilinear, whereas the inviscid case is quasilinear, and the inviscid Burgers' equation is famous as a conservation law that exhibits a finite-time shock (e. g., [132, Section 3.4.1]). In fact, we can replace $\upsilon\partial_x^2$ in (1.3) by $-\upsilon\Lambda^\gamma$, where Λ is the fractional Laplacian (see (3.10)), so that the resulting equation is quasilinear when $\gamma \in [0, 1]$, and it is known that a finite-time shock persists as long as $\gamma \in [0, 1)$ even if

$\upsilon > 0$ (e. g., [6, 128, 220]). We will also discuss the higher-dimensional Burgers' equation in (6.53).

1.1.3 Surface quasi-geostrophic equations

Introduced by Charney [77], the 2D SQG equations model the motion of buoyancy or potential temperature $\theta : \mathbb{R}_+ \times U \to \mathbb{R}$ on a horizontal fluid boundary and have applications in oceanography, meteorology, and geophysics. Denoting by $\mathcal{R}$ the Riesz transform vector (see Definition 3.4.3) and by Λ a fractional Laplacian (see (3.10)), the SQG equations can be written as

$$\partial_t \theta + u \cdot \nabla \theta + \upsilon \Lambda^\gamma \theta = 0, \quad u = \mathcal{R}^\perp \theta, \quad \gamma \in [0, 2]. \tag{1.4}$$

The SQG equations are semilinear if $\gamma > 1$ and $\upsilon > 0$ and quasilinear if $\gamma \in [0, 1]$ and $\upsilon \geq 0$.

Constantin, Majda, and Tabak [100] pointed out the resemblance between the vorticity formulation (1.2) in the 2D case and (1.4). In fact, applying the gradient operator to (1.4) results in

$$\partial_t \nabla \theta + (u \cdot \nabla) \nabla \theta + \upsilon \Lambda^\gamma \nabla \theta = -\nabla u \cdot \nabla \theta,$$

which resembles (1.2) even in the 3D case. In [100] the authors verified the similarities of these solutions numerically. Subsequently, Kiselev [217, eq. (1.3)] introduced the following generalized SQG equations:

$$\partial_t \theta + u \cdot \nabla \theta + \upsilon \Lambda^{\gamma_1} \theta = 0, \quad u = \Lambda^{1-\gamma_2} \mathcal{R}^\perp \theta, \quad \gamma_1 \in [0, 2], \gamma_2 \in [1, 2], \tag{1.5}$$

although the modified SQG (MQG) equations were introduced previously by Constantin, Iyer, and Wu [98] when $\gamma_1 \in (0, 1]$ and $\gamma_2 = 2 - \gamma_1$. Notice that when $\gamma_2 = 2$, (1.5) recovers the SQG equations (1.4).

Finally, we also mention that when v satisfies $-\nabla^\perp \cdot v = \theta$, (1.4) transforms to the following momentum SQG equations:

$$\partial_t v + (u \cdot \nabla) v - (\nabla v)^T \cdot u + \nabla \pi + \upsilon \Lambda^\gamma v = 0, \quad u = \Lambda v, \quad \nabla \cdot v = 0. \tag{1.6}$$

Indeed, it can be directly verified that

$$-\nabla^\perp \cdot ((u \cdot \nabla) v - (\nabla v)^T \cdot u) \overset{(4.21)}{=} -\nabla^\perp \cdot (u^\perp (\nabla^\perp \cdot v)) = \operatorname{div}(u\theta),$$

so that (1.4) is recovered; we refer to [46] for details about derivations of the momentum SQG equations through the Euler–Poincaré principle. We will return to the momentum SQG equations (1.6) in Section 4.5.1, where we study their generalized version (see (4.20)).

1.1.4 Boussinesq system

Instead of toy models of the Navier–Stokes equations, the Navier–Stokes equations can be coupled to form a system. We start with a transport-diffusion equation that results in the Boussinesq system

$$\partial_t u + (u \cdot \nabla)u + \nabla\pi = \upsilon\Delta u + \theta e_d, \quad \nabla \cdot u = 0, \tag{1.7a}$$

$$\partial_t \theta + u \cdot \nabla\theta = \kappa\Delta\theta, \tag{1.7b}$$

where $u : \mathbb{R}_+ \times U \to \mathbb{R}^d$, $\pi : \mathbb{R}_+ \times U \to \mathbb{R}$, and $\theta : \mathbb{R}_+ \times U \to \mathbb{R}$ represent the velocity field, pressure field, and temperature, respectively.

In applications, ocean circulation consists of turbulent motions on a wide range of scales from a few centimeters to thousands of kilometers that interact continuously with one another, whereas atmospheric turbulence refers to small-scale irregular air motions that vary in both speed and direction due to wind. The Boussinesq system arises as a model of interaction of gravity and rotation of the earth with density variations about a reference state (e. g., [258, Chapter 1.1]). Additionally, subtracting pure conduction solution from the temperature scalar field transforms the Boussinesq system into the Bénard problem of thermohydraulics (see, e. g., [319, p. 133]). Finally, the 2D Boussinesq system corresponds to the 3D axisymmetric swirling flows (see, e. g., [259, Sec. 5.4.1]).

1.1.5 MHD system

If we couple the Navier–Stokes equations with the Maxwell equation from electromagnetism instead of a transport-diffusion equation, as we did in (1.7), then we obtain the MHD system

$$\partial_t u + (u \cdot \nabla)u + \nabla\pi = \upsilon\Delta u + (b \cdot \nabla)b, \quad \nabla \cdot u = 0, \tag{1.8a}$$

$$\partial_t b + \nabla \times (b \times u) = \eta\Delta b, \tag{1.8b}$$

where $b : \mathbb{R}_+ \times U \to \mathbb{R}^d$ and $\eta \geq 0$ represent the magnetic vector field and resistivity, respectively. When $b \equiv 0$, the MHD system reduces to the Navier–Stokes equations. We refer to the inviscid and nonresistive case as the ideal MHD system. If the initial magnetic velocity field is divergence-free, then since $\nabla \cdot \nabla \times F = 0$ for all F, the divergence-free property will propagate for all $t \geq 0$; consequently, thanks to (2.1c), in the literature, (1.8b) often takes the form

$$\partial_t b + (u \cdot \nabla)b - (b \cdot \nabla)u = \eta\Delta b.$$

We can define $j \triangleq \nabla \times b$ as the current density, and obtain an equivalent form by replacing $(b \cdot \nabla)b$ by $j \times b$, which is called the Lorentz force.

Led by the pioneering work of Alfvén [5], the study of MHD is concerned with properties of electrically conducting fluids and has wide applications in astrophysics, geophysics, and plasma physics.

1.1.6 Hall-MHD and electron MHD systems

In addition to the notations of the MHD system, denoting by $h \geq 0$ the Hall coefficient, we can introduce the Hall-MHD system

$$\partial_t u + (u \cdot \nabla)u + \nabla \pi = \upsilon \Delta u + j \times b, \quad \nabla \cdot u = 0, \tag{1.9a}$$

$$\partial_t b + \nabla \times (b \times u) + h\nabla \times (j \times b) = \eta \Delta b. \tag{1.9b}$$

When $h = 0$, the Hall-MHD system reduces to the MHD system (1.8). On the other hand, if $u \equiv 0$, then the Hall-MHD system reduces to the electron MHD system, which governs the electron's self-induced magnetic field (see [329]):

$$\partial_t b + h\nabla \times (j \times b) = \eta \Delta b. \tag{1.10}$$

Introduced by Lighthill [240], the Hall-MHD system has various applications such as the study of the sun [55] and star formation [328]. The Hall-MHD and the electron MHD systems are both quasilinear even in the presence of diffusion. Because the Hall term decouples in the 2D case, the $2\frac{1}{2}$-D Hall-MHD system has been studied by physicists (e. g., [127]), the case in which the system is solved by a pair of $(u_1, u_2, u_3)(t, x_1, x_2)$ and $(b_1, b_2, b_3)(t, x_1, x_2)$ (see [259, Section 2.3.1] on the $2\frac{1}{2}$-D Navier–Stokes and Euler equations).

1.1.7 Hyperbolic Navier–Stokes equations

As a last model of fluid mechanics, we introduce the hyperbolic Navier–Stokes equations (e. g., [35, eq. (1.6)])

$$\gamma \partial_t^2 u + \partial_t u + (u \cdot \nabla)u + \nabla \pi = \upsilon \Delta u, \quad \nabla \cdot u = 0, \tag{1.11}$$

where $\gamma \geq 0$. The case $\gamma = 0$ reduces (1.11) to (1.1). More than half a century ago, Cattaneo [62, 63] and Vernotte [323] proposed replacing the heat operator with a damped wave operator to make the propagation speed of heat transfer finite. This idea was subsequently extended by others (e. g., [60, 110]).

One way to quickly understand the origin of this model is to consider delayed stress. We can write the Navier–Stokes equations (1.1) as

$$F(t) = -\nabla \cdot (\pi \,\mathrm{Id} - \upsilon \nabla u)(t), \quad F(t) \triangleq \partial_t u + (u \cdot \nabla)u,$$

where $\pi\,\mathrm{Id} - \upsilon\nabla u$ is the sum of pressure and viscous stress representing Cauchy stress. We can then replace $F(t)$ by $F(t+\gamma)$ to take into account the delay and apply Taylor expansion to approximate $F(t+\gamma)$ by $(1+\gamma\partial_t)F$, which leads to

$$\gamma\partial_t(\partial_t u + (u\cdot\nabla)u) + (\partial_t u + (u\cdot\nabla)u) = -\nabla\cdot(\pi\,\mathrm{Id} - \upsilon\nabla u). \tag{1.12}$$

Dropping $\gamma\partial_t((u\cdot\nabla)u)$ for simplicity reduces (1.12) to (1.11).

1.2 SPDEs in mathematical physics

In this section, we introduce various SPDEs in mathematical physics, namely the KPZ equation, parabolic Anderson model, stochastic Yang–Mills equation, Φ^4 model, and Cahn–Hilliard equation. For simplicity of presentation, we focus on the spatial domain of $\mathbb{T}^d$ and denote all random forces by ξ, with more details given later in Chapters 4 and 8.

1.2.1 Kardar–Parisi–Zhang equations

The KPZ equation, originally derived by Kardar, Parisi, and Zhang [210] describes growing interface and has found much attraction because surface growth appears, e. g., in paper wetting, crystal growth, an even expansion of the universe. Letting $h:\mathbb{R}_+\times\mathbb{T}\to\mathbb{R}$ represent the interface height, the 1D KPZ equation can be written as

$$\partial_t h = \upsilon\partial_x^2 h + \lambda|\partial_x h|^2 + \xi, \tag{1.13}$$

where $\lambda > 0$ represents coupling strength. Differentiating (1.13) in space and defining $u \triangleq -\partial_x h$, we see that the resulting PDE is a variant of the Burgers' equation (1.3), and many researchers have taken advantage of this relationship (e. g., [28, eq. (B.1)]).

1.2.2 Parabolic Anderson model

The parabolic Anderson model has applications in random potential, directed polymers, trapping of random paths, and branching processes in random medium (see, e. g., [7, 162, 230] and references therein). The SPDE consists of

$$\partial_t u = \upsilon\Delta u + \xi u \tag{1.14}$$

solved by $u:\mathbb{R}_+\times\mathbb{T}^d\mapsto\mathbb{R}$.

1.2.3 Stochastic Yang–Mills equation

To introduce the Yang–Mills equation and explain stochastic quantization, we fix a gauge group G, a closed subgroup of the group of unitary matrices of order N for simplicity. We also denote by $\mathfrak{g}$ the Lie algebra of G, specifically the subspace of the vector space of all $N \times N$ skew-Hermitian matrices and $A \triangleq \sum_{j=1}^{d} A_j dx_j$ on $\mathbb{T}^d$, the G connection form. Additionally, we denote the curvature form F of A in coordinates as

$$F_{j,k}(x) = \frac{\partial A_k}{\partial x_j} - \frac{\partial A_j}{\partial x_k} + [A_j(x), A_k(x)], \tag{1.15}$$

where $[B, C] \triangleq BC - CB$ is the Lie bracket. Denoting by d_A the covariant derivative associated with A, d_A^* its adjoint, and F_A the curvature 2-form of A, the Yang–Mills equation can now be written as

$$d_A^* F_A = 0.$$

A part of the Yang–Mills problem (cf. [59]) consists of rigorously constructing the following Yang–Mills measure in the 4D case:

$$d\mu_{YM}(A) \triangleq Z^{-1} e^{\int_{\mathbb{T}^d} \sum_{j,k=1}^{d} \mathrm{Tr}(F_{j,k}(x)^2) dx} dA, \tag{1.16}$$

where Z is a normalizing constant, and dA is infinite-dimensional Lebesgue measure on the space of G connection forms on $\mathbb{T}^d$; equivalently, we can define the Yang–Mills action by

$$S_{YM}(A) \triangleq \int_{\mathbb{T}^d} |F_A(x)|^2 dx,$$

so that $d\mu_{YM}(A) = Z^{-1} e^{-S_{YM}(A)} dA$. Here the immediate warning sign is that there is no Lebesgue measure in the infinite-dimensional setting (see Lemma 2.2.2). We refer to [78, 152] and [295, Chapter 6] for more detail and also to [277] on the Osterwalder–Schrader axioms.

The idea of stochastic quantization initiated by Parisi and Wu [279] in 1981 is to not directly construct $e^{-S_{YM}(A)}$ but consider the Langevin equation

$$\partial_t A = -d_A^* F_A + \xi \quad \text{or, equivalently,} \quad \partial_t A = -\frac{\delta S}{\delta A} + \xi. \tag{1.17}$$

A simple example is a Gaussian free field in case $S(\Phi) = \int \frac{1}{2} |\nabla \Phi|^2 dx$, so that $-\frac{\delta S}{\delta \Phi} = \Delta \Phi$, which implies $\partial_t \Phi = \Delta \Phi + \xi$. Analogous computation in the case of the stochastic Yang–Mills equations leads to, in coordinates, the equations

$$\partial_t A_i = \Delta A_i + \xi_i + [A_j, 2\partial_j A_i - \partial_i A_j + [A_j, A_i]], \tag{1.18}$$

which is a parabolic SPDE with a Lie-algebra-valued solution. For a precise description with more details, we refer to [74, eq. (1.7)].

It was observed in [279] that if μ_t is the law of $A(t)$, then μ_t tends to an invariant measure $\mu_\infty(dA) \propto e^{-S(A)}dA$ as $t \to \infty$.

Remark 1.2.1. We observe that the nonlinear terms of (1.18) consist informally of cubic and Burgers' types, specifically $[A_j, [A_j, A_i]]$ and $[A_j, 2\partial_j A_i - \partial_i A_j]$, respectively.

1.2.4 Φ^4 model

In the context of nonconservative Glauber dynamics, a special case of (1.17) gives the Φ^4 model

$$\partial_t \Phi + \Phi^3 = \Delta\Phi + \xi \tag{1.19}$$

solved by $\Phi : \mathbb{R}_+ \times \mathbb{T}^d \mapsto \mathbb{R}$ (e. g., [279, eq. (4.1)] and [152, Section 20.1]).

1.2.5 Cahn–Hilliard equation

Cahn and Hilliard [53] originally proposed the following equation:

$$\partial_t u = \kappa \Delta \mu \tag{1.20}$$

where u is the order parameter that varies continuously through the diffuse interface separating the pure states from -1 to 1, $\kappa > 0$ is the mobility constant, $\mu \triangleq -\alpha\Delta u + f(u)$ is the chemical potential, $\alpha > 0$ is the surface tension coefficient, and $f = F'$, where typically $F(s) = \frac{1}{4}(s^2 - 1)^2$, so that

$$f(s) = s^3 - s. \tag{1.21}$$

This leads (1.20) to

$$\partial_t u = \kappa\Delta(-\alpha\Delta u + u^3 - u). \tag{1.22}$$

In contrast to the Glauber dynamics (1.17), the Kawasaki dynamics $\partial_t A = \nabla \cdot (\nabla \frac{\delta S}{\delta A}) + \nabla \cdot \xi$ from [213] is conservative, and the following Cahn–Hilliard equation forced by divergence of the noise in a matrix form appeared in [168]:

$$\partial_t u + \Delta^2 u = \Delta u^3 + \operatorname{div} \xi.$$

2 Preliminaries on measure theory and functional analysis

Although the focus of this monograph is applications of harmonic and stochastic analysis to PDEs, measure theory and functional analysis play vital roles, which motivates us to devote this chapter for their review. We recall vector calculus, useful identities, and inequalities in Section 2.1 and prove the absence of Lebesgue measure in infinite-dimensional space in Section 2.2. Then we recall useful definitions and facts concerning Lebesgue spaces, operator theory, mollifiers, and Steklov average in Sections 2.3, 2.4, 2.5, and 2.6, respectively. Finally, as an application, we derive the energy identity for the solutions to the Euler and Navier–Stokes equations (1.1) and state the Navier–Stokes problem on $\mathbb{T}^3$ in Section 2.7.

2.1 Vector calculus, useful identities and inequalities

The following vector calculus identities come in handy, especially when analyzing the Riesz transform (see Definition 3.4.3) and the Hall-MHD system. For all $\Xi, \Psi, \Theta \in \mathbb{R}^3$, we have

$$\nabla \times (\nabla \times \Xi) = \nabla(\nabla \cdot \Xi) - \Delta \Xi, \tag{2.1a}$$

$$(\Xi \times \Psi) \cdot \Xi = 0, \tag{2.1b}$$

$$\nabla \times (\Theta \times \Psi) = \Theta(\nabla \cdot \Psi) - \Psi(\nabla \cdot \Theta) + (\Psi \cdot \nabla)\Theta - (\Theta \cdot \nabla)\Psi. \tag{2.1c}$$

Polar coordinates are useful to estimate higher-dimensional integrals.

Definition 2.1.1 (Polar coordinates). Let $f : \mathbb{R}^d \to \mathbb{R}$ be continuous and measurable. Then $\int_{\mathbb{R}^d} f(x)dx = \int_0^\infty (\int_{\partial B(x_0,r)} f dS)dr$ for all $x_0 \in \mathbb{R}^d$. In particular, $\frac{d}{dr}(\int_{B(x_0,r)} f dx) = \int_{\partial B(x_0,r)} f dS$ for all $r > 0$.

Lemma 2.1.2 ([347, Lemma 2.2]). *For all* $r \in \mathbb{C} \setminus \{1\}$,

$$\sum_{j=n}^{m} r^j = \frac{r^{m+1} - r^n}{r - 1}.$$

We emphasize that this partial sum formula does not require $|r| < 1$, which is needed for convergence. For completeness, we include the proof of Lemma 2.1.2 in Appendix A.1.

An immediate corollary of Lemma 2.1.2 in case $|r| < 1$, which will be useful for us upon studying homogeneous Besov spaces later (Proposition 3.2.8), is the following:

https://doi.org/10.1515/9783111349985-003

Corollary 2.1.3.

1. *For all* $m, n \in \mathbb{Z}$,

$$\sum_{j=-\infty}^{m} r^j \lesssim r^m \quad \text{if } |r| > 1, \quad \text{and} \quad \sum_{j=n}^{\infty} r^j \lesssim r^n \quad \text{if } |r| < 1. \tag{2.2}$$

2. *For all* $m, n \in \mathbb{R}$,

$$\sum_{j=-\infty}^{m} r^j \lesssim r^{m+1} \quad \text{if } r > 1, \quad \text{and} \quad \sum_{j=n}^{\infty} r^j \lesssim r^{n-1} \quad \text{if } r \in (0,1). \tag{2.3}$$

It is well known that $(a+b)^p \le (2\max\{a,b\})^p \le 2^p(a^p + b^p)$ for all $a, b \ge 0$ and $p \in [0,\infty)$. Although this suffices in most circumstances, we record the following improvements.

Lemma 2.1.4. *Let* $a, b \ge 0$. *Then*

$$(a+b)^p \le \begin{cases} 2^{p-1}(a^p + b^p) & \text{if } p \in [1,\infty), \\ a^p + b^p & \text{if } p \in [0,1). \end{cases} \tag{2.4}$$

For completeness, we give the proof of Lemma 2.1.4 in Appendix A.2.

Lemma 2.1.5 (Young's inequality). *Let* $\alpha, \beta \ge 0$, *and let* $p, q \in [1,\infty]$ *be such that* $\frac{1}{p} + \frac{1}{q} = 1$. *Then*

$$\alpha\beta \le \frac{\alpha^p}{p} + \frac{\beta^q}{q}.$$

Lemma 2.1.6 (Gronwall's inequality). *Let* ξ *be a nonnegative function in* $L^1([0,T])$ *such that for some* $C_1, C_2 \ge 0$, $\xi(t) \le C_1 \int_0^t \xi(s)ds + C_2$ *for a. e.* $t \in [0,T]$. *Then* $\xi(t) \le C_2(1 + C_1 t e^{C_1 t})$ *for a. e.* $t \in [0,T]$.

We refer to, e. g., [132, Appendix B] for proofs and some variations of Lemmas 2.1.5 and 2.1.6.

2.2 Absence of Lebesgue measure in infinite-dimensional space and atoms

The absence of Lebesgue measure in an infinite-dimensional space is the very first step to understand the difficulty of the Yang–Mills problem, which was briefly discussed in Section 1.2.3. To prove such a claim, we recall the following Riesz' lemma (see, e. g., [225, Lemma 2.5-4]).

Lemma 2.2.1 (Riesz' lemma). *Let Y and Z be subspaces of a normed space X of any dimension, and suppose that Y is a closed proper subset of Z. Then for every $\theta \in (0,1)$, there exists $z \in Z$ such that $\|z\|_X = 1$ and $\|z - y\|_X \geq \theta$ for all $y \in Y$.*

We follow the approach of Alexanderian [4] to deduce the desired conclusion that is more general than in the case of Lebesgue measure.

Lemma 2.2.2. *Let X be an infinite-dimensional separable normed linear vector space. Then X does not admit any nontrivial translation-invariant positive Borel measure μ that is finite on open balls.*

Proof of Lemma 2.2.2. We take $y_1 \in \mathbb{S}^\infty \triangleq \{x \in X : \|x\|_X = 1\}$ and define $W_1 \triangleq \operatorname{span}\{y_1\}$. By Riesz' lemma there is $y_2 \in \mathbb{S}^\infty$ such that $\|y_2 - w\|_X \geq \frac{1}{2}$ for all $w \in W_1$. We can repeat this with $W_2 \triangleq \operatorname{span}\{y_1, y_2\}$ to find $y_3 \in \mathbb{S}^\infty$ such that $\|y_3 - w\|_X \geq \frac{1}{2}$ for all $w \in W_2$. Consequently, for all $n \in \mathbb{N}$, we have $y_n \in \mathbb{S}^\infty$ that satisfies $\|y_n - y_m\|_X \geq \frac{1}{2}$ for all $m \in \{1, \ldots, n-1\}$. It follows that $\{B(\frac{1}{2}y_n, \frac{1}{8})\}_{n=1}^\infty$ consists of mutually disjoint balls within $B(0,1)$. Now by translation invariance we have $\mu(B(x, \frac{1}{8})) = C \geq 0$ for all $x \in X$. If $C > 0$, then $\mu(B(0,1)) \geq \sum_{n=1}^\infty C = \infty$. This contradicts the hypothesis that $\mu(B(0,1))$ is finite. On the other hand, if $C = 0$, then because X is separable, there are $\{q_j\}_{j=1}^\infty \subset X$ such that $\|x - q_j\|_X < \frac{1}{8}$. Consequently, $\mu(X) \leq \sum_{j=1}^\infty \mu(B(q_j, \frac{1}{8})) \leq 0$. □

The following notions of atoms are needed in Section 8.3.2.

Definition 2.2.3 ([38, p. 108]). Let $(X, \mathcal{M}, \mu)$ be a measure space. A set $A \in \mathcal{M}$ is called an atom if $\mu(A) > 0$ and for all $B \in \mathcal{M}$ such that $B \subset A$, we have $\mu(B) = 0$ or $\mu(A \setminus B) = 0$. The measure space called nonatomic if there are no atoms.

2.3 Lebesgue spaces

The following come from [132, 147], to which we refer the readers for details.

Definition 2.3.1. For an open set $U \subset \mathbb{R}^d$ and $p \in [1, \infty]$, we define $L^p(U; \mathbb{R}) \triangleq \{f : U \to \mathbb{R} : f$ is measurable, and $\|f\|_{L^p(U)} < \infty\}$, where

$$\|f\|_{L^p(U)} \triangleq \begin{cases} (\int_U |f(x)|^p d\mu(x))^{\frac{1}{p}} & \text{if } p \in [1, \infty), \\ \operatorname{ess\,sup}_{x \in U} |f(x)| & \text{if } p = \infty. \end{cases} \tag{2.5}$$

When we need not emphasize the range $\mathbb{R}$ in $L^p(U; \mathbb{R})$, we abbreviate the latter by $L^p(U)$. We define $L^p_{\text{loc}}(U) \triangleq \{f : U \to \mathbb{R}$ such that $f \in L^p(V)\ \forall\, V \Subset U\}$. As special cases of $L^p(\mathbb{R})$ with counting measure, we define $l^p(\mathbb{Z})$ analogously with its norm defined by

$$\|f\|_{l^r(\mathbb{Z})} \triangleq \begin{cases} (\sum_{j \in \mathbb{Z}} |f_j|^r)^{\frac{1}{r}} & \text{if } r \in [1, \infty), \\ \sup_{j \in \mathbb{Z}} |f_j| & \text{if } r = \infty. \end{cases} \tag{2.6}$$

Definition 2.3.2. Let $U \subset \mathbb{R}^d$ be an open set.

(a) For $s \in \mathbb{N}_0$, we define $C^s(U)$ as the space of functions with all continuous derivatives of order s, endowed with the norm

$$\|f\|_{C^s} \triangleq \sum_{|\alpha| \le s} \|D^\alpha f\|_{C(\overline{U})}, \quad \text{where } \|f\|_{C(\overline{U})} \triangleq \sup_{x \in U} |f(x)|. \tag{2.7}$$

Then we define

$$C_c^s(U) \triangleq \{f \in C^s(U) : f \text{ has compact support}\}, \tag{2.8a}$$
$$C_b(U) \triangleq \{f \in C(U) : f \text{ is bounded}\}, \tag{2.8b}$$
$$C^\infty(U) \triangleq \cap_{s=0}^\infty C^s(U), \quad \text{and} \quad C_c^\infty(U) \triangleq \cap_{s=0}^\infty C_c^s(U). \tag{2.8c}$$

Additionally in case $U = \mathbb{R}^d$, we define

$$C_0^\infty(\mathbb{R}^d) \triangleq \left\{f \in C^\infty(\mathbb{R}^d) : \lim_{|x| \to \infty} |f(x)| = 0\right\}. \tag{2.9}$$

(b) We define the $\dot{C}^{s,\gamma}(U)$-Hölder seminorm for $s \in \mathbb{N}_0$ and $\gamma \in (0, 1]$ by

$$[f]_{\dot{C}^s} \triangleq \max_{|\alpha|=s} \|D^\alpha f\|_C, \tag{2.10a}$$
$$[f]_{\dot{C}^{s,\gamma}(\overline{U})} \triangleq \sum_{|\alpha|=s} \sup_{x,y \in U: x \ne y} \frac{|D^\alpha f(x) - D^\alpha f(y)|}{|x-y|^\gamma}, \tag{2.10b}$$

which are homogeneous. To make it a norm, we define the $C^{s,\gamma}$-Hölder norm by

$$\|f\|_{C^{s,\gamma}(\overline{U})} \triangleq \|f\|_{C^s} + [f]_{\dot{C}^{s,\gamma}(\overline{U})}. \tag{2.11}$$

For $s \in \mathbb{R}_+ \setminus \mathbb{N}$, we will write $\|\cdot\|_{C^s} \triangleq \|\cdot\|_{C^{\lfloor s \rfloor, s - \lfloor s \rfloor}}$ and refer to functions with finite such norms as s-Hölder continuous.

(c) Finally, we say that a function is s-locally Hölder continuous in U if it is s-Hölder continuous in K for all $K \Subset U$.

The following inequalities are independent of spatial dimensions.

Lemma 2.3.3. *Let (X, μ) be a measure space, and let $p, q, r \in [1, \infty]$.*

(a) *Hölder's inequality:*

$$\|fg\|_{L^r} \le \|f\|_{L^p} \|g\|_{L^q} \quad \text{if } \frac{1}{r} = \frac{1}{p} + \frac{1}{q}. \tag{2.12}$$

(b) *Interpolation inequality:*

$$\|f\|_{L^q} \lesssim \|f\|_{L^p}^{\lambda} \|f\|_{L^r}^{1-\lambda} \quad \text{if } 1 \le p \le q \le r \le \infty, \ \frac{1}{q} = \frac{\lambda}{p} + \frac{1-\lambda}{r}, \tag{2.13a}$$

$$[f]_{\dot{C}^{0,s}} \lesssim \|f\|_{C^0}^{1-\frac{s}{r}} [f]_{\dot{C}^{0,r}}^{\frac{s}{r}} \quad \text{if } 0 \le s \le r \le 1. \tag{2.13b}$$

We refer to [147, Section 6.1] for proofs of (2.12)–(2.13a) and [44, eq. (128)] for the proof of (2.13b). For $p \in [1, \infty]$, we refer to p' such that $\frac{1}{p} + \frac{1}{p'} = 1$ as the Hölder's conjugate of p.

The reverse Hölder inequality (e. g., [12, Lemma 1.2]) can be very useful.

Lemma 2.3.4 (Reverse Hölder's inequality). *Let (X, μ) be a measurable space, let $p \in [1, \infty]$, and let f be a measurable function. If*

$$\sup_{\|g\|_{L^{p'}} \le 1} \int_X |f(x)g(x)| d\mu(x) < \infty, \tag{2.14}$$

then $\|f\|_{L^p} = \sup_{\|g\|_{L^{p'}} \le 1} \int_X f(x)g(x)d\mu(x)$, so that $f \in L^p(X)$.

It is well known that due to Hölder's inequality, as long as $\mu(U) < \infty$, $\|f\|_{L^q} \le (\mu(U))^{\frac{1}{q}-\frac{1}{p}} \|f\|_{L^p}$ for all $p, q \in [1, \infty]$ such that $q \le p$ (e. g., [147, Proposition 6.12]). In contrast, the inclusion of l^p-spaces is clear as stated in the following lemma, for which we refer to, e. g., [147, Proposition 6.11] for a proof.

Lemma 2.3.5. *For all $1 \le p \le q \le \infty$, $f = \{f_j\}_{j \in \mathbb{Z}}$ satisfies*

$$\|f\|_{l^q} \le \|f\|_{l^p}. \tag{2.15}$$

For further details on Lebesgue and Hölder spaces, we refer respectively to [147, Chapter 6] and [132, Section 5.1].

For functions f and g, we denote their convolution by $f * g$ and state the Young inequality for convolutions; the proof can be found in [147, Theorem 8.9(a)] and [12, Lemma 1.4].

Lemma 2.3.6.

(a) *(Young's inequality for convolutions) For $p, q \in [1, \infty]$, $f \in L^p$, and $g \in L^q$,*

$$\|f * g\|_{L^r} \le \|f\|_{L^p} \|g\|_{L^q} \quad \text{if } \frac{1}{r} + 1 = \frac{1}{p} + \frac{1}{q}. \tag{2.16}$$

(b) $\operatorname{supp}(f * g) \subset \operatorname{supp}(f) + \operatorname{supp}(g)$.

We refer to [147, Chapter 8] for further details.

For convenience, we recall the following definition.

Definition 2.3.7. An additive function satisfies $f(x+y) = f(x) + f(y)$, whereas a superadditive function satisfies $f(x+y) \geq f(x) + f(y)$ for all x, y.

Finally, we recall the following useful definition and result.

Definition 2.3.8. Given a topological space X and $E \subset X$, a partition of unity on E is a collection $\{h_\alpha\}_{\alpha\in A}$ of functions in $C(X;[0,1])$ such that each $x \in X$ has a neighborhood on which only finitely many h_α are nonzero, and $\sum_{\alpha\in A} h_\alpha(x) = 1$ for all $x \in E$. A partition of unity $\{h_\alpha\}_{\alpha\in A}$ is subordinate to an open cover $\mathcal{U}$ of E if for each $\alpha \in A$, there exists $U \in \mathcal{U}$ such that $\operatorname{supp} h_\alpha \subset U$.

Lemma 2.3.9 ([147, Proposition 4.41]). *Let X be a locally compact Hausdorff space, let K be a compact subset of X, and let $\{U_j\}_{j=1}^n$ be an open cover of K. Then there exists a partition of unity on K subordinate to $\{U_j\}_{j=1}^n$ consisting of compactly supported functions.*

2.4 Operator theory

Let us start by clarifying two distinct notions of duality.

Definition 2.4.1 (Dual space and Gelfand dual).

(a) For a normed linear vector space X, we denote by X^* its dual space, the space of all continuous linear functionals on X.

(b) It is well known that every Hilbert space H can be canonically identified with its dual H^* ([147, Theorem 5.25]). Let $V \subset H$ be a linear subspace that is dense in H with continuous embedding. Then every functional $l \in H^*$ restricts to a continuous linear functional on V, defining a canonical embedding $T : H^* \to V^*$ by $\langle T(l), v\rangle_{V^*\times V} = \langle l, v\rangle_{H^*\times H}$. This yields

$$V \subset H \cong H^* \subset V', \tag{2.17}$$

sometimes called a Gelfand triple, in which case H is called a pivot space, and such Gelfand dual V' is compatible with the inner product on H in the sense that $\langle f, g\rangle_{V'\times V} = (f, g)_H$ for all $f \in H$, $v \in V$ (see, e. g., [36, Remark 3, p. 136] for details).

Definition 2.4.2. Let X and Y be Banach spaces.

(a) If $X^{**} = X$, then X is called a reflexive space.

(b) A linear operator $T : X \to Y$ is bounded if there exists a constant $C \geq 0$ such that $\|T(x)\|_Y \leq C\|x\|_X$ for all $x \in X$.

(c) A linear operator $T : X \to Y$ is continuous if for all $\epsilon > 0$, there exists $\delta > 0$ such that $\|x_1 - x_2\|_X \leq \delta$ implies $\|T(x_1) - T(x_2)\|_Y \leq \epsilon$. A linear operator is bounded if and only if it is continuous. We denote by $L(X, Y)$ the space of all bounded linear operators from X to Y and $L(X) \triangleq L(X, X)$.

(d) An operator $T \in L(X, Y)$ (e. g., embedding) is compact if for all $\{x_n\} \subset X$ bounded, $\{Tx_n\} \subset Y$ has a convergent subsequence in Y.
(e) Let H be a Hilbert space, and let $T \in L(H)$. Then its adjoint T^* is a unique operator in $L(H)$ such that $\langle Tx, y\rangle = \langle x, T^*y\rangle$ for all $x, y \in H$. The operator T is called self-adjoint if $T = T^*$. Additionally, if $T : D(T) \subset H \to H$ is linear with dense domain and $\langle Tx, y\rangle = \langle x, Ty\rangle$ for all $x, y \in D(T)$, then we say that T is symmetric.

We will comment on the self-adjointness of an operator later in Remark 3.1.13.

Definition 2.4.3. Let X be a Banach space, and let X^* be its dual.
(a) (Strong convergence, a. k. a. convergence in norm). We say that a family $\{u_k\}_{k=1}^{\infty} \subset X$ converges to $u \in X$ and write $u_k \to u$ as $k \to \infty$ if $\lim_{k\to\infty} \|u_k - u\|_X = 0$.
(b) (Weak convergence) We say that $\{u_k\}_{k=1}^{\infty} \subset X$ converges to $u \in X$ weakly and write $u_k \overset{w}{\rightharpoonup} u$ if

$$\langle u_k, F\rangle \to \langle u, F\rangle \quad \text{as } k \to \infty \ \forall F \in X^*. \tag{2.18}$$

(c) (Weak*-convergence) We say that $\{u_k\}_{k=1}^{\infty} \subset X^*$ converges to $u \in X^*$ and write $u_k \overset{w^*}{\rightharpoonup} u$ if

$$\langle x, u_k\rangle \to \langle x, u\rangle \quad \text{as } k \to \infty \ \forall x \in X. \tag{2.19}$$

Definition 2.4.4 (Weak continuity). Let X be a Banach space, and let X^* be its dual. A function $\phi : [0, T] \to X$ is said to be weakly continuous if the map $t \mapsto \langle l, \phi(t)\rangle$ is continuous on $[0, T]$ for all $l \in X^*$. In such a case, we write $\phi \in C_{\text{weak}}([0, T]; Y)$.

Lemma 2.4.5.
(a) ([287, p. 200]) *Let X, Y, Z be Banach spaces, and let $T \in L(X, Y)$ and $S \in L(Y, Z)$. If T or S is compact, then ST is compact.*
(b) (*Weak compactness* [132, p. 639]) *Let X be a reflexive Banach space, and let $\{u_k\}_{k=1}^{\infty} \subset X$ be a bounded sequence. Then there exist a subsequence $\{u_{k_j}\}_{j=1}^{\infty} \subset \{u_k\}_{k=1}^{\infty}$ and $u \in X$ such that $u_{k_j} \overset{w}{\rightharpoonup} u$ as $j \to \infty$.*
(c) (*Banach–Alaoglu theorem* [147, *Theorem IV.21 on p.* 115]) *Let X^* be the dual of a Banach space X. Then the unit ball in X^* is compact in the weak*-topology.*
(d) ([287, *Theorem VI.11 on p.* 199]) *A compact operator maps weakly convergent sequences into norm convergent sequences.*

Lemma 2.4.6 ([302, Theorem 5], [303, Lemma 4 (i), (ii), and (iv)]). *Let X, Y, and E be Banach spaces such that $X \Subset E \subset Y$. For $q \in [1, \infty]$, $s \in (0, 1)$, and $T > 0$, denote*

$$N^{s,q}(0, T; Y) \triangleq \Big\{v \in L^q(0, T; Y) : \sup_{\zeta>0} \zeta^{-s} \|v(t + \zeta) - v(t)\|_{L^q(0,T-\zeta;Y)} < \infty\Big\}.$$

Then we have the following embeddings:

$$L^q(0,T;X) \cap \{v : \partial_t v \in L^1(0,T;Y)\} \Subset L^q(0,T;E), \tag{2.20a}$$
$$L^\infty(0,T;X) \cap \{v : \partial_t v \in L^r(0,T;Y)\} \Subset C([0,T];E) \quad \forall r \in (1,\infty], \tag{2.20b}$$
$$L^q(0,T;X) \cap N^{s,q}(0,T;Y) \Subset L^q(0,T;E). \tag{2.20c}$$

Lemma 2.4.7 ([141, Theorems 2.1–2.2]).
(a) *Let B, B_0, B_1 be Banach spaces such that B_0 and B_1 are reflexive and $B_0 \Subset B \subset B_1$, and let $p \in (1,\infty)$ and $\gamma \in (0,1)$. Then $L^p(0,T;B_0) \cap W^{\gamma,p}(0,T;B_1) \Subset L^p(0,T;B)$.*
(b) *Let B_1 and B_2 be Banach spaces such that $B_1 \Subset B_2$, and let $\gamma \in (0,1)$ and $p > 1$ satisfy $\gamma p > 1$. Then $W^{\gamma,p}(0,T;B_1) \Subset C([0,T];B_2)$.*

Lemma 2.4.8 (Hilbert–Schmidt theorem [287, Theorem VI.16 on p. 203]). *Let H be a Hilbert space, and let B be a self-adjoint compact operator on H. Then there exists a complete o. n. b. $\{f_k\}_{k\in\mathbb{N}}$ for H and $\{\lambda_k\}_{k\in\mathbb{N}}$ such that $Bf_k = \lambda_k f_k$ for all $k \in \mathbb{N}$ and $\lim_{k\to\infty} \lambda_k = 0$.*

Lemma 2.4.9.
(a) (*Lions–Magenes theorem* [318, *Lemma* 1.2 *on pp.* 260–261]) *Let V, H, V' be Hilbert spaces such that $V \hookrightarrow H \hookrightarrow V'$ with continuous embeddings (recall* (2.17)*). If $u \in L^2(0,T;V)$ and $\frac{\partial u}{\partial t} \in L^2(0,T;V')$, then $u \in C([0,T];H)$.*
(b) ([318, *Lemma* 1.4 *on p.* 178]) *Let X and Y be Banach spaces such that $X \subset Y$ with continuous embedding. If $\phi \in L^\infty(0,T;X) \cap C_{weak}([0,T];Y)$, then $\phi \in C_{weak}([0,T];X)$ (recall Definition* 2.4.4*).*

The following notions are especially important in the study of SPDEs and are taken from [118, Appendix C] (see also [287, Section VI.6]).

Definition 2.4.10. Let H be s separable Hilbert space.
(a) An operator $T \in L(H)$ is nonnegative if $\langle Tx, x\rangle \geq 0$ for all $x \in H$.
(b) For any o. n. b. $\{e_k\}_{k=1}^\infty$ of H and any nonnegative $T \in L(H)$, define the trace of T as $\operatorname{Tr} T \triangleq \sum_{k=1}^\infty \langle Te_k, e_k\rangle$. By [118, Propositions C.1 and C.3] Tr is well-defined independently of the choice of $\{e_k\}_{k=1}^\infty$, and we define the space of trace-class operators as

$$L_1(H) \triangleq \{T \in L(H) : \|T\|_{L_1} \triangleq \operatorname{Tr} T < \infty\}. \tag{2.21}$$

(c) Additionally, if F is another separable Hilbert space and $\{e_k\}_{k=1}^\infty$ is a complete o. n. b. of H, then $T \in L(H,F)$ belongs to the space of all Hilbert–Schmidt operators $L_2(H,F)$ if

$$\|T\|_{L_2(H,F)}^2 \triangleq \sum_{k=1}^\infty \|Te_k\|_F^2 < \infty. \tag{2.22}$$

We denote $L_2(H) \triangleq L_2(H,H)$. By the definition of trace (Definition 2.4.10(b)) we can write

$$\|T\|^2_{L_2(H,F)} \overset{(2.22)}{=} \sum_{k=1}^{\infty} \|Te_k\|^2_F = \sum_{k=1}^{\infty} \langle Te_k, Te_k\rangle = \sum_{k=1}^{\infty} \langle TT^* e_k, e_k\rangle = \mathrm{Tr}(TT^*), \tag{2.23}$$

where T^* is the adjoint of T.

Lemma 2.4.11. *Let H be a separable Hilbert space.*

1. ([287, *Theorem VI.22*]) $\|T\|_{L(H)} \le \|T\|_{L_2(H)} \le \|T\|_{L_1(H)}$ *for* $T \in L_1(H)$. *Moreover,*

$$\|T\|_{L_2(H)} = \|T^*\|_{L_2(H)}, \tag{2.24}$$

where T^* is the adjoint of T.

2. ([287, *p.* 218] *and* [118, *Proposition C.4*])
   (a) $\|TS\|_{L_1(H)} \le \|T\|_{L(H)}\|S\|_{L_1(H)}$ *for* $T \in L(H)$ *and* $S \in L_1(H)$.
   (b) $\|TS\|_{L_2(H)} \le \|T\|_{L(H)}\|S\|_{L_2(H)}$ *for* $T \in L(H)$ *and* $S \in L_2(H)$.
   (c) $\|TS\|_{L_1(H)} \le \|T\|_{L_2(H)}\|S\|_{L_2(H)}$ *for* $T \in L_2(H)$ *and* $S \in L_2(H)$.

We refer to [280] for definitions and notions of semigroups.

Lemma 2.4.12 ([117, Proposition A.1.1]). *Let H be a Hilbert space, and let $A : D(A) \to H$, where $D(A)$ denotes the domain of A. Suppose that A is an infinitesimal generator of an analytic semigroup $S(t)$, $t \ge 0$. Suppose that there exist $M > 0$ and $w > 0$ such that $\|S(t)\|_{L(H)} \le Me^{-wt}$ for all $t \ge 0$. For $\alpha, \gamma \in (0,1)$ and $p > 1$, define the bounded linear operator on $L^p(0,T;H)$ by*

$$R_{\alpha,\gamma}\psi(t) \triangleq \int_0^t (t-\sigma)^{\alpha-1}(-A)^{\gamma}S(t-\sigma)\psi(\sigma)d\sigma, \quad \psi \in L^p(0,T;H).$$

(a) *If $\gamma > 0$ and $\alpha > \gamma + 1 + \frac{1}{p}$, then $R_{\alpha,\gamma}$ is a bounded operator from $L^p(0,T;H)$ into $C^{\alpha-\gamma-\frac{1}{p}}([0,T];D((-A)^{\gamma}))$, where $D((-A)^{\gamma})$ is the domain of $(-A)^{\gamma}$.*
(b) *If $\gamma = 0$ and $\alpha > \frac{1}{p}$, then for all $\delta \in (0, \alpha - \frac{1}{p})$, $R_{\alpha,\gamma}$ is a bounded operator from $L^p(0,T;H)$ into $C^{\delta}([0,T];H)$.*

2.5 Mollifier

Mollification, convoluting potentially a nondifferentiable function with a mollifier, is a standard operation, which can be used to prove, e. g., local solution theory to nonlinear

PDEs on $\mathbb{R}^d$ (see also Remark 2.5.2 concerning $\mathbb{T}^d$). Informally stated, we fix a nonnegative smooth function ρ supported in $B_1(0)$ with mass one, e. g.,

$$\rho(x) = \begin{cases} C\exp(\frac{1}{|x|^2-1}) & \text{if } |x| < 1, \\ 0 & \text{otherwise,} \end{cases}$$

where $C^{-1} \triangleq \int_{\mathbb{R}^d} \exp(\frac{1}{|x|^2-1})dx$ (see [132, Appendix C]). Then we rescale it by $\rho_\epsilon(x) \triangleq \frac{1}{\epsilon^d}\rho(\frac{x}{\epsilon})$ for $\epsilon > 0$. It follows that $\rho_\epsilon \in C^\infty(\mathbb{R}^d)$, $\int_{\mathbb{R}^d} \rho_\epsilon(x)dx = 1$, and $\operatorname{supp} \rho_\epsilon \subset B_\epsilon(0)$. The main purpose of mollification is to approximate f by

$$\mathcal{T}_\epsilon f \triangleq f_\epsilon \triangleq f * \rho_\epsilon. \tag{2.25}$$

The following properties of $\mathcal{T}_\epsilon$ are well known, and their proofs can be found in [132, Theorem 6 in Appendix C], [259, Lemma 3.5], and [147, Theorem 8.14].

Lemma 2.5.1. *Let $f \in L^1_{loc}(\mathbb{R}^d)$ and define $\mathcal{T}_\epsilon f = f_\epsilon$ by (2.25). Then we have the following:*
(a) $f_\epsilon \in C^\infty(\mathbb{R}^d)$.
(b) *$f_\epsilon \to f$ for a. e. x as $\epsilon \to 0$.*
(c) *If $f \in C(\mathbb{R}^d)$, then $f_\epsilon \to f$ uniformly on compact subsets of $\mathbb{R}^d$.*
(d) *If $f \in L^p_{loc}(\mathbb{R}^d)$ for $p \in [1, \infty)$, then $f_\epsilon \to f$ in $L^p_{loc}(\mathbb{R}^d)$. Similarly, if $f \in L^p(\mathbb{R}^d)$ for $p \in [1, \infty)$, then $f_\epsilon \to f$ in $L^p(\mathbb{R}^d)$.*
(e) $D^\alpha \mathcal{T}_\epsilon f = \mathcal{T}_\epsilon D^\alpha f$.
(f) *$\mathcal{T}_\epsilon$ is self-adjoint.*

We note that Lemma 2.5.1(d) is false in the case $p = \infty$; consider, e. g., $d = 1$ and $f = 1_{(0,\infty)} \in L^\infty(\mathbb{R}^d)$.

Commutativity in (e) and self-adjoint property in (f) are easy to explain via Fourier transforms (see Remarks 3.1.12–3.1.13). We postpone other useful estimates on mollifiers to Lemma 3.1.24 after introducing Sobolev spaces. Mollifiers can be used for a comprehensive proof of the local solution theory of the Euler and Navier–Stokes equations (1.1); see [259, Chapter 3, Sections 3.1–3.2]. We will sketch the proof in Section 3.9.

Remark 2.5.2. We briefly explain the extension of mollifiers on $\mathbb{R}^d$ to periodic functions on $\mathbb{T}^d$ for later use, e. g., in convex integration and the study of singular SPDEs. First, we can start from the same ρ and $\rho_\epsilon(x) = \frac{1}{\epsilon^d}\rho(\frac{x}{\epsilon})$. Second, for $x \in \mathbb{T}^d = [0,1]^d$, we define

$$\tilde{\rho}_\epsilon(x) \triangleq \sum_{k\in\mathbb{Z}^d} \rho_\epsilon(x+k) \triangleq \sum_{k\in\mathbb{Z}^d} \epsilon^{-d}\rho\left(\frac{x+k}{\epsilon}\right).$$

For fixed $\epsilon > 0$, this series only has finitely many nonzero terms due to the compact support of ρ. Consequently, $\tilde{\rho}_\epsilon$ is smooth and, additionally, periodic. Third, this allows us to define $f_\epsilon(x) \triangleq (f * \tilde{\rho}_\epsilon)(x) = \int_{\mathbb{T}^d} f(y)\tilde{\rho}_\epsilon(x-y)dy$ for $f \in L^1(\mathbb{T}^d)$.

The following notions of tensor products of Hilbert spaces will be useful to understand covariance operator, to be defined in (4.2a).

Definition 2.5.3 (Tensor products of Hilbert spaces). Let H_1 and H_2 be Hilbert spaces. For $\phi_1 \in H_1$ and $\phi_2 \in H_2$, we denote by $\phi_1 \otimes \phi_2$ the conjugate bilinear form acting on $H_1 \times H_2$ by

$$(\phi_1 \otimes \phi_2)\langle \psi_1, \psi_2 \rangle = (\phi_1, \psi_1)(\phi_2, \psi_2). \tag{2.26}$$

Let $\mathcal{E}$ be the set of all finite linear combinations of such conjugate bilinear forms and define their inner product by

$$(\phi \otimes \psi, \eta \otimes \mu)_{\mathcal{E}} \triangleq (\phi, \eta)(\psi, \mu) \tag{2.27}$$

and extend it to $\mathcal{E}$ by linearity. The tensor product $H_1 \otimes H_2$ is the completion of $\mathcal{E}$ under the inner product $(\cdot, \cdot)_{\mathcal{E}}$ in (2.27).

2.6 Steklov average

We will use the Steklov average in Section 9.3.1, and we refer the interested readers to [20, 73] for details.

Definition 2.6.1 (Steklov average). Let $I \subset \mathbb{R}$ be an interval, let E be a measurable set over $\mathbb{T}^d$, and let $q, r \in [1, \infty]$ and $h > 0$. For $v \in L^r(I; L^q(E))$, we define the Steklov average as

$$v_h(t, \cdot) \triangleq \frac{1}{h} \int_{t-h}^{t} \tilde{v}(s, \cdot) ds, \quad \text{where } \tilde{v}(t, \cdot) \triangleq \begin{cases} v(t, \cdot) & \text{for } t \in I, \\ 0 & \text{for } t \in \mathbb{R} \setminus I. \end{cases} \tag{2.28}$$

Lemma 2.6.2. *Let $I \subset \mathbb{R}$ be an interval, let E be a measurable set over $\mathbb{T}^d$, and let $q, r \in [1, \infty]$ and $h > 0$. For $v \in L^r(I; L^q(E))$, define the Steklov average by (2.28). Then it satisfies the following properties:*
(a) $\|v_h\|_{L^q(I;L^q(E))} \leq \|v\|_{L^q(I;L^q(E))}$, *and* $(\partial_t v_h)(t, x) = \frac{\tilde{v}(t,x) - \tilde{v}(t-h,x)}{h}$.
(b) $v_h \in C(I; L^q(E)) \cap L^\infty(I; L^q(E))$.
(c) $v_h \to v$ *in* $L^r(I; L^q(E))$ *as* $h \searrow 0$.

2.7 Energy identity and the Navier–Stokes problem

As an application of Lebesgue spaces introduced in Definition 2.3.1, we derive the energy identity of the solutions to the Euler and Navier–Stokes equations (1.1) and informally state the Navier–Stokes problem. Let u be a smooth solution to (1.1) on $\mathbb{R}^d$ with decay at infinity, take $L^2(\mathbb{R}^d)$-inner products on (1.1) with u, and use the fact that

$$\int_{\mathbb{R}^d} (u \cdot \nabla) u \cdot u dx = \frac{1}{2} \int_{\mathbb{R}^d} (u \cdot \nabla)|u|^2 dx = -\frac{1}{2} \int_{\mathbb{R}^d} (\nabla \cdot u)|u|^2 dx = 0 \tag{2.29}$$

due to the divergence-free property of u. This leads to the energy identity

$$\|u(t)\|_{L^2}^2 + 2v \int_0^t \|\nabla u\|_{L^2}^2 d\tau = \|u(0)\|_{L^2}^2. \tag{2.30}$$

Identity (2.30) shows that the solution to the Euler equations ((1.1) with $v = 0$) conserves the kinetic energy L_x^2-norm. For a comprehensive list of conserved quantities for the Euler equations, we refer to [259, Section 1.7]. We recall the Millennium Prize problem on the Navier–Stokes equations (cf. [59]).

Statement 1 (Informally stated Navier–Stokes problem on $\mathbb{T}^3$).
(a) For any smooth initial data $u^{\text{in}} \in C^\infty(\mathbb{T}^3)$ such that $\nabla \cdot u^{\text{in}} = 0$, prove that there exists a unique smooth solution $u(t)$, $t > 0$, such that $u(0) = u^{\text{in}}$ and $\sup_{t>0} \|u(t)\|_{L^2}^2 \leq C < \infty$.
(b) Or prove that there exists some smooth initial data $u^{\text{in}} \in C^\infty(\mathbb{T}^3)$ such that $\nabla \cdot u^{\text{in}} = 0$ and the solution u that satisfies $u(0) = u^{\text{in}}$ has a finite-time blowup (singularity).

As we will see in Section 3.9, initial data only needs to be in H_x^s for $s > \frac{5}{2}$ (see Definition 3.1.19) to guarantee that a local-in-time unique solution exists, which also means that the negative direction of the Navier–Stokes problem requires finding initial data such that the corresponding solution exists, is unique, smooths up to some time $T > 0$, and only then blows up.

3 Preliminaries on harmonic analysis

This chapter is relatively comprehensive due to its importance for later discussions. We introduce Fourier analysis and Littlewood–Paley theory in Sections 3.1 and 3.2, respectively. We recall useful interpolation inequalities in Section 3.3 and the Riesz transform and Brezis–Wainger-type inequalities in Section 3.4. We discuss BMO and Hardy spaces in Section 3.5. Then we introduce Bony's paraproducts and remainder and discuss consequences such as product estimates in Section 3.6. We introduce anisotropic analogues of Littlewood–Paley theory, Besov spaces, Bony's paraproducts, and their consequences in Section 3.7. Section 3.8 consists of important commutator estimates.

We conclude this chapter with two sections on applications. Section 3.9 provides a proof of local well-posedness and blowup criterion through an example of the Euler equations using mollifiers. Section 3.10 discusses Galerkin approximation with an example of the Cahn–Hilliard equation (1.22) in detail; a sketch in the case of the SQG equations (1.4) is given as well. Strictly speaking, Galerkin approximation is a classical technique in PDEs that does not directly require advanced harmonic analysis; however, because this chapter on harmonic analysis is the first time we introduce the Fourier transform (see (3.1)) and Sobolev spaces (Definitions 3.1.16 and 3.1.19), we include Galerkin approximation in this chapter instead of Chapter 2.

3.1 Fourier analysis

The following materials are partially from [12, 307, 147], to which we refer the interested readers for details.

Definition 3.1.1. For $f \in L^1(\mathbb{R}^d)$, we define its Fourier transform as

$$\mathcal{F}f(\xi) \triangleq \hat{f}(\xi) \triangleq \int_{\mathbb{R}^d} f(x)e^{-ix\cdot\xi}dx. \tag{3.1}$$

It is readily verified that $\mathcal{F} : L^1(\mathbb{R}^d) \to L^\infty(\mathbb{R}^d)$ is a bounded linear operator.

Definition 3.1.2. The Schwartz space $\mathcal{S}(\mathbb{R}^d)$ is defined as

$$\mathcal{S}(\mathbb{R}^d) \triangleq \{u \in C^\infty(\mathbb{R}^d) : \forall k \in \mathbb{N}, \|u\|_{k,\mathcal{S}} < \infty\},$$
$$\text{where } \|u\|_{k,\mathcal{S}} \triangleq \sup_{|\alpha|\le k, x\in\mathbb{R}^d} (1+|x|)^k |D^\alpha u(x)|,$$

where $\alpha \in \mathbb{N}_0^d$ is a multiindex. We define the space of tempered distributions as $\mathcal{S}'(\mathbb{R}^d)$.

Next, we denote the translation and dilation operators by

$$\tau_a f(x) \triangleq f(x-a),\ a \in \mathbb{R}^d, \quad \text{and} \quad \delta_\lambda f(x) = f(\lambda x),\ \lambda > 0. \tag{3.2}$$

https://doi.org/10.1515/9783111349985-004

Let us recall for convenience that

$$f \quad \text{is homogeneous of degree } k \text{ if} \quad f(\lambda x) = \lambda^k f(x) \quad \forall x, \forall s \neq 0. \tag{3.3}$$

Remark 3.1.3. When f is not a function but a tempered distribution, $\delta_\lambda f$ is understood as follows: $\langle \delta_\lambda f, \phi\rangle = \langle f, \lambda^{-d}\delta_{\lambda^{-1}}\phi\rangle$ for all $\phi \in \mathcal{S}(\mathbb{R}^d)$.

Remark 3.1.4. We recall ([12, Section 1.2.2]) that $\mathcal{F}$ can be extended from $L^1(\mathbb{R}^d)$-functions to tempered distributions via

$$\langle {}^t\mathcal{F}f, \phi\rangle \triangleq \langle f, \mathcal{F}\phi\rangle \quad \text{for } f \in \mathcal{S}'(\mathbb{R}^d),\ \phi \in \mathcal{S}(\mathbb{R}^d), \tag{3.4}$$

where ${}^t\mathcal{F}|_{L^1(\mathbb{R}^d)} = \mathcal{F}|_{L^1(\mathbb{R}^d)}$, so that we will further write $\mathcal{F}$ instead of ${}^t\mathcal{F}$. Of relevance, we point out that (3.4) is familiar with the well-known identity

$$\int_{\mathbb{R}^d} \hat{f}(x)g(x)dx = \int_{\mathbb{R}^d} f(x)\hat{g}(x)dx \quad \forall f, g \in L^1(\mathbb{R}^d). \tag{3.5}$$

The Plancherel theorem is a valuable tool, which, e. g., allows us to shift our estimates to the frequency side, although at the possible expense of having to come back (see Remark 3.5.5).

Theorem 3.1.5 (Plancherel theorem). *$\mathcal{F}$ is an automorphism of $\mathcal{S}'(\mathbb{R}^d)$ with inverse*

$$\mathcal{F}^{-1}f(x) \triangleq (2\pi)^{-d}\int_{\mathbb{R}^d} f(\xi)e^{ix\cdot\xi}d\xi. \tag{3.6}$$

Moreover, $\mathcal{F}$ is an automorphism on $L^2(\mathbb{R}^d)$ such that

$$\|\hat{f}\|_{L^2} = (2\pi)^{\frac{d}{2}}\|f\|_{L^2} \quad \forall f \in L^2(\mathbb{R}^d).$$

Definition 3.1.6 (Fourier symbols, a. k. a. multipliers, [307, p. 94]). Let m be a bounded measurable function on $\mathbb{R}^d$. We can define the linear operator T_m with domain $L^2(\mathbb{R}^d) \cap L^p(\mathbb{R}^d)$, $p \in [1, \infty]$, by

$$\mathcal{F}(T_m f)(\xi) = m(\xi)\hat{f}(\xi), \quad f \in L^2(\mathbb{R}^d) \cap L^p(\mathbb{R}^d). \tag{3.7}$$

Then we say that m is a Fourier symbol (multiplier) for $L^p(\mathbb{R}^d)$ if there exists $A \geq 0$ such that

$$\|T_m(f)\|_{L^p} \leq A\|f\|_{L^p} \quad \forall f \in L^2(\mathbb{R}^d) \cap L^p(\mathbb{R}^d), \tag{3.8}$$

where A is independent of f. We define $\mathcal{M}_p$ for $p \in [1, \infty]$ as the class of multipliers with norm

$$\|m\|_{\mathcal{M}_p} \triangleq \inf\{A \geq 0 : \|T_m(f)\|_{L^p} \leq A\|f\|_{L^p}\ \forall f \in L^2(\mathbb{R}^d) \cap L^p(\mathbb{R}^d)\}.$$

The notion of multiplier may be extended to the case of unbounded m, e. g., $m(\xi) = |\xi|^2$ for $T_m = -\Delta$.

A variant of [307, Theorem 3] is known as the Mihlin–Hörmander multiplier theorem (see [154, Theorem 6.2.7, p. 445] or [23, Theorem 6.1.6, p. 135]).

Lemma 3.1.7 (Multiplier theorem). *Suppose that* $p \in (1,\infty)$, $k \in \mathbb{N}$, $k > \frac{d}{2}$, *and* $m \in C^k(\mathbb{R}^d \setminus \{0\})$. *Assume that*

$$\left|D_\xi^\alpha m(\xi)\right| \lesssim |\xi|^{-|\alpha|} \quad \forall \alpha \in \mathbb{N}_0^d \text{ such that } |\alpha| \le k.$$

Then $m \in \mathcal{M}_p$; *in particular,* T_m *defined by* (3.7) *satisfies* $\|T_m f\|_{L^p} \lesssim_p \|f\|_{L^p}$.

Another powerful result of relevance is the Calderón–Zygmund theory. To explain, let us first give a definition of the Calderón–Zygmund operator.

Definition 3.1.8. An operator $T \in L(L^2(\mathbb{R}^d))$ is called a Calderón–Zygmund operator if there exists a function $K(x,y)$ defined on $\mathbb{R}^d \times \mathbb{R}^d \setminus \{(x,x) \in \mathbb{R}^d \times \mathbb{R}^d\}$ satisfying the following:

(a) There is a constant $C_0 > 0$ such that $|K(x,y)| \le \frac{C_0}{|x-y|^d}$ for all $x, y \in \mathbb{R}^d$.

(b) There are constants $\epsilon > 0$ and $C_1 > 0$ such that $|K(x+z,y) - K(x,y)| \le \frac{C_1|z|^\epsilon}{|x-y|^{d+\epsilon}}$ for all $x,y \in \mathbb{R}^d$ and $z \in B(0, \frac{1}{2}|x-y|)$.

(c) There are constants $\epsilon > 0$ and $C_1 > 0$ such that $|K(x,y+z) - K(x,y)| \le \frac{C_1|z|^\epsilon}{|x-y|^{d+\epsilon}}$ for all $x,y \in \mathbb{R}^d$ and $z \in B(0, \frac{1}{2}|x-y|)$.

(d) For $f, g \in L^2(\mathbb{R}^d;\mathbb{C})$ with compact support, $\operatorname{supp} f \cap \operatorname{supp} g = \emptyset$ implies $\langle Tf, g\rangle = \int_{\mathbb{R}^d}\int_{\mathbb{R}^d} K(x,y)f(y)\overline{g}(x)dxdy$.

The proof of the following statement of Calderón–Zygmund theory can be found in [236, Theorem 6.8]; $\mathcal{H}^1(\mathbb{R}^d)$ and $\mathrm{BMO}(\mathbb{R}^d)$ therein are respectively the Hardy space and the space of bounded mean oscillations (see Definition 3.5.1).

Lemma 3.1.9 (Calderón–Zygmund). *Let T be a Calderón–Zygmund operator. Then T is bounded from $L^p(\mathbb{R}^d)$ to $L^p(\mathbb{R}^d)$ for all $p \in (1,\infty)$, from $\mathcal{H}^1(\mathbb{R}^d)$ to $L^1(\mathbb{R}^d)$, and from $L^\infty(\mathbb{R}^d)$ to* $\mathrm{BMO}(\mathbb{R}^d)$.

To compare Lemmas 3.1.7 and 3.1.9, we recall the following fact from [307, Proposition 2, p. 28].

Lemma 3.1.10. *Let $T \in L(L^2(\mathbb{R}^d))$. Then T commutes with translation operator τ_a from* (3.2) *if and only if there exists a Fourier multiplier $m(\xi)$ such that $\mathcal{F}(Tf)(\xi) = m(\xi)\hat{f}(\xi)$ for all $f \in L^2(\mathbb{R}^d)$, in which case $\|T\|_{L(L^2(\mathbb{R}^d))} = \|m\|_{L^\infty}$.*

Lemma 3.1.10 makes it clear that the hypothesis of multiplier theorem (Lemma 3.1.7) requires T to commute with translation, whereas Calderón–Zygmund theory (Lemma 3.1.9) does not require such a hypothesis. In most of our applications, the multiplier theorem suffices and is simpler to apply due to its hypothesis on Fourier frequency, but Calderón–Zygmund theory is more general.

We now record some useful properties of the Fourier transform.

Lemma 3.1.11 ([12, Proposition 1.24]). *Let $(u, \theta) \in \mathcal{S}'(\mathbb{R}^d) \times \mathcal{S}(\mathbb{R}^d)$, let α be a multi-index, and let $\lambda > 0$ and $a \in \mathbb{R}^d$. Then*

$$(iD)^{\alpha}\hat{u} = \mathcal{F}(x^{\alpha}u), \quad (i\xi)^{\alpha}\hat{u} = \mathcal{F}(D^{\alpha}u), \tag{3.9a}$$

$$e^{-ia\cdot\xi}\hat{u} = \mathcal{F}(\tau_a u), \quad \tau_a\hat{u} = \mathcal{F}(e^{ix\cdot a}u), \tag{3.9b}$$

$$\mathcal{F}(\delta_{\lambda}u) = \lambda^{-d}\delta_{\lambda^{-1}}\hat{u}, \quad \mathcal{F}(u * \theta) = \hat{\theta}\hat{u}. \tag{3.9c}$$

Remark 3.1.12. The commutativity of operators is sometimes easy to understand upon seeing it through the Fourier frequency side. As an example, we come back to Lemma 2.5.1(e):

$$\mathcal{F}(D^{\alpha}\mathcal{J}_{\epsilon}f)(\xi) = (i\xi)^{\alpha}\hat{\rho}_{\epsilon}(\xi)\hat{f}(\xi) = \hat{\rho}_{\epsilon}(\xi)\mathcal{F}(D^{\alpha}f)(\xi) = \mathcal{F}(\mathcal{J}_{\epsilon}D^{\alpha}f)(\xi),$$

and thus applying $\mathcal{F}^{-1}$ shows that $D^{\alpha}\mathcal{J}_{\epsilon}f = \mathcal{J}_{\epsilon}D^{\alpha}f$.

Remark 3.1.13.

(a) It is immediate that if T is an operator with convolution kernel K that is radial, then T is self-adjoint because

$$\begin{aligned}\int_{\mathbb{R}^d}(Tf)(x)g(x)dx &= \int_{\mathbb{R}^d}(K * f)(x)g(x)dx \\ &= \int_{\mathbb{R}^d}f(y)(K * f)(y)dy = \int_{\mathbb{R}^d}f(y)Tf(y)dy.\end{aligned}$$

(b) Using the Fourier transform, we can also observe that whether an operator is self-adjoint or not also follows from whether the Fourier symbol is radial or not, because $\mathcal{F}(Tf)(\xi) = m(\xi)\mathcal{F}f(\xi)$ implies $\mathcal{F}^{-1}(Tf)(x) = m(-x)\mathcal{F}^{-1}f(x)$, so that

$$\int_{\mathbb{R}^d}(Tf)(x)g(x)dx \overset{(3.5)}{=} \int_{\mathbb{R}^d}m(\xi)\hat{f}(\xi)\mathcal{F}^{-1}g(\xi)d\xi \overset{(3.5)}{=} \int_{\mathbb{R}^d}f(x)Tg(x)dx.$$

This particularly implies that the fractional Laplacian, defined next in (3.10), is self-adjoint. Concerning the mollification operator $\mathcal{J}_{\epsilon}$ in (2.25), it is useful to recall that the Fourier transform of a radial function remains radial ([309, Corollary 2.3]). Consequently, we can choose ρ to be radial, so that ρ_{ϵ} becomes radial, which in turn implies that $\hat{\rho}_{\epsilon}$ is radial, and thus $\mathcal{J}_{\epsilon}$ is self-adjoint, as claimed in Lemma 2.5.1(f).

Next, let us define the fractional Laplacian Λ^{β} by

$$\mathcal{F}(\Lambda^{\beta}f)(\xi) = |\xi|^{\beta}\hat{f}(\xi), \quad \beta \in \mathbb{R}. \tag{3.10}$$

It follows from Remark 3.1.13(b) that Λ^β is self-adjoint. A "fractional chain rule" plays its role ubiquitously, and we state it here.

Proposition 3.1.14 (Fractional chain rule). *Let* $\lambda > 0$, *let* δ_λ *denote the dilation operator from* (3.2), *and let* $f \in \mathcal{S}'(\mathbb{R}^d)$. *Then*

$$\Lambda^\beta \delta_\lambda f = \lambda^\beta \delta_\lambda \Lambda^\beta f. \tag{3.11}$$

Proof. The proof follows from the following identity: for all $\phi \in \mathcal{S}(\mathbb{R}^d)$,

$$\mathcal{F}(\Lambda^\beta \delta_\lambda \phi)(\xi) \overset{(3.10)(3.9c)}{=} \left|\frac{\xi}{\lambda}\right|^\beta \frac{\lambda^\beta}{\lambda^d}\hat{\phi}\left(\frac{\xi}{\lambda}\right) \overset{(3.10)}{=} \frac{\lambda^\beta}{\lambda^d}\widehat{\Lambda^\beta \phi}\left(\frac{\xi}{\lambda}\right) \overset{(3.9c)}{=} \lambda^\beta \mathcal{F}(\delta_\lambda \Lambda^\beta \phi)(\xi). \qquad \square$$

Definition 3.1.15. The support of a function f is defined as $\operatorname{supp} f \triangleq \overline{\{x : f(x) \neq 0\}}$. We say that $T \in \mathcal{S}'(\mathbb{R}^d)$ vanishes on an open set $U \subset \mathbb{R}^d$ if $T\phi \equiv 0$ for all $\phi \in \mathcal{S}(\mathbb{R}^d)$ such that $\operatorname{supp} \phi \subset U$, which leads to the following definition of the support of a tempered distribution ([147, p. 284], [287, p. 139]):

$$\operatorname{supp} T \triangleq \mathbb{R}^d \setminus \bigcup \{U \subset \mathbb{R}^d : U \text{ open}, T\phi = 0 \ \forall \phi \in \mathcal{S}(\mathbb{R}^d) \text{ such that } \operatorname{supp} \phi \subset U\}.$$

Example 1. We define the Dirac delta distribution at the origin δ_0 by

$$\delta_0 \phi \triangleq \langle \delta_0, \phi \rangle \triangleq \phi(0) \quad \forall \phi \in C_c^\infty(\mathbb{R}^d). \tag{3.12}$$

Let $U \subset \mathbb{R}^d \setminus \{0\}$ be an open set, and let $\phi \in C_c^\infty(\mathbb{R}^d)$ be supported in U. Then it follows that $\langle \delta_0, \phi \rangle = 0$, $\langle D^\alpha \delta_0, \phi \rangle = 0$ for all multiindices $\alpha \in \mathbb{N}_0^d$ of any order, and, consequently, $\operatorname{supp} \delta_0 = \{0\}$ and $\operatorname{supp} D^\alpha \delta_0 = \{0\}$. Additionally, for all $\phi \in C_c^\infty(\mathbb{R}^d)$, $\langle \mathcal{F}\delta_0, \phi \rangle = \langle 1, \phi \rangle$ and $\langle \mathcal{F} D^\alpha \delta_0, \phi \rangle = \langle (ix)^\alpha, \phi \rangle$, implying that, importantly for our later discussions, the Fourier transforms of nonzero polynomials have support only at the origin. An immediate corollary of this fact is that $\Lambda^s C = 0$ for all $C \in \mathbb{R}$ and $s > 0$, as expected.

Definition 3.1.16. Let $s \in \mathbb{R}$. The homogeneous Sobolev space $\dot{H}^s(\mathbb{R}^d)$ is defined as

$$\dot{H}^s(\mathbb{R}^d) \triangleq \left\{u \in \mathcal{S}'(\mathbb{R}^d) : \hat{u} \in L^1_{\text{loc}}(\mathbb{R}^d),\ \|u\|^2_{\dot{H}^s} \triangleq \int_{\mathbb{R}^d} |\xi|^{2s} |\hat{u}(\xi)|^2 d\xi < \infty\right\}. \tag{3.13}$$

It is always good to know when a space is a Hilbert space, and the following result informs us of this in the case of $\dot{H}^s(\mathbb{R}^d)$.

Lemma 3.1.17 ([12, Proposition 1.34]). $\dot{H}^s(\mathbb{R}^d)$ *is a Hilbert space if and only if* $s < \frac{d}{2}$.

Lemma 3.1.18 (Sobolev embedding [12, Theorem 1.38 and Corollary 1.39]). *The following embeddings are continuous:*

$$\dot{H}^s(\mathbb{R}^d) \hookrightarrow L^{\frac{2d}{d-2s}}(\mathbb{R}^d) \quad for\ s \in \left[0, \frac{d}{2}\right), \tag{3.14a}$$

$$L^{\frac{2d}{d-2s}}(\mathbb{R}^d) \hookrightarrow \dot{H}^s(\mathbb{R}^d) \quad for\ s \in \left(-\frac{d}{2}, 0\right]. \tag{3.14b}$$

Definition 3.1.19. Let $s \in \mathbb{R}$. The inhomogeneous Sobolev space $H^s(\mathbb{R}^d)$ is defined as

$$H^s(\mathbb{R}^d) \triangleq \Big\{ u \in \mathcal{S}'(\mathbb{R}^d) : \tag{3.15}$$

$$\hat{u} \in L^2_{\mathrm{loc}}(\mathbb{R}^d),\ \|u\|^2_{H^s} \triangleq \int_{\mathbb{R}^d} (1 + |\xi|^2)^s |\hat{u}(\xi)|^2 d\xi < \infty \Big\}.$$

Remark 3.1.20. In contrast to the homogeneous case stated in Lemma 3.1.17, $H^s(\mathbb{R}^d)$ is a Hilbert space for all $s \in \mathbb{R}$ ([12, p. 38]).

Definition 3.1.21. Analogously on $\mathbb{T}^d = (\mathbb{R}/\mathbb{Z})^d$, we can define inhomogeneous and homogeneous Sobolev spaces. We restrict ourselves to the case of mean-zero functions in the homogeneous case, where they have these corresponding norms for all $s \in \mathbb{R}$:

$$\|u\|_{H^s(\mathbb{T}^d)} \triangleq \left(\sum_{n \in \mathbb{Z}^d} (1 + 4\pi^2 |n|^2)^s |\hat{u}(n)|^2 \right)^{\frac{1}{2}}, \tag{3.16a}$$

$$\|u\|_{\dot{H}^s(\mathbb{T}^d)} \triangleq \left(\sum_{n \in \mathbb{Z}^d \setminus \{0\}} |n|^{2s} |\hat{u}(n)|^2 \right)^{\frac{1}{2}}. \tag{3.16b}$$

We refer to [22] for details.

Remark 3.1.22. Conveniently, $\dot{H}^s(\mathbb{T}^d)$ is a Hilbert space for all $s \in \mathbb{R}$ in contrast to the case of $\dot{H}^s(\mathbb{R}^d)$ due to the absence of the issue at the origin. Informally, this can be readily explained as follows. Upon inspecting the proof of Lemma 3.1.17 in [12, Proposition 1.34], we see that the restriction of $s < \frac{d}{2}$ comes about when showing that $\hat{u} \in L^2(\mathbb{R}^d; |\xi|^{2s} d\xi)$ is in $L^1_{\mathrm{loc}}(\mathbb{R}^d)$ in accordance with (3.13), which requires

$$\int_{B(0,1)} |\hat{u}(\xi)| d\xi \leq \left(\int_{B(0,1)} |\xi|^{2s} |\hat{u}(\xi)|^2 d\xi \right)^{\frac{1}{2}} \left(\int_0^1 r^{-2s} d\alpha(d) r^{d-1} dr \right)^{\frac{1}{2}} < \infty,$$

where we used the polar coordinates from Definition 2.1.1. In case of mean-zero functions in $\dot{H}^s(\mathbb{T}^d)$, the singularity at the origin is absent, and thus $\dot{H}^s(\mathbb{T}^d)$ is a Hilbert space for all $s \in \mathbb{R}$.

As another example of application of polar coordinates, we show that the Sobolev embedding (3.14a) fails at the endpoint case.

Example 2. By (3.14a) in case $d = 2$, we have $\dot{H}^s(\mathbb{R}^2) \hookrightarrow L^{\frac{2}{1-s}}(\mathbb{R}^2)$ for all $s \in [0,1)$. To show that $\dot{H}^1(\mathbb{R}^2) \hookrightarrow L^\infty(\mathbb{R}^2)$ fails, we consider $f(x) = \ln(\ln(1+\frac{1}{|x|}))$. Because $\lim_{|x|\to 0^+} \ln(\ln(1+\frac{1}{|x|})) = +\infty$, $f \notin L^\infty(\mathbb{R}^2)$. On the other hand, we can write

$$\|f\|_{\dot{H}^1}^2 = \left(\int_0^1 + \int_1^\infty\right) 2\alpha(2) \left|\frac{1}{\ln(1+\frac{1}{r})}\right|^2 \frac{1}{(1+r)^2 r} dr$$

and compute

$$\int_0^1 \left|\frac{1}{\ln(1+\frac{1}{r})}\right|^2 \frac{1}{(1+r)^2 r} dr \le \frac{1}{\ln(2)},$$

$$\int_1^\infty \left|\frac{1}{\ln(1+\frac{1}{r})}\right|^2 \frac{1}{(1+r)^2 r} dr = \int_0^{\ln(2)} \frac{1}{z^2} \frac{e^z - 1}{e^z} dz < \infty,$$

to conclude that $f \in \dot{H}^1(\mathbb{R}^2)$, which implies that $\dot{H}^1(\mathbb{R}^2) \not\hookrightarrow L^\infty(\mathbb{R}^2)$.

Next, we recall the notion of weak derivatives and classical general inhomogeneous Sobolev spaces $W^{s,p}$ for all $s \in \mathbb{N}_0$ and $p \in [1,\infty]$, where informally stated, all weak derivatives up to order s lie in L^p; the homogeneous case requires only that all weak derivatives of order s lie in L^p (e. g., [132, Section 5.2]). Analogously to $\dot{H}^s$ and H^s in Definitions 3.1.16–3.1.19, we consider the following generalization of $s \in \mathbb{R}_{\ge 0}$ using fractional Laplacian (3.10).

Definition 3.1.23 ([211, Appendix]). We define the seminorms and norms of the homogeneous and inhomogeneous Sobolev spaces $\dot{W}^{s,p}$ and $W^{s,p}$ for all $s \in [0,\infty)$ and $p \in [1,\infty]$ by

$$\|f\|_{\dot{W}^{s,p}} \triangleq \|\Lambda^s f\|_{L^p} \quad \text{and} \quad \|f\|_{W^{s,p}} \triangleq \|f\|_{L^p} + \|f\|_{\dot{W}^{s,p}}. \tag{3.17}$$

We give some useful properties of mollified functions on such generalized Sobolev and Hölder spaces; they can be further generalized in various ways.

Lemma 3.1.24. *Define f_ϵ by (2.25) for $\epsilon \in (0,1)$. Then:*

(a) *For all $s \in \mathbb{R}$ and $a \ge 0$,*

$$\|f_\epsilon\|_{H^{s+a}(\mathbb{R}^d)} \lesssim \epsilon^{-a}\|f\|_{H^s(\mathbb{R}^d)}, \quad \|f_\epsilon - f\|_{H^{s-a}(\mathbb{R}^d)} \lesssim \epsilon^a \|f\|_{H^s(\mathbb{R}^d)}; \tag{3.18}$$

(b) *For all $s \in \mathbb{N}_0$ and $a \in (0,1]$,*

$$\|f_\epsilon\|_{C^{s+a}(\mathbb{R}^d)} \lesssim \epsilon^{-a}\|f\|_{C^s(\mathbb{R}^d)}, \quad \|f_\epsilon - f\|_{C^s(\mathbb{R}^d)} \lesssim \epsilon^a \|f\|_{C^{s+a}(\mathbb{R}^d)}; \tag{3.19}$$

(c) *For all $p \in [1,\infty]$,*

$$\|f_\epsilon\|_{C(\mathbb{R}^d)} \lesssim \epsilon^{-\frac{d}{p}}\|f\|_{L^p(\mathbb{R}^d)}; \tag{3.20}$$

(d) *For all* $s \geq 0$ *and* $\alpha \in (0,1]$,

$$\|(fg)_\epsilon - f_\epsilon g_\epsilon\|_{C^s(\mathbb{R}^d)} \lesssim l^{2\alpha - s}\|f\|_{C^\alpha(\mathbb{R}^d)}\|g\|_{C^\alpha(\mathbb{R}^d)}. \tag{3.21}$$

Proof. A particular case of (3.18) can be found in [259, Lemma 3.5] and can be proven as follows:

$$\|f_\epsilon\|^2_{H^{s+\alpha}(\mathbb{R}^d)} \lesssim \sup_{\xi \in \mathbb{R}^d}(1+|\xi|^2)^\alpha |\hat{\rho}_\epsilon(\xi)|^2 \|f\|^2_{H^s(\mathbb{R}^d)} \lesssim \epsilon^{-\alpha}\|f\|_{H^s(\mathbb{R}^d)},$$

$$\|f_\epsilon - f\|^2_{H^{s-\alpha}(\mathbb{R}^d)} = \int_{\mathbb{R}^d} \frac{(\hat{\rho}(\epsilon\xi) - \hat{\rho}(0))^2}{(1+|\xi|^2)^\alpha}(1+|\xi|^2)^s |\hat{f}(\xi)|^2 d\xi \lesssim \epsilon^\alpha \|f\|_{H^s(\mathbb{R}^d)}.$$

Similarly, a particular case of (3.19) can be found in [105, Lemma 1] and can be proven by the following observations in case $s = 0$: for the first inequality, $\|f * \rho_\epsilon\|_{C^0} \lesssim \|f\|_{L^\infty} \lesssim \epsilon^{-\alpha}\|f\|_{L^\infty}$, and

$$\frac{|f_\epsilon(x) - f_\epsilon(y)|}{|x-y|^\alpha} \leq \begin{cases} \frac{2\|f_\epsilon\|_{L^\infty}}{\epsilon^\alpha} \lesssim \epsilon^{-\alpha}\|f\|_{L^\infty} & \text{if } |x-y| > \epsilon, \\ \|f\|_{L^\infty}\|\nabla \rho_\epsilon\|_{L^1}|x-y|^{1-\alpha} \lesssim \epsilon^{-\alpha}\|f\|_{L^\infty} & \text{if } |x-y| \leq \epsilon, \end{cases}$$

whereas for the second inequality, we can estimate

$$|(f_\epsilon - f)(x)| \lesssim \int_{\mathbb{R}^d} \frac{1}{\epsilon^d}\rho\left(\frac{y}{\epsilon}\right)|y|^\alpha dy \|f\|_{C^\alpha} \lesssim \epsilon^\alpha \|f\|_{C^\alpha}.$$

To prove (3.20), for p', a Hölder conjugate of p, we compute

$$\|f_\epsilon\|_{C(\mathbb{R}^d)} \leq \sup_{x \in \mathbb{R}^d}\|f(x-\cdot)\|_{L^p(\mathbb{R}^d)}\left(\int_{\mathbb{R}^d}\left|\frac{1}{\epsilon^d}\rho\left(\frac{y}{\epsilon}\right)\right|^{p'} dy\right)^{\frac{1}{p'}} \approx \epsilon^{-\frac{d}{p}}\|f\|_{L^p(\mathbb{R}^d)}.$$

Finally, the proof of (3.21) can be found in [105, Lemma 1]. □

For various other properties of $\dot{H}^s(\mathbb{R}^d)$ and $H^s(\mathbb{R}^d)$, we refer to [12, Sections 1.3–1.4].

3.2 Littlewood–Paley theory

First, let us define

$$\theta(\lambda D)u \triangleq \mathcal{F}^{-1}(\theta(\lambda \cdot)\mathcal{F}(u)) \quad \text{for } \lambda > 0, \tag{3.22}$$

recall $\mathcal{S}'(\mathbb{R}^d)$ from Definition 3.1.2, and additionally define

$$\mathcal{S}'_h(\mathbb{R}^d) \triangleq \left\{u \in \mathcal{S}'(\mathbb{R}^d) : \lim_{\lambda\to\infty}\|\theta(\lambda D)u\|_{L^\infty} = 0 \; \forall \theta \in C_c^\infty(\mathbb{R}^d)\right\}. \tag{3.23}$$

The following Bernstein's inequality is arguably the backbone of Littlewood–Paley theory and harmonic analysis on PDEs (e. g., [12, Lemma 2.1] and [236, Proposition 3.2]).

Lemma 3.2.1 (Bernstein's inequality). *Fix an annulus* $C(0, r_1, r_2) \triangleq \{\xi \in \mathbb{R}^d : 0 < r_1 \leq |\xi| \leq r_2\}$ *and a ball* $B_{r_1}(0) \triangleq \{\xi \in \mathbb{R}^d : |\xi| \leq r_1\}$. *Then for all* $k \geq 0$, *there exists a constant* $C(k) \geq 0$ *such that for all* $1 \leq p \leq q \leq \infty$ *and* $u \in L^p(\mathbb{R}^d)$, *we have:*
(a) $\operatorname{supp} \hat{u} \subset B_{r_1\lambda}(0)$ *implies*

$$\|\Lambda^k u\|_{L^q(\mathbb{R}^d)} \leq C(k)\lambda^{k+d(\frac{1}{p}-\frac{1}{q})}\|u\|_{L^p(\mathbb{R}^d)}; \tag{3.24}$$

(b) $\operatorname{supp} \hat{u} \subset C(0, r_1\lambda, r_2\lambda)$ *implies*

$$C(k)^{-1}\lambda^k\|u\|_{L^p(\mathbb{R}^d)} \leq \|\Lambda^k u\|_{L^p(\mathbb{R}^d)} \leq C(k)\lambda^k\|u\|_{L^p(\mathbb{R}^d)}. \tag{3.25}$$

In most textbooks, k is taken to be in $\mathbb{N}_0$; such a statement with fractional Laplacian can be found in [334, Proposition 2.3], although without proof. We follow the same line of reasoning in [12, Lemma 2.1] and [79, Lemma 2.1.1], but with fractional chain rule thanks to Proposition 3.1.14, and sketch its proof.

Proof of Lemma 3.2.1.
(a) Let $\phi \in C_0^\infty(\mathbb{R}^d)$ be such that

$$\phi = \begin{cases} 1 & \text{in } B_{r_1}(0), \\ 0 & \text{in } \mathbb{R}^d \setminus B_{r_1+\epsilon}(0), \end{cases} \tag{3.26}$$

for any $\epsilon > 0$ and interpolates smoothly in between. Because $\operatorname{supp} \hat{u} \subset B_{r_1\lambda}(0)$, this implies

$$\hat{u}(\xi) = \hat{u}(\xi)\phi\left(\frac{\xi}{\lambda}\right). \tag{3.27}$$

We observe that

$$\mathcal{F}^{-1}(\delta_{\lambda^{-1}}\phi)(x) = \lambda^d\Phi(\lambda x) = \lambda^d\delta_\lambda\Phi(x) \tag{3.28}$$

if we define $\Phi \triangleq \mathcal{F}^{-1}\phi$, where δ_λ is from (3.2). It follows that $u(x) = \lambda^d(u * \delta_\lambda\Phi)(x)$, so that using

$$\mathcal{F}(\Lambda^k(f_1 * f_2))(\xi) = \mathcal{F}(f_1 * \Lambda^k f_2)(\xi)$$

due to (3.9c), we get $\Lambda^k(f_1 * f_2) = f_1 * \Lambda^k f_2$, and via Young's inequality for convolution (2.16), we can estimate

$$\|\Lambda^k u\|_{L^q} \overset{(3.11)}{=} \lambda^{d+k}\|u * \delta_\lambda\Lambda^k\Phi\|_{L^q} \leq \lambda^{k+d(\frac{1}{p}-\frac{1}{q})}\|u\|_{L^p}\|\Lambda^k\Phi\|_{L^{\frac{pq}{p+pq-q}}}. \tag{3.29}$$

We can further estimate via interpolation inequality (2.13a) and Young's inequality to conclude the proof as follows: since $\phi \in C_c^\infty(\mathbb{R}^d)$,

$$\begin{aligned}\|\Lambda^k \Phi\|_{L^{\frac{pq}{p+pq-q}}} &\leq \left(\frac{q-p}{pq}\right)\|\Lambda^k \Phi\|_{L^\infty} + \left(\frac{p+pq-q}{pq}\right)\|\Lambda^k \Phi\|_{L^1} \\ &\overset{(3.28)(3.9a)}{\lesssim} \int_{\mathbb{R}^d} |(1-\Delta)^d(|\xi|^k \phi)| d\xi = C(k). \end{aligned} \tag{3.30}$$

(b) We may assume that $k > 0$ because (3.25) is clear for $k = 0$. Moreover, the second inequality of (3.25) follows from (3.24); thus we focus only on the first inequality of (3.25). Let $\{\theta_i\}_{i=1}^d \subset C^\infty(\mathbb{S}^{d-1})$ be such that

$$\operatorname{supp} \theta_i \subset \{\xi : \xi_i \neq 0\} \quad \forall\, i \in \{1, \ldots, d\}, \quad \sum_{i=1}^d \theta_i = 1. \tag{3.31}$$

Define a homogeneous function of degree 0 (recall (3.3)) that is equal to θ_i on $\mathbb{S}^{d-1}$, and relabel it as θ_i. Now take $\phi \in C_c^\infty(\mathbb{R}^d)$ such that $\phi \approx 1$ near $C(0, r_1, r_2)$ while $\phi \equiv 0$ on $B_{\frac{r_1}{2}}(0)$, and then define $g \triangleq (2\pi)^d \mathcal{F}^{-1}\phi$. Similarly to (3.27), we can compute

$$\hat{u}(\xi) \overset{(3.31)}{=} \sum_{i=1}^d \theta_i(\xi)|\xi|^{-k}\mathcal{F}(\Lambda^k u)(\xi)\phi\left(\frac{\xi}{\lambda}\right) = \lambda^{-k}\sum_{i=1}^d \phi_{i,k}\left(\frac{\xi}{\lambda}\right)\mathcal{F}(\Lambda^k u)(\xi), \tag{3.32}$$

where

$$\phi_{i,k}(\xi) \triangleq |\xi|^{-k}\theta_i(\xi)\phi(\xi). \tag{3.33}$$

As $\phi \in C_c^\infty(\mathbb{R}^d)$, we have $\phi_{i,k} \in \mathcal{S}(\mathbb{R}^d)$, so that we can define

$$g_{i,k} \triangleq \mathcal{F}^{-1}\phi_{i,k} \tag{3.34}$$

and using (3.32) and (3.34), compute

$$\lambda^k u(x) = \mathcal{F}^{-1}\left(\sum_{i=1}^d \phi_{i,k}\left(\frac{\cdot}{\lambda}\right)\mathcal{F}(\Lambda^k u)\right)(x) = \sum_{i=1}^d \lambda^d(\delta_\lambda g_{i,k})(x) * \Lambda^k u(x). \tag{3.35}$$

Thus similarly to (3.30), using Young's inequality for convolution, we compute

$$\lambda^k \|u\|_{L^p} \overset{(3.35)}{\leq} \sum_{i=1}^d \lambda^d \|\delta_\lambda g_{i,k}\|_{L^1}\|\Lambda^k u\|_{L^p} \leq C(k)\|\Lambda^k u\|_{L^p}.$$

□

We now state a dyadic partition of unity on the frequency side.

Theorem 3.2.2 (Dyadic partition of unity). *Let*

$$\mathcal{C} \triangleq \left\{ \xi \in \mathbb{R}^d : \frac{3}{4} \leq |\xi| \leq \frac{8}{3} \right\} \tag{3.36}$$

(sometimes referred to as "dyadic shells"). Then there exist radial functions

$$\chi, \phi \in C_c^\infty(\mathbb{R}^d) \quad \text{such that} \quad \operatorname{supp}\chi \subset B_{\frac{4}{3}}(0), \quad \operatorname{supp}\phi \subset \mathcal{C}, \quad \chi, \phi \in [0,1] \tag{3.37}$$

such that χ, *together with*

$$\phi_j(\xi) \triangleq \phi(2^{-j}\xi), \tag{3.38}$$

satisfies

$$\chi(\xi) + \sum_{j\geq 0} \phi_j(\xi) = 1 \quad \forall \xi \in \mathbb{R}^d, \tag{3.39a}$$

$$\sum_{j\in\mathbb{Z}} \phi_j(\xi) = 1 \quad \forall \xi \in \mathbb{R}^d \setminus \{0\}, \tag{3.39b}$$

as well as the following property (sometimes referred to as the "almost orthogonality"):

$$|j - j'| \geq 2 \quad \text{implies} \quad \operatorname{supp}\phi_j \cap \operatorname{supp}\phi_{j'} = \emptyset, \tag{3.40a}$$

$$j \geq 1 \quad \text{implies} \quad \operatorname{supp}\chi \cap \operatorname{supp}\phi_j = \emptyset. \tag{3.40b}$$

The proof can be found in [12, Proposition 2.10] (see also [308, Chapter VI, Section 4]). With such χ and ϕ fixed, we can define inhomogeneous and homogeneous Littlewood–Paley operators Δ_j and $\dot{\Delta}_j$, followed by their corresponding low-frequency cut-off operators S_j and $\dot{S}_j$, respectively, as

$$\Delta_j u \triangleq \begin{cases} 0 & \text{if } j \leq -2, \\ (\mathcal{F}^{-1}\chi) * u & \text{if } j = -1, \\ (\mathcal{F}^{-1}\phi_j) * u & \text{if } j \geq 0, \end{cases} \tag{3.41a}$$

$$\dot{\Delta}_j u \triangleq (\mathcal{F}^{-1}\phi_j) * u \quad \forall j \in \mathbb{Z}, \tag{3.41b}$$

$$S_j u \triangleq \sum_{j' \leq j-1} \Delta_{j'} u, \quad \dot{S}_j u \triangleq \chi(2^{-j}D)u \tag{3.41c}$$

(recall (3.22)). Multiplying (3.39a)–(3.39b) by $\hat{u}$ gives us informally

$$\chi(\xi)\hat{u}(\xi) + \sum_{j\geq 0} \phi_j(\xi)\hat{u}(\xi) = \hat{u}(\xi) \quad \forall \xi \in \mathbb{R}^d, \tag{3.42a}$$

$$\sum_{j\in\mathbb{Z}} \phi_j(\xi)\hat{u}(\xi) = \hat{u}(\xi) \quad \forall \xi \in \mathbb{R}^d \setminus \{0\}. \tag{3.42b}$$

We now want to take the Fourier inverse and we can do so in the inhomogeneous case for all $u \in \mathcal{S}'(\mathbb{R}^d)$ (see [12, Proposition 2.12]). The following caveat is in order.

Remark 3.2.3. For the homogeneous case, Example 1 indicates that we certainly cannot do so for nonzero polynomials P because their Fourier transforms have support only at the origin. Fortunately, $\mathcal{S}_h'(\mathbb{R}^d)$ defined in (3.23) eliminates P because for all $\theta \in C_c^\infty(\mathbb{R}^d)$ such that $\theta(0) = 1$ and $\lambda > 0$, we can write $\theta(\lambda D)P \overset{(3.22)}{=} \mathcal{F}^{-1}(\theta(\lambda\cdot)\mathcal{F}(P)) = P$, so that $\lim_{\lambda\to\infty}\|\theta(\lambda D)P\|_{L^\infty} \neq 0$, implying that $P \notin \mathcal{S}_h'(\mathbb{R}^d)$.

Consequently, we can deduce

$$\Delta_{-1}u + \sum_{j\geq 0}\Delta_j u = u \quad \forall\, u \in \mathcal{S}'(\mathbb{R}^d), \quad \sum_{j\in\mathbb{Z}}\dot{\Delta}_j u = u \quad \forall\, u \in \mathcal{S}_h'(\mathbb{R}^d) \tag{3.43}$$

(see [12, Propositions 2.13–2.14]). For convenience later, let us define

$$\Psi \triangleq \mathcal{F}^{-1}\chi, \quad \Phi \triangleq \mathcal{F}^{-1}\phi, \quad \Phi_j(x) \triangleq 2^{jd}\Phi(2^j x) \tag{3.44}$$

which leads to

$$(\mathcal{F}^{-1}\phi_j)(x) = (2\pi)^{-d}\Phi_j(x). \tag{3.45}$$

Applications of Young's inequality for convolution leads to the following facts about the operators Δ_j, $\dot{\Delta}_j$, S_j, and $\dot{S}_j$ in (3.41).

Proposition 3.2.4. *All of $\Delta_j, \dot{\Delta}_j, S_j, \dot{S}_j, j \in \mathbb{Z}$ in (3.41) are linear bounded operators from $L^p(\mathbb{R}^d)$ to $L^p(\mathbb{R}^d)$ for all $p \in [1,\infty]$ with their operator norms independent of p or j.*

Variations of Bernstein's inequality (Lemma 3.2.1) have been developed, e. g., the following by Chen, Miao, and Zhang [85] within the dyadic shells.

Lemma 3.2.5 ([85, Theorem 1.1]). *Let $p \in [2,\infty)$ and $\alpha \in [0,1]$. Then there exist two positive constants c_p and C_p such that for all $f \in \mathcal{S}'(\mathbb{R}^d)$ and $j \in \mathbb{Z}$,*

$$c_p 2^{\frac{2\alpha j}{p}}\|\dot{\Delta}_j f\|_{L^p} \leq \|\Lambda^\alpha(|\dot{\Delta}_j f|^{\frac{p}{2}})\|_{L^2}^{\frac{2}{p}} \leq C_p 2^{\frac{2\alpha j}{p}}\|\dot{\Delta}_j f\|_{L^p}.$$

We are now in a position to define the homogeneous and inhomogeneous Besov spaces $\dot{B}^s_{p,r}(\mathbb{R}^d)$ and $B^s_{p,r}(\mathbb{R}^d)$.

Definition 3.2.6. Let $s \in \mathbb{R}$ and $p, r \in [1,\infty]$. The homogeneous and inhomogeneous Besov spaces are respectively defined as

$$\dot{B}^s_{p,r}(\mathbb{R}^d) \triangleq \{u \in \mathcal{S}_h'(\mathbb{R}^d) : \|u\|_{\dot{B}^s_{p,r}} < \infty\}, \quad \|u\|_{\dot{B}^s_{p,r}} \triangleq \|2^{js}\|\dot{\Delta}_j u\|_{L^p}\|_{l^r(\mathbb{Z})}, \tag{3.46a}$$

$$B^s_{p,r}(\mathbb{R}^d) \triangleq \{u \in \mathcal{S}'(\mathbb{R}^d) : \|u\|_{B^s_{p,r}} < \infty\}, \quad \|u\|_{B^s_{p,r}} \triangleq \|2^{js}\|\Delta_j u\|_{L^p}\|_{l^r(\mathbb{Z})}. \tag{3.46b}$$

Lemma 3.2.7 ([12, Sections 2.3 and 2.7]).
(a) *For al $s \in \mathbb{R}$, the seminorms $\|\cdot\|_{\dot{H}^s}$ and $\|\cdot\|_{\dot{B}^s_{2,2}}$ are equivalent, and $\dot{H}^s(\mathbb{R}^d) \subset \dot{B}^s_{2,2}(\mathbb{R}^d)$; on the other hand, $\dot{H}^s(\mathbb{R}^d) = \dot{B}^s_{2,2}(\mathbb{R}^d)$ for $s < \frac{d}{2}$.*
(b) *For all $s \in \mathbb{R}$, the norms $\|\cdot\|_{H^s}$ and $\|\cdot\|_{B^s_{2,2}}$ are equivalent, and $H^s(\mathbb{R}^d) = B^s_{2,2}(\mathbb{R}^d)$.*

In the subsequent discussion of singular SPDEs, the relation between Besov and Hölder spaces are of importance because the solutions to singular SPDEs are often distribution-valued and $B^\alpha_{\infty,\infty}$ is the natural way to extend the classical Hölder spaces C^α to negative exponents.

Proposition 3.2.8. *If $\alpha \in \mathbb{R}_+ \setminus \mathbb{N}$, then for $U = \mathbb{R}^d$ or $\mathbb{T}^d$,*
(a) $\dot{B}^\alpha_{\infty,\infty}(U) = \dot{C}^{\lfloor\alpha\rfloor,\alpha-\lfloor\alpha\rfloor}(U)$,
(b) $B^\alpha_{\infty,\infty}(U) = C^{\lfloor\alpha\rfloor,\alpha-\lfloor\alpha\rfloor}(U)$.

Our proof shows that if $\alpha \in \mathbb{N}$, then $\dot{B}^\alpha_{\infty,\infty}(U) \supset \dot{C}^{\lfloor\alpha\rfloor,\alpha-\lfloor\alpha\rfloor}(U)$ and $B^\alpha_{\infty,\infty}(U) \supset C^{\lfloor\alpha\rfloor,\alpha-\lfloor\alpha\rfloor}(U)$.

The first statement in the case $U = \mathbb{R}^d$ can be found in [12, Section 2.3, specifically Example on p. 63 and Theorem 2.36], whereas the second statement can be found in [12, Section 2.7, specifically Example on p. 99]. We provide below a simple proof of Proposition 3.2.8 in the case $U = \mathbb{R}^d$, which is thanks to the author's discussion with Prof. David Ullrich; intuitively, the uniformity over $x, y \in U$ in the definition of Hölder norm (or seminorm) from (2.10) is played by the uniformity over j in $l^\infty(\mathbb{Z})$ in the definition of Besov norm (or seminorm) from (3.46). The trick of splitting up the frequencies and optimizing over the threshold in our proof below will also be repeatedly used (e. g., in the proofs of Theorems 3.3.1 and 3.3.3).

Proof of Proposition 3.2.8. We provide the proof in the case $U = \mathbb{R}^d$ and $\alpha \in (0,1)$ because the case $\alpha > 1$ can be proven analogously.
(a) We first prove that $\dot{C}^{0,\alpha}(\mathbb{R}^d) \subset \dot{B}^\alpha_{\infty,\infty}(\mathbb{R}^d)$. We recall $\Phi_j = (2\pi)^d \mathcal{F}^{-1}\phi_j$ from (3.45) and observe that $\int_{\mathbb{R}^d} \Phi_j(y)dy = \phi(0) = 0$ because $\operatorname{supp}\phi \subset \mathcal{C}$ due to (3.37). Thus, assuming that $f \in \dot{C}^{0,\alpha}(\mathbb{R}^d)$, for all $j \in \mathbb{Z}$, we have

$$\begin{aligned} |\dot{\Delta}_j f(x)| &\overset{(3.41b)(3.44)}{=} (2\pi)^{-d} \left| \int_{\mathbb{R}^d} \Phi_j(y)(f(x-y) - f(x))dy \right| \\ &\overset{(3.44)}{\lesssim} \int_{\mathbb{R}^d} 2^{jd} |\Phi(2^j y)||y|^\alpha dy \approx 2^{-j\alpha} \int_{\mathbb{R}^d} |\Phi(z)||z|^\alpha dz \lesssim 2^{-j\alpha}, \end{aligned} \tag{3.47}$$

where we point out that this part of the proof goes through even if $\alpha = 1$. Consequently, definition from (3.46a) shows $\|f\|_{\dot{B}^\alpha_{\infty,\infty}} = \sup_{j\in\mathbb{Z}} 2^{j\alpha} \operatorname{esssup}|\Delta_j f(x)| \lesssim 1$, allowing us to conclude that $\dot{C}^{0,\alpha}(\mathbb{R}^d) \subset \dot{B}^\alpha_{\infty,\infty}(\mathbb{R}^d)$.

Next, we prove that $\dot{C}^{0,\alpha}(\mathbb{R}^d) \supset \dot{B}^\alpha_{\infty,\infty}(\mathbb{R}^d)$. Suppose that $f \in \dot{B}^\alpha_{\infty,\infty}(\mathbb{R}^d)$ and write using (3.43), for M to be chosen subsequently,

$$f \triangleq I_1 + I_2 \quad \text{with} \quad I_1 \triangleq \sum_{j=-\infty}^{M} \dot{\Delta}_j f, \quad I_2 \triangleq \sum_{j=M+1}^{\infty} \dot{\Delta}_j f. \tag{3.48}$$

Since $1 - \alpha > 0$, by Bernstein's inequality (Lemma 3.2.1) we can estimate

$$|I_1(x+h) - I_1(x)| \lesssim \sum_{j=-\infty}^{M} \|\nabla \dot{\Delta}_j f\|_{L^\infty} |h| \overset{(2.2)}{\lesssim} |h| 2^{M(1-\alpha)},$$

$$|I_2(x+h) - I_2(x)| \leq 2 \sum_{j=M+1}^{\infty} \|\dot{\Delta}_j f\|_{L^\infty} \lesssim \sum_{j=M+1}^{\infty} 2^{-j\alpha} \overset{(2.2)}{\lesssim} 2^{-M\alpha}.$$

To optimize, we match the bounds by choosing $M \approx \log_2(|h|^{-1})$ to conclude that $|f(x+h) - f(x)| \lesssim |h|^\alpha$, and hence $\dot{B}^\alpha_{\infty,\infty}(\mathbb{R}^d) \subset \dot{C}^{0,\alpha}(\mathbb{R}^d)$.

(b) First, we prove that $C^{0,\alpha}(\mathbb{R}^d) \subset B^\alpha_{\infty,\infty}(\mathbb{R}^d)$. Suppose that $f \in C^{0,\alpha}(\mathbb{R}^d)$. In this inhomogeneous case, estimate (3.47) goes through for all $j \geq 0$ but not for $j = -1$ because $\int_{\mathbb{R}^d} \Psi(y)dy \neq 0$. Nevertheless, because $f \in C^{0,\alpha}(\mathbb{R}^d)$, for the fixed α, we can bound $|\Delta_{-1} f(x)| \lesssim \|\Psi\|_{L^1} \|f\|_{L^\infty} \lesssim 1 \lesssim 2^\alpha$, and the rest of the proof goes through to show that $f \in B^\alpha_{\infty,\infty}(\mathbb{R}^d)$.

Next, we prove that $C^{0,\alpha}(\mathbb{R}^d) \supset B^\alpha_{\infty,\infty}(\mathbb{R}^d)$. In the case $|h| \in (0,1]$, the proof of $B^\alpha_{\infty,\infty}(\mathbb{R}^d) \subset C^{0,\alpha}(\mathbb{R}^d)$ goes through identically to give us $\dot{B}^\alpha_{\infty,\infty}(\mathbb{R}^d) \subset \dot{C}^{0,\alpha}(\mathbb{R}^d)$ because the only difference is resolved due to $\sum_{j=-\infty}^{M} 2^{j(1-\alpha)} \approx 2^{M(1-\alpha)} \approx \sum_{j=-1}^{M} 2^{j(1-\alpha)}$ by (2.2). In the case $|h| \geq 1$, we are just not guaranteed to be able to find $M \approx \log_2(|h|^{-1}) \in \mathbb{N}_0 \cup \{-1\}$. However, in this case, we can just bound

$$|f(x+h) - f(x)| \leq 2\left(\|\Psi * f\|_{L^\infty} + \sum_{j=0}^{\infty} \|\Delta_j f\|_{L^\infty} \right) \lesssim 2^\alpha + \sum_{j=0}^{\infty} 2^{-j\alpha} \lesssim 1 \lesssim |h|^\alpha$$

since $|h| \geq 1$. □

Remark 3.2.9. We note that [12, Theorem 2.37] sheds light in the homogeneous case with $\alpha = 1$ and [12, Remark 2.38] describes the relationship between $\dot{B}^1_{\infty,\infty}(\mathbb{R}^d)$ with the Zygmund class of functions f that particularly satisfy

$$|f(x+y) + f(x-y) - 2f(x)| \leq C|y| \quad \forall x, y \in \mathbb{R}^d. \tag{3.49}$$

Additionally, $C^\alpha(\mathbb{R}^d) \subsetneq B^\alpha_{\infty,\infty}(\mathbb{R}^d)$ if $\alpha \in \mathbb{N}_0$ and $C^{\alpha-1,1}(\mathbb{R}^d) \subsetneq B^\alpha_{\infty,\infty}(\mathbb{R}^d)$ if $\alpha \in \mathbb{N}$ (see [12, Example on p. 99]).

Definition 3.2.10. For subsequent sections, let us formally define the Besov–Hölder space

$$\mathcal{C}^\alpha \triangleq B^\alpha_{\infty,\infty}, \quad \alpha \in \mathbb{R}. \tag{3.50}$$

Lemma 3.2.11. *We have the following continuous embeddings:*
(a) ([12, *Theorems* 2.40–2.41])

$$\begin{cases} \dot{B}^0_{p,p}(\mathbb{R}^d) \hookrightarrow L^p(\mathbb{R}^d) & \text{if } p \in [1,2], \\ L^p(\mathbb{R}^d) \hookrightarrow \dot{B}^0_{p,2}(\mathbb{R}^d) & \text{if } p \in (1,2], \end{cases} \tag{3.51a}$$

$$\begin{cases} \dot{B}^0_{p,2}(\mathbb{R}^d) \hookrightarrow L^p(\mathbb{R}^d) & \text{if } p \in [2,\infty), \\ L^p(\mathbb{R}^d) \hookrightarrow \dot{B}^0_{p,p}(\mathbb{R}^d) & \text{if } p \in [2,\infty). \end{cases} \tag{3.51b}$$

(b) ([23, *Theorem* 6.4.4])

$$B^0_{p,p}(\mathbb{R}^d) \hookrightarrow L^p(\mathbb{R}^d) \hookrightarrow B^0_{p,2}(\mathbb{R}^d) \quad \text{if } p \in (1,2], \tag{3.52a}$$

$$B^0_{p,2}(\mathbb{R}^d) \hookrightarrow L^p(\mathbb{R}^d) \hookrightarrow B^0_{p,p}(\mathbb{R}^d) \quad \text{if } p \in [2,\infty). \tag{3.52b}$$

They can be generalized to $\dot{W}^{s,p}(\mathbb{R}^d)$ and $W^{s,p}(\mathbb{R}^d)$ as well; however, we only present the case $s = 0$ for simplicity and practical applications.

Let us comment on one of the endpoint cases $p = \infty$ in (3.51b). It is easy to check that $\|\dot{\Delta}_j f\|_{L^\infty} \lesssim \|f\|_{L^\infty}$ by Proposition 3.2.4, so that $\|f\|_{\dot{B}^0_{\infty,\infty}} \lesssim \|f\|_{L^\infty}$. This suggests that $L^\infty(\mathbb{R}^d) \hookrightarrow \dot{B}^0_{\infty,\infty}(\mathbb{R}^d)$. Unfortunately, we realize that, e. g., $1 \in L^\infty(\mathbb{R}^d)$, but we know that $1 \notin \mathcal{S}'_h(\mathbb{R}^d)$ from Remark 3.2.3, and thus $1 \notin \dot{B}^0_{\infty,\infty}(\mathbb{R}^d)$. With this in mind, the following extension of Lemma 3.2.11 in a particular case can be useful. Its proof is based on application of Young's inequality (or Proposition 3.2.4).

Lemma 3.2.12. *For all* $p \in [1,\infty]$,

$$\mathcal{S}'_h(\mathbb{R}^d) \cap L^p(\mathbb{R}^d) \hookrightarrow \dot{B}^0_{p,\infty}(\mathbb{R}^d) \quad \text{and} \quad L^p(\mathbb{R}^d) \hookrightarrow B^0_{p,\infty}(\mathbb{R}^d)$$

continuously.

Remark 3.2.13. Although we will focus on the Besov spaces, we refer to [322, Theorem 2.3.2] for the Triebel–Lizorkin spaces $F^s_{p,r}$, where it is informally stated that, although $\|u\|_{B^s_{p,r}} = \|2^{js}\|\Delta_j u\|_{L^p}\|_{l^r(\mathbb{Z})}$ in (3.46b), the Triebel–Lizorkin analogue would be $\|u\|_{F^s_{p,r}} = \|2^{js}\|\Delta_j u\|_{l^r(\mathbb{Z})}\|_{L^p}$ and enjoys that for all $s \in \mathbb{R}$,

$$\begin{aligned} B^s_{p,r}(\mathbb{R}^d) \subset F^s_{p,r}(\mathbb{R}^d) \subset B^s_{p,p}(\mathbb{R}^d) \quad & \text{if } 1 < r \le p < \infty, \\ B^s_{p,p}(\mathbb{R}^d) \subset F^s_{p,r}(\mathbb{R}^d) \subset B^s_{p,r}(\mathbb{R}^d) \quad & \text{if } 1 < p \le r < \infty, \\ B^s_{p,p}(\mathbb{R}^d) = F^s_{p,p}(\mathbb{R}^d) \quad & \text{if } p \in [1,\infty), \\ F^s_{\infty,\infty}(\mathbb{R}^d) \subsetneq B^s_{\infty,\infty}(\mathbb{R}^d), & \end{aligned}$$

where the last $\subsetneq$ is hinted by $\sup_{j\ge -1} 2^{js}\|\Delta_j f\|_{L^\infty(\mathbb{R}^d)} \le \|\sup_{j\ge -1} 2^{js}|\Delta_j f|\|_{L^\infty(\mathbb{R}^d)}$.

Analogously to the Sobolev embedding (Lemma 3.1.18), we now state Besov embeddings, which are more general ([12, Proposition 2.20]).

Lemma 3.2.14 (Besov embedding theorem). *Let* $1 \le p_1 \le p_2 \le \infty$, $1 \le r_1 \le r_2 \le \infty$, *and* $s \in \mathbb{R}$. *Then we have the following continuous embeddings:*

$$\dot{B}^{s}_{p_1,r_1}(\mathbb{R}^d) \hookrightarrow \dot{B}^{s-d(\frac{1}{p_1}-\frac{1}{p_2})}_{p_2,r_2}(\mathbb{R}^d), \tag{3.53a}$$

$$B^{s}_{p_1,r_1}(\mathbb{R}^d) \hookrightarrow B^{s-d(\frac{1}{p_1}-\frac{1}{p_2})}_{p_2,r_2}(\mathbb{R}^d). \tag{3.53b}$$

The following straightforward observation will useful from time to time.

Lemma 3.2.15. *For all* $p \in [1, \infty]$, $1 \le r_1 \le r_2 \le \infty$, $s \in \mathbb{R}$, *and* $\epsilon > 0$, *we have the following embeddings:*

$$\dot{B}^{s}_{p,r_1}(\mathbb{R}^d) \hookrightarrow \dot{B}^{s}_{p,r_2}(\mathbb{R}^d) \quad \textit{and} \quad B^{s}_{p,r_1}(\mathbb{R}^d) \hookrightarrow B^{s}_{p,r_2}(\mathbb{R}^d), \tag{3.54a}$$

$$B^{s+\epsilon}_{p,\infty}(\mathbb{R}^d) \hookrightarrow B^{s}_{p,1}(\mathbb{R}^d). \tag{3.54b}$$

Proof. Embedding (3.54a) follows from Lemma 2.3.5, and embedding (3.54b) from

$$\|f\|_{B^{s}_{p,1}} \le \|2^{-j\epsilon}\|_{l^1(\mathbb{N}_0\cup\{-1\})}\|2^{j(s+\epsilon)}\|\Delta_j f\|_{L^p}\|_{l^\infty(\mathbb{N}_0\cup\{-1\})} \lesssim \|f\|_{B^{s+\epsilon}_{p,\infty}}. \tag{3.55}$$

Note that (3.55) shows that it cannot be extended to the homogeneous case. □

Sometimes the following fact or its proof is also useful.

Lemma 3.2.16. *Let* $p, q \in [1, \infty]$. *Suppose that*

$$s \begin{cases} \ge 0 & \textit{if } q = \infty, \\ > 0 & \textit{if } q < \infty. \end{cases}$$

Then

$$\mathcal{S}'_h(\mathbb{R}^d) \cap B^{s}_{p,q}(\mathbb{R}^d) \hookrightarrow \dot{B}^{s}_{p,q}(\mathbb{R}^d) \tag{3.56}$$

continuously.

Proof. The claim follows once we prove $\|f\|_{\dot{B}^{s}_{p,q}} \lesssim \|f\|_{B^{s}_{p,q}}$ for all $f \in \mathcal{S}'_h(\mathbb{R}^d) \cap B^{s}_{p,q}(\mathbb{R}^d)$; the reason for the need to intersection with $\mathcal{S}'_h(\mathbb{R}^d)$ is the same as that explained for Lemma 3.2.12. First, we split

$$\|f\|_{\dot{B}^{s}_{p,q}} \lesssim \sum_{i=1}^{3} \mathrm{II}_i, \tag{3.57}$$

where

$$\mathrm{II}_1 \triangleq \big\|2^{js}\|\dot\Delta_j f\|_{L^p}\big\|_{l^q(j\le -2)}, \tag{3.58a}$$
$$\mathrm{II}_2 \triangleq 2^{-s}\|\dot\Delta_{-1} f\|_{L^p}, \tag{3.58b}$$
$$\mathrm{II}_3 \triangleq \big\|2^{js}\|\dot\Delta_j f\|_{L^p}\big\|_{l^q(j\ge 0)}. \tag{3.58c}$$

For all $j \le -2$, we can write $\dot\Delta_j f = \mathcal{F}^{-1}\chi * \dot\Delta_j f$ and bound

$$\mathrm{II}_1 = \big\|2^{js}\|\dot\Delta_j \Delta_{-1} f\|_{L^p}\big\|_{l^q(j\in\{-2,-1,0\})} \lesssim \|\Delta_{-1} f\|_{L^p} \tag{3.59}$$

due to (3.41) and (3.40) and Proposition 3.2.4. For II_2, we can write $\dot\Delta_{-1} = \sum_{j\ge -1}\dot\Delta_{-1}\Delta_j$ and use (3.40) and Proposition 3.2.4 to bound it by $\|\Delta_{-1} f\|_{L^p} + \|\Delta_0 f\|_{L^p}$. Therefore

$$\|f\|_{\dot B^s_{p,q}} \overset{(3.57)}{\lesssim} \sum_{i=1}^{3} \mathrm{II}_i \lesssim \|\Delta_{-1} f\|_{L^p} + \|\Delta_0 f\|_{L^p} + \mathrm{II}_3 \approx \big\|2^{js}\|\Delta_j f\|_{L^p}\big\|_{l^q(j\ge -1)},$$

which completes the proof. □

3.3 Interpolation inequalities

We now discuss the Gagliardo–Nirenberg inequality, which is akin to the interpolation inequality (2.13a) but involving derivatives. For a comprehensive classical study with derivatives of only integers, we refer to [260, Chapter 3]. We give two examples, more for the purpose of elaborating on cheap tricks to deduce such original inequalities rather than their sharpness.

Theorem 3.3.1 (Gagliardo–Nirenberg inequality I). *Let*

$$1 \le p < q \le \infty, \quad 1 \le r \le q, \quad s > d\left(\frac{1}{r} - \frac{1}{q}\right), \tag{3.60}$$

and let $f \in \mathcal{S}'_h(\mathbb{R}^d) \cap L^p(\mathbb{R}^d) \cap \dot W^{s,r}(\mathbb{R}^d)$. *Then*

$$\|f\|_{L^q} \lesssim \|f\|_{L^p}^{\epsilon}\|f\|_{\dot W^{s,r}}^{1-\epsilon}, \quad \epsilon \in (0,1), \quad \frac{1}{q} = \epsilon\left(\frac{1}{p}\right) + (1-\epsilon)\left(\frac{1}{r} - \frac{s}{d}\right). \tag{3.61}$$

Remark 3.3.2. We can allow the case $p = q \in [1,\infty]$, which is the trivial case of $\epsilon = 1$ and $\|f\|_{L^q} \le \|f\|_{L^p}$. Moreover, in the case $q < \infty$, we can allow $s = d(\frac{1}{r} - \frac{1}{q})$ and $\epsilon = 0$, which is the Sobolev embedding theorem (Lemma 3.1.18).

Proof of Theorem 3.3.1. First, we observe that the case $f \equiv 0$ is trivial; thus we assume that $f \not\equiv 0$. Using (3.42b), we compute

$$\|f\|_{L^q} \lesssim \sum_{j=-\infty}^{M} 2^{jd(\frac{1}{p}-\frac{1}{q})}\|\dot\Delta_j f\|_{L^p} + \sum_{j=M}^{\infty} 2^{jd(\frac{1}{r}-\frac{1}{q})}2^{-js}\|\dot\Delta_j \Lambda^s f\|_{L^r},$$

where we used Bernstein's inequality (Lemma 3.2.1) and the hypothesis that $p < q$ and $r \le q$ by (3.60). We can continue to bound by Lemma 3.2.12 and (2.3):

$$\begin{aligned}\|f\|_{L^q} &\lesssim \left(\sum_{j=-\infty}^{M} 2^{jd(\frac{1}{p}-\frac{1}{q})}\right)\|f\|_{\dot{B}^0_{p,\infty}} + \left(\sum_{j=M+1}^{\infty} 2^{j(d(\frac{1}{r}-\frac{1}{q})-s)}\right)\|\Lambda^s f\|_{\dot{B}^0_{r,\infty}} \\ &\lesssim 2^{\mathrm{M}d(\frac{1}{p}-\frac{1}{q})}\|f\|_{L^p} + 2^{M(d(\frac{1}{r}-\frac{1}{q})-s)}\|f\|_{\dot{W}^{s,r}}.\end{aligned}$$

Notice that if, e. g., we estimate

$$\sum_{j=-\infty}^{M} 2^{jd(\frac{1}{p}-\frac{1}{q})}\|\dot{\Delta}_j f\|_{L^p} \le \left(\sum_{j=-\infty}^{M} 2^{jdp'(\frac{1}{p}-\frac{1}{q})}\right)^{\frac{1}{p'}} \|f\|_{\dot{B}^0_{p,p}} \overset{(2.2)}{\lesssim} 2^{\mathrm{M}d(\frac{1}{p}-\frac{1}{q})}\|f\|_{\dot{B}^0_{p,p}},$$

where p' is the Hölder conjugate of p, and rely on (3.51b), then we will need $2 \le p$ and similarly $2 \le r$. We now match the two bounds to optimize via $M \approx \frac{1}{d(\frac{1}{p}-\frac{1}{r})+s}\log_2(\frac{\|f\|_{\dot{W}^{s,r}}}{\|f\|_{L^p}})$ and deduce (3.61). □

The proof of Theorem 3.3.3 can be generalized in various ways, and we present one example.

Theorem 3.3.3 (Gagliardo–Nirenberg inequality II). *Let*

$$1 \le p \le q \le \infty, \quad 1 \le r \le q \quad s > d\left(\frac{1}{r} - \frac{1}{q}\right) + 1, \tag{3.62}$$

and let $f \in \mathcal{S}_h'(\mathbb{R}^d) \cap L^p(\mathbb{R}^d) \cap \dot{W}^{s,r}(\mathbb{R}^d)$. *Then*

$$\|\nabla f\|_{L^q} \lesssim \|f\|_{L^p}^{\epsilon}\|f\|_{\dot{W}^{s,r}}^{1-\epsilon} \quad \epsilon \in (0,1), \quad \frac{1}{q} - \frac{1}{d} = \epsilon\left(\frac{1}{p}\right) + (1-\epsilon)\left(\frac{1}{r} - \frac{s}{d}\right). \tag{3.63}$$

We notice that in the Gagliardo–Nirenberg inequality I, we needed $p < q$, but now we can allow $p \le q$.

Proof of Theorem 3.3.3. Again, we observe that the case $f \equiv 0$ is trivial; thus we assume that $f \not\equiv 0$. By (3.42b), Bernstein's inequality (Lemma 3.2.1), Lemma 3.2.12, and (2.3), we compute

$$\begin{aligned}\|\nabla f\|_{L^q} &\lesssim \sum_{j=-\infty}^{M} 2^{jd(\frac{1}{p}-\frac{1}{q})+j}\|\dot{\Delta}_j f\|_{L^p} + \sum_{j=M}^{\infty} 2^{jd(\frac{1}{r}-\frac{1}{q})}2^{j(1-s)}\|\dot{\Delta}_j\Lambda^s f\|_{L^r} \\ &\lesssim 2^{M(d(\frac{1}{p}-\frac{1}{q})+1)}\|f\|_{L^p} + 2^{M(d(\frac{1}{r}-\frac{1}{q})+1-s)}\|f\|_{\dot{W}^{s,r}}.\end{aligned}$$

We match the bounds to optimize with $M \approx \frac{1}{d(\frac{1}{p}-\frac{1}{r})+s}\log_2(\frac{\|f\|_{\dot{W}^{s,r}}}{\|f\|_{L^p}})$ and deduce (3.63). □

Remark 3.3.4. We notice that the Sobolev embedding (Lemma 3.1.18) and Bernstein's inequality (Lemma 3.2.1) depend on spatial dimension, and thus the Gagliardo–Nirenberg inequalities (Theorems 3.3.1–3.3.3) depend on spatial dimension. This is not always the case. First, as a special case of Theorem 3.3.3 with $q = p = r = s = 2$, we can deduce the following Sobolev interpolation inequality, which is independent of spatial dimension:

$$\|\nabla f\|_{L^2} \lesssim \|f\|_{L^2}^{\frac{1}{2}} \|\Delta f\|_{L^2}^{\frac{1}{2}}. \tag{3.64}$$

The multiplicative constant can be sharpened to be one; this can be verified via Plancherel theorem (Theorem 3.1.5).

More generally, we record [12, Theorem 2.44] here. For $p, r \in (1, \infty]$ and $\sigma, s \in (0, \infty)$ such that $\sigma < s$,

$$\|u\|_{\dot{W}^{\sigma,q}} \lesssim \|u\|_{L^p}^{\epsilon} \|u\|_{\dot{W}^{s,r}}^{1-\epsilon}, \quad \epsilon = 1 - \frac{\sigma}{s}, \quad \frac{1}{q} = \epsilon\left(\frac{1}{p}\right) + (1-\epsilon)\left(\frac{1}{r}\right). \tag{3.65}$$

We observe that ϵ, q, p, r do not depend on d. Indeed, this is because $\epsilon = 1 - \frac{\sigma}{s}$ is fixed, whereas Theorems 3.3.1 and 3.3.3 allowed freedom over $\epsilon \in (0, 1)$; e. g., if we choose $\epsilon = 1 - \frac{1}{s}$ in (3.63), then the condition $\frac{1}{q} - \frac{1}{d} = \epsilon(\frac{1}{p}) + (1-\epsilon)(\frac{1}{r} - \frac{s}{d})$ therein becomes $\frac{1}{q} = (1 - \frac{1}{s})(\frac{1}{p}) + \frac{1}{s}(\frac{1}{r})$, independent of d.

To give another example, we have [367, eq. (271)]

$$\|f\|_{L^4} \overset{(3.52b)}{\lesssim} \|f\|_{B^0_{4,2}} \lesssim \left(\sum_{m \geq -1} \|\Delta_m f\|_{L^2} \|\Delta_m f\|_{L^\infty}\right)^{\frac{1}{2}} \lesssim \|f\|_{H^{\frac{1}{4}}}^{\frac{1}{2}} \|f\|_{B^{-\frac{1}{4}}_{\infty,2}}^{\frac{1}{2}}, \tag{3.66}$$

which is independent of spatial dimension because the estimates depend only on Hölder's and interpolation inequalities (2.13a).

3.4 Riesz transform, Brezis–Wainger-type inequalities

Before we introduce the Riesz transform vector, which plays a crucial role in analysis of PDEs in fluid mechanics, it is convenient to recall the Riesz potential and its properties.

Definition 3.4.1 (Riesz potential). For $\alpha \in (0, d)$, the Riesz potential is defined as

$$(I_\alpha f)(x) \triangleq \frac{1}{\gamma(\alpha)} \int_{\mathbb{R}^d} |x - y|^{-d+\alpha} f(y) dy \quad \text{where} \quad \gamma(\alpha) \triangleq \pi^{\frac{d}{2}} 2^{\alpha} \frac{\Gamma(\frac{\alpha}{2})}{\Gamma(\frac{d}{2} - \frac{\alpha}{2})}. \tag{3.67}$$

The Riesz potential has various important properties; here we limit ourselves to the following result and refer the interested readers to [307, Chapter V] for further details.

Lemma 3.4.2 ([307, Lemma 1(b), Chapter V]). *Let* $\alpha \in (0, d)$. *Then* $\mathcal{F}(I_\alpha f)(\xi) = C(\alpha)|\xi|^{-\alpha}\hat{f}(\xi)$ *in the sense that for all* $f, g \in \mathcal{S}(\mathbb{R}^d)$,

$$\int_{\mathbb{R}^d} I_\alpha(f)(\xi)\overline{g}(\xi)d\xi = C(\alpha)\int_{\mathbb{R}^d} |\xi|^{-\alpha}\hat{f}(\xi)\hat{\overline{g}}(\xi)d\xi.$$

Definition 3.4.3 (Riesz transform). The Riesz transform vector $\mathcal{R} = (\mathcal{R}_1, \dots, \mathcal{R}_d)$ is the singular integral operator defined by

$$\widehat{\mathcal{R}f}(\xi) \triangleq i\frac{\xi}{|\xi|}\hat{f}(\xi) \quad \text{or} \quad \mathcal{R}_j f(x) \triangleq C(d)\lim_{\epsilon\to 0}\int_{\mathbb{R}^d\setminus B_\epsilon(x)} \frac{x_j - y_j}{|x-y|^{d+1}} f(y)dy. \tag{3.68}$$

The two definitions in (3.68) are equivalent; e. g., by observing that $\frac{x_j}{|x|^{d+1}} = \frac{1}{1-d}\partial_{x_j}|x|^{1-d}$, we can informally compute, using (3.9a) and Lemma 3.4.2,

$$\int_{\mathbb{R}^d}\left(\frac{x_j}{|x|^{d+1}} * f\right)(x)\hat{\phi}(x)dx = C(d)\int_{\mathbb{R}^d}\frac{i\xi_j}{|\xi|}\hat{f}(\xi)\phi(\xi)d\xi,$$

which clarifies (3.68). Thanks to Lemma 3.1.7 or 3.1.9, the Riesz transform satisfies

$$\|\mathcal{R}f\|_{L^p} \le C(p)\|f\|_{L^p} \quad \forall p \in (1, \infty). \tag{3.69}$$

For relevance, we also mention that the Hilbert transform $\mathcal{H}$ is a scalar version of the Riesz transform defined by $\widehat{\mathcal{H}f}(\xi) = i\,\mathrm{sgn}\,(\xi)\hat{f}(\xi)$. For further details, we refer to, e. g., [307, Chapters II and III].

In case the frequency support is bounded away from the origin, (3.69) may be extended to $p \in [1, \infty]$ as follows.

Lemma 3.4.4. *The Riesz transform is bounded in* $\dot{B}^s_{p,r}(\mathbb{R}^d)$ *for all* $s \in \mathbb{R}$ *and* $p, r \in [1, \infty]$, *with a bound independent of* p, r, s.

Proof. We recall $\mathcal{C}$ from (3.36), which implies that ϕ from (3.37) with $\operatorname{supp}\phi \subset \mathcal{C}$ is compactly supported away from the origin. Thus we can define $m_k(\xi) \triangleq \frac{i\xi_k}{|\xi|}$ as the Fourier symbol of $\mathcal{R}_k$ and then $a_k(\xi) \triangleq m_k(\xi)\phi(\xi)$ to see that $a_k \in C_c^\infty(\mathbb{R}^d)$. Consequently, $L_k \triangleq \mathcal{F}^{-1}a_k \in \mathcal{S}(\mathbb{R}^d) \subset L^1(\mathbb{R}^d)$. Now we can take advantage of the fact that the Fourier symbol of $\mathcal{R}_k$, namely $m_k(\xi)$, is homogeneous of degree 0 (recall (3.3)) and write

$$\mathcal{F}(\mathcal{R}_k\dot{\Delta}_j f)(\xi) \overset{(3.68)(3.41b)}{=} i\frac{\xi_k}{|\xi|}\phi_j(\xi)\hat{f}(\xi) \overset{(3.38)}{=} a_k(2^{-j}\xi)\hat{f}(\xi). \tag{3.70}$$

Due to (3.9c), we have $\mathcal{F}(2^{jd}L_k(2^j\cdot))(\xi) = a_k(2^{-j}\xi)$, and therefore $\mathcal{R}_k\dot{\Delta}_j f = 2^{jd}L_k(2^j\cdot) * \dot{\Delta}_j f$. Therefore, for all $j \in \mathbb{Z}$, $k \in \{1, \dots, d\}$, and $p \in [1, \infty]$,

$$\|\dot{\Delta}_j\mathcal{R}_k f\|_{L^p} \lesssim \|\dot{\Delta}_j f\|_{L^p} \tag{3.71}$$

by Young's inequality for convolution, and, consequently, $\|\mathcal{R}_k f\|_{\dot{B}^s_{p,r}} \lesssim \|2^{js}\|\dot{\Delta}_j f\|_{L^p}\|_{l^r} \approx \|f\|_{\dot{B}^s_{p,r}}$. □

Example 3. Besides the explicit appearance of $\mathcal{R}^{\perp}\theta = \nabla^{\perp}(-\Delta)^{-\frac{1}{2}}\theta$ in the SQG equations (1.4), the utility of the Riesz transform in fluid mechanics is ubiquitous. For example, if f is divergence-free so that $\sum_{l=1}^{d}\xi_l\hat{f}_l(\xi) = 0$, then for $\xi \neq 0$, we can write

$$\xi_j\hat{f}_k(\xi) = \sum_{l=1}^{d}\frac{\xi_j\xi_l}{|\xi|^2}\xi_l\hat{f}_k(\xi) - \frac{\xi_j\xi_k}{|\xi|^2}\xi_l\hat{f}_l(\xi) = \sum_{l=1}^{d}\frac{\xi_j\xi_l}{|\xi|^2}(\xi_l\hat{f}_k - \xi_k\hat{f}_l)(\xi),$$

so that taking the Fourier inverse gives

$$\partial_j f_k = -\sum_{l=1}^{d}\mathcal{R}_j\mathcal{R}_l(\partial_l f_k - \partial_k f_l), \tag{3.72}$$

to which we can apply (3.69) or Lemma 3.4.4, depending on the frequency support of f. Consequently, we can deduce the following inequality (3.73), which is stated in [79, Theorem 3.1.1] modulo the explicit constant $\frac{p^2}{p-1}$: if f is smooth and $\nabla\cdot f = 0$, then for all $p \in (1,\infty)$, there exists a constant $C \geq 0$ such that

$$\|\nabla f\|_{L^p} \leq C\frac{p^2}{p-1}\|\nabla\times f\|_{L^p}, \tag{3.73}$$

where the curl operator in the case $d \geq 4$ was defined by (2). In the case $p = 2$, (3.73) immediately follows as an equality with $C = 1$ thanks to (2.1a), which leads to

$$\|\nabla f\|_{L^2}^2 = -\int_{\mathbb{R}^d}(\Delta f)\cdot f dx = \int_{\mathbb{R}^d}\nabla\times(\nabla\times f)\cdot f dx = \|\nabla\times f\|_{L^2}^2. \tag{3.74}$$

We note that (3.73) can be used to bound ∇u in terms of vorticity $\omega = \nabla\times u$ in (1.2).

Finally, we also mention that if (u,π) solves (1.1), then taking the divergence and applying the divergence-free property lead to

$$\operatorname{div}\operatorname{div}(u\otimes u) = -\Delta\pi,$$

so that $\pi = (-\Delta)^{-1}\operatorname{div}\operatorname{div}(u\otimes u)$, allowing us to express π in terms of the double Riesz transform applied on u.

Now we give another example of application of the Riesz transform.

Lemma 3.4.5. *For all $p \in (1,\infty)$, there exist constants $C_1(p), C_2(p) \geq 0$ such that*

$$C_1(p)\|\Lambda f\|_{L^p} \leq \|\nabla f\|_{L^p} \leq C_2(p)\|\Lambda f\|_{L^p} \quad \forall f \in W^{1,p}(\mathbb{R}^d). \tag{3.75}$$

The author learned this proof from Prof. David Ullrich in the case $d = 2$; in fact, there is similarity of the following proof to that of Lemma 3.2.1(b), e. g., how singularity is carefully avoided in (3.32).

Proof of Lemma 3.4.5. First, because $\widehat{\partial_j f}(\xi) = \widehat{\mathcal{R}_j \Lambda f}(\xi)$ for all $j \in \{1, \dots, d\}$, $\partial_j f = \mathcal{R}_j \Lambda f$, and thus (3.69) gives us the second inequality in (3.75). For the first inequality, we can cover $\mathbb{S}^{d-1} \subset \mathbb{R}^d$ by $\{U_j\}_{j=1}^d$, $U_j \triangleq \{x \in \mathbb{S}^{d-1} : |x_j| > \frac{1}{d}\}$. Then by Lemma 2.3.9 there exists a partition of unity on $\mathbb{S}^{d-1}$ subordinate to $\{U_j\}_{j=1}^d$ consisting of continuous functions $\phi_j : \mathbb{S}^{d-1} \to [0,1], j \in \{1, \dots, d\}$, such that $\operatorname{supp} \phi_j \subset U_j$ and $\sum_{j=1}^d \phi_j(\xi) = 1$ for all $\xi \in \mathbb{S}^{d-1}$. Then we define $h_j(\xi) \triangleq \phi_j(\frac{\xi}{|\xi|}), j \in \{1, \dots, d\}$, for all $\xi \in \mathbb{R}^d \setminus \{0\}$, so that $h_j \in C^\infty(\mathbb{R}^d \setminus \{0\})$ and $\sum_{j=1}^d h_j = 1$ on $\mathbb{R}^d \setminus \{0\}$. It follows that $\psi_j(\xi) \triangleq h_j(\xi)\frac{|\xi|}{\xi_j}$ is well-defined and $\sum_{j=1}^d \psi_j(\xi)\xi_j = |\xi|$. Thus, defining the operator T_j by the Fourier symbol $m_j(\xi) = i\psi_j(\xi)$, we deduce that $\Lambda f = -\sum_{j=1}^d T_j(\partial_j f)$. By Lemma 3.1.7 we conclude that $\|\Lambda f\|_{L^p} \lesssim_p \|\nabla f\|_{L^p}$. □

Various other properties of fractional Laplacian were investigated; we state the following positivity lemma from [106], which is useful in handling the diffusive term in an L_x^p-estimate, and we refer the interested readers to references therein.

Lemma 3.4.6 (Positivity lemma, [106, Lemma 2.5]). *For all $\alpha \in [0,2]$, $U = \mathbb{R}^d$ or $\mathbb{T}^d$, and $p \in [1, \infty)$, we have $\int_U \Lambda^\alpha \theta |\theta|^{p-2}\theta dx \geq 0$ for all $\theta \in W^{\alpha,p}(U)$.*

We saw in (3.73) that even for smooth divergence-free functions, we have $\|\nabla f\|_{L^p} \lesssim_p \|\nabla \times f\|_{L^p}$ only for $p \in (1, \infty)$. The classical Brezis–Wainger inequality [37] states that it can be extended to the case $p = \infty$ at the expense of multiplication by a logarithmic function of $H^s(\mathbb{R}^d)$-norm for $s > \frac{d}{2}$ (see (3.76) for a precise statement). It turns out that Littlewood–Paley theory readily allows us to improve this in two different ways, as we will show in Lemma 3.4.7. First, instead of $\|\nabla \times f\|_{L^\infty}$, we will bound by $\|\nabla \times f\|_{\dot{B}^0_{\infty,\infty}}$, recalling that $\|\cdot\|_{\dot{B}^0_{\infty,\infty}} \lesssim \|\cdot\|_{L^\infty}$ by Lemma 3.2.12. Second, instead of bound by $\log_2(2 + \|f\|_{H^s})$, we will bound by $\log_2(2 + \|f\|_{B^s_{2,\infty}})$, recalling that $\|\cdot\|_{B^s_{2,\infty}} \lesssim \|\cdot\|_{B^s_{2,2}} \approx \|\cdot\|_{H^s}$ due to Lemma 2.3.5 and Lemma 3.2.7(b).

Lemma 3.4.7 (Variant of Brezis–Wainger inequality, [339, Lemma 2.3]). *Let $d \geq 2$ and $s > 1 + \frac{d}{2}$. Then, for smooth f such that $\nabla \cdot f = 0$,*

$$\|\nabla f\|_{L^\infty} \lesssim \|f\|_{L^2} + \|\nabla \times f\|_{\dot{B}^0_{\infty,\infty}} \log_2(2 + \|f\|_{B^s_{2,\infty}}) + 1, \tag{3.76}$$

where the curl operator in the case $d \geq 4$ was defined in (2).

Proof. Without loss of generality, we assume that $s \in (1 + \frac{d}{2}, 2 + \frac{d}{2}]$. Because (3.41) shows that $\dot{\Delta}_j = \Delta_j$ for all $j \in \mathbb{N}_0$, we can rely on (3.71) to estimate

$$
\begin{aligned}
\|\nabla f\|_{L^\infty} &\overset{(3.43)}{\lesssim} \|\Delta_{-1}\nabla f\|_{L^\infty} + \sum_{j=0}^{M-1}\|\Delta_j \nabla f\|_{L^\infty} + \sum_{j=M}^{\infty}\|\Delta_j\nabla f\|_{L^\infty} \\
&\overset{(3.72)(3.24)(3.71)}{\lesssim} \|f\|_{L^2} + \|\nabla\times f\|_{B^0_{\infty,\infty}} M + \sum_{j=M}^{\infty} 2^{j(\frac{d}{2}+1-s)}2^{j(s-1)}\|\Delta_j\nabla f\|_{L^2} \\
&\overset{(2.2)}{\lesssim} \|f\|_{L^2} + \|\nabla\times f\|_{\dot{B}^0_{\infty,\infty}} M + 2^{M(\frac{d}{2}+1-s)}\|f\|_{B^s_{2,\infty}}.
\end{aligned}
$$

Choosing $M \approx (\frac{1}{s-1-\frac{d}{2}})\log_2(2+\|u\|_{B^s_{2,\infty}})$ leads to (3.76). □

Lemma 3.4.7 is a variant of Brezis–Wainger inequality [37] and allows us to immediately deduce the classical Beale–Kato–Majda criterion due to [17] in Theorem 3.9.1, in fact, a slight improvement, because $\|\cdot\|_{\dot{B}^0_{\infty,\infty}} \lesssim \|\cdot\|_{L^\infty}$ by Lemma 3.2.12. We postpone this until the introduction of commutator estimates in Section 3.8.

Applications of Littlewood–Paley theory can allow us to prove another variant of Brezis–Wainger inequality, which overcomes the barrier of $\dot{H}^{\frac{d}{2}}(\mathbb{R}^d) \not\subset L^\infty(\mathbb{R}^d)$ (recall Example 2). Again, the classical Brezis–Wainger-type inequality would require multiplying by a logarithmic function of $\|\cdot\|_{W^{s,p}}$-norm for $s > \frac{d}{p}$. We again improve it in two ways in Lemma 3.4.8. First, instead of bound by $\|\cdot\|_{\dot{H}^{\frac{d}{2}}}$, we will bound by $\|\cdot\|_{\dot{B}^{\frac{d}{2}}_{2,\infty}}$, recalling that $\|\cdot\|_{\dot{B}^{\frac{d}{2}}_{2,\infty}} \lesssim \|\cdot\|_{\dot{B}^{\frac{d}{2}}_{2,2}} \approx \|\cdot\|_{\dot{H}^{\frac{d}{2}}}$ due to Lemma 2.3.5 and Lemma 3.2.7(a). Second, instead of bound by $\|\cdot\|_{W^{s,p}}$-norm, we bound by $\|f\|_{B^s_{p,\infty}}$, recalling that $\|f\|_{B^s_{p,\infty}} \lesssim \|f\|_{B^s_{p,p}} \lesssim \|f\|_{W^{s,p}}$ due to Lemma 2.3.5 and (3.52b) for $p \in [2,\infty)$.

Lemma 3.4.8 (Variant of Brezis–Wainger inequality, [342, Lemma 2.7]). *Let $p \in [2,\infty)$ and $s > \frac{d}{p}$. Then, for smooth f,*

$$
\|f\|_{L^\infty} \lesssim \|f\|_{L^2} + \|f\|_{\dot{B}^{\frac{d}{2}}_{2,\infty}} \log_2(2+\|f\|_{B^s_{p,\infty}}) + 1.
$$

The proof is similar to that of Lemma 3.4.7 and can be found in [342, Section 4.1]; for completeness, we leave it to Appendix A.3. We observe that the hypothesis of Lemma 3.4.8 does not require f to be divergence-free, in contrast to Lemma 3.4.7, which used the divergence-free property to apply (3.72).

We end this section by introducing the Leray projection operator $\mathcal{P}_L$, which we define in $\mathbb{T}^d$ as

$$
\mathcal{P}_L f(x) \triangleq \sum_{k\in\mathbb{Z}^d\setminus\{0\}} e^{i2\pi k\cdot x}\left(\mathrm{Id} - \frac{k\otimes k}{|k|^2}\right)\hat{f}(k) \tag{3.77}
$$

for all mean-zero f. We remark how $\frac{k\otimes k}{|k|^2}$ has a similar structure as the double Riesz transform, and thus (3.69) and Lemma 3.4.4 apply to $\mathcal{P}_L$.

3.5 BMO and Hardy spaces

Besides the Brezis–Wainger-type inequality, another classical technique to overcome the threshold of the Sobolev embedding is the duality of BMO and Hardy spaces, previewed earlier in Lemma 3.1.9, which we briefly explain now.

Definition 3.5.1.

(a) The space $\mathrm{BMO}(\mathbb{R}^d)$ of bounded mean oscillation is defined as $\mathrm{BMO}(\mathbb{R}^d) \triangleq \{f \in L^1_{\mathrm{loc}}(\mathbb{R}^d) : \|f\|_{\mathrm{BMO}} < \infty\}$, where

$$\|f\|_{\mathrm{BMO}} \triangleq \sup_B \frac{1}{|B|} \int_B |f(x) - f_B| dx, \quad f_B \triangleq \frac{1}{|B|} \int_B f(x) dx, \tag{3.78}$$

and B are balls in $\mathbb{R}^d$.

(b) The Hardy space $\mathcal{H}^1(\mathbb{R}^d)$ consists of all $f = \sum_{j\in\mathbb{N}} \lambda_j a_j$ with a_j and λ_j satisfying

$$\operatorname{supp} a_j \subset B_{R_j}(x_j), \quad \|a_j\|_{L^\infty} \leq R_j^{-d}, \quad \int_{\mathbb{R}^d} a_j(x) dx = 0, \quad \sum_{j\in\mathbb{N}} |\lambda_j| < \infty.$$

Such functions a_j are called atoms carried by the ball $B_{R_j}(x_j)$. The space $\mathcal{H}^1(\mathbb{R}^d)$ equipped with norm $\|\cdot\|_{\mathcal{H}^1}$ is a Banach space that is continuously embedded in $L^1(\mathbb{R}^d)$, where

$$\|f\|_{\mathcal{H}^1} \triangleq \inf\left\{\sum_{j\in\mathbb{N}} |\lambda_j| : f = \sum_j \lambda_j a_j\right\}.$$

Remark 3.5.2.

(a) As $\|f\|_{\mathrm{BMO}} = 0$ for all $f \equiv c \in \mathbb{R}$, $\|\cdot\|_{\mathrm{BMO}}$ is only a seminorm.

(b) $L^\infty(\mathbb{R}^d) \subset \mathrm{BMO}(\mathbb{R}^d)$ because $\frac{1}{|B|}\int_B |f(x) - f_B| dx \leq 2\|f\|_{L^\infty}$. The converse is false; in particular, $f(x) = \ln(|x|)1_{(0,\infty)}(x)$ is not in $L^\infty(\mathbb{R}^d)$ but is in $\mathrm{BMO}(\mathbb{R}^d)$ ([308, p. 140]). Therefore $L^\infty(\mathbb{R}^d) \subsetneq \mathrm{BMO}(\mathbb{R}^d)$. On the other hand, $\mathrm{BMO}(\mathbb{R}^d) \subset L^p_{\mathrm{loc}}(\mathbb{R}^d)$ for all $p \in [1, \infty)$ ([308, p. 144]). In fact, $(\mathcal{H}^1(\mathbb{R}^d))^* = \mathrm{BMO}(\mathbb{R}^d)$ ([236, p. 54]).

Recall from Example 2 that $\dot{H}^{\frac{d}{2}}(\mathbb{R}^d) \subset L^\infty(\mathbb{R}^d)$ is "logarithmically" false. The fact that $L^\infty(\mathbb{R}^d) \subsetneq \mathrm{BMO}(\mathbb{R}^d) \subset L^p_{\mathrm{loc}}(\mathbb{R}^d)$ for all $p < \infty$ while BMO-norm is smaller than L^∞-norm raises the possibility that BMO-norm may be bounded by the $\dot{H}^{\frac{d}{2}}$-norm for appropriate functions, and that is the claim of the following lemma.

Lemma 3.5.3 ([12, Theorem 1.48]). *The embedding $L^1_{\mathrm{loc}}(\mathbb{R}^d) \cap \dot{H}^{\frac{d}{2}}(\mathbb{R}^d) \hookrightarrow \mathrm{BMO}(\mathbb{R}^d)$ holds continuously, i. e., there exists $C \geq 0$ such that*

$$\|f\|_{\mathrm{BMO}} \leq C\|f\|_{\dot{H}^{\frac{d}{2}}} \quad \forall f \in L^1_{\mathrm{loc}}(\mathbb{R}^d) \cap \dot{H}^{\frac{d}{2}}(\mathbb{R}^d). \tag{3.79}$$

Of relevance, the following classical div-curl lemma, originally due to Coifman, Lions, Meyer, and Semmes [94], has valuable applications.

Lemma 3.5.4 (Div-curl lemma, [236, Theorem 12.1]). *Let $f \in L^2(\mathbb{R}^d)$, and let $g = \nabla G$ for $G \in \dot{H}^1(\mathbb{R}^d)$. Then $f \cdot g \in \mathcal{H}^1(\mathbb{R}^d)$, and*

$$\|f \cdot g\|_{\mathcal{H}^1} \lesssim \|f\|_{L^2}\|g\|_{L^2}. \tag{3.80}$$

Example 4. A standard example of an application of the div-curl lemma is the following estimate on the vorticity-stretching term in (1.2) of the 3D Navier–Stokes equations. Specifically,

$$\int_{\mathbb{R}^3} (\omega \cdot \nabla)u \cdot \omega dx$$

cannot be estimated by $\|\omega\|_{L^2}\|\nabla u\|_{L^2}\|\omega\|_{\dot{H}^{\frac{3}{2}}}$ through Hölder's inequality and Sobolev embedding (recall Example 2). However, we can readily achieve the claim by

$$\left|\int_{\mathbb{R}^3} (\omega \cdot \nabla)u \cdot \omega dx\right| \lesssim \|(\omega \cdot \nabla)u\|_{\mathcal{H}^1}\|\omega\|_{\mathrm{BMO}} \overset{(3.79)(3.80)}{\lesssim} \|\omega\|_{L^2}\|\nabla u\|_{L^2}\|\omega\|_{\dot{H}^{\frac{3}{2}}}.$$

Before we proceed to a next section, we give another example where Besov space techniques can overcome the critical case, at a small expense as usual.

Remark 3.5.5. In various occasions, Plancherel theorem (Theorem 3.1.5) attempts us to carry over the estimate to the Fourier side; however, it may come at the cost of subsequently having to estimate $\|\hat{f}\|_{L^1(\mathbb{R}^d)}$. The most straightforward estimate using the polar coordinates gives us, for all $\epsilon > 0$,

$$\|\hat{f}\|_{L^1} \lesssim \left(\int_1^\infty \frac{1}{r^{d+2\epsilon}} r^{d-1} dr\right)^{\frac{1}{2}} \|f\|_{H^{\frac{d}{2}+\epsilon}} \approx \|f\|_{H^{\frac{d}{2}+\epsilon}}. \tag{3.81}$$

We point out two improvements. The first one is interpolation in the same spirit as in the proofs of Theorems 3.3.1 and 3.3.3. For λ to be chosen subsequently, using Plancherel theorem (Theorem 3.1.5), we estimate, for any $s > \frac{d}{2}$,

$$\begin{aligned}
\|\hat{f}\|_{L^1} &= \int_{B_\lambda(0)} |\hat{f}(\xi)| d\xi + \int_{\mathbb{R}^d \setminus B_\lambda(0)} |\hat{f}(\xi)| \frac{|\xi|^s}{|\xi|^s} d\xi \\
&\lesssim \left(\int_0^\lambda r^{d-1} dr\right)^{\frac{1}{2}} \|f\|_{L^2} + \left(\int_\lambda^\infty \frac{1}{r^{2s}} r^{d-1} dr\right)^{\frac{1}{2}} \|f\|_{\dot{H}^s} \lesssim \lambda^{\frac{d}{2}} \|f\|_{L^2} + \lambda^{\frac{d}{2}-s} \|f\|_{\dot{H}^s}.
\end{aligned}$$

We match the two bounds and choose $\lambda \approx \|f\|_{\dot{H}^s}^{\frac{1}{s}} \|f\|_{L^2}^{-\frac{1}{s}}$. This leads to

$$\|\hat{f}\|_{L^1} \lesssim \|f\|_{L^2}^{1-\frac{d}{2s}} \|f\|_{\dot{H}^s}^{\frac{d}{2s}}, \tag{3.82}$$

which implies (3.81) via Young's inequality and hence is slightly better than (3.81). However, we observe that the restriction $s > \frac{d}{2}$ remains in (3.81). The second improvement uses Besov spaces and actually achieves the case $s = \frac{d}{2}$ at the expense of worsening the summability condition. We write $\hat{f} = (\chi + \sum_{j\geq 0}\phi_j)\hat{f}$ using (3.42a) and estimate

$$\|\hat{f}\|_{L^1(\mathbb{R}^d)} \lesssim |\mathrm{supp}\,\chi|^{\frac{1}{2}}\|\chi\hat{f}\|_{L^2(\mathbb{R}^d)} + \sum_{j\geq 0}|\mathrm{supp}\,\phi_j|^{\frac{1}{2}}\|\phi_j\hat{f}\|_{L^2(\mathbb{R}^d)}$$

$$\overset{(3.38)(3.37)(3.36)}{\approx} \|\Delta_{-1}f\|_{L^2} + \sum_{j\geq 0}2^{\frac{jd}{2}}\|\Delta_j f\|_{L^2(\mathbb{R}^d)} \approx \|f\|_{B^{\frac{d}{2}}_{2,1}},$$

which is a clear improvement considering Lemma 3.2.15 and that $B^s_{2,2} = H^s$.

3.6 Bony's paraproducts and remainder and product estimates

Estimates on products are inevitable for nonlinear PDEs and can be difficult when they involve derivatives. Toward this goal, we recall S_j from (3.41c) and define the inhomogeneous Bony paraproducts and remainder (a. k. a. resonant term) respectively as

$$f \prec g \triangleq \sum_{i\geq -1} S_{i-1}f\Delta_i g \quad \text{and} \quad f \circ g \triangleq \sum_{i\geq -1}\sum_{j:|j|\leq 1}\Delta_i f\Delta_{i+j}g, \tag{3.83a}$$

$$fg = f \prec g + f \succ g + f \circ g, \tag{3.83b}$$

where $f \succ g \triangleq g \prec f$. We will later see in Lemma 8.1.4 that the resonant term $f \circ g$ can make products become ill-defined.

The homogeneous Bony's paraproducts are analogous, just with homogeneous counterparts in (3.41), specifically $\dot{\Delta}_i$, $\dot{S}_{i-1}$, and we refer to [12, Sections 2.6.1 and 2.8.1] for details. For a subsequent purpose when we introduce anisotropic analogues in Section 3.7, let us introduce $T(f,g)$ and $R(f,g)$ as

$$T(f,g) \triangleq f \prec g \quad \text{and} \quad R(f,g) \triangleq f \circ g. \tag{3.84}$$

It can be verified ([12, p. 87]) and is useful to know that there exist $N_1, N_2 \in \mathbb{N}$ such that

$$\Delta_m(f \prec g) = \sum_{i:|i-m|\leq N_1}(S_{i-1}f)\Delta_i g, \tag{3.85a}$$

$$\Delta_m(f \circ g) = \Delta_m \sum_{i\geq m-N_2}\sum_{j:|j|\leq 1}\Delta_i f\Delta_{i+j}g. \tag{3.85b}$$

The following is a classical product estimate in Sobolev and Hölder spaces.

Lemma 3.6.1 (Product estimate).
(a) *([211, Lemma A.2] and also [79]) Let* $f \in W^{\delta,p_1}(\mathbb{R}^d) \cap L^{q_2}(\mathbb{R}^d)$ *and* $g \in W^{\delta,p_2}(\mathbb{R}^d) \cap L^{q_1}(\mathbb{R}^d)$*, where* $\delta \geq 0$, $1 < p_k < \infty$, $1 < q_k \leq \infty$, $\frac{1}{p_k} + \frac{1}{q_k} = \frac{1}{p}$, $k \in \{1,2\}$*. Then*

$$\|fg\|_{\dot{W}^{\delta,p}} \lesssim \|f\|_{\dot{W}^{\delta,p_1}} \|g\|_{L^{q_1}} + \|f\|_{L^{q_2}} \|g\|_{\dot{W}^{\delta,p_2}}. \tag{3.86}$$

(b) *([44, eq. (127)]) If* $r \in [0,1)$*, then*

$$[fg]_{\dot{C}^{0,r}} \lesssim [f]_{\dot{C}^{0,r}} \|g\|_{C^0} + \|f\|_{C^0} [g]_{\dot{C}^{0,r}}. \tag{3.87}$$

The heuristic behind Lemma 3.6.1 is that the most singular terms are the extreme cases where all the derivatives fall into one term as all other terms may be interpolated between the two.

The following product estimate is also classical and can be proven using Bony's paraproducts.

Lemma 3.6.2 ([366, Lemma 3.2]). *Let* $\sigma_1, \sigma_2 < \frac{d}{2}$ *satisfy* $\sigma_1 + \sigma_2 > 0$*. Then, for all* $f \in \dot{H}^{\sigma_1}(\mathbb{R}^d)$ *and* $g \in \dot{H}^{\sigma_2}(\mathbb{R}^d)$,

$$\|fg\|_{\dot{H}^{\sigma_1+\sigma_2-\frac{d}{2}}} \lesssim_{\sigma_1,\sigma_2} \|f\|_{\dot{H}^{\sigma_1}} \|g\|_{\dot{H}^{\sigma_2}}. \tag{3.88}$$

Some generalization is certainly possible, and we give the following example to explain the idea of the proof of Lemma 3.6.2.

Lemma 3.6.3. *Let* $p_3 \in [2,\infty)$ *and* $p_1, p_2 \in [p_3, \infty)$ *satisfy*

$$p_1 \leq \frac{p_2 p_3}{p_2 - p_3}, \quad p_2 \leq \frac{p_1 p_3}{p_1 - p_3}, \tag{3.89}$$

and let σ_1, σ_2 *satisfy*

$$\sigma_1, \sigma_2 < d\left(\frac{p_1 p_3 + p_2 p_3 - p_1 p_2}{p_1 p_2 p_3}\right), \quad \sigma_1 + \sigma_2 > 0, \tag{3.90}$$

where in the case $p_2 = p_3$, $\frac{p_2 p_3}{p_2 - p_3}$ *is understood as* ∞ *and similarly in the case* $p_1 = p_3$*. Then, for all* $f \in \dot{B}^{\sigma_1}_{p_1,\infty}(\mathbb{R}^d)$ *and* $g \in \dot{B}^{\sigma_2}_{p_2,\infty}(\mathbb{R}^d)$,

$$\|fg\|_{\dot{B}^{\sigma_1+\sigma_2-d(\frac{p_1p_3+p_2p_3-p_1p_2}{p_1p_2p_3})}_{p_3,\infty}} \lesssim \|f\|_{\dot{B}^{\sigma_1}_{p_1,\infty}} \|g\|_{\dot{B}^{\sigma_2}_{p_2,\infty}}.$$

Proof. First, we estimate by Proposition 3.2.4, when N_1 is from (3.85a) and because $p_1 \leq \frac{p_2 p_3}{p_2 - p_3}$ from (3.89),

$$\|\dot{\Delta}_k(f \prec g)\|_{L^{p_3}} \overset{(3.85a)}{\lesssim} \sum_{j:|j-k|\leq N_1} \|\dot{S}_{j-1} f \dot{\Delta}_j g\|_{L^{p_3}}$$
$$\overset{(3.24)}{\lesssim} \sum_{j:|j-k|\leq N_1} \sum_{l\leq j-2} 2^{dl(\frac{p_2p_3-p_1p_2+p_1p_3}{p_1p_2p_3})} \|\dot{\Delta}_l f\|_{L^{p_1}} \|\dot{\Delta}_j g\|_{L^{p_2}}.$$

This implies by $d(\frac{p_1p_3+p_2p_3-p_1p_2}{p_1p_2p_3}) - \sigma_1 > 0$ from (3.90) that

$$2^{k(\sigma_1+\sigma_2-d(\frac{p_2p_3+p_1p_3-p_1p_2}{p_1p_2p_3})}\|\dot{\Delta}_k(f \prec g)\|_{L^{p_3}} \lesssim \|f\|_{\dot{B}^{\sigma_1}_{p_1,\infty}} 2^{k\sigma_2}\|\dot{\Delta}_k g\|_{L^{p_2}}.$$

This allows us to conclude that

$$\|2^{k(\sigma_1+\sigma_2-d(\frac{p_1p_3+p_2p_3-p_1p_2}{p_1p_2p_3}))}\|\dot{\Delta}_k(f \prec g)\|_{L^{p_3}}\|_{l^\infty} \lesssim \|f\|_{\dot{B}^{\sigma_1}_{p_1,\infty}} \|g\|_{\dot{B}^{\sigma_2}_{p_2,\infty}}. \tag{3.91}$$

Estimates on $g \prec f$ are analogous, and we can deduce

$$\|2^{k(\sigma_1+\sigma_2-d(\frac{p_1p_3+p_2p_3-p_1p_2}{p_1p_2p_3}))}\|\dot{\Delta}_k(g \prec f)\|_{L^{p_3}}\|_{l^\infty} \lesssim \|g\|_{\dot{B}^{\sigma_2}_{p_2,\infty}} \|f\|_{\dot{B}^{\sigma_1}_{p_1,\infty}}. \tag{3.92}$$

Finally, by Proposition 3.2.4 and $p_3 \geq \frac{p_1p_2}{p_1+p_2}$, which follows from (3.89),

$$\|\dot{\Delta}_k(f \circ g)\|_{L^{p_3}} \overset{(3.24)}{\lesssim} 2^{\mathrm{kd}(\frac{p_2+p_1}{p_1p_2} - \frac{1}{p_3})} \left\|\dot{\Delta}_k \sum_{i=-1}^{1} \sum_{j>k-N_2} \dot{\Delta}_j f \dot{\Delta}_{j+i} g\right\|_{L^{\frac{p_1p_2}{p_1+p_2}}}$$
$$\lesssim 2^{\mathrm{kd}(\frac{p_2p_3+p_1p_3-p_1p_2}{p_1p_2p_3})} \sum_{i=-1}^{1} \sum_{j>k-N_2} \|\dot{\Delta}_j f\|_{L^{p_1}} \|\dot{\Delta}_{j+i} g\|_{L^{p_2}}.$$

This implies

$$2^{k(\sigma_1+\sigma_2-d(\frac{p_1p_3+p_2p_3-p_1p_2}{p_1p_2p_3}))}\|\dot{\Delta}_k(f \circ g)\|_{L^{p_3}}$$
$$\lesssim \sum_{i=-1}^{1} 2^{-|k|(\sigma_1+\sigma_2)} *_k (2^{k(\sigma_1+\sigma_2)}\|\dot{\Delta}_k f\|_{L^{p_1}} \|\dot{\Delta}_{k+i} g\|_{L^{p_2}}).$$

Hence, since $\sigma_1 + \sigma_2 > 0$ from (3.90),

$$\|2^{k(\sigma_1+\sigma_2-d(\frac{p_1p_3+p_2p_3-p_1p_2}{p_1p_2p_3}))}\|\dot{\Delta}_k(f \circ g)\|_{L^{p_3}}\|_{l^\infty} \lesssim \|f\|_{\dot{W}^{s_1,p_1}} \|g\|_{\dot{W}^{s_2,p_2}}. \tag{3.93}$$

Combining (3.91), (3.92), and (3.93) concludes the proof. □

We leave further consequences of Bony's paraproducts in Chapter 8 upon discussion of singular SPDEs.

3.7 Anisotropic Littlewood–Paley theory

Due to the divergence-free property of the solution to the Navier–Stokes equations (1.1), analysis can sometimes benefit from considering anisotropic setting (see, e. g., [80]). For example, in the 3D case, we can define for $s, s' \in \mathbb{R}$ the anisotropic homogeneous Sobolev spaces $\dot{H}^{s,s'}$ that consist of $f \in \mathcal{S}'(\mathbb{R}^3)$ such that

$$\|f\|^2_{\dot{H}^{s,s'}} \triangleq \int_{\mathbb{R}^3} |\xi_h|^{2s} |\xi_3|^{2s'} |\hat{f}(\xi)|^2 \, d\xi < \infty, \quad \xi_h \triangleq (\xi_1, \xi_2, 0)$$

([195]). Moreover, with the same ϕ and χ from (3.37), we can define the anisotropic homogeneous Littlewood-Paley operator and low-frequency cutoff operators as

$$\dot{\Delta}^h_k f \triangleq \mathcal{F}^{-1}(\phi(2^{-k}|\xi_h|)\hat{f}), \quad \dot{S}^h_k f \triangleq \mathcal{F}^{-1}(\chi(2^{-k}|\xi_h|)\hat{f}),$$
$$\dot{\Delta}^v_l f \triangleq \mathcal{F}^{-1}(\phi(2^{-l}|\xi_3|)\hat{f}), \quad \dot{S}^v_l f \triangleq \mathcal{F}^{-1}(\chi(2^{-l}|\xi_3|)\hat{f}),$$

and analogously in the inhomogeneous case. Analogues of Bernstein's inequality (Lemma 3.2.1) also exist as follows.

Lemma 3.7.1 (Anisotropic Bernstein's inequality). *Let $\mathcal{B}_h$ and $\mathcal{B}_v$ be balls in $\mathbb{R}^2_h$ and $\mathbb{R}_v$, and let $\mathcal{C}_h$ and $\mathcal{C}_v$) be annuli in $\mathbb{R}^2_h$ and $\mathbb{R}_v$, respectively. Then, for all $1 \le p_2 \le p_1 \le \infty$ and $1 \le q_2 \le q_1 \le \infty$,*

$$\|D^\alpha_h f\|_{L^{p_1}_h(L^{q_1}_v)} \lesssim 2^{k(|\alpha|+2(\frac{1}{p_2}-\frac{1}{p_1}))} \|f\|_{L^{p_2}_h(L^{q_1}_v)} \quad \text{if } \operatorname{supp}\hat{f} \subset 2^k\mathcal{B}_h,$$
$$\|\partial^\beta_3 f\|_{L^{p_1}_h(L^{q_1}_v)} \lesssim 2^{l(|\beta|+(\frac{1}{q_2}-\frac{1}{q_1}))} \|f\|_{L^{p_1}_h(L^{q_2}_v)} \quad \text{if } \operatorname{supp}\hat{f} \subset 2^l\mathcal{B}_v,$$
$$\|f\|_{L^{p_1}_h(L^{q_1}_v)} \lesssim 2^{-kN} \sup_{|\alpha|=N} \|\nabla^\alpha_h f\|_{L^{p_1}_h(L^{q_1}_v)} \quad \text{if } \operatorname{supp}\hat{f} \subset 2^k\mathcal{C}_h,$$
$$\|f\|_{L^{p_1}_h(L^{q_1}_v)} \lesssim 2^{-lN} \|\partial^N_3 f\|_{L^{p_1}_h(L^{q_1}_v)} \quad \text{if } \operatorname{supp}\hat{f} \subset 2^l\mathcal{C}_v.$$

Anisotropic Besov space $(\dot{B}^{s_1}_{p,q_1})_h(\dot{B}^{s_2}_{p,q_2})_v$ denotes roughly the subspace of $\mathcal{S}'(\mathbb{R}^3)$ such that each its element f satisfies

$$\|f\|_{(\dot{B}^{s_1}_{p,q_1})_h(\dot{B}^{s_2}_{p,q_2})_v} \triangleq \left(\sum_{k\in\mathbb{Z}} 2^{q_1 k s_1}\left(\sum_{l\in\mathbb{Z}} 2^{q_2 l s_2} \|\dot{\Delta}^h_k \dot{\Delta}^v_l f\|^{q_2}_{L^p}\right)^{\frac{q_1}{q_2}}\right)^{\frac{1}{q_1}} < \infty.$$

We can also introduce horizontal and vertical Bony's paraproduct decompositions. The idea is to split the product into nine pieces

$$fg = (T^h(f,g) + T^h(g,f) + R^h(f,g))(T^v(f,g) + T^v(g,f) + R^v(f,g)),$$
$$T^h(f,g) \triangleq \sum_m \dot{S}^h_{m-1} f \dot{\Delta}^h_m g, \quad R^h(f,g) \triangleq \sum_m \sum_{j:|j|\le 1} \dot{\Delta}^h_m f \dot{\Delta}^h_{m+j} g,$$
$$T^v(f,g) \triangleq \sum_n \dot{S}^v_{n-1} f \dot{\Delta}^v_n g, \quad R^v(f,g) \triangleq \sum_n \sum_{j:|j|\le 1} \dot{\Delta}^v_n f \dot{\Delta}^v_{n+j} g$$

([196]), so that e. g.

$$R^h(f,g)T^v(f,g) = \sum_{m,n\in\mathbb{Z}} \sum_{j:|j|\le 1} \dot{\Delta}_m^h \dot{S}_{n-1}^v f \dot{\Delta}_{m+j}^h \dot{\Delta}_n^v g,$$

where we recall the notations $T(f,g)$ and $R(f,g)$ from (3.84). Using such tools, the following anisotropic analogues of (3.88) can be proven ([343, eqs. (94)–(95)]):

$$\|fg\|_{\dot{H}^{s-1,\sigma_1+\sigma_2-\frac{1}{2}}} \lesssim \|f\|_{(\dot{B}^1_{2,1})_h(\dot{B}^{\sigma_1}_{2,1})_v} \|g\|_{\dot{H}^{s-1,\sigma_2}},$$

where $\sigma_1 < \frac{1}{2}, \sigma_2 < \frac{1}{2}, \sigma_1 + \sigma_2 > 0, 2 > s > 0$, and

$$\|fg\|_{\dot{H}^{s_1+s_2-1,\sigma-\frac{1}{2}}} \lesssim \|f\|_{(\dot{B}^{s_1}_{2,2})_h(\dot{B}^{\frac{1}{2}}_{2,1})_v} \|g\|_{\dot{H}^{s_2,\sigma-\frac{1}{2}}}$$

where $s_1 < 1, s_2 < 1, s_1 + s_2 > 0, 1 > \sigma > 0$.

3.8 Commutator estimates

Commutator estimates play a crucial role in advanced estimates of nonlinear terms. To give the basic idea, we start with [259, Lemma 3.4]: for all $s \in \mathbb{N}$,

$$\sum_{|\gamma|\le s} \|D^\gamma(fg) - (D^\gamma f)g\|_{L^2} \lesssim (\|f\|_{H^{s-1}}\|\nabla g\|_{L^\infty} + \|f\|_{L^\infty}\|g\|_{H^s}), \tag{3.94}$$

where we can write $D^\gamma(fg) - (D^\gamma f)g = [D^\gamma, g]f$ with $[B,C] \triangleq BC - CB$ as a commutator of operators (cf. the Lie bracket in (1.15)). The heuristic of (3.94) is as follows. Because $s \in \mathbb{N}$, we can apply the Leibniz rule and distribute derivatives. Then the worst terms that arise from $D^\gamma(fg)$ are $(D^\gamma f)g$ and $f(D^\gamma g)$ because any other case can be bounded by a sum of such, thanks to the Gagliardo–Nirenberg inequalities from Theorems 3.3.1 and 3.3.3. Because $(D^\gamma f)g$ is subtracted out in (3.94), the worst terms we get instead are $(D^\eta f)\nabla g$ for $|\eta| \le s-1$ and $f(D^\gamma g)$, as the bound in (3.94) indicates.

The intuition behind (3.94) was clear because the Leibniz rule was directly applicable since $s \in \mathbb{N}$. In the fractional Laplacian case, the Leibniz rule no longer holds, but the following classical Kato–Ponce commutator estimate remarkably justifies our intuition anyhow.

Lemma 3.8.1 (Kato–Ponce commutator estimate, [212] and [84, Lemma 2.2]). *Let* $p, p_2, p_3 \in (1,\infty)$ *and* $p_1, p_4 \in [1,\infty]$ *satisfy* $\frac{1}{p} = \frac{1}{p_1} + \frac{1}{p_2} = \frac{1}{p_3} + \frac{1}{p_4}$, *and let* $s > 0$. *Then for smooth* $f \in W^{1,p_1} \cap W^{s,p_3}$ *and* $g \in W^{s-1,p_2} \cap L^{p_4}$, *we have*

$$\|\Lambda^s(fg) - f\Lambda^s g\|_{L^p} \lesssim \|\nabla f\|_{L^{p_1}}\|\Lambda^{s-1} g\|_{L^{p_2}} + \|\Lambda^s f\|_{L^{p_3}}\|g\|_{L^{p_4}}. \tag{3.95}$$

For other variants, we refer to [64, Proposition 2.1], [367, Remark 2.2(c)], and [367, eq. (169)].

The following is an interesting generalization of (3.95), which arises in our discussions in Section 9.1.

Lemma 3.8.2 ([238, Remark 1.3], originally [214]). *Let $p, p_1, p_2 \in (1,\infty)$ satisfy $\frac{1}{p} = \frac{1}{p_1} + \frac{1}{p_2}$, and let $s, s_1, s_2 \in [0,1)$ satisfy $s = s_1 + s_2$. If $f \in W^{s_1,p_1}$ and $g \in W^{s_2,p_2}$ are smooth, then*

$$\|\Lambda^s(fg) - f\Lambda^s g - g\Lambda^s f\|_{L^p} \lesssim_{s,s_1,s_2,p_1,p_2} \|\Lambda^{s_1} f\|_{L^{p_1}} \|\Lambda^{s_2} g\|_{L^{p_2}}. \tag{3.96}$$

Its extended version can be found in [112, eq. (1.5)] by D'Ancona, where $s \in (0,2)$ is allowed as long as $\frac{2d}{d+2s_j} < p_j < \infty$ for both $j \in \{1,2\}$.

Intuition behind (3.96) is that although the Kato–Ponce commutator estimate (3.95) starts with $f\Lambda^s g$ subtracted out, (3.96) starts with the other worse term $g\Lambda^s f$ subtracted out, allowing us to distribute derivatives completely up to convenience, at the expense of some restrictions such as $s \in (0,2)$. Then a natural question is whether we can extend such an estimate of $\|\Lambda^s(fg) - f\Lambda^s g - g\Lambda^s f\|_{L^p}$ for $s \ge 2$ beyond [112]. In fact, if we do not need complete flexibility in (3.96) and just wish to avoid many derivatives falling on f or g, then it is not so hard to do. Let us explain our heuristics by realizing that in the case $s = 2$, the worst-case scenario of all the derivatives falling on either f or g can be trivially avoided as

$$\Delta(fg) - \Delta fg - f\Delta g = 2\nabla f \cdot \nabla g. \tag{3.97}$$

Suggested by this observation, we can prove the following proposition, which was suggested by Prof. In-Jee Jeong upon the author's discussion with him.

Proposition 3.8.3. *Let $p, p_2, p_3 \in (1,\infty)$, let $p_1, p_4 \in [1,\infty]$ satisfy $\frac{1}{p} = \frac{1}{p_1} + \frac{1}{p_2} = \frac{1}{p_3} + \frac{1}{p_4}$, and let $s \in \mathbb{R}$ satisfy $s > 2$. If $f \in \cap_{j=1}^2 \dot{W}^{j,p_1} \cap \dot{W}^{s-j,p_3}$ and $g \in \cap_{j=1}^2 \dot{W}^{s-j,p_2} \cap \dot{W}^{j,p_4}$ are smooth, then*

$$\begin{aligned}&\|\Lambda^s(fg) - f\Lambda^s g - g\Lambda^s f\|_{L^p}\\ &\lesssim \sum_{j=1}^{2} \|f\|_{\dot{W}^{j,p_1}} \|g\|_{\dot{W}^{s-j,p_2}} + \|f\|_{\dot{W}^{s-j,p_3}} \|g\|_{\dot{W}^{j,p_4}}.\end{aligned} \tag{3.98}$$

Proof. Observe that by writing $\Lambda^2 = -\Delta$ we obtain by (3.97)

$$\begin{aligned}\Lambda^s(fg) - f(\Lambda^s g) - (\Lambda^s f)g &= \Lambda^{s-2}((-\Delta f)g) - (\Lambda^{s-2}(-\Delta)f)g\\ &\quad + \Lambda^{s-2}(f(-\Delta)g) - f(\Lambda^{s-2}(-\Delta)g) - 2\Lambda^{s-2}(\nabla f \cdot \nabla g).\end{aligned}$$

This allows us to apply the classical Kato–Ponce commutator estimate to the first two terms and the product estimate on the third to deduce

$$
\begin{aligned}
&\|\Lambda^s(fg) - f(\Lambda^s g) - (\Lambda^s f)g\|_{L^p} \\
&\overset{(3.95)(3.86)}{\lesssim} \|\Delta f\|_{L^{p_1}} \|\Lambda^{s-2} g\|_{L^{p_2}} + \|\nabla g\|_{L^{p_4}} \|\Lambda^{s-3}\Delta f\|_{L^{p_3}} \\
&\quad + \|\Lambda^{s-2} f\|_{L^{p_3}} \|\Delta g\|_{L^{p_4}} + \|\nabla f\|_{L^{p_1}} \|\Lambda^{s-3}\Delta g\|_{L^{p_2}} \\
&\quad + \|\Lambda^{s-2}\nabla f\|_{L^{p_3}} \|\nabla g\|_{L^{p_4}} + \|\nabla f\|_{L^{p_1}} \|\Lambda^{s-2}\nabla g\|_{L^{p_2}} \\
&\lesssim \sum_{j=1}^{2} \|f\|_{\dot{W}^{j,p_1}} \|g\|_{\dot{W}^{s-j,p_2}} + \|f\|_{\dot{W}^{s-j,p_3}} \|g\|_{\dot{W}^{j,p_4}},
\end{aligned}
$$

where in the last inequality, we used Lemma 3.4.5 and the hypothesis that $p_2, p_3 \in (1, \infty)$. □

The following generalized Calderón commutator estimate is useful in the analysis of the momentum SQG equations (1.6).

Proposition 3.8.4 (Calderón commutator). *Let $p \in (1, \infty)$.*

(a) *([46, Lemma A.5], [262, Lemma 2.2], and [236, Theorem 10.3])) If $\phi \in W^{1,\infty}(\mathbb{T}^2)$, then for all mean-zero $f \in L^p(\mathbb{T}^2)$,*

$$\|[\Lambda, \phi]f\|_{L^p} \lesssim_p \|\phi\|_{W^{1,\infty}} \|f\|_{L^p}. \tag{3.99}$$

Moreover, if $s \in [0,1]$ and $\phi \in W^{2,\infty}(\mathbb{T}^2)$, then for all mean-zero $f \in H^s(\mathbb{T}^2)$,

$$\|[\Lambda, \phi]f\|_{H^s} \lesssim_s \|\phi\|_{W^{2,\infty}} \|f\|_{H^s}. \tag{3.100}$$

(b) *([362, Proposition 5.3]) Let $\gamma \in (1,2)$. If $\phi \in C^{2-\gamma}(\mathbb{T}^2)$ satisfies $\Lambda^{2-\gamma}\phi \in L^\infty(\mathbb{T}^2)$, then for all mean-zero $f \in L^p(\mathbb{T}^2)$,*

$$\|[\Lambda^{2-\gamma}, \phi]f\|_{L^p} \lesssim_p (\|\phi\|_{C^{2-\gamma}} + \|\Lambda^{2-\gamma}\phi\|_{L^\infty}) \|f\|_{L^p}. \tag{3.101}$$

Moreover, if $s \in [0,1]$ and $\phi \in C^{1,2-\gamma}(\mathbb{T}^2)$ satisfies $\Lambda^{2-\gamma}\phi \in W^{1,\infty}(\mathbb{T}^2)$, then for all mean-zero $f \in H^s(\mathbb{T}^2)$,

$$\|[\Lambda^{2-\gamma}, \phi]f\|_{H^s} \lesssim_s (\|\phi\|_{C^{1,2-\gamma}} + \|\Lambda^{2-\gamma_2}\phi\|_{W^{1,\infty}}) \|f\|_{H^s}. \tag{3.102}$$

We recall the Besov–Hölder space $\mathcal{C}^\alpha = B^\alpha_{\infty,\infty}$, $\alpha \in \mathbb{R}$, from Definition 3.2.10 and record one more commutator estimate, which is widely used in the theory of paracontrolled distributions applied to singular SPDEs.

Lemma 3.8.5 ([159, Lemma 2.4]). *Let $\alpha \in (0,1)$ and $\beta, \gamma \in \mathbb{R}$ be such that $\beta + \gamma < 0$ and $\alpha + \beta + \gamma > 0$. Define the trilinear operator*

$$\mathcal{C}(f,g,h) \triangleq (f \prec g) \circ h - f(g \circ h) \tag{3.103}$$

for smooth functions on $\mathbb{T}^d$. This trilinear operator can be extended to the product space $\mathcal{C}^\alpha(\mathbb{T}^d) \times \mathcal{C}^\beta(\mathbb{T}^d) \times \mathcal{C}^\gamma(\mathbb{T}^d)$, and

$$\|\mathcal{C}(f, g.h)\|_{\mathcal{C}^{\alpha+\beta+\gamma}} \lesssim \|f\|_{\mathcal{C}^\alpha}\|g\|_{\mathcal{C}^\beta}\|h\|_{\mathcal{C}^\gamma} \tag{3.104}$$

for all $f \in \mathcal{C}^\alpha(\mathbb{T}^d)$, $g \in \mathcal{C}^\beta(\mathbb{T}^d)$, and $h \in \mathcal{C}^\gamma(\mathbb{T}^d)$.

We refer to [7, Proposition 4.3] and [162, Proposition A.2] for some variations involving the Sobolev space instead of the Besov–Hölder space $\mathcal{C}^\alpha = B^\alpha_{\infty,\infty}$, $\alpha \in \mathbb{R}$.

3.9 Local well-posedness and blowup criterion of the Euler equations using mollifiers

The purpose of this section and Section 3.10 is to give examples of applications of our discussions thus far. Here we sketch the proof of local well-posedness of the deterministic Euler and Navier–Stokes equations in $H^s(\mathbb{R}^d)$ for $s \in \mathbb{N}$ such that $s > 1 + \frac{d}{2}$ following [259, Chapter 3, Sections 3.1–3.2]. Improvement toward any $s \in \mathbb{R}$ such that $s > 1 + \frac{d}{2}$ can be readily achieved by applying (3.95) instead of (3.94) at the step of (3.109). In the case of $\mathbb{T}^d$, a typical approach is the Galerkin approximation to construct a global-in-time weak solution. In relevance to the convex integration solution, which we will discuss later in Chapters 6–7, we will elaborate more on the Galerkin approximation in Section 3.10 and keep short this sketch of the proof using mollifiers.

For $U = \mathbb{T}^d$ or $\mathbb{R}^d$, we define

$$H^s_\sigma \triangleq \{f \in H^s(U) : \nabla \cdot f = 0\} \quad \text{for } s \geq 0, \tag{3.105}$$

and similarly $\dot{H}^s_\sigma$; the case in which $s < 0$ is understood as a Gelfand dual via (2.17). We fix the same ρ in Section 2.5 and recall $\mathcal{J}_\epsilon$ from (2.25). Then given $u^{\text{in}} \in H^s_\sigma$ with $s > 1 + \frac{d}{2}$, for $\epsilon > 0$, we approximate (1.1) by

$$\partial_t u^\epsilon + \mathcal{J}_\epsilon((\mathcal{J}_\epsilon u^\epsilon) \cdot \nabla \mathcal{J}_\epsilon u^\epsilon) + \nabla \pi^\epsilon = \upsilon \mathcal{J}_\epsilon^2 \Delta u^\epsilon, \quad \nabla \cdot u^\epsilon = 0, \quad u^\epsilon|_{t=0} = u^{\text{in}}, \tag{3.106}$$

where we emphasize that $f_\epsilon = f * \eta_\epsilon$ and the superindices of u^ϵ and π^ϵ are intended to indicate the dependence on $\epsilon > 0$. Using the Picard theorem ([259, Theorem 3.1]), we can show that for all $\epsilon > 0$ fixed, there exists a unique solution $u^\epsilon \in C^1([0, T^\epsilon); H^s_\sigma(\mathbb{R}^d))$ for some $T^\epsilon > 0$. We can directly compute

$$\frac{1}{2}\frac{d}{dt}\|u^\epsilon\|^2_{L^2} + \upsilon\|\nabla \mathcal{J}_\epsilon u^\epsilon\|^2_{L^2} = 0 \tag{3.107}$$

identically to (2.29)–(2.30). Now we apply D^j for $j \in \mathbb{N}^d_0$ such that $|j| \leq s$ to (3.106) and take the $L^2(\mathbb{R}^d)$-inner products with $D^j u^\epsilon$ to deduce

$$\frac{1}{2}\frac{d}{dt}\|u^\epsilon\|_{H^s}^2 + \nu\|\mathcal{J}_\epsilon \nabla u^\epsilon\|_{H^s}^2 = -\sum_{|j|\le s}\int_{\mathbb{R}^d} D^j((\mathcal{J}_\epsilon u^\epsilon)\cdot\nabla\mathcal{J}_\epsilon u^\epsilon)\cdot D^j\mathcal{J}_\epsilon u^\epsilon dx,$$

where we used the fact that $\mathcal{J}_\epsilon$ is self-adjoint and commutes with D^j by Lemma 2.5.1(e)–(f). We observe that when the Leibniz rule is applied and the highest order derivative falls on $\nabla\mathcal{J}_\epsilon u^\epsilon$, it vanishes due to the divergence-free property of $\mathcal{J}_\epsilon u^\epsilon$:

$$\int_{\mathbb{R}^d}((\mathcal{J}_\epsilon u^\epsilon\cdot\nabla)D^j\mathcal{J}_\epsilon u^\epsilon)\cdot D^j\mathcal{J}_\epsilon u^\epsilon dx = 0. \tag{3.108}$$

Thus we can rely on (3.94) and $H^s(\mathbb{R}^d)\hookrightarrow W^{1,\infty}(\mathbb{R}^d)$ to deduce

$$\begin{aligned}\frac{1}{2}\frac{d}{dt}\|u^\epsilon\|_{H^s}^2 + \nu\|\mathcal{J}_\epsilon\nabla u^\epsilon\|_{H^s}^2 &\lesssim (\|\nabla\mathcal{J}_\epsilon u^\epsilon\|_{H^{s-1}}\|\nabla\mathcal{J}_\epsilon u^\epsilon\|_{L^\infty} \\ &\quad + \|\nabla\mathcal{J}_\epsilon u^\epsilon\|_{L^\infty}\|\mathcal{J}_\epsilon u^\epsilon\|_{H^s})\|\mathcal{J}_\epsilon u^\epsilon\|_{H^s} \lesssim \|u^\epsilon\|_{H^s}^3.\end{aligned} \tag{3.109}$$

After dividing by $\|u^\epsilon\|_{H^s}$, we can estimate

$$\frac{d}{dt}\|u^\epsilon\|_{H^s} \le C_s\|u^\epsilon\|_{H^s}^2 \overset{(3.18)}{\le} \epsilon^{-2s}C_s\|u^\epsilon\|_{L^2}^2 \overset{(3.107)}{\le} \epsilon^{-2s}C_s\|u^{\text{in}}\|_{L^2}^2.$$

Thus the continuation of an autonomous ODE on a Banach space ([259, Theorem 3.3]) shows that for all $\epsilon > 0$, there exists a unique solution $u^\epsilon \in C^1([0,\infty); H_\sigma^s)$. On the other hand, from (3.109), for all $T \in (0, \frac{1}{C_s\|u^{\text{in}}\|_{H^s}})$ independent of $\epsilon > 0$, we can deduce a solution up to T that satisfies

$$\sup_{t\in[0,T]}\|u^\epsilon\|_{H^s} \le \frac{\|u^{\text{in}}\|_{H^s}}{1 - C_s T\|u^{\text{in}}\|_{H^s}}. \tag{3.110}$$

Lastly, we can show that $\{u^\epsilon\}_{\epsilon>0}$ is Cauchy in $C([0,T]; L^2(\mathbb{R}^d))$ and interpolate with (3.110) to deduce that it is Cauchy in $C([0,T]; H^{s'}(\mathbb{R}^d))$ for all $s' \in [0,s)$ to conclude the proof of the existence of a local-in-time solution. We refer the interested readers to [259, Chapter 3, Sections 3.1–3.2] for further details.

Having constructed a local-in-time unique solution, we are ready to prove the aforementioned Beale–Kato–Majda criterion [17] using (3.76); as already stressed, it is an improved version because $\|\cdot\|_{\dot{B}^0_{\infty,\infty}} \lesssim \|\cdot\|_{L^\infty}$ by Lemma 3.2.12.

Theorem 3.9.1 (Beale–Kato–Majda criterion for the 3D Euler equations [17]). *Let u be a smooth solution to the 3D Euler equations on $[0,T]$. If $\int_0^T\|\omega\|_{\dot{B}^0_{\infty,\infty}} dt < \infty$, then the solution can be extended beyond T.*

Proof. We know the local solution theory in $H^3(\mathbb{R}^3)$ and (3.109) shows that

$$\frac{1}{2}\frac{d}{dt}\|u\|_{H^3}^2 \lesssim \|\nabla u\|_{L^\infty}\|u\|_{H^3}^2.$$

Applying (3.76) and then dividing by $\|u\|_{H^3}$ give us

$$\frac{d}{dt}\|u\|_{H^3} \lesssim (1+\|\omega\|_{\dot{B}^0_{\infty,\infty}})\ln(e+\|u\|_{H^3})\|u\|_{H^3}. \tag{3.111}$$

□

Another classical result, the global well-posedness for small initial data for damped Euler equations, follows immediately from our previous computations, and it will be of relevance in Section 7.1. Thus we give it here.

Proposition 3.9.2. *Consider the following damped Euler equations: for $\alpha > 0$,*

$$\partial_t u + (u\cdot\nabla)u + \nabla\pi + \alpha u = 0, \quad \nabla\cdot u = 0. \tag{3.112}$$

Then there exists $\gamma = \gamma(\alpha) > 0$ such that for all $u^{\mathrm{in}} \in H^s(\mathbb{R}^d)$, $s > 1+\frac{d}{2}$, such that $\|u^{\mathrm{in}}\|_{H^s} \le \gamma$, there exists a unique solution u to (3.112) for all $t > 0$ such that $u|_{t=0} = u^{\mathrm{in}}$. Additionally, $u \in L^\infty(0,\infty;H^s(\mathbb{R}^d))$; in fact, $\|u(t)\|_{H^s} \le \|u^{\mathrm{in}}\|_{H^s}$ for all $t \ge 0$.

Proof. By (3.95) and Sobolev embedding of $H^s(\mathbb{R}^d) \hookrightarrow \dot{W}^{1,\infty}(\mathbb{R}^d)$ we know that there exists a universal constant $C > 0$ such that

$$\|\Lambda^s((f\cdot\nabla)f) - f\cdot\Lambda^s\nabla f\|_{L^2} \le C\|f\|_{H^s}^2.$$

We let $\gamma \triangleq \frac{\alpha}{2C}$. The same computations as in (3.109), but using

$$\int_{\mathbb{R}^d}(u\cdot\nabla)\Lambda^s u\cdot\Lambda^s u\,dx = 0$$

instead of (3.108) and (3.95) instead of (3.94) gives us

$$\frac{1}{2}\frac{d}{dt}\|u\|_{\dot{H}^s}^2 + \alpha\|u\|_{\dot{H}^s}^2 \le C\|u\|_{H^s}^3.$$

Adding the energy inequality of $\frac{1}{2}\frac{d}{dt}\|u\|_{L^2}^2 + \alpha\|u\|_{L^2}^2 = 0$, which can be obtained identically to (2.29)–(2.30), and dividing by $\|u\|_{H^s}$ lead to

$$\frac{d}{dt}\|u\|_{H^s} + \alpha\|u\|_{H^s} \le C\|u\|_{H^s}^2 \tag{3.113}$$

and thus

$$\|u(t)\|_{H^s} \le \frac{\|u^{\mathrm{in}}\|_{H^s}}{1-Ct\|u^{\mathrm{in}}\|_{H^s}}, \tag{3.114}$$

identically to (3.110). This implies that $\|u(t)\|_{H^s} \le 2\gamma$ for all $t \in [0,\frac{1}{2C\gamma})$. Applying this to (3.113) finally allows us to conclude

$$\frac{d}{dt}\|u\|_{H^s} \le -\alpha\|u\|_{H^s} + C\|u\|_{H^s}^2 \le (-\alpha + 2\gamma C)\|u\|_{H^s} = 0.$$

Repeating this procedure from $t = \frac{1}{2C\gamma}$ allows us to reach any $T > 0$. □

As we will revisit in Section 7.2, such a global well-posedness result for small initial data requires some damping.

3.10 Galerkin approximation

In Section 3.9, we described the application of mollifiers to establish local-in-time solution theory to the Euler and Navier–Stokes equations following [259, Chapter 3]. We also refer to [138] for the Fourier truncation approach in the deterministic case. Informally stated, one works on the Fourier frequency restriction to $B_N(0)$, attain bounds independently of N, and take $N \nearrow +\infty$; this approach can be adapted to the stochastic case as well ([368]). We devote this section to details on another classical technique, specifically Galerkin approximation, to construct a global-in-time weak solution through an example of the Cahn–Hilliard equation (1.22).

3.10.1 Example of Cahn–Hilliard equation (1.22)

Here we give an example of classical Galerkin approximation in the case of Cahn–Hilliard equation (1.22) on $\mathbb{T}^d$. We follow the approach of [265, Chapter 3], although our aim is to seek the a weak solution with low regularity (cf. [265, Theorem 3.2 on p. 37]). First, we recall (e. g., from [319, p. 56]; cf. the Rellich–Kondrachov compactness theorem from [132, p. 272]) that

$$H^s(\mathbb{T}^d) \Subset H^{s-\epsilon}(\mathbb{T}^d) \quad \forall \epsilon > 0 \quad \text{and} \quad \forall s \in \mathbb{R}. \tag{3.115}$$

Integrating (1.22) over $\mathbb{T}^d$ and time gives us conservation of mass $⨍_{\mathbb{T}^d} u(t,x)dx = ⨍_{\mathbb{T}^d} u(0,x)dx$; because we can always consider $\overline{u}(0,t) = u(0,x) - ⨍_{\mathbb{T}^d} u(0,y)dy$, throughout this section, w. l. o. g. we assume that the initial data and consequently the solution are mean-zero.

Definition 3.10.1. We define

$$H \triangleq \dot{L}^2(\mathbb{T}^d) \triangleq \left\{ f \in L^2(\mathbb{T}^d) : ⨍_{\mathbb{T}^d} f(x)dx = 0 \right\}, \quad V \triangleq \dot{H}^1(\mathbb{T}^d), \tag{3.116}$$

where V' is the Gelfand dual of V (recall Definition 2.4.1(b)). We denote by $(\cdot,\cdot)$ and $|\cdot|$ respectively the inner product and norm of H and define

$$A \triangleq -\Delta : V \to V' \quad \text{by} \quad \langle Au, v\rangle = (\nabla u, \nabla v) \quad \text{with} \quad D(A) \triangleq H^2(\mathbb{T}^d). \tag{3.117}$$

We have $f(s) = s^3 - s$ from (1.21), so that $f'(s) = 3s^2 - 1 \geq -1$, which is called the dissipative property.

Now $A^{-1} = (-\Delta)^{-1} : H \to H$ is self-adjoint. Moreover, $A^{-1} : H \to V$ is continuous, and $V \hookrightarrow H$ is compact due to (3.115), and thus $A^{-1} : H \to H$ is compact by Lemma 2.4.5(a). Therefore by the Hilbert–Schmidt theorem (Lemma 2.4.8) we can find $\{\mu_j\}_{j\in\mathbb{N}} \subset \mathbb{R}_{>0}$ and o. n. b. $\{w_j\}_{j\in\mathbb{N}}$ of H such that

$$A^{-1}w_j = \mu_j w_j, \quad \lim_{j\to\infty} \mu_j = 0 \quad \forall j \in \mathbb{N}. \tag{3.118}$$

We can define

$$\lambda_j \triangleq \mu_j^{-1} \tag{3.119}$$

and see that (w_j, λ_j) for all $j \in \mathbb{N}$ forms pairs of eigenvectors and eigenvalues of A such that $0 < \lambda_1 \leq \lambda_2 \leq \cdots$ and $\lim_{j\to\infty} \lambda_j = \infty$. We apply A^{-1} to (1.22) to write

$$A^{-1}\partial_t u + Au + f(u) = 0. \tag{3.120}$$

We write a formal definition of a weak solution.

Definition 3.10.2. Given $u^{\text{in}} \in \dot{H}^{-1}(\mathbb{T}^d)$, we say that u is a weak solution to (1.22) if $u \in L^3(0,T;L^3(\mathbb{T}^d))$ satisfies the weak formulation: for a. e. $t \in (0,T]$,

$$\begin{aligned}(A^{-1}u, v)(t) - (A^{-1}u^{\text{in}}, v(0)) - \int_0^t (A^{-1}u(s), \partial_s v(s))ds \\ + \int_0^t (A^{\frac{1}{2}}u, A^{\frac{1}{2}}v)(s) + (f(u), v)(s)ds = 0\end{aligned} \tag{3.121}$$

for all $v \in C^\infty([0,T); V)$.

By the denseness of $\{\sum_{j=1}^N \phi_j(t)v_j : \phi_j \in C_c^\infty([0,T)), v_j \in V, N \in \mathbb{N}\}$ in $C_c^\infty([0,T); V)$, it suffices to look for a weak solution that satisfies

$$\begin{aligned}(A^{-1}u(t), v)\phi(t) - (A^{-1}u^{\text{in}}, v)\phi(0) - \int_0^t (A^{-1}u(s), v)\partial_s\phi(s)ds \\ + \int_0^t ((A^{\frac{1}{2}}u(s), A^{\frac{1}{2}}v) + (f(u(s), v))\phi(s)ds = 0\end{aligned} \tag{3.122}$$

for a. e. $t \in (0,T]$ and all $v \in V$ and $\phi \in C_c^\infty([0,T))$.

Theorem 3.10.3.

1. *(Existence)*
   (a) *In the case* $d = 1$, *for all* $u^{\text{in}} \in \dot{H}^{-1}(\mathbb{T})$, *there exists a weak solution to* (1.22) *such that*

$$u \in L^\infty(0,T;V') \cap L^2(0,T;V) \cap L^4(0,T;L^4(\mathbb{T})). \tag{3.123}$$

If $u^{\text{in}} \in L^2(\mathbb{T})$, *then*

$$u \in L^\infty(0,T;H) \cap L^2(0,T;H^2(\mathbb{T})). \tag{3.124}$$

(b) *In the case* $d \in \{2,3,4\}$, *for all* $u^{\text{in}} \in L^2(\mathbb{T}^d)$, *there exists a weak solution* u *to* (1.22) *that satisfies* (3.124).

2. *In all cases* $d \in \{1,2,3,4\}$, *the weak solution with regularity* (3.124) *is unique.*

In the case $d \in \{2,3,4\}$, we have $u \in L^4(0,T;L^4(\mathbb{T}^d))$ too; however, we did not mention it because it follows from (3.124). As Theorem 3.10.3(2) shows, the uniqueness requires regularity (3.124), for which the initial data must be in L_x^2. Nevertheless, Theorem 3.10.3(1) shows that for the existence, the 1D case requires only $u^{\text{in}} \in \dot{H}^{-1}(\mathbb{T})$. The proof consists of estimates that we state formally as Proposition 3.10.4; we prove the existence and uniqueness thereafter.

First, let P_m be the projection onto $\operatorname{span}\{w_j\}_{j=1}^m$. We consider an approximation by searching for

$$u_m(t,x) = \sum_{i=1}^m u_{i,m}(t) w_i(x) \tag{3.125}$$

in $\operatorname{span}\{w_j\}_{j=1}^m$ such that

$$\left(\frac{d}{dt}A^{-1}u_m(t), v\right) + (A^{\frac{1}{2}}u_m, A^{\frac{1}{2}}v) + (P_m f(u_m)(t), v) = 0 \quad \forall v \in \operatorname{span}\{w_j\}_{j=1}^m, \tag{3.126a}$$

$$u_m(0) = P_m u^{\text{in}}, \tag{3.126b}$$

so that $u_m(0) \to u^{\text{in}}$ as $m \to \infty$ in appropriate initial data space. We now start our estimates.

Proposition 3.10.4.

1. *In the case* $d = 1$, *the solution to* (3.126), *under the hypothesis of Theorem* 3.10.3(a), *has the regularity*

$$u_m \in L^\infty(0,T;V') \cap L^2(0,T;V) \cap L^4(0,T;L^4(\mathbb{T})), \tag{3.127a}$$

$$\partial_t u_m \in L^{\frac{4}{3}}(0,T;H^{-3}(\mathbb{T})), \tag{3.127b}$$

$$u_m \in L^3(0,T;W^{\epsilon,3}(\mathbb{T})) \quad \forall \epsilon \in \left(0, \frac{1}{6}\right], \tag{3.127c}$$

if $u^{\text{in}} \in \dot{H}^{-1}(\mathbb{T})$. *In the case* $u^{\text{in}} \in L^2(\mathbb{T})$, *the solution satisfies*

$$u_m \in L^\infty(0,T;H) \cap L^2(0,T;H^2(\mathbb{T})). \tag{3.128}$$

2. *In the case $d \in \{2,3,4\}$, the solution to* (3.126), *under the hypothesis of Theorem* 3.10.3(b), *has the regularity of* (3.127a), (3.127b), *and* (3.128), *as well as*

$$u_m \in L^3(0,T;W^{\epsilon,3}(\mathbb{T}^d)) \quad \forall \epsilon \in \left(0, \frac{8-d}{6}\right]. \tag{3.129}$$

Proof. (1) We take $v = w_i$ in (3.126a), multiply by $u_{i,m}(t)$, and sum over $i \in \{1,\dots,m\}$ to deduce

$$\frac{1}{2}\frac{d}{dt}\|u_m\|^2_{\dot{H}^{-1}} + \|u_m\|^2_{\dot{H}^1} + (f(u_m),u_m) = 0. \tag{3.130}$$

As $f(s)s \geq \frac{1}{2}s^4 - \frac{1}{2}$ by (1.21), we deduce

$$\frac{1}{2}\|u_m(t)\|^2_{\dot{H}^{-1}} + \int_0^t \|u_m(s)\|^2_{\dot{H}^1} + \frac{1}{2}\|u_m(s)\|^4_{L^4} ds \leq \frac{1}{2}\|u^{\text{in}}\|^2_{\dot{H}^{-1}} + \frac{t}{2}, \tag{3.131}$$

and thus integrating in time gives (3.127a). Next, we take $v = w_i$ in (3.126a), multiply by $\lambda_i^{-2}\frac{du_{im}}{dt}$, and sum over $i \in \{1,\dots,d\}$ to obtain

$$\|\partial_t u_m\|^2_{H^{-3}} \lesssim \|\partial_t u_m\|_{H^{-3}}[\|\nabla u_m\|_{L^2} + \|A^{-\frac{1}{2}}u_m^3\|_{L^2} + \|A^{-\frac{1}{2}}u_m\|_{L^2}]. \tag{3.132}$$

Now $\|A^{-\frac{1}{2}}u_m^3\|_{L^2} \lesssim \|u_m\|^3_{L^4}$ due to the Sobolev embedding of $L^{\frac{4}{3}}(\mathbb{T}^d) \hookrightarrow H^{-1}(\mathbb{T}^d)$, which is valid for all $d \in \{1,2,3,4\}$. Applying this to (3.132) and relying on (3.127a) allow us to deduce (3.127b). Lastly, (3.127c) is needed to deduce strong convergence in $L^3(0,T;L^3(\mathbb{T}))$ afterward. Thus we interpolate and estimate for $\epsilon \in (0,\frac{1}{6}]$ only for the case $d = 1$:

$$\|u_m\|^3_{L^3_T W^{\epsilon,3}} \lesssim \|u_m\|^{\frac{5-6\epsilon}{4}}_{L^\infty_T \dot{H}^{-1}} \int_0^T \|u_m\|^{\frac{7+6\epsilon}{4}}_{\dot{H}^1} dt \overset{(3.127a)}{\lesssim} 1.$$

Note that this computation would not work for $d \in \{2,3,4\}$. In general, we end up with $\int_0^T \|u_m\|^{\frac{3}{2}+\frac{6\epsilon+d}{4}}_{\dot{H}^1} dt$, which is not necessarily finite because $\frac{3}{2} + \frac{6\epsilon+d}{4} > 2$ for all $d \geq 2$, no matter how small we take $\epsilon > 0$.

Next, taking $v = w_i$ in (3.126a), multiplying by $\lambda_i u_{i,m}$, and summing over $i = 1,\dots,m$, we obtain, using (1.21) and (3.64), that for all $d \in \{1,2,3,4\}$,

$$\frac{1}{2}\frac{d}{dt}\|u_m\|^2_{L^2} + \|u_m\|^2_{\dot{H}^2} \leq \frac{1}{2}\|u_m\|^2_{\dot{H}^2} + \frac{1}{2}\|u_m\|^2_{L^2}. \tag{3.133}$$

This implies (3.128). This completes the proof of the first part.
(2) Our verification of (3.127a), (3.127b), and (3.128) is valid for all $d \in \{1,2,3,4\}$. Thus we interpolate to estimate for $\epsilon \in (0,\frac{8-d}{6}]$ and deduce (3.129) as follows:

$$\|u_m\|^3_{L^3_T W^{\epsilon,3}} \lesssim \|u_m\|^{\frac{12-6\epsilon-d}{4}}_{L^\infty_T L^2} \int_0^T \|u_m\|^{\frac{6\epsilon+d}{4}}_{\dot{H}^2} dt \overset{(3.128)}{\lesssim} 1. \qquad \square$$

Proof of Theorem 3.10.3. (1) Using Proposition 3.10.4, we first prove the existence of a weak solution. We can use (2.20a), which implies that

$$L^3(0,T;W^{\epsilon,3}(\mathbb{T}^d)) \cap \{\phi : \partial_t\phi \in L^{\frac{4}{3}}(0,T;H^{-3}(\mathbb{T}^d)) \Subset L^3(0,T;L^3(\mathbb{T}^d)),$$

where we uses (3.127c) in the case $d = 1$ and (3.127c) in the case $d \in \{2,3,4\}$.

Thus we deduce, along with weak compactness and Banach–Alaoglu theorem from Lemma 2.4.5(b), relabeling subsequence by u_m if necessary for simplicity,

$$u_m \overset{w^*}{\rightharpoonup} u \text{ in } L^\infty(0,T;H^{-1}(\mathbb{T}^d)), \tag{3.134a}$$

$$u_m \overset{w}{\rightharpoonup} u \text{ in } L^2(0,T;H^1(\mathbb{T}^d)), \tag{3.134b}$$

$$u_m \to u \text{ in } L^3(0,T;L^3(\mathbb{T}^d)). \tag{3.134c}$$

Multiplying (3.126a) by $\phi \in C_c^\infty([0,T))$ and integrating in time over $[t_0,t]$, we have

$$(A^{-1}u_m(t),v)\phi(t) - (A^{-1}u_m(t_0),v)\phi(t_0) - \int_{t_0}^t (A^{-1}u_m(s),v)\partial_s\phi(s)ds$$
$$+ \int_{t_0}^t (A^{\frac{1}{2}}u_m, A^{\frac{1}{2}}v)\phi(s)ds + \int_{t_0}^t (f(u_m)(s),v)\phi(s)ds = 0.$$

We know that $u_m \overset{w}{\rightharpoonup} u$ in $L^2(0,T;H^1(\mathbb{T}^d))$; thus there exists a subsequence, which we relabel as u_m, such that for a. e. $t \in [0,T]$, $u_m(t) \overset{w}{\rightharpoonup} u$ in $H^1(\mathbb{T}^d)$ ([147, Corollary 2.32]). As $H^1(\mathbb{T}^d) \Subset L^2(\mathbb{T}^d)$ by (3.115), we obtain $u_m(t) \to u(t)$ strongly in $L^2(\mathbb{T}^d)$ for a. e. $t \in [0,T]$ by Lemma 2.4.5(c). This leads to appropriate convergence of $(A^{-1}u_m(t),v)\phi(t) - (A^{-1}u_m(t_0),v)\phi(t_0)$ for a. e. t, t_0 such that $t \geq t_0$, whereas the convergence of linear terms follows from (3.134b), and the convergence of nonlinear term follows from (3.134c).
(2) We now prove part (2) of Theorem 3.10.3, namely the uniqueness of the weak solution constructed in part (1). We assume that for both $j \in \{1,2\}$, u_j distributionally satisfies

$$(\partial_t A^{-1}u_j(t),v) + (A^{\frac{1}{2}}u_j, A^{\frac{1}{2}}v) + (f(u_j),v) = 0 \quad \forall v \in V \quad \text{and for a. e. } t > 0,$$

starting from $u_1(0) = u_2(0) = u^{\text{in}}$. Thus we can let $w \triangleq u_1 - u_2$ and use the assumption that $u_1, u_2 \in L_T^2 H_x^2 \cap L_T^4 L_x^4$ due to (3.127a) and (3.128) to deduce

$$\frac{1}{2}\frac{d}{dt}\|w\|_{\dot{H}^{-1}}^2 + \|w\|_{\dot{H}^1}^2 = -(f(u_1) - f(u_2)) \cdot w. \tag{3.135}$$

Using $-(f(u_1) - f(u_2))(u_1 - u_2) \leq |u_1 - u_2|^2$, we get

$$\frac{1}{2}\frac{d}{dt}\|w\|_{\dot{H}^{-1}}^2 + \|w\|_{\dot{H}^1}^2 \leq \|u_1 - u_2\|_{L^2}^2 \overset{(3.64)}{\leq} \frac{1}{2}\|u_1 - u_2\|_{\dot{H}^{-1}}^2 + \frac{1}{2}\|u_1 - u_2\|_{\dot{H}^1}^2$$

and conclude the proof of uniqueness. □

3.10.2 Galerkin approximation on other PDEs, e. g., Euler, Navier–Stokes, and SQG equations

In Section 3.10.1, we sketched Galerkin approximation with an example of Cahn–Hilliard equation (1.22). Here we elaborate on generalizations to the Euler, Navier–Stokes equations (1.1), and SQG equations (1.4). First, in the case of Euler/Navier–Stokes equations (1.1), we consider two types of solutions, for clarity with spatial domain $\mathbb{T}^d$.

Definition 3.10.5.

(a) For all $\nu \geq 0$, $u \in C_t^0 L_x^2$ is an analytically weak solution to (1.1) if for all t, $u(t,\cdot)$ has zero mean, is weakly divergence-free, and distributionally satisfies (1.1), i. e., for all $\psi \in C_c^\infty(\mathbb{T}^d)$ and $\phi \in \{\Phi \in C_c^\infty(\mathbb{R} \times \mathbb{T}^d) : \nabla \cdot \Phi = 0\}$,

$$\int_{\mathbb{T}^d} (u(t,x) \cdot \nabla)\psi(x)dx = 0, \tag{3.136a}$$

$$\int_{\mathbb{R}} \int_{\mathbb{T}^d} u \cdot (\partial_t \phi + (u \cdot \nabla)\phi + \nu\Delta\phi)dxdt = 0, \tag{3.136b}$$

with appropriate modification in case of initial value problem.

(b) In the case $\nu > 0$, $u \in C_{\text{weak}}^0([0,T]; L^2(\mathbb{T}^d)) \cap L^2(0,T; \dot{H}^1(\mathbb{T}^d))$ is a Leray–Hopf weak solution to (1.1) if $u(t,\cdot)$ has zero mean for all $t \in [0,T]$, is weakly divergence-free, and distributionally satisfies (1.1) (i. e., (3.136)) and additionally the energy inequality

$$\frac{1}{2}\|u(t)\|_{L^2}^2 + \nu \int_0^t \|u(s)\|_{\dot{H}^1}^2 ds \leq \frac{1}{2}\|u^{\text{in}}\|_{L^2}^2. \tag{3.137}$$

The construction of a Leray–Hopf weak solution to the Navier–Stokes equations via Galerkin approximation dates back to [193, 237] (see, e. g., [97, Chapter 8]) and is analogous to the example of the Cahn–Hilliard equation in Section 3.10.1. Classical regularity criteria for the Leray–Hopf weak solutions, which amount to $u \in L_t^r L_x^p$ for $\frac{d}{p} + \frac{2}{r} \leq 1$, include, but are not limited to, [131, 284, 298]. For analogous regularity criteria for weak solutions, we refer to discussions in [48, Section 3.2] and references therein. Historically, Ladyzhenskaya [231] conjectured the nonuniqueness of a Leray–Hopf weak solution to the 3D Navier–Stokes equations.

We emphasize that the weak solution to the Euler equations cannot be constructed from $L^2(\mathbb{T}^d)$ initial data by Galerkin approximation due to the lack of diffusion and thus compactness. Only in the 2D case and if the initial data is in $H^1(\mathbb{T}^2)$, taking $L^2(\mathbb{T}^2)$-inner products of (1.2) with ω and observing that

$$\int_{\mathbb{T}^2} (u \cdot \nabla)\omega\omega dx = 0 \tag{3.138}$$

lead to an $H^1(\mathbb{T}^2)$-bound of $u(t)$, which then allows us to use compactness to prove the global existence of a weak solution to the 2D Euler equations on $\mathbb{T}^2$ (e. g., [29], which even treats the case with random force). We mention that using the vector calculus identity (2.1a), so that $\Delta u = -\nabla\times\omega$ and $\nabla\times((u\cdot\nabla)u) = (u\cdot\nabla)\omega$ in the 2D case, we see that (3.138) is equivalent to

$$\int_{\mathbb{T}^2} (u\cdot\nabla)u\cdot\Delta u dx = 0, \tag{3.139}$$

at least for smooth solutions. In fact, multiplying (1.2) in the 2D case with $|\omega|^{p-2}\omega$ for $p \geq 2$ and integrating over $\mathbb{T}^2$ or $\mathbb{R}^2$ lead to

$$\|\omega(t)\|_{L^\infty} \leq \|\omega(0)\|_{L^\infty}, \tag{3.140}$$

and it is known since the pioneering work of Yudovich [371] that the 2D Euler equations admit a unique global-in-time weak solution starting from bounded vorticity (see, e. g., [259, Theorem 8.2] for details). However, if the initial data is in $L^2(\mathbb{T}^2)$, then there is no account of the successful application of the Galerkin approximation to the Euler equations.

Remarkably, for the SQG equations (1.4), the Galerkin approximation can be applied even in the inviscid case to deduce the global existence of a weak solution, as shown by Resnick [289].

Proposition 3.10.6 ([289, Theorem 2.1]). *Given any initial data $\theta^{\text{in}} \in L^2(\mathbb{T}^2)$, there exists a global-in-time weak solution to the inviscid SQG equations ((1.4) with $\nu = 0$); i. e., for all $T > 0$ and $\phi \in C_c^\infty([0,T)\times\mathbb{T}^2)$,*

$$\begin{aligned}\int_{\mathbb{T}^2} \theta(t,x)\phi(t,x)dx &= \int_{\mathbb{T}^2} \theta^{\text{in}}(x)\phi(t,x)dx \\ &\quad + \int_0^t\int_{\mathbb{T}^2} u(s,x)\cdot\nabla\phi(s,x)\theta(s,x)dxds.\end{aligned}$$

We mention that Marchand [262] improved this result to $\theta^{\text{in}} \in L_x^p$ for $p > \frac{4}{3}$. Such a result is relevant to our subsequent discussions on momentum SQG equations in Section 4.5.1, and its proof relies on commutator; thus, we sketch its proof.

Proof of Proposition 3.10.6. Since $u = \mathcal{R}^\perp\theta$ is divergence-free, we define a stream function ψ such that $u = \nabla^\perp\psi$. We let P_n be the orthogonal projection in $L^2(\mathbb{T}^2)$ onto the span of $\{e^{imx}\}_{0<|m|\leq n}$, $n \in \mathbb{N}$, and for m such that $|m| \leq n$, we have

$$\partial_t\theta_m + P_m(u_m\cdot\nabla\theta_m) = 0, \quad u_m = \nabla^\perp\psi_m, \quad \theta_m(0) = P_m\theta^{\text{in}}. \tag{3.141}$$

We observe that

$$\psi_m = -(-\Delta)^{-1}\nabla^\perp\cdot u_m = -(-\Delta)^{-\frac{1}{2}}\theta_m. \tag{3.142}$$

This is in fact the key we will come back to later. Informally, $u_m = \nabla^\perp \psi_m$, and thus u_m and ψ_m have "opposite signs" (i. e., oddness of $i\xi^\perp$), but because $u_m = \mathcal{R}^\perp\theta_m$ and θ_m have opposite signs (i. e., oddness of $\frac{i\xi^\perp}{|\xi|}$), this makes $\theta_m = -(-\Delta)^{\frac{1}{2}}\psi_m$ and ψ_m have the same signs (i. e., evenness of $-|\xi|$).

Now analogous computations to (2.29)–(2.30) give us

$$\|\theta_m(t)\|_{L^2} \leq \|\theta^{\text{in}}\|_{L^2}. \tag{3.143}$$

In addition, we can multiply (3.141) by $(-\Delta)^{-\frac{1}{2}}\theta_m$ to obtain

$$\|\theta_m(t)\|_{\dot{H}^{-\frac{1}{2}}} \leq \|\theta^{\text{in}}\|_{\dot{H}^{-\frac{1}{2}}}. \tag{3.144}$$

Using Remark 3.1.20 and $H^{2+\epsilon}(\mathbb{T}^2) \hookrightarrow W^{1,\infty}(\mathbb{T}^2)$ for all $\epsilon > 0$, we deduce

$$\|\partial_t\theta_m(t)\|_{H^{-(2+\epsilon)}} \overset{(3.141)}{=} \sup_{\|\varphi\|_{H^{2+\epsilon}}\leq 1}\left|\int_{\mathbb{T}^2} u_m \cdot \nabla P_m\varphi\theta_m dx\right| \overset{(3.143)}{\lesssim} \|\theta^{\text{in}}\|_{L^2}^2. \tag{3.145}$$

Via $L^2(\mathbb{T}^2) \overset{(3.115)}{\Subset} H^{-1}(\mathbb{T}^2) \overset{(3.115)}{\Subset} H^{-(2+\epsilon)}(\mathbb{T}^2)$ and (2.20b) we deduce

$$\theta_m \overset{w^*}{\rightharpoonup} \theta \text{ in } L^\infty(0,T;L^2(\mathbb{T}^2)), \quad u_m \overset{w^*}{\rightharpoonup} u \text{ in } L^\infty(0,T;L^2(\mathbb{T}^2)), \tag{3.146a}$$

$$\theta_m \overset{w}{\rightharpoonup} \theta \text{ in } L^2(0,T;L^2(\mathbb{T}^2)), \tag{3.146b}$$

$$\theta_m \to \theta \text{ in } C([0,T];H^{-1}(\mathbb{T}^2)), \quad \psi_m \to \psi \text{ in } C([0,T];L^2(\mathbb{T}^2)), \tag{3.146c}$$

as $m \to \infty$. From the proof in the case of the Cahn–Hilliard equation (1.22) we know that we need to gain a derivative from $\dot{H}^{-1}(\mathbb{T}^2)$ somehow to successfully pass to the limit in the nonlinear term, and we rely on the approach of commutator, as illustrated already in (3.108)–(3.109). Recalling that $\theta = -\Lambda\psi$ and $u = \nabla^\perp\psi$, let us compute, for $\phi \in C_c^\infty([0,T)\times\mathbb{T}^2)$,

$$\int_{\mathbb{T}^2} u\cdot\nabla\phi\theta dx = \int_{\mathbb{T}^2}\Lambda(\psi\nabla\phi)\cdot u dx \overset{(3.142)}{=} \int_{\mathbb{T}^2} u\cdot[\Lambda,\nabla\phi]\psi - u\cdot\nabla\phi\theta dx,$$

and therefore

$$\int_{\mathbb{T}^2}(u\cdot\nabla)\phi\theta dx = \frac{1}{2}\int_{\mathbb{T}^2} u\cdot[\Lambda,\nabla\phi]\psi dx. \tag{3.147}$$

To show that the nonlinear term converges appropriately, by (3.147) we can write

$$\int_0^t\int_{\mathbb{T}^2} u\cdot\nabla\phi\theta dxds - \int_0^t\int_{\mathbb{T}^2}(u_m\cdot\nabla P_m\phi)\theta_m dxds = \frac{1}{2}\sum_{k=1}^3 \mathrm{III}_k,$$

where

$$\mathrm{III}_1 \triangleq \int_0^t \int_{\mathbb{T}^2} (u - u_m) \cdot [\Lambda, \nabla \phi] \psi dx ds, \tag{3.148a}$$

$$\mathrm{III}_2 \triangleq \int_0^t \int_{\mathbb{T}^2} u_m \cdot [\Lambda, \nabla(\mathrm{Id} - P_m)\phi] \psi dx ds, \tag{3.148b}$$

$$\mathrm{III}_3 \triangleq \int_0^t \int_{\mathbb{T}^2} u_m \cdot [\Lambda, \nabla P_m \phi](\psi - \psi_m) dx ds. \tag{3.148c}$$

It follows that

$$\|[\Lambda, \nabla\phi]\psi\|_{L_x^2} \overset{(3.100)}{\lesssim} \|\phi\|_{W_x^{2,\infty}} \|\psi\|_{L_x^2} \overset{(3.146c)}{\lesssim} 1, \tag{3.149}$$

so that $\mathrm{III}_1 \to 0$ as $m \to \infty$ due to (3.146b). Additionally,

$$\mathrm{III}_2 \overset{(3.143)(3.100)}{\lesssim} \int_0^t \|(\mathrm{Id} - P_m)\phi\|_{W_x^{2,\infty}} \|\psi\|_{L_x^2} ds \to 0$$

and

$$\mathrm{III}_3 \overset{(3.143)(3.100)}{\lesssim} \int_0^t \|P_m \phi\|_{W_x^{2,\infty}} \|\psi - \psi_m\|_{L_x^2} ds \overset{(3.146c)}{\to} 0$$

as $m \nearrow +\infty$. □

Remark 3.10.7. Toward convex integration to be discussed in Chapters 6–7, we point out that in the classical approaches, to construct solution such as the Galerkin approximation, mollifier, or Fourier truncation, the approximation solution already satisfies the energy inequality (see (3.131) and (3.143)–(3.144)), and thus so does the limit. As we will see, a convex integration scheme deliberately constructs a solution that disobeys the energy inequality, and hence its approximation consists of the original PDE with an error vanishing in the limit.

Toward convex integration on singular SPDEs to be discussed in Chapters 8–9, concerning energy inequalities, it is clear from the proofs of Theorem 3.10.3 and Proposition 3.10.6 that the case with external force f can be allowed as long as it is sufficiently smooth, e. g., $L^2(0, T; \dot{H}^{-1}(\mathbb{T}^3))$ in the case of the 3D Navier–Stokes equations ([97, Theorem on p. 71]), so that

$$\int_0^t \int_{\mathbb{T}^3} f \cdot u dx \leq \frac{1}{2} \int_0^t \|u(s)\|_{\dot{H}^1}^2 ds + \frac{1}{2} \int_0^t \|f(s)\|_{\dot{H}^{-1}}^2 ds,$$

and the energy inequality (3.137) can be modified appropriately. The external forces in singular SPDEs are quite rough, far worse than $L^2(0, T; \dot{H}^{-1}(\mathbb{T}^d))$, which makes convex integration approach seem more fit, considering that this approach, again, typically constructs a solution that does not obey the classical energy inequality anyway. We will come back to this issue in Section 9.2.

4 Preliminaries on stochastic analysis

This chapter is devoted to a review on basic stochastic analysis. As described in Preface, this chapter introduces basic Itô calculus, whereas Chapter 8 will describe preliminaries on singular SPDEs. Section 4.1 starts with fundamentals of probability theory, and Section 4.2 sketches the construction of Itô's integrals. Section 4.3 describes the notions of uniqueness of solutions to SPDEs. Section 4.4 gives preliminaries on probabilistic Galerkin approximation. Section 4.5 gives details of probabilistic Galerkin approximation through an example of momentum SQG equations (1.6). Section 4.6 includes further preliminaries for determining the regularity of the solution to the stochastic heat equation (see Lemma 7.1.1), Wiener chaos decomposition, invariant measure, and ergodicity.

4.1 Fundamentals of probability theory

A probability space is $(\Omega, \mathcal{F}, \mathbb{P})$, where Ω represents a sample space, $\mathcal{F}$ its σ-algebra, and $\mathbb{P}$ a probability measure. A random variable X is an $\mathcal{F}$-measurable function from Ω to $\mathbb{R}^N, N \in \mathbb{N}$, or to $\overline{\mathbb{R}^N}$ if needed. Let us collect the basic definitions.

Definition 4.1.1.

(a) For a random variable X, its induced probability measure $\mathcal{L}(X) : \mathcal{B}_{\mathbb{R}^N} \to [0,1]$ is defined by $\mathcal{L}(X)(A) \triangleq \mathbb{P}(\{X^{-1}(A)\})$ for $A \in \mathcal{B}_{\mathbb{R}^N}$; such $\mathcal{L}(X)$ is called the probability law or distribution of X.

(b) For Borel-measurable $f : \mathbb{R}^d \to \mathbb{R}$, we define the mathematical expectation of $f(X)$ w. r. t. $\mathbb{P}$ as $\mathbb{E}[f(X)] \triangleq \int_\Omega f(X(\omega))d\mathbb{P}(\omega)$. We write $\mathbb{E}^{\mathbb{P}}$ to emphasize its dependence on $\mathbb{P}$.

(c) Let E be a separable Banach space, and let X be an E-valued random variable. If $\int_\Omega \|X(\omega)\|_E d\mathbb{P}(\omega) < \infty$, then we call X Bochner integrable and refer to $\int_\Omega X d\mathbb{P}$ as Bochner's integral.

(d) A stochastic process is a collection $\{X_t\}_{t\in I}$ for some temporal interval I, e. g., $[0,\infty)$. For fixed $t \in I$, the mapping $\omega \mapsto X_t(\omega)$ is a random variable, whereas for fixed $\omega \in \Omega$, the mapping $t \mapsto X_t(\omega)$ is called a path of X_t. We say that a process $\{X_t\}_{t\in I}$ is centered if $\mathbb{E}[X_t] = 0$ for all $t \in I$.

(e) Given a stochastic process $\{X_t\}_{t\in I}$, a stochastic process $\{Y_t\}_{t\in I}$ is a version (a. k. a. modification) of $\{X_t\}_{t\in I}$ if $\mathbb{P}(\{X_t = Y_t\}) = 1$ for all $t \in I$.

(f) A stochastic process $\{X_t\}_{t\in I}$ is called Gaussian if for all $k \in \mathbb{N}$ and $t_1, \ldots, t_k \in I$ such that $t_1 < t_2 < \cdots < t_k$, $(X_{t_1}, X_{t_2}, \ldots, X_{t_k})$ has a joint normal distribution. If the law of $(X_{t+t_1}, X_{t+t_2}, \ldots, X_{t+t_k})$ does not depend on t, then we say that the process is stationary.

https://doi.org/10.1515/9783111349985-005

Upon discussing a stochastic process hereafter, we assume that $I = [0,\infty)$ unless otherwise mentioned. We can write X_t or even X instead of $\{X_t\}_{t\in I}$ and, if convenient, write $X_t(\omega) = X(t,\omega)$. We sometimes call a stochastic process simply a process.

Definition 4.1.2.

(a) A filtration $\{\mathcal{F}_t\}_{t\geq 0}$ is a nondecreasing family of σ-subalgebras of $\mathcal{F}$. The quadruple $(\Omega,\mathcal{F},\mathbb{P},\{\mathcal{F}_t\}_{t\geq 0})$ is called a filtered probability space.

(b) $\{\mathcal{F}_t\}_{t\geq 0}$ is said to be normal if it satisfies the "usual conditions:"
- $\mathcal{F}_0$ is complete;
- $\{\mathcal{F}_t\}_{t\geq 0}$ is right-continuous, i. e., $\mathcal{F}_t = \mathcal{F}_{t+} \triangleq \cap_{s>t}\mathcal{F}_s$.

A filtered probability space $(\Omega,\mathcal{F},\mathbb{P},\{\mathcal{F}_t\}_{t\geq 0})$ that satisfies the usual conditions is called a stochastic basis.

(c) Given a filtration $\{\mathcal{F}_t\}_{t\geq 0}$, a stochastic process $\{X_t\}_{t\geq 0}$ is said to be $\mathcal{F}_t$-adapted if for all $t \geq 0$, the mapping $\omega \mapsto X_t(\omega)$ is $\mathcal{F}_t$-measurable.

(d) A stochastic process $\{X_t\}_{t\geq 0}$ is said to be progressively measurable if for all $t \in [0,\infty)$ and $A \in \mathcal{B}_{\mathbb{R}^N}$,

$$\{(s,\omega) \in [0,t]\times\Omega : X_s(\omega) \in A\} \in \mathcal{B}([0,t]) \otimes \mathcal{F}_t,$$

where $\mathcal{B}([0,t]) \otimes \mathcal{F}_t$ denotes the product σ-algebra.

(e) Let E be a separable Banach space. We define $\mathcal{P}_\infty$ as the σ-algebra on $[0,\infty)\times\Omega$ generated by

$$\{\{0\}\times F \text{ and } (s,t]\times F, \text{ where } F \in \mathcal{F}_s,\ 0 \leq s < t < \infty\}.$$

The σ-algebra $\mathcal{P}_\infty$ is called the predictable σ-algebra, and its elements are called predictable sets. Any measurable mapping from $([0,+\infty)\times\Omega,\mathcal{P}_\infty)$ into $(E,\mathcal{B}_E)$ is called a predictable process.

(f) A process $\{X_t\}_{t\geq 0}$ is stochastically continuous at t_0 if for all $\epsilon > 0$ and $\delta > 0$, there exists $\rho > 0$ such that $\mathbb{P}(\{\|X_t - X_{t_0}\| \geq \epsilon\}) \leq \delta$ for all $t \in [t_0-\rho, t_0+\rho]$. A process is stochastically continuous on an interval I if it is stochastically continuous at every $t_0 \in I$; additionally, it is stochastically uniformly continuous on I if ρ can be chosen independently of t_0. Finally, we say that $\{X_t\}_{t\geq 0}$ is continuous if its paths $X(\cdot,\omega)$ are continuous $\mathbb{P}$-a. s.; analogously, $\{X_t\}_{t\geq 0}$ is α-Hölder continuous, $\alpha \in (0,1]$, if its paths $X(\cdot,\omega)$ are α-Hölder continuous $\mathbb{P}$-a. s. (recall Definition 2.3.2).

(g) Given $(\Omega,\mathcal{F},\mathbb{P})$, if a stochastic process $\{X_t\}_{t\geq 0}$ satisfies $\mathbb{E}[|X_t|] < \infty$ for all $t \geq 0$, then we say that it is integrable and lies in $L^1(\Omega)$. More generally, extending Definition 2.3.1, we can define the Banach space $L^p(\Omega;E)$ as the set of all equivalence classes of E-valued random variables, equipped with the norm

$$\|X\|_{L^p_\omega} \triangleq \begin{cases} (\mathbb{E}[\|X\|_E^p])^{\frac{1}{p}} & \text{if } p \in [1,\infty), \\ \operatorname{esssup}_{\omega\in\Omega}\|X(\omega)\|_E & \text{if } p = \infty. \end{cases} \tag{4.1}$$

(h) Let H be a Hilbert space with an inner product $(\cdot,\cdot)$, and let $X, Y \in L^2(\Omega; H)$. Then we define the covariance operator of X and the correlation operator of (X, Y) by

$$\mathrm{Cov}(X) \triangleq \mathbb{E}[(X - \mathbb{E}[X]) \otimes (X - \mathbb{E}[X])], \tag{4.2a}$$

$$\mathrm{Cor}(X, Y) \triangleq \mathbb{E}[(X - \mathbb{E}[X]) \otimes (Y - \mathbb{E}[Y])], \tag{4.2b}$$

where $a \otimes b$ for $a, b \in H$ is understood as the linear operator defined by $(a \otimes b)h \triangleq a\langle b, h\rangle_H$ for all $h \in H$.

To be precise, we recall the notion of tensor product of Hilbert spaces from Definition 2.5.3 and understand the precise meaning of (4.2a) as follows: for all $h, g \in H$, due to (2.26),

$$\mathbb{E}[(X - \mathbb{E}[X]) \otimes (X - \mathbb{E}[X])]\langle h, g\rangle = \mathbb{E}[(X - \mathbb{E}[X], h)(X - \mathbb{E}[X], g)]. \tag{4.3}$$

The operator (4.2b) can be understood similarly. The following fact will be useful in Section 4.4.

Lemma 4.1.3. *Let H be a Hilbert space with an inner product $(\cdot,\cdot)$, let $X, Y \in L^2(\Omega; H)$, and let $L : H \to H$ be a bounded linear operator on H. Then*

$$Cov(LX) = LCov(X)L^*$$

where L^* is the adjoint of L.

Proof. For all $h, g \in H$, using (4.2b) we see that

$$\langle \mathrm{Cov}(LX)h, g\rangle_H = \langle \mathrm{Cov}(X)L^*h, L^*g\rangle_H = \langle L\mathrm{Cov}(X)L^*h, g\rangle_H. \qquad \square$$

As a consequence of the Radon–Nikodym theorem ([32, p. 422]), we can construct the conditional expectation.

Definition 4.1.4. Let $(\Omega, \mathcal{F}, \mathbb{P})$ be a probability space, and let $X : \Omega \to \mathbb{R}^N$ be in $L^1(\Omega)$. If $\mathcal{H} \subset \mathcal{F}$ is a σ-algebra, then the conditional expectation of X given $\mathcal{H}$, denoted $\mathbb{E}[X|\mathcal{H}]$, is the $\mathbb{P}|_{\mathcal{H}}$-a. s. unique function from Ω to $\mathbb{R}^N$ such that
(a) $\mathbb{E}[X|\mathcal{H}]$ is $\mathcal{H}$-measurable,
(b) $\int_H \mathbb{E}[X|\mathcal{H}]d\mathbb{P} = \int_H Xd\mathbb{P}\ \forall H \in \mathcal{H}$.

The proof of the following properties of the conditional expectation can be found in [275, Appendix B].

Proposition 4.1.5. *Let $X, Y : \Omega \to \mathbb{R}^N$ be integrable random variables. Then we have the following properties:*
(a) $\mathbb{E}[aX + bY|\mathcal{H}] = a\mathbb{E}[X|\mathcal{H}] + b\mathbb{E}[Y|\mathcal{H}]$ *for all $a, b \in \mathbb{R}$;*
(b) $\mathbb{E}[\mathbb{E}[X|\mathcal{H}]] = \mathbb{E}[X]$;

(c) $\mathbb{E}[X|\mathcal{H}] = X$ *if* X *is* $\mathcal{H}$*-measurable;*
(d) $\mathbb{E}[X|\mathcal{H}] = \mathbb{E}[X]$ *if* X *is independent of* $\mathcal{H}$*;*
(e) $\mathbb{E}[Y \cdot X|\mathcal{H}] = Y \cdot \mathbb{E}[X|\mathcal{H}]$ *if* Y *is* $\mathcal{H}$*-measurable.*

Definition 4.1.6.

(a) $\{X_t\}_{t\geq 0}$ is called a martingale if $\mathbb{E}[X_t|\mathcal{F}_s] = X_s$ $\mathbb{P}$-a. s. for all $t \geq s$. It is a submartingale or a supermartingale if the equality is replaced by $\geq$ or $\leq$, respectively.
(b) Given $(\Omega, \mathcal{F}, \{\mathcal{F}_t\}_{t\geq 0})$, a random variable $\tau : \Omega \to [0, \infty]$ is called an $\{\mathcal{F}_t\}_{t\geq 0}$-stopping time if $\{\omega : \tau(\omega) \leq t\} \in \mathcal{F}_t$ for all $t \in [0, \infty)$. We emphasize that the definition of a stopping time does not require a probability measure.
(c) Given $(\Omega, \mathcal{F}, \{\mathcal{F}_t\}_{t\geq 0}, \mathbb{P})$, an $\{\mathcal{F}_t\}_{t\geq 0}$-adapted stochastic process $\{X_t\}_{t\geq 0}$ is an $\{\mathcal{F}_t\}_{t\geq 0}$-local martingale if there exists an increasing family of stopping times $\{\tau_n\}_{n\in\mathbb{N}}$ (i. e., $\tau_n \nearrow \infty$ $\mathbb{P}$-a. s.) such that the stopped process $X^{\tau_n}(\cdot) \triangleq X(\tau_n \wedge \cdot)$ is an $\{\mathcal{F}_t\}_{t\geq 0}$-martingale for all $n \in \mathbb{N}$.
(d) Given $(\Omega, \mathcal{F}, \{\mathcal{F}_t\}_{t\geq 0}, \mathbb{P})$, an $\{\mathcal{F}_t\}_{t\geq 0}$-adapted continuous stochastic process $\{X_t\}_{t\geq 0}$ is an increasing process if for a. a. $\omega \in \Omega$, its path $t \mapsto X_t(\omega)$ is an increasing function and $X_0(\omega) = 0$. It is called a process of bounded variation if can be written as a difference of two increasing processes. Finally, it is a semimartingale if it can be written as the sum of a local martingale and a process of bounded variation.

Definition 4.1.7.

(a) Given $(\Omega, \mathcal{F}, \{\mathcal{F}_t\}_{t\geq 0}, \mathbb{P})$, $T > 0$, and a separable Banach space E, we define $\mathcal{M}_T^2(E)$ as the space of all E-valued continuous square-integrable martingales M such that $M(0, \cdot) = 0$ $\mathbb{P}$-a. s.; by [118, Proposition 3.10], $\mathcal{M}_T^2(E)$ is a Banach space with norm $\|M\|_{\mathcal{M}_T^2(E)} \triangleq (\mathbb{E}[\sup_{t\in[0,T]} \|M_t\|_E^2])^{\frac{1}{2}}$. If $M \in \mathcal{M}_T^2(\mathbb{R})$, then there exists a unique increasing $\{\mathcal{F}_t\}_{t\geq 0}$-adapted process $\langle\langle M\rangle\rangle_t$ starting from $t = 0$ such that the process $M(t)^2 - \langle\langle M\rangle\rangle_t$ is a continuous martingale for all $t \geq 0$. The process $\langle\langle M\rangle\rangle_t$ is called a quadratic variation of $\{M\}_{t\geq 0}$.
(b) If $M_1, M_2 \in \mathcal{M}_T^2(\mathbb{R})$, then the process

$$\langle\langle M_1, M_2\rangle\rangle_t = \frac{1}{4}[\langle\langle (M_1 + M_2)\rangle\rangle_t - \langle\langle (M_1 - M_2)\rangle\rangle_t] \tag{4.4}$$

is called the cross quadratic variation of $\{M_1(t)\}_{t\geq 0}$ and $\{M_2(t)\}_{t\geq 0}$; it is the unique $\{\mathcal{F}_t\}_{t\geq 0}$-adapted process with paths of bounded variation such that $M_1(t)M_2(t) - \langle\langle M_1, M_2\rangle\rangle_t$ is a continuous martingale.

Remark 4.1.8. For generality beyond Definition 4.1.7, we mention that if $\{X_t\}_{t\geq 0}$ is a continuous stochastic process, then its pth variation process for $p \in (0, \infty)$ is defined as

$$\langle\langle X\rangle\rangle_t^{(p)} \triangleq \lim_{\Delta t_k \to 0} \sum_{t_k \leq t} |X_{t_{k+1}}(\omega) - X_{t_k}(\omega)|^p, \tag{4.5}$$

where the limit is taken in $\mathbb{P}$, and $0 = t_1 < t_2 < \cdots < t_n = t$, $\Delta t_k \triangleq t_{k+1} - t_k$ ([275, p. 19]). This generalizes the cases of bounded variation ($p = 1$) and quadratic variation ($p = 2$). This will be revisited in Section 8.2 when discussing Young's integral.

4.2 Wiener process and stochastic integrals

In this section, we sketch the ideas of stochastic integrals so that we can later write down the definition of SPDEs.

Definition 4.2.1. Given $(\Omega, \mathcal{F}, \{\mathcal{F}_t\}_{t\geq 0}, \mathbb{P})$, an $\mathbb{R}^N$-valued stochastic process $\{W_t\}_{t\geq 0}$ is an $\{\mathcal{F}_t\}_{t\geq 0}$-Wiener process if

1. its paths are continuous $\mathbb{P}$-a. s.,
2. it is $\{\mathcal{F}_t\}_{t\geq 0}$-adapted,
3. $W(0,\cdot) = 0$ $\mathbb{P}$-a. s.,
4. $W_t - W_s$ satisfy the following:
   (a) they have the normal distribution $\mathcal{N}(0, (t-s)\,\mathrm{Id})$,
   (b) are independent of $\mathcal{F}_s$ for all $0 \leq s \leq t < \infty$.

Remark 4.2.2 ([209, Remark 2.12 on p. 56]). The paths of a Wiener process are α-Hölder continuous for $\alpha \in (0, \frac{1}{2})$ $\mathbb{P}$-a. s. but are not locally $\frac{1}{2}$-Hölder continuous anywhere on $[0,\infty)$ ([209, pp. 113–114]). Additionally, its covariance $\mathbb{E}[W(t)W(s)] = \min\{t, s\}$ ([275, p. 13]).

Lemma 4.2.3 (Lévy's characterization theorem, [118, Proposition 3.11]). *Recall $\mathcal{M}_T^2(\mathbb{R})$ from Definition 4.1.7. If $M \in \mathcal{M}_T^2(\mathbb{R})$ and $\langle\langle M\rangle\rangle_t = t$ for all $t \in [0, T]$, then $\{M_t\}_{t\in[0,T]}$ is an $\{\mathcal{F}_t\}_{t\in[0,T]}$-adapted Wiener process with increments $M_t - M_s$, $t > s$, that are independent of $\mathcal{F}_s$.*

Definition 4.2.4. Let U and H be separable Hilbert spaces, and let Q be a trace-class nonnegative operator on H (recall Definition 2.4.10(a)–(b)).

(a) An U-valued stochastic process $\{W_t\}_{t\geq 0}$ is called a Q-Wiener process if
   (i) $W(0,\cdot) = 0$ $\mathbb{P}$-a. s.,
   (ii) W_t has continuous paths $\mathbb{P}$-a. s.,
   (iii) W_t has independent increments,
   (iv) $\mathcal{L}(W_t - W_s) = \mathcal{N}(0, (t-s)Q)$ for all $t \geq s \geq 0$.
(b) A Q-Wiener process $\{W_t\}_{t\geq 0}$ is called a Q-Wiener process w. r. t. $\{\mathcal{F}_t\}_{t\geq 0}$ if it additionally satisfies the following:
   (i) W_t is $\mathcal{F}_t$-measurable for all $t \geq 0$,
   (ii) $W_{t+h} - W_t$ is independent of $\mathcal{F}_t$ for all $t \geq 0$ and $h \geq 0$.

Lemma 4.2.5 ([118, Proposition 4.3]). *Given $(\Omega, \mathcal{F}, \mathbb{P})$, a separable Hilbert space U, and a trace-class nonnegative operator Q (recall Definition 2.4.10(a)–(b)), let $\{W_t\}_{t\geq 0}$ be a*

Q-Wiener process. Then there exists a complete o. n. b. $\{e_j\}_{j\geq 0}$ for U and a bounded sequence of nonnegative real numbers $\{\lambda_j\}_{j\in\mathbb{N}}$ such that

$$Qe_j = \lambda_j e_j \quad \forall j \in \mathbb{N}. \tag{4.6}$$

Additionally, $\{W_t\}_{t\geq 0}$ is a Gaussian process on U that satisfies $\mathbb{E}[W_t] = 0$ and $\mathrm{Cov}(W_t) = tQ$ for all $t \geq 0$ (recall (4.2a)). Finally, for all $t \geq 0$, $W_t = \sum_{j=1}^{\infty} \sqrt{\lambda_j}\beta_j(t)e_j$, where $\beta_j(t) \triangleq \lambda_j^{-\frac{1}{2}}\langle W_t, e_j\rangle$ for all $j \in \mathbb{N}$ are $\mathbb{R}$-valued Wiener processes that are mutually independent on $(\Omega, \mathcal{F}, \mathbb{P})$, and the series converges in $L^2(\Omega)$.

Lemma 4.2.5 particularly states that if Q is trace-class, then a Q-Wiener process $\{W_t\}_{t\geq 0}$ satisfies $\mathrm{Cov}(W_t) = tQ$ (see also [118, Proposition 2.16]).

Definition 4.2.6. Under the hypotheses of Definition 4.2.4, a Q-Wiener process is called a cylindrical Wiener process if $Q = \mathrm{Id}$.

Given $(\Omega, \mathcal{F}, \{\mathcal{F}_t\}_{t\geq 0}, \mathbb{P})$ and a Q-Wiener process $\{W_t\}_{t\geq 0}$ w. r. t. $\{\mathcal{F}_t\}_{t\geq 0}$ (recall Definition 4.2.4(b)), we are now ready to introduce the stochastic integral w. r. t. W_t. Intuitively, the starting point is the approach of Riemann–Stieltjes integral $\int_a^b f(t)dW_t$, which goes through the standard procedure of partitioning the time interval. An immediate problem, of course, is that such an approach alone requires W_t be monotone over $[a, b]$, but the paths of Wiener process have no such property. This is why while the classical approach defines an integral precisely when the limiting value is independent of the choice of left and right endpoints or midpoints of the partition intervals, stochastic integrals depend on such a choice made a priori. For example, [275, Example 3.1.1] offers a simple example of $\int_0^T W(t)dW(t)$, which gives completely different values under mathematical expectation depending on the choice of left and right endpoints; to be precise, the left endpoint gives 0, whereas the right endpoint gives T.

We now start giving the main ideas of the construction of Itô's integral, the stochastic integral with the choice of left endpoint, referring to [118, Section 4.2] for complete details. First, for fixed $T \in (0, \infty)$, an $L(U, H)$-valued process $\{\Phi_t\}_{t\geq 0}$ is called an elementary process if it takes the form

$$\Phi(t) \triangleq \sum_{m=0}^{k-1} \Phi_{t_m}\chi_{[t_m, t_{m+1})}(t) \tag{4.7}$$

where Φ_{t_m} are $\mathcal{F}_{t_m}$-measurable, and $0 = t_0 < t_1 < \cdots < t_k = T$. For such an elementary process, we can introduce a natural definition of its stochastic integral as

$$\int_0^t \Phi(s)dW(s) \triangleq \sum_{m=0}^{k-1} \Phi(t_m)(W_{t_{m+1}\wedge t} - W_{t_m\wedge t}). \tag{4.8}$$

We introduce the following subspace of U.

Definition 4.2.7. Let H and U be separable Hilbert spaces, the latter with an inner product $\langle\cdot,\cdot\rangle_H$, and Q any trace-class nonnegative operator on U.

(a) We define

$$U_0 \triangleq Q^{\frac{1}{2}}U \quad \text{with } \langle u,v\rangle_{U_0} \triangleq \langle Q^{-\frac{1}{2}}u, Q^{-\frac{1}{2}}v\rangle_U \quad \forall u,v \in U, \tag{4.9}$$

which is a Hilbert space itself.

(b) Recall the space of Hilbert–Schmidt operators $L_2(U,H)$ from Definition 2.4.10(c). Considering (4.6), we see that if $\{e_j\}_{j=1}^{\infty}$, $\{f_j\}_{j=1}^{\infty}$, and $\{g_j\}_{j=1}^{\infty}$ with $g_j \triangleq \lambda_j^{\frac{1}{2}} e_j$ are o.n.b.s of U, H, and U_0, respectively, then we can define another space $L_2(U_0,H)$ of Hilbert–Schmidt operators from U_0 to H equipped with the norm

$$\|\Psi\|^2_{L_2(U_0,H)} \triangleq \sum_{h,k=1}^{\infty} |\langle \Psi g_h, f_k\rangle|^2 \overset{(4.6)}{=} \mathrm{Tr}[(\Psi Q^{\frac{1}{2}})(\Psi Q^{\frac{1}{2}})^*]. \tag{4.10}$$

The space $L_2(U_0,H)$ is a separable Hilbert space.

(c) We say that an $L_2(U_0,H)$-valued process Φ is stochastically integrable on $[0,T]$ if it is predictable and

$$\mathbb{P}\left(\left\{\int_0^T \|\Phi(s)\|^2_{L_2(U_0,H)}ds < \infty\right\}\right) = 1. \tag{4.11}$$

The Itô's isometry below informally serves as a bridge between the stochastic integral w. r. t. W_s and an ordinary integral w. r. t. time s.

Lemma 4.2.8 (Itô's isometry, [118, Proposition 4.20]). *Under the hypotheses of Definition 4.2.7, if $\{\Phi_t\}_{t\geq 0}$ is an elementary process such that $\mathbb{E}[\int_0^t \|\Phi(s)\|^2_{L_2(U_0,H)}ds] < \infty$, then $\int_0^t \Phi(s)dW(s)$ is a continuous square-integrable H-valued martingale on $[0,T]$ such that*

$$\mathbb{E}\left[\left|\int_0^t \Phi(s)dW(s)\right|^2\right] = \mathbb{E}\left[\int_0^t \|\Phi(s)\|^2_{L_2(U_0,H)}ds\right] \quad \forall t \in [0,T]. \tag{4.12}$$

Relying on the Itô's isometry from Lemma 4.2.8 and suitable approximations in $L^2(\Omega; L^2(0,t;L_2(U_0,H)))$ by elementary processes (see [118, Proposition 4.22] for details), we are now able to define a stochastic integral for any predictable process $\Phi \in L^2(\Omega;L^2(0,t;L_2(U_0,H)))$. Such a stochastic integral is a continuous square-integrable martingale, and the Itô's isometry from Lemma 4.2.8 continues to hold. It is easy to see that $\Phi \in L^2(\Omega;L^2(0,t;L_2(U_0,H)))$ satisfies (4.11) by Chebyshev's inequality ([147, Theorem 6.17]), and [118, pp. 99–100] is devoted to show that a stochastic integral can be extended to $L_2(U_0,H)$-valued stochastically integrable processes (i. e., (4.11)).

Now that a stochastic integral is constructed, we may force any PDE by $\Phi(t)dW(t)$ and understand its weak formulation with $\int_0^t \Phi(s)dW(s)$ (see (4.13)) or mild formulation with $\int_0^t e^{(t-s)\Delta}\Phi(s)dW(s)$ if diffusion consists of $-\Delta$.

The following Itô's formula, of which many variants exist ([118, Theorem 4.32], also (4.14)), is an informal Taylor expansion that provides a stochastic chain rule.

Lemma 4.2.9 (Itô's formula). *Under the hypotheses of Definition 4.2.7, suppose that Φ is an $L_2(U_0, H)$-valued stochastically integrable process on $[0, T]$, ϕ is an H-valued $\{\mathcal{F}_t\}_{t\geq 0}$-adapted Bochner-integrable process on $[0, T]$ $\mathbb{P}$-a. s., and $X(0)$ is an $\mathcal{F}_0$-measurable H-valued random variable that satisfies*

$$X(t) = X(0) + \int_0^t \phi(s)ds + \int_0^t \Phi(s)dW(s), \quad \forall t \in [0, T]. \tag{4.13}$$

Suppose that $F : [0, T] \times H \to \mathbb{R}$ has all of $\partial_t F$, $\partial_x F$, $\partial_x^2 F$ uniformly continuous on bounded subsets of $[0, T] \times H$. Then $\mathbb{P}$-a. s., for all $t \in [0, T]$,

$$\begin{aligned} F(t, X(t)) = F(0, X(0)) &+ \int_0^t \langle \partial_s F(s, X(s)), \phi(s)\rangle ds + \int_0^t \langle \partial_x F(s, X(s)), \phi(s)\rangle ds \\ &+ \int_0^t \langle \partial_x F(s, X(s)), \Phi(s)dW(s)\rangle \\ &+ \frac{1}{2}\int_0^t \operatorname{Tr}[\partial_x^2 F(s, X(s))(\Phi(s)Q^{\frac{1}{2}})(\Phi(s)Q^{\frac{1}{2}})^*]ds. \end{aligned}$$

Remark 4.2.10. Thus far, we have focused exclusively on the Itô's integral that fixed the left endpoint convention upon construction of the stochastic integral. Another popular choice is the midpoint, which leads to the Stratonovich integral, often denoted as $\int_0^t \Phi(s) \circ dW(s)$. We refer the interested readers to, e. g., [209, Definition 3.13] and [227, Theorem 2.3.1], especially to the Itô–Stratonovich correction (e. g., [211, Theorem 2.3.5]), and we will come back in Section 7.1 to briefly discuss such a case. We will also very briefly discuss the fractional Brownian motion in Example 12.

SPDEs have a long history; e. g., the theory of hydrodynamic fluctuations was pioneered by Landau and Lifschitz [233] in 1957. Besides the motivation to study turbulence, there are various reasons to force the PDEs in fluid mechanics and mathematical physics by random noise and consider their SPDE analogues. For example, effects from an external force such as wind in an ocean are unpredictable. An addition of the noise turns a model to become probabilistic, allowing us to close modeling without having to consider extra terms. Finally, fluid in microscopic scales display collisions of molecules, suggesting chaotic motion rather than a function of only space and time.

Broadly stated, the random force in an SPDE is one of the following two types.

Definition 4.2.11. We say that an SPDE with random force ΦdW has an additive noise if Φ is independent of the solution; in contrast, we say that it has a multiplicative noise if Φ depends on the solution.

Remark 4.2.12. A popular random force that we only describe briefly here is the Lévy noise. For example, in the case of multiplicative noise for the Navier–Stokes equations with the solution $u \in H^s(\mathbb{R}^d)$, in addition to the cylindrical Wiener process $\Phi(u(t))dW(t)$, we can define $\lambda(dz)$ as a σ-finite Lévy measure on $(\mathbb{Z}, \mathcal{B}_{\mathbb{Z}})$ with an associated Poisson random measure $\mathcal{N}(dt, dz)$ and the compensated Poisson random measure by $\tilde{\mathcal{N}}(dt, dz) \triangleq \mathcal{N}(dt, dz) - \lambda(dz)dt$ that satisfies $\mathbb{E}[\mathcal{N}(dt, dz)] = \lambda(dz)dt$ (e. g., [11, p. 105]). We furthermore define the jump coefficients $\gamma : H^s(\mathbb{R}^d) \times \mathbb{Z} \to H^s(\mathbb{R}^d)$ so that a Lévy noise in this case can be written as

$$\Phi(u(t))dW(t) + \int_{\mathbb{Z}} \gamma(u(t-), z)\tilde{\mathcal{N}}(dt, dz).$$

We refer to [11, 261, 293, 368] for further details.

Remark 4.2.13. We constructed a stochastic integral w. r. t. a U-valued Q-Wiener process $\{W_t\}_{t\geq 0}$. By Definition 4.2.4(a) we see that $\{W_t\}_{t\geq 0}$ satisfies $W(0, \cdot) = 0$ $\mathbb{P}$-a. s. and has continuous paths, independent increments, and $\mathcal{L}(W_t - W_s) = \mathcal{N}(0, (t-s)Q)$ for all $t \geq s \geq 0$. It follows that $\{W_t\}_{t\geq 0}$ satisfies

$$\mathbb{E}[W(t)|\mathcal{F}_s] = \mathbb{E}[W(s) + W(t) - W(s)|\mathcal{F}_s] = W(s) + \mathbb{E}[W(t) - W(s)] = W(s)$$

due to Proposition 4.1.5. This shows that $\{W_t\}_{t\geq 0}$ is a martingale and thus trivially a continuous $\{\mathcal{F}_t\}_{t\geq 0}$-semimartingale due to Definition 4.1.6. In case $X = \{X_t\}_{t\in[0,\infty)}$ is a continuous semimartingale of the form $X = M_t + B_t$ where $\{M_t\}$ is local martingale and B is of bounded variation, then Itô's formula [209, Theorem 3.6] states that for $f : \mathbb{R} \to \mathbb{R}$ of class C^2,

$$df(X_t) = f'(X_t)dX_t + \frac{1}{2}f''(X_t)d\langle\langle M\rangle\rangle_t. \tag{4.14}$$

There is a more general theory of stochastic integrals w. r. t. continuous semi-martingales. We refer to [209, Definitions 2.9, 2.23, and 3.1 in Chapter 3] and [290] for details.

The correlation of a random noise may be categorized into the following three types.

Definition 4.2.14. Given a probability space $(\Omega, \mathcal{F}, \mathbb{P})$, a Gaussian field $\xi = (\xi_1, \ldots, \xi_d)$ with covariance

$$\mathbb{E}[\xi_i(t,x)\xi_j(s,y)] = 1_{\{i=j\}}\delta(t-s)\prod_{l=1}^{d}\delta(x_l - y_l) \quad \forall\, i,j \in \{1,\ldots,d\} \tag{4.15}$$

is called a space-time white noise (STWN). As we will describe lalter in Lemma 8.1.3, for all $d \in \mathbb{N}$, such a noise is distribution-valued, and hence it is more accurate to describe (4.15) as

$$\mathbb{E}[\xi_i(\phi)\xi_j(\psi)] = 1_{\{i=j\}}\int_{\mathbb{R}\times\mathbb{T}^d} \phi(t,x)\psi(t,x)dxdt \quad \forall\, i,j \in \{1,\ldots,d\}$$

for all $\phi, \psi \in \mathcal{S}(\mathbb{R} \times \mathbb{T}^d)$ in the case of spatial domain $\mathbb{T}^d$. In case the product of a Dirac delta distribution in space and that in time in (4.15) are reduced to the first or the second, the corresponding random noise is said to be white in time or white in space, respectively.

We will primarily focus on the case where the noise is white in time in the rest of this chapter and Chapter 7. As we will see in Lemma 8.1.3, the cases of the white-in-space noise or STWN induce roughness, which will be the topic of singular SPDEs in Chapters 8–9.

4.3 Uniqueness of solutions to SPDEs

In this section, we discuss the uniqueness issue of solutions to SPDEs. For clarity, we choose to state the following definition for the Navier–Stokes equations (1.1) analogously to (4.20), starting from given deterministic initial data u^{in}:

$$du + ((u \cdot \nabla)u + \nabla\pi - \nu\Delta u)dt = G(u)dW, \quad \nabla \cdot u = 0, \quad u|_{t=0} = u^{\text{in}}. \tag{4.16}$$

Definition 4.3.1.

(a) The uniqueness in law holds for (4.16) if for any solutions (u, W) and $(\tilde{u}, \tilde{W})$ with the same initial distributions, possibly defined on different probability spaces, the law of (u, W) coincides with that of $(\tilde{u}, \tilde{W})$. The pathwise uniqueness holds for (4.16) if for any solutions (u, W) and $(\tilde{u}, W)$ with common initial data defined on the same probability space $(\Omega, \mathcal{F}, \mathbb{P})$, $u(t) = \tilde{u}(t)$ for all t $\mathbb{P}$-a. s.

(b) A process u is a probabilistically strong solution to (4.16) if it is adapted to the canonical left-continuous filtration generated by the a priori fixed W and augmented by all the $\mathbb{P}$-negligible sets. If it is not a probabilistically strong solution, then u is called a probabilistically weak solution.

Example 5 (Tanaka's counterexample, [87, Example 2.2]). Consider the SDE

$$du(t) = \operatorname{sgn}(u(t))dW(t), \quad u|_{t=0} = 0, \quad \text{where } \operatorname{sgn}(x) \triangleq \begin{cases} 1 & \text{if } x > 0, \\ -1 & \text{if } x \le 0. \end{cases} \tag{4.17}$$

If (u, W) is a solution to (4.17), then $u(t) = \int_0^t \operatorname{sgn}(u(s))dW(s)$, so that $u(t)$ is a continuous martingale with quadratic variation t. By Levy's characterization theorem (Lemma 4.2.3) this implies that $u(t)$ must be Brownian motion, and thus we have the uniqueness in law. On the other hand, if for fixed $W(t)$, $u(t)$ is a solution, then $-u(t)$ is also a solution, indicating the lack of pathwise uniqueness. In conclusion, we see that the uniqueness in law does not imply the pathwise uniqueness.

On the other hand, the pathwise uniqueness implies the uniqueness in law, as the following classical result states.

Lemma 4.3.2 (Yamada–Watanabe theorem [330, 338], also [87, Proposition 1.6]).
(a) *The pathwise uniqueness implies the uniqueness in law.*
(b) *The existence of a probabilistically weak solution, together with pathwise uniqueness, implies the existence of a probabilistically strong solution.*

The following Cherny's theorem provides a way to deduce the pathwise uniqueness.

Lemma 4.3.3 (Cherny's theorem [87]). *The existence of a probabilistically strong solution, together with uniqueness in law, implies the pathwise uniqueness of the solution.*

For extension of Cherny's theorem to the infinite-dimensional setting, we refer to [186, Appendix C].

4.4 Preliminaries on probabilistic Galerkin approximation

In Section 4.5, we will provide details of probabilistic Galerkin approximation through an example of the momentum SQG equations (1.6), and the following preliminaries will be needed for that purpose.

Definition 4.4.1.
(a) A family Λ of probability measures on a separable Banach space E is tight if for all $\epsilon > 0$, there exists a compact set $K_\epsilon \subset E$ such that $\mu(K_\epsilon) \geq 1 - \epsilon$ for all $\mu \in \Lambda$.
(b) A family of measures $\{\mu_n\}_{n=1}^{\infty}$ on $(E, \mathcal{B}_E)$ for a separable Banach space E is said to be weakly convergent to μ on $(E, \mathcal{B}_E)$ if for every $\phi \in C_b(E)$ (recall (2.8b)), $\lim_{n\to\infty} \int_E \phi(x)\mu_n(dx) = \int_E \phi(x)\mu(dx)$.
(c) A family Λ of measures is relatively compact if for all $\{\mu_n\}_{n=1}^{\infty} \subset \Lambda$, there exist a measure μ and a subsequence $\{\mu_{n_k}\}_{k=1}^{\infty}$ that weakly converges to μ on $(E, \mathcal{B}_E)$.

Lemma 4.4.2 (Prokhorov's theorem, [118, Theorem 2.3 on p. 31]). *Let E be a complete separable metric space. A set $\Lambda = \{\mu_n\}_{n=1}^{\infty}$ of probability measures on $(E, \mathcal{B}_E)$ is relatively compact if and only if it is tight.*

Lemma 4.4.3 (Skorokhod's theorem, [118, Theorem 2.4 on p. 32]). *Let E be a complete separable metric space. For every family of probability measures $\{\mu_n\}_{n=1}^{\infty}$ on $\mathcal{B}_E$ that weakly converges to a probability measure μ, there exists a probability space $(\Omega, \mathcal{F}, \mathbb{P})$ and a family of random variables X and $\{X_n\}_{n=1}^{\infty}$ such that (i) $\mathcal{L}(X) = \mu$, $\mathcal{L}(X_n) = \mu_n$, and (ii) $\lim_{n\to\infty} X_n = X$ $\mathbb{P}$-a. s.*

The following Burkholder–Davis–Gundy inequality is crucial in probabilistic Galerkin approximation ([33, Proposition 2.3.8], [209, Theorem 3.28], [227, Theorem 2.3.12]).

Lemma 4.4.4 (Burkholder–Davis–Gundy inequality). *Let H and U be Hilbert spaces, let Q be a trace-class nonnegative operator on U, and let U_0 be defined by (4.9). Suppose that Φ is an $L_2(U_0, H)$-valued process that is progressively measurable w. r. t. $\{\mathcal{F}_t\}_{t\geq 0}$ and satisfies $\mathbb{E}[(\int_0^T \|\Phi(s)\|^2_{L_2(U_0,H)} ds)^{\frac{p}{2}}] < \infty$. Then for every $p \in (0,\infty)$, there exists a constant $C_p > 0$ such that*

$$\mathbb{E}\left[\sup_{t\in[0,T]}\left\|\int_0^t \Phi(s)dW(s)\right\|_H^p\right] \leq C_p \mathbb{E}\left[\left(\int_0^T \|\Phi(s)\|^2_{L_2(U_0,H)} ds\right)^{\frac{p}{2}}\right]. \tag{4.18}$$

Remark 4.4.5. The following variant can be found in [118, Theorem 4.36]. Let U be a Hilbert space, let H be a separable Hilbert space, let Q be a trace-class nonnegative operator on U, and let U_0 be defined by (4.9). Suppose that Φ is an $L_2(U_0, H)$-valued random process that is stochastically integrable on $[0,T]$ and satisfies $\mathbb{E}[\int_0^T \|\Phi(s)\|^2_{L_2(U_0,H)} ds] < \infty$. Then, for all $p \in (0,\infty)$, there exists a constant $C_p > 0$ such that

$$\mathbb{E}\left[\sup_{t\in[0,T]}\left\|\int_0^t \Phi(s)dW(s)\right\|_H^p\right] \leq C_p \left(\mathbb{E}\left[\int_0^T \|\Phi(s)\|^2_{L_2(U_0,H)} ds\right]\right)^{\frac{p}{2}}. \tag{4.19}$$

Lemma 4.4.6 ([141, Lemma 2.1], [33, Lemma 2.3.9]). *Let U be a Hilbert space, let H be a separable Hilbert space, let Q be a trace-class nonnegative operator on U, and let U_0 be defined by (4.9). Suppose that $\Phi \in L^p(\Omega \times [0,T]; L_2(U_0,H))$ is a progressively measurable process. Then for all $p \in [2,\infty)$ and $\gamma \in (0,\frac{1}{2})$, there exists a constant $C_{p,\gamma} > 0$ such that*

$$\mathbb{E}\left[\left\|\int_0^t \Phi(s)dW(s)\right\|^p_{W^{\gamma,p}(0,T;H)}\right] \leq C_{p,\gamma}\mathbb{E}\left[\int_0^T \|\Phi\|^p_{L_2(U_0,H)} dt\right].$$

Lemma 4.4.7 (De la Vallée-Poussin's theorem, [263, Theorem T22 on p. 19]). *Let $(X,\mathcal{F},\mu)$ be a measure space, and let $\{X_\alpha\}_{\alpha\in A}$ be a family of measurable functions on X. Then the family $\{X_\alpha\}_{\alpha\in A} \subset L^1(\mu)$ is uniformly integrable if and only if there exists an increasing convex function $f(\tau) \geq 0$ such that $\lim_{\tau\to\infty}\frac{f(\tau)}{\tau} = \infty$ and $\sup_{\alpha\in A}\int_X f(|X_\alpha|)d\mu < \infty$.*

Lemma 4.4.8 (Vitali's convergence theorem, [294, p. 133]). *Let $(X,\mathcal{F},\mu)$ be a measure space. If $\mu(X) < \infty$, $\{f_n\}_{n=1}^\infty$ is uniformly integrable, $f_n(x) \to f(x)$ a. e. as $n \to \infty$, and $|f(x)| < \infty$ a. e., then $f \in L^1(\mu)$, and $\lim_{n\to\infty}\int_X |f_n - f| d\mu = 0$.*

Lemma 4.4.9 (Martingale representation theorem, [118, p. 222]). *Let H, U be separable Hilbert spaces, let Q be a symmetric bounded positive definite operator on U, let U_0 be defined by (4.9), $T > 0$, and let Φ be a predictable $L_2(U_0,H)$-valued process. Suppose that $M \in \mathcal{M}_T^2(H)$ (recall Definition 4.1.7) satisfies*

$$\langle\langle M\rangle\rangle_t = \int_0^t (\Phi(s)Q^{\frac{1}{2}})(\Phi(s)Q^{\frac{1}{2}})^* ds \quad \forall t \in [0,T].$$

Then there exist a filtered probability space $(\tilde{\Omega}, \tilde{\mathcal{F}}, \{\tilde{\mathcal{F}}_t\}_{t\geq 0}, \tilde{\mathbb{P}})$ *and a Q-Wiener process (recall Definition 4.2.4) W that is U-valued, defined on* $(\Omega \times \tilde{\Omega}, \mathcal{F} \times \tilde{\mathcal{F}}, \mathbb{P} \times \tilde{\mathbb{P}})$*, and adapted to* $\{\mathcal{F}_t \times \tilde{\mathcal{F}}_t\}_{t\geq 0}$ *such that*

$$M(t, \omega, \tilde{\omega}) = \int_0^t \Phi(s, \omega, \tilde{\omega}) dW(s, \omega, \tilde{\omega}) \quad \forall t \in [0, T], \quad (\omega, \tilde{\omega}) \in \Omega \times \tilde{\Omega},$$

where

$$M(t, \omega, \tilde{\omega}) = M(t, \omega), \quad \Phi(t, \omega, \tilde{\omega}) = \Phi(t, \omega) \quad \forall t \in [0, T], \quad \forall (\omega, \tilde{\omega}) \in \Omega \times \tilde{\Omega}.$$

In the following section, we construct a martingale solution and formally state its definition.

Definition 4.4.10. In the context of this monograph, a martingale solution to an SPDE refers to a probabilistically weak and analytically weak solution, unless specifically upgraded to probabilistically or analytically strong solution.

The flexibility of a probabilistically weak solution allows us to not fix a probability space or a Wiener process a priori, so that by application of Prokhorov and Skorokhod theorems (Lemmas 4.4.2 and 4.4.3) we can construct a martingale solution. Because it is analytically weak, the random force will result in a multiplication by a smooth cut-off function and integrating over space-time, which is a martingale (see (4.29) and (4.68)).

Finally, for discussions in Section 4.5.2, let us briefly introduce the following notions of selection.

Definition 4.4.11. Given a set Φ, let $\Upsilon \triangleq \{\gamma : [0, \infty) \to \Phi\}$ and then $S \triangleq \{S(\phi)\}_{\phi\in\Phi}$, where $S(\phi) \triangleq \{\gamma \in \Upsilon : \gamma(0) = \phi\}$. A selection of the family S is a map $u : [0, \infty) \times \Phi \to \Phi$ such that $u(t, \phi) \in S(\phi)$ for every $\phi \in \Phi$; i. e., a selection picks a solution for every initial data ϕ.

4.5 Probabilistic Galerkin approximation

In this section, we prove the global existence of a martingale solution to the generalized momentum SQG equations forced by multiplicative noise (4.20) (see Theorem 4.5.5) as an example of probabilistic Galerkin approximation. We hope that the similarities and differences with Section 3.10 can shed light on how to treat SPDEs.

The construction of global-in-time probabilistically and analytically weak solution for the Navier–Stokes equations with white-in-time noise goes back, at least, to Bensoussan and Temam [21], who worked on the system ω-wise, and the energy inequality did not involve mathematical expectation. We follow the approach of Flandoli and Gatarek [141], who obtained uniform energy bounds using the Burkholder–Davis–Gundy inequality (Lemma 4.4.4) and [118, Chapter 8] and relied on the martingale representation theorem (Lemma 4.4.9).

4.5.1 Example of momentum SQG equations (1.6)

In this section, we briefly review the literature on the SQG (1.4) and the momentum SQG (1.6) equations. The generalized SQG equations (1.5), and thus the SQG equations (1.4), have the rescaling property that if $\theta(t,x)$ is a solution, then so is $\theta_\lambda(t,x) \triangleq \lambda^{\gamma_1+\gamma_2-2}\theta(\lambda^{\gamma_1}t,\lambda x)$ for all $\lambda \in \mathbb{R}_+$. Moreover, multiplying (1.5) by $|\theta|^{p-2}\theta$ for $p \geq 2$, and integrating over $\mathbb{T}^2$, and relying on the positivity lemma (Lemma 3.4.6) lead to a uniform $L^\infty(\mathbb{T}^2)$-bound by that of initial data, similarly to the smooth solution to the 2D Euler equations (recall (3.140)). With this in mind, the cases $\gamma_1 + \gamma_2 \in (2,4]$, $\gamma_1 + \gamma_2 = 2$, and $\gamma_1 + \gamma_2 \in (1,2)$, are referred to respectively as the $L^\infty(\mathbb{T}^2)$-subcritical, critical, and supercritical cases, and the global regularity issue of the SQG equations asks if the global-in-time weak solution constructed by Resnick in Proposition 3.10.6, which continues to hold with any diffusion, remains smooth for all time $t > 0$. In the $L^\infty(\mathbb{T}^2)$-subcritical case, Constantin and Wu [102] settled its global regularity theory. In the $L^\infty(\mathbb{T}^2)$-critical case, Caffarelli and Vasseur [52] and Kiselev, Nazarov, and Volberg [221] accomplished breakthrough works to prove that the global regularity continues to hold (see also [101, 219, 218]). Although this problem remains unsolved in the $L^\infty(\mathbb{T}^2)$-supercritical case, some properties of the solution were studied in various works (e. g., [103, 104]).

The mathematical theory of the stochastic SQG equations was launched by Zhu [373] and Zhu [377] in their PhD theses. Their results include (i) the existence of a probabilistically martingale solution proven via Galerkin approximation if $\gamma_1 > 0$, $\gamma_2 = 1$ in [377, Theorem 4.3.2]; (ii) the existence of a pathwise unique and consequently probabilistically strong solution starting from $\theta^{\text{in}} \in L^p(\mathbb{T}^2)$ for $p > \frac{2}{\gamma_1 - 1}$ where $\gamma_1 > 1$ and $\gamma_2 = 1$ in [377, Theorem 4.4.4]; (iii) the existence of a unique invariant measure in [373, Theorem 4.3.10]. We also refer to [45, 291, 356] and references therein for more results on the stochastic SQG equations. We mention that the noise was white only in time in all these works.

The generalized momentum SQG equations forced by multiplicative noise (cf. (1.6)) of our main interest in this section take the following form:

$$dv + ((u \cdot \nabla)v - (\nabla v)^T \cdot u + \nabla \pi + \Lambda^{\gamma_1} v)dt = G(v)dW, \tag{4.20a}$$

$$u = \Lambda^{2-\gamma_2} v, \quad \nabla \cdot v = 0, \tag{4.20b}$$

where $\gamma_1 \in (0,2]$, $\gamma_2 \in [1,2]$, and $G(v)dW$ is a cylindrical Wiener process to be elaborated later (see (4.26)).

Remark 4.5.1. To the best of the author's knowledge, this is the first time the generalized momentum SQG equations have been studied even in the deterministic case, although the momentum SQG equations (equation (4.20) with $u = \Lambda v$) and the generalized SQG equations (1.5) (the equation solved by $\theta = -\nabla^\perp \cdot v$) have received a lot of attention. In terms of technicality, as the proof of Proposition 3.10.6 showed, we will need to rely on the Calderón commutator estimates from Proposition 3.8.4, and the extension of Proposition 3.8.4(a) to the general case stated in Proposition 3.8.4(b) was proven only recently in [362].

A key identity that will play the crucial role in our analysis is

$$(f \cdot \nabla)g - (\nabla g)^T \cdot f = f^{\perp}(\nabla^{\perp} \cdot g), \tag{4.21}$$

which is valid for any vector f and g, not necessarily $f = \Lambda^{2-\gamma_2} g$, and no divergence-free property is needed. Due to this identity, we see that the solution to (4.20) is mean-zero for all $t > 0$ if so initially. We also see that upon $\dot{H}^{1-\frac{\gamma_2}{2}}(\mathbb{T}^2)$-estimate of v in (4.20), the nonlinear term vanishes since

$$\int_{\mathbb{T}^2} (\Lambda^{2-\gamma_2} v)^{\perp} (\nabla^{\perp} \cdot v) \cdot (\Lambda^{2-\gamma_2} v) dx = 0. \tag{4.22}$$

Let us give another key identity for one of the nonlinear terms of (4.20). Its case where $\gamma_2 = 1$ has appeared in various previous works (e. g., [46, Definition 2.1]). We present it in a general form.

Lemma 4.5.2. *Suppose that v and ϕ are smooth, $u = \Lambda^{2-\gamma_2} v$ for $\gamma_2 \in [1, 2]$, and ϕ is divergence-free. Then*

$$-\int_{\mathbb{T}^2} (\nabla v)^T \cdot u \cdot \phi dx = \frac{1}{2} \sum_{i,j=1}^{2} \int_{\mathbb{T}^2} \partial_i v_j [\Lambda^{2-\gamma_2}, \phi_i] v_j dx. \tag{4.23}$$

We emphasize that for this identity, v does not have to be divergence-free.

Proof. We can integrate by parts and compute

$$\begin{aligned} -\int_{\mathbb{T}^2} (\nabla v)^T \cdot u \cdot \phi dx &= -\sum_{i,j=1}^{2} \int_{\mathbb{T}^2} \partial_i v_j \phi_i \Lambda^{2-\gamma_2} v_j dx \\ &= \sum_{i,j=1}^{2} \int_{\mathbb{T}^2} (\partial_i v_j [\Lambda^{2-\gamma_2}, \phi_i] v_j + \partial_i v_j \phi_i \Lambda^{2-\gamma_2} v_j) dx. \end{aligned}$$

Subtracting $\sum_{i,j=1}^{2} \int_{\mathbb{T}^2} \partial_i v_j \phi_i \Lambda^{2-\gamma_2} v_j dx$ from both sides gives us

$$-\int_{\mathbb{T}^2} (\nabla v)^T \cdot u \cdot \phi dx = -\sum_{i,j=1}^{2} \int_{\mathbb{T}^2} \partial_i v_j \phi_i \Lambda^{2-\gamma_2} v_j dx = \frac{1}{2} \sum_{i,j=1}^{2} \int_{\mathbb{T}^2} \partial_i v_j [\Lambda^{2-\gamma_2}, \phi_i] v_j dx. \qquad \square$$

Remark 4.5.3. For (4.23) to hold, we need that $u = \Lambda^{2-\gamma_2} v$. For example, if we replace u by $\Lambda^{2-\gamma_2} f$, then the identical proof leads to

$$
\begin{aligned}
-\int_{\mathbb{T}^2} (\nabla v)^T \cdot \Lambda^{2-\gamma_2} f \cdot \phi dx &= -\sum_{i,j=1}^{2} \int_{\mathbb{T}^2} \partial_i v_j \phi_i \Lambda^{2-\gamma_2} f_j dx \\
&= \sum_{i,j=1}^{2} \int_{\mathbb{T}^2} \partial_i f_j [\Lambda^{2-\gamma_2}, \phi_i] v_j + \partial_i f_j \phi_i \Lambda^{2-\gamma_2} v_j dx,
\end{aligned}
$$

for which subtracting $\sum_{i,j=1}^{2} \int_{\mathbb{T}^2} \partial_i f_j \phi_i \Lambda^{2-\gamma_2} v_j dx$ is meaningless due to a mismatch with $-\sum_{i,j=1}^{2} \int_{\mathbb{T}^2} \partial_i v_j \phi_i \Lambda^{2-\gamma_2} f_j dx$. We will discuss this issue in (4.61).

We now start with functional setups. We define

$$
\begin{aligned}
\mathcal{V} \triangleq \Big\{ \phi : \phi \text{ is } \mathbb{R}^2\text{-valued trigonometric polynomials on } \mathbb{T}^2, \\
\nabla \cdot \phi = 0, \fint_{\mathbb{T}^2} \phi dx = 0 \Big\},
\end{aligned} \tag{4.24a}
$$

$$
V \triangleq \text{closure of } \mathcal{V} \text{ in } \dot{H}^{1-\frac{\gamma_2}{2}+\frac{\gamma_1}{2}}(\mathbb{T}^2), \quad H \triangleq \text{closure of } \mathcal{V} \text{ in } \dot{H}^{1-\frac{\gamma_2}{2}}(\mathbb{T}^2). \tag{4.24b}
$$

Following the standard practice, we denote an inner product and a norm of H by $(\cdot,\cdot)$ and $|\cdot|$, respectively. We observe that H is a Hilbert space by Remark 3.1.22. Recalling the Leray projection from (3.77), we define

$$
B(\phi, \psi) \triangleq \mathcal{P}_L((\Lambda^{2-\gamma_2}\phi) \cdot \nabla\psi - (\nabla\psi)^T \cdot \Lambda^{2-\gamma_2}\phi), \quad B(\phi) \triangleq B(\phi, \phi), \tag{4.25a}
$$

$$
A = \Lambda^{2-\gamma_2+\gamma_1} \quad \text{with domain } D(A) \triangleq \dot{H}^{2-\gamma_2+\gamma_1}(\mathbb{T}^2) \cap V. \tag{4.25b}
$$

Let Z be a separable Hilbert space, and let $\{W_t\}_{t\geq 0}$ be the cylindrical Wiener process defined on $(\Omega, \mathcal{F}, \{\mathcal{F}_t\}_{t\geq 0}, \mathbb{P})$ and valued in Z. We recall the notation $L_2(H, F)$ from Definition 2.4.10 and define

$$
G(v(t))dW(t) \triangleq \mathcal{P}_L \sum_{i=1}^{\infty} c_i(v(t))d\beta_i(t), \tag{4.26}
$$

where $\{\beta_i\}_{i=1}^{\infty}$ is a family of 1D i. i. d. Wiener processes, and $\{c_i\}_{i=1}^{\infty} \subset L(V, H)$. If $\{e_i\}_{i=1}^{\infty}$ is a basis for H, this implies that $G(v)e_i = c_i v$, so that

$$
\|G(v)\|_{L_2(Z,H)} \overset{(2.22)}{=} \left(\sum_{i=1}^{\infty} \|G(v)e_i\|_H^2 \right)^{\frac{1}{2}} = \left(\sum_{i=1}^{\infty} \|c_i v\|_H^2 \right)^{\frac{1}{2}}. \tag{4.27}
$$

We recall the notation $\dot{H}_\sigma^s$ from (3.105) and formally define a martingale solution to (4.20).

Definition 4.5.4. Given an initial data $v^{\text{in}} \in H$, for all $T > 0$, a martingale solution to (4.20) on $[0, T]$ consists of $(\Omega, \mathcal{F}, \{\mathcal{F}_t\}_{t\in[0,T]}, \mathbb{P})$, a Z-valued cylindrical Wiener process $\{W_t\}_{t\geq 0}$, and a progressively measurable process v that satisfies

$$
v \in C_{\text{weak}}([0, T]; H) \cap L^\infty(0, T; H) \cap L^2(0, T; V) \quad \mathbb{P}\text{-a. s.}, \tag{4.28}
$$

$$\int_{\mathbb{T}^2} (v(t) - v^{\mathrm{in}}) \cdot \phi dx \tag{4.29}$$
$$- \int_0^t \left(\sum_{i,j=1}^{2} \langle \Lambda^{2-\gamma_2} v_j, \partial_j \phi_i v_i \rangle_{\dot{H}^{-1+\frac{\gamma_2}{2}} \times \dot{H}^{1-\frac{\gamma_2}{2}}} - \frac{1}{2} \sum_{i,j=1}^{2} \langle \partial_i v_j, [\Lambda^{2-\gamma_2}, \phi_i] v_j \rangle_{\dot{H}^{-\frac{\gamma_2}{2}} \times \dot{H}^{\frac{\gamma_2}{2}}}$$
$$+ \int_{\mathbb{T}^2} v \cdot \Lambda^{\gamma_1} \phi \right) ds = \int_{\mathbb{T}^2} \int_0^t G(v(s)) dW(s) \cdot \phi dx$$

$\mathbb{P}$-a. s. for all $\phi \in \dot{H}_\sigma^\beta$, $\beta > 4 - \gamma_2$.

Theorem 4.5.5. *Let $\gamma_1 \in (0,2)$, $\gamma_2 \in [1, 1 + \frac{\gamma_1}{2}]$, and $v^{\mathrm{in}} \in H$, $T > 0$, let $\{c_i\}_{i=1}^\infty$ of* (4.26) *satisfy*

$$\sum_{i=1}^{\infty} \|c_i \phi\|^2_{\dot{H}^{1-\frac{\gamma_2}{2}}} \lesssim \|\phi\|^2_{\dot{H}^{1-\frac{\gamma_2}{2}}}, \quad \|G(\psi)\|^2_{L_2(Z, \dot{H}_\sigma^{-\alpha})} \lesssim 1 + \|\psi\|^2_{\dot{H}^{1-\frac{\gamma_2}{2}}} \tag{4.30}$$

for

$$\alpha > 4 - \gamma_2, \tag{4.31}$$

and let G of (4.26) *extend to a Lipschitz continuous mapping from H to $L_2(Z, \dot{H}_\sigma^{-\alpha})$. Then there exists a martingale solution to the momentum SQG equations* (4.20) *over* $[0, T]$ *according to Definition* 4.5.4.

Remark 4.5.6. The uniqueness of the martingale solution to the momentum SQG equations (4.20) can be investigated, e. g., using the approach of [93]. We refer to this discussion in [262, Section 8] in the deterministic case and leave it for future works, because our main concern in this book is comparison with solutions constructed via convex integration, to be discussed in Chapters 6–7.

We refer to Remark 7.3.6 for comments concerning $\gamma_2 \in [1, 1 + \frac{\gamma_1}{2}]$ in the hypothesis of Theorem 4.5.5.

Our proof is partially inspired by [42]. The parts of our poof of Theorem 4.5.5 that has similarities to the Galerkin approximation described in Section 3.10 will only be sketched.

We take an o. n. b. $\{e_j\}_{j=1}^\infty$ of H and denote by P_m the orthogonal projection onto $\mathrm{span}\{e_j\}_{j=1}^m$. On $(\overline{\Omega}, \overline{\mathcal{F}}, \overline{\mathbb{P}})$ with $\overline{W}$ and the mathematical expectation $\overline{\mathbb{E}}$ w. r. t. $\overline{\mathbb{P}}$, we look for $v^m \in \mathrm{span}\{e_j\}_{j=1}^m$ that satisfies

$$dv^m + (P_m B(v^m) + \Lambda^{\gamma_1} v^m) dt = P_m G(v^m(t)) d\overline{W}(t), \quad v^m(0) = P_m v^{\mathrm{in}}, \tag{4.32}$$

so that $v^m(0) \to v^{\mathrm{in}}$ strongly in H as $m \to \infty$, where

$$G(v^m(t)) d\overline{W}(t) = \mathcal{P}_L \sum_{i=1}^{\infty} c_i v^m(t) d\overline{\beta}_i(t) \tag{4.33}$$

(cf. (4.26)), so that for all $j \in \{1, \ldots, m\}$,

$$d(v^m, e_j) + (B(v^m) + \Lambda^{\gamma_1} v^m, e_j)dt = (G(v^m(t)), e_j)d\overline{W}. \tag{4.34}$$

By the standard theory of SDEs (e. g., [305, Theorem 2 on p. 121]) we know that the solution v^m exists locally on $[0, T_m]$, $T_m \le T$ possibly.

Proposition 4.5.7. *Under the hypotheses of Theorem* 4.5.5, *for all* $p \in [1, \infty)$, v^m *satisfies*

$$v^m \in L^p(\overline{\Omega}; L^\infty(0, T; H)) \cap L^p(\overline{\Omega}; L^2(0, T; V)). \tag{4.35}$$

Proof. Such estimates are standard. Thus we only mention that using Itô's formula (Lemma 4.2.9) on (4.34) and taking advantage of (4.22), for all $p \in [4, \infty)$, we can deduce

$$\begin{aligned} &\sup_{\tau \in [0, t\wedge\tau_N]} |v^m(\tau)|^{2p} + p^2 \left(\int_0^{t\wedge\tau_N} |v^m|^{p-2} \|v^m\|_V^2 d\tau \right)^2 \lesssim_p |v^m(0)|^{2p} \\ &+ \left(\int_0^{t\wedge\tau_N} |v^m|^{p-2} \sum_{i=1}^{\infty} |c_i v^m|^2 d\tau \right)^2 + \left(\sup_{\tau \in [0, t\wedge\tau_N]} \left| \int_0^{\tau} |v^m|^{p-2} (G(v^m), v^m) d\overline{W}(\delta) \right| \right)^2, \end{aligned} \tag{4.36}$$

where

$$\tau_N \triangleq \begin{cases} \inf\{\tau > 0 : |v^m(\tau)| \ge N\} & \text{if } \{\overline{\omega} \in \overline{\Omega} : |v^m(\tau)|^2 \ge N^2\} = \emptyset, \\ +\infty & \text{otherwise,} \end{cases} \tag{4.37}$$

is an $(\overline{\mathcal{F}}_t)_{t\ge 0}$-stopping time. Taking mathematical expectation $\overline{\mathbb{E}}$ and applying (4.30) and the Burkholder–Davis–Gundy inequality to handle the noise term lead to the desired claim; we leave the details to Appendix A.4. □

We recall the notions of Gelfand dual in (2.17) and state the following result concerning a family of laws.

Proposition 4.5.8. *Under the hypotheses of Theorem* 4.5.5, $\{\mathcal{L}(v^m)\}_{m\in\mathbb{N}}$ *is tight in* $L^2(0, T; H) \cap C([0, T]; \dot{H}_\sigma^{-a})$ *for* a *that satisfies* (4.31).

The fact that $\{\mathcal{L}(v^m)\}_{m\in\mathbb{N}}$ is tight in $C([0, T]; \dot{H}_\sigma^{-a})$ for $a > 4 - \gamma_2$, combined with Lemma 2.4.9(b), leads to the fact that $v \in C_{\text{weak}}([0, T]; \dot{H}_\sigma^{1-\frac{\gamma_2}{2}})$ $\mathbb{P}$-a. s., as claimed in (4.28).

Proof of Proposition 4.5.8. Because we already know that $v^m \in L^p(0, T; V)$ $\overline{\mathbb{P}}$-a. s. from Proposition 4.5.7, to show the tightness in $L^2(0, T; H)$, according to Lemma 2.4.7(a), it suffices to show that $v^m \in W^{\gamma,2}(0, T; \dot{H}_\sigma^{-a})$ for some $\gamma \in (0, \frac{1}{2})$ $\overline{\mathbb{P}}$-a. s. To this end, from (4.34) we can directly compute

$$\|v^m\|_{W^{\gamma,2}(0,T;\dot{H}_\sigma^{-a})} \le \sum_{i=1}^{4} \mathrm{IV}_i, \tag{4.38}$$

where

$$\text{IV}_1 \triangleq \|v^m(0)\|_{W^{1,2}(0,T;\dot{H}^{-\alpha})}, \tag{4.39a}$$

$$\text{IV}_2 \triangleq \left\| \sup_{\|e_j\|_{\dot{H}^\alpha_\sigma}\le 1} \left| \int_0^t \int_{\mathbb{T}^2} [(\Lambda^{2-\gamma_2} v^m \cdot \nabla) v^m - (\nabla v^m)^T \cdot \Lambda^{2-\gamma_2} v^m] \cdot e_j dx d\tau \right| \right\|_{W^{1,1}(0,T)}, \tag{4.39b}$$

$$\text{IV}_3 \triangleq \left\| \sup_{\|e_j\|_{\dot{H}^\alpha_\sigma}\le 1} \left| \int_0^t \int_{\mathbb{T}^2} \Lambda^{\gamma_1} v^m \cdot e_j dx d\tau \right| \right\|_{W^{1,2}(0,T)}, \tag{4.39c}$$

$$\text{IV}_4 \triangleq \left\| \sup_{\|e_j\|_{\dot{H}^\alpha_\sigma}\le 1} \left| \int_{\mathbb{T}^2} \int_0^t G(v^m(\tau)) d\overline{W}(\tau) e_j dx \right| \right\|_{W^{\gamma,2}(0,T)}, \tag{4.39d}$$

due to $W^{1,1}(0,T) \subset W^{\gamma,2}(0,T)$ since $\gamma \in (0, \frac{1}{2})$. We only mention that to handle the difficult term IV_2 that is special for the momentum SQG equations, we compute

$$\overline{\mathbb{E}}[\text{IV}_2^2] \overset{(4.23)}{\lesssim} \sum_{k=1}^{4} \text{IV}_{2,k}^2, \tag{4.40}$$

where

$$\text{IV}_{2,1}^2 \triangleq \overline{\mathbb{E}}\left[\left\| \sup_{\|e_j\|_{\dot{H}^\alpha_\sigma}\le 1} \left| \int_0^t \sum_{i,l=1}^{2} \int_{\mathbb{T}^2} \Lambda^{2-\gamma_2} v_l^m \partial_l e_{j,i} v_i^m dx d\tau \right|^2 \right\|_{L^2(0,T)} \right], \tag{4.41a}$$

$$\text{IV}_{2,2}^2 \triangleq \overline{\mathbb{E}}\left[\left\| \sup_{\|e_j\|_{\dot{H}^\alpha_\sigma}\le 1} \left| \sum_{i,l=1}^{2} \int_{\mathbb{T}^2} \Lambda^{2-\gamma_2} v_l^m \partial_l e_{j,i} v_i^m dx \right|^2 \right\|_{L^2(0,T)} \right], \tag{4.41b}$$

$$\text{IV}_{2,3}^2 \triangleq \overline{\mathbb{E}}\left[\left\| \sup_{\|e_j\|_{\dot{H}^\alpha_\sigma}\le 1} \left| \int_0^t \sum_{i,l=1}^{2} \int_{\mathbb{T}^2} \partial_i v_l^m [\Lambda^{2-\gamma_2}, e_{j,i}] v_l^m dx d\tau \right|^2 \right\|_{L^1(0,T)} \right], \tag{4.41c}$$

$$\text{IV}_{2,4}^2 \triangleq \overline{\mathbb{E}}\left[\left\| \sup_{\|e_j\|_{\dot{H}^\alpha_\sigma}\le 1} \left| \sum_{i,l=1}^{2} \int_{\mathbb{T}^2} \partial_i v_l^m [\Lambda^{2-\gamma_2}, e_{j,i}] v_l^m dx \right|^2 \right\|_{L^1(0,T)} \right]. \tag{4.41d}$$

Using (3.100), (3.102), and (4.31), we estimate

$$\|[\Lambda^{2-\gamma_2}, e_j] v^m\|_{\dot{H}^{1-\frac{\gamma_2}{2}}} \lesssim \|e_j\|_{H^\alpha} \|v^m\|_{\dot{H}^{1-\frac{\gamma_2}{2}}}. \tag{4.42}$$

This leads to

$$\begin{aligned} \text{IV}_{2,4}^2 &\lesssim \overline{\mathbb{E}}[\sup_{\|e_j\|_{\dot{H}^\alpha_\sigma}\le 1} \| |\nabla v^m\|_{\dot{H}^{\frac{\gamma_2}{2}-1}} \|[\Lambda^{2-\gamma_2}, e_j] v^m\|_{\dot{H}^{1-\frac{\gamma_2}{2}}} |^2 \|_{L^1(0,T)} \\ &\lesssim \left(\overline{\mathbb{E}}\left[\left(\int_0^T \|v^m\|_{\dot{H}^{1-\frac{\gamma_2}{2}+\frac{\gamma_1}{2}}}^2 d\tau \right)^2 \right] \right)^{\frac{1}{2}} (\overline{\mathbb{E}}[\|v^m\|_{L_T^\infty \dot{H}^{1-\frac{\gamma_2}{2}}}^4])^{\frac{1}{2}} \overset{(4.35)}{\lesssim} 1. \end{aligned} \tag{4.43}$$

Concerning IV_4 from (4.39d), using Lemma 4.4.6, we can estimate

$$\begin{aligned}\overline{\mathbb{E}}[\mathrm{IV}_4^2] &\lesssim \overline{\mathbb{E}}\left[\left\|\int_0^t G(v^m(\tau))d\overline{W}(\tau)\Lambda^{2-\gamma_2}e_j|\right\|^2_{W^{\gamma,2}(0,T;\dot{H}_\sigma^{\frac{3\gamma_2}{2}-5})}\right] \\ &\overset{(4.30)}{\lesssim}{}_{\gamma}\overline{\mathbb{E}}\left[\int_0^T 1+\|v^m(t)\|^2_{\dot{H}^{1-\frac{\gamma_2}{2}}}dt\right]\overset{(4.35)}{\lesssim} 1. \end{aligned} \tag{4.44}$$

Likewise, it suffices to rely on Lemma 2.4.7(b) to deduce the tightness in $C([0,T];\dot{H}_\sigma^{-a})$ for $a > 4-\gamma_2$. □

Thanks to Proposition 4.5.8, Prokhorov's theorem (Lemma 4.4.2) implies that $\{\mathcal{L}(v^m)\}_{m\in\mathbb{N}}$ is relatively compact, so that there exists a subsequence, which we still relabel as $\{\mathcal{L}(v^m)\}_{m\in\mathbb{N}}$, that converges weakly in $L^2(0,T;H)\cap C([0,T];\dot{H}_\sigma^{-a})$ for $a > 4-\gamma_2$. By Skorokhod's theorem (Lemma 4.4.3) this gives us $(\Omega,\mathcal{F},\{\mathcal{F}_t\}_{t\in[0,T]},\mathbb{P})$ and $L^2(0,T;H)\cap C([0,T];\dot{H}_\sigma^{-a})$-valued processes $\hat{v}$ and $\hat{v}^m$ defined on $(\Omega,\mathcal{F},\{\mathcal{F}_t\}_{t\in[0,T]},\mathbb{P})$ such that

$$\mathcal{L}(\hat{v}^m) = \mathcal{L}(v^m), \tag{4.45a}$$

$$\hat{v}^m \to \hat{v} \quad \text{as } m\to\infty \quad \text{in } L^2(0,T;H)\cap C([0,T];\dot{H}_\sigma^{-a})\ \mathbb{P}\text{-a.s.} \tag{4.45b}$$

Due to (4.45a) and Proposition 4.5.7, for all $m\in\mathbb{N}$ and $p\in[1,\infty)$,

$$\hat{v}^m \in L^p(\Omega;L^\infty(0,T;H))\cap L^p(\Omega;L^2(0,T;V)). \tag{4.46}$$

Thus by the weak compactness and Banach–Alaoglu theorem (Lemma 2.4.5(b)–(c)), we obtain that for all $p\in[1,\infty)$,

$$\begin{aligned}&\hat{v}^m \overset{w^*}{\rightharpoonup} \hat{v} \text{ weak}^* \text{ in } L^p(\Omega;L^\infty(0,T;H)),\\ &\hat{v}^m \overset{w}{\rightharpoonup} \hat{v} \quad \text{weakly in } L^p(\Omega;L^2(0,T;V)).\end{aligned}$$

For all $t\in[0,T]$, we define

$$M_m(t) \triangleq v^m(t) - v^m(0) + \int_0^t [P_m B(v^m(\tau)) + \Lambda^{\gamma_1}v^m(\tau)]d\tau, \tag{4.47a}$$

$$\hat{M}_m(t) \triangleq \hat{v}^m(t) - \hat{v}^m(0) + \int_0^t [P_m B(\hat{v}^m(\tau)) + \Lambda^{\gamma_1}\hat{v}^m(\tau)]d\tau. \tag{4.47b}$$

After integrating (4.32) over $[0,t]$ so that

$$M_m(t) = \int_0^t P_m G(v^m(\tau))d\overline{W}(\tau), \tag{4.48}$$

we see that M_m with paths in $C([0,T];H)$ is a square-integrable martingale w. r. t. $\mathcal{F}(\{v^m(\tau)\}_{\tau\in[0,T]})$ with quadratic variation

$$\langle\langle M_m\rangle\rangle_t = \int_0^t P_m G(v^m(\tau))G(v^m(\tau))^* P_m d\tau, \tag{4.49}$$

where $*$ denotes the adjoint from Definition 2.4.2(e) (cf. [209, eq. (2.19) on p. 138]), and we also recall Lemma 4.1.3; additionally, here we consider $\sigma(\{v^m(\tau) : \tau \in [0,T]\})$ as the σ-algebra generated by $\{v^m(\tau)\}_{\tau\in[0,T]}$ and $\mathcal{F}(\{v^m(\tau)\}_{\tau\in[0,T]})$ as the filtration generated by v^m. Because $\hat{M}_m$ in (4.47b) and M_m in (4.47a) are defined respectively in terms of $\hat{v}^m$ and v^m, by (4.45a) we have

$$\mathcal{L}(\hat{M}_m) = \mathcal{L}(M_m). \tag{4.50}$$

Moreover, for all $s \in [0,t]$, $t \in [0,T]$, we let φ be a continuous linear mapping on $L^2(0,s;H)\cap C([0,s];\dot{H}_\sigma^{-a})$, $a > 4-\gamma_2$, so that $\varphi(v^m)$ is well-defined. Then, for $w \in \dot{H}_\sigma^a(\mathbb{T}^2)$, we compute

$$\begin{aligned}
&\mathbb{E}[(\hat{M}_m(t) - \hat{M}_m(s), w)\varphi(\hat{v}^m|_{[0,s]})] \\
&\quad = \overline{\mathbb{E}}[(M_m(s) - M_m(s), w)\varphi(v^m|_{[0,s]})] = 0, && (4.51)\\
&\mathbb{E}[\|\hat{M}_m(t)\|^2_{\dot{H}^{1-\frac{\gamma_2}{2}}}] = \overline{\mathbb{E}}[\|M_m(t)\|^2_{\dot{H}^{1-\frac{\gamma_2}{2}}}] \lesssim \overline{\mathbb{E}}\Big[\sup_{t\in[0,T]}\|v^m(t)\|^2_{\dot{H}^{1-\frac{\gamma_2}{2}}}\Big] \lesssim 1, && (4.52)
\end{aligned}$$

where $\mathbb{E}$ is the mathematical expectation w. r. t. $\mathbb{P}$. Consequently, $\hat{M}_m$ with paths in $C([0,T];H)$ is a square-integrable martingale w. r. t. $\mathcal{F}(\{\hat{v}^m(\tau)\}_{\tau\in[0,t]})$ with quadratic variation

$$\langle\langle \hat{M}_m\rangle\rangle_t = \int_0^t P_m G(\hat{v}^m(\tau))G(\hat{v}^m(\tau))^* P_m d\tau. \tag{4.53}$$

Similarly, for $w, z \in \dot{H}_\sigma^a$,

$$\begin{aligned}
\mathbb{E}\Bigg[\Bigg(&(\hat{M}_m(t), w)(\hat{M}_m(t), z) - (\hat{M}_m(s), w)(\hat{M}_m(s), z) \\
&- \int_s^t (G(\hat{v}^m(\tau))^* P_m w, G(\hat{v}^m(\tau))^* P_m z)d\tau\Bigg)\varphi(\hat{v}^m|_{[0,s]})\Bigg] = 0 && (4.54)
\end{aligned}$$

by (4.49) (recall Definition 4.1.7 (b)), so that

$$\begin{aligned}
\mathbb{E}\Bigg[&\Bigg(\Bigg(\hat{v}^m(t) - \hat{v}^m(0) + \int_0^t [P_m B(\hat{v}^m) + \Lambda^{\gamma_1}\hat{v}^m]d\tau\Bigg), w\Bigg) \\
&\times \Bigg(\Bigg(\hat{v}^m(t) - \hat{v}^m(0) + \int_0^t [P_m B(\hat{v}^m) + \Lambda^{\gamma_1}\hat{v}^m]d\tau\Bigg), z\Bigg)
\end{aligned}$$

$$- \left(\left(\hat{v}^m(s) - \hat{v}^m(0) + \int_0^s [P_m B(\hat{v}^m) + \Lambda^{\gamma_1}\hat{v}^m] d\tau \right), w \right)$$

$$\times \left(\left(\hat{v}^m(s) - \hat{v}^m(0) + \int_0^s [P_m B(\hat{v}^m) + \Lambda^{\gamma_1}\hat{v}^m] d\tau \right), z \right)$$

$$- \int_s^t (G(\hat{v}^m(\tau))^* P_m w, G(\hat{v}^m(\tau))^* P_m z) d\tau \Bigg) \varphi(\hat{v}^m|_{[0,s]}) \Bigg] = 0. \tag{4.55}$$

Next, we define

$$g_m \triangleq \int_0^t \int_{\mathbb{T}^2} P_m \mathcal{P}_L((\Lambda^{2-\gamma_2}\hat{v}^m) \cdot \nabla \hat{v}^m - (\nabla \hat{v}^m)^T \cdot \Lambda^{2-\gamma_2}\hat{v}^m) w dx d\tau \tag{4.56a}$$

$$\times \int_0^t \int_{\mathbb{T}^2} P_m \mathcal{P}_L((\Lambda^{2-\gamma_2}\hat{v}^m) \cdot \nabla \hat{v}^m - (\nabla \hat{v}^m)^T \cdot \Lambda^{2-\gamma_2}\hat{v}^m) z dx d\tau \varphi(\hat{v}^m|_{[0,s]}),$$

$$g \triangleq \int_0^t \int_{\mathbb{T}^2} \mathcal{P}_L((\Lambda^{2-\gamma_2}\hat{v}) \cdot \nabla \hat{v} - (\nabla \hat{v})^T \cdot \Lambda^{2-\gamma_2}\hat{v}) w dx d\tau \tag{4.56b}$$

$$\times \int_0^t \int_{\mathbb{T}^2} \mathcal{P}_L((\Lambda^{2-\gamma_2}\hat{v}) \cdot \nabla \hat{v} - (\nabla \hat{v})^T \cdot \Lambda^{2-\gamma_2}\hat{v}) z dx d\tau \varphi(\hat{v}|_{[0,s]}).$$

By a combination of de la Vallée-Poussin theorem (Lemma 4.4.7) and Vitali's convergence theorem (Lemma 4.4.8) we see that

$$\hat{v}^m \to \hat{v} \quad \text{as} \quad m \to \infty \quad \text{strongly in } L^2(\Omega; L^2(0,T;H)), \tag{4.57}$$

and thus relabeling subsequence if necessary, we obtain

$$\hat{v}^m \to \hat{v} \quad \text{as } m \to \infty \quad \text{in } H \quad \text{for a. e. } (\omega, t). \tag{4.58}$$

Similarly to (4.43), for all $r \in (1, \infty)$, we can compute, $\mathbb{P}$-a. s.,

$$|g_m|^r \overset{(4.56a)(4.23)}{=} \Bigg| \Bigg[\int_0^t \sum_{i,j=1}^2 \int_{\mathbb{T}^2} \Lambda^{2-\gamma_2}\hat{v}_j^m \partial_j P_m w_i \hat{v}_i^m dx d\tau$$

$$- \frac{1}{2} \sum_{i,j=1}^2 \int_{\mathbb{T}^2} \partial_i \hat{v}_j^m [\Lambda^{2-\gamma_2}, P_m w_i] \hat{v}_j^m dx d\tau \Bigg]$$

$$\times \Bigg[\int_0^t \sum_{i,j=1}^2 \int_{\mathbb{T}^2} \Lambda^{2-\gamma_2}\hat{v}_j^m \partial_j P_m z_i \hat{v}_i^m dx d\tau$$

$$- \frac{1}{2} \sum_{i,j=1}^2 \int_{\mathbb{T}^2} \partial_i \hat{v}_j^m [\Lambda^{2-\gamma_2}, P_m z_i] \hat{v}_j^m dx d\tau \Bigg] \varphi(\hat{v}^m \Big|_{[0,s]}) \Bigg|^r$$

$$\lesssim \sup_{t \in [0,T]} \|\hat{v}^m(t)\|_{\dot{H}^{1-\frac{\gamma_2}{2}}}^{4r} + \left(\int_0^T \|\hat{v}^m\|_{\dot{H}^{1-\frac{\gamma_2}{2}+\frac{\gamma_1}{2}}}^2 d\tau \right)^{2r} \lesssim 1$$

due to (4.42), Proposition 4.5.7, and (4.45a). Consequently, $\{g_m\}_{m\in\mathbb{N}}$ is uniformly integrable by the de la Vallée-Poussin theorem (Lemma 4.4.7). Moreover, (4.45b) implies that for a. e. $\omega \in \Omega$,

$$|\varphi(\hat{v}^m|_{[0,s]}(\omega)) - \varphi(\hat{v}|_{[0,s]}(\omega))| \to 0 \tag{4.59}$$

as $m \to \infty$. Thus, to prove that $g_m \to g$ $\mathbb{P}$-a. s. as $m \to \infty$, it suffices to focus on

$$\int_0^t \int_{\mathbb{T}^2} P_m \mathcal{P}_L((\Lambda^{2-\gamma_2}\hat{v}^m)\cdot\nabla\hat{v}^m - (\nabla\hat{v}^m)^T\cdot\Lambda^{2-\gamma_2}\hat{v}^m)$$
$$- \mathcal{P}_L((\Lambda^{2-\gamma_2}\hat{v})\cdot\nabla\hat{v} - (\nabla\hat{v})^T\cdot\Lambda^{2-\gamma_2}\hat{v})wdxdt = \sum_{i=1}^4 V_i,$$

where we use (4.23) to deduce

$$V_1 \triangleq \int_0^t \int_{\mathbb{T}^2} (\Lambda^{2-\gamma_2}\hat{v}^m)\cdot\nabla\hat{v}^m(P_m - \mathrm{Id})wdxd\tau, \tag{4.60a}$$

$$V_2 \triangleq \int_0^t \int_{\mathbb{T}^2} [(\Lambda^{2-\gamma_2}(\hat{v}^m - \hat{v}))\cdot\nabla\hat{v}^m + \Lambda^{2-\gamma_2}\hat{v}\cdot\nabla(\hat{v}^m - \hat{v})]\cdot wdxd\tau, \tag{4.60b}$$

$$V_3 \triangleq \frac{1}{2}\int_0^t \sum_{i,j=1}^2 \int_{\mathbb{T}^2} \partial_i\hat{v}_j^m[\Lambda^{2-\gamma_2}, (P_m - \mathrm{Id})w_i]\hat{v}_j^m dxd\tau, \tag{4.60c}$$

$$V_4 \triangleq -\int_0^t \int_{\mathbb{T}^2} [(\nabla\hat{v}^m)^T\cdot\Lambda^{2-\gamma_2}\hat{v}^m - (\nabla\hat{v})^T\cdot\Lambda^{2-\gamma_2}\hat{v}]\cdot wdxd\tau. \tag{4.60d}$$

We can show by (4.60a) that $V_1 \to 0$ as $m \to \infty$ $\mathbb{P}$-a. s. by Hölder's inequality and that $\lim_{m\to\infty}\|(P_m - \mathrm{Id})w\|_{\dot{H}^{4-\gamma_2}} = 0$. Next, we estimate V_2 from (4.60b) by

$$\begin{aligned}
V_2 \overset{(3.86)}{\lesssim} & \int_0^t \|\hat{v}^m - \hat{v}\|_{\dot{H}^{1-\frac{\gamma_2}{2}}} (\|\Lambda^{4-\frac{3\gamma_2}{2}} w\|_{L^{\frac{4}{2-\gamma_2}}} \|\hat{v}^m\|_{L^{\frac{4}{\gamma_2}}} \\
& + \|\nabla w\|_{L^\infty} \|\Lambda^{1-\frac{\gamma_2}{2}} \hat{v}^m\|_{L^2}) \\
& + \|\hat{v}\|_{\dot{H}^{1-\frac{\gamma_2}{2}}} (\|\Lambda^{4-\frac{3\gamma_2}{2}} w\|_{L^{\frac{4}{2-\gamma_2}}} \|\hat{v}^m - \hat{v}\|_{L^{\frac{4}{\gamma_2}}} \\
& + \|\nabla w\|_{L^\infty} \|\Lambda^{1-\frac{\gamma_2}{2}} (\hat{v}^m - \hat{v})\|_{L^2}) d\tau \\
\lesssim & \sup_{t\in[0,T]} (\|\hat{v}^m\|_{\dot{H}^{1-\frac{\gamma_2}{2}}} + \|\hat{v}\|_{\dot{H}^{1-\frac{\gamma_2}{2}}})(t) \int_0^t \|\hat{v}^m - \hat{v}\|_{\dot{H}^{1-\frac{\gamma_2}{2}}} d\tau \to 0
\end{aligned}$$

$\mathbb{P}$-a. s. as $m \to \infty$. Next, V_3 in (4.60c) can be shown to vanish as $m \to \infty$ $\mathbb{P}$-a. s. using (3.100) and (3.102), which implies

$$\|[\Lambda^{2-\gamma_2}, (P_m - \mathrm{Id})w]\hat{v}^m\|_{\dot{H}^{1-\frac{\gamma_2}{2}}} \lesssim \|(P_m - \mathrm{Id})w\|_{H^a} \|\hat{v}^m\|_{\dot{H}^{1-\frac{\gamma_2}{2}}}$$

analogously to (4.42).

Lastly, to treat V_4 in (4.60d), by Remark 4.5.3 we can rewrite it as

$$V_4 = -\int_0^t \int_{\mathbb{T}^2} [(\nabla(\hat{v}^m - \hat{v}))^T \cdot \Lambda^{2-\gamma_2}\hat{v}^m + (\nabla\hat{v})^T \cdot \Lambda^{2-\gamma_2}(\hat{v}^m - \hat{v})] \cdot w dx d\tau,$$

but then (4.23) becomes inapplicable. Instead, we can first use (4.23) and then split (4.60d) as

$$V_4 = V_{4,1} + V_{4,2}, \tag{4.61}$$

where

$$V_{4,1} \triangleq \frac{1}{2}\int_0^t \sum_{i,j=1}^2 \int_{\mathbb{T}^2} \partial_i(\hat{v}_j^m - \hat{v}_j)[\Lambda^{2-\gamma_2}, w_i]\hat{v}_j^m dx d\tau, \tag{4.62a}$$

$$V_{4,2} \triangleq \frac{1}{2}\int_0^t \sum_{i,j=1}^2 \int_{\mathbb{T}^2} \partial_i\hat{v}_j[\Lambda^{2-\gamma_2}, w_i](\hat{v}_j^m - \hat{v}_j) dx d\tau. \tag{4.62b}$$

Then separately using (3.100) and (3.102), we can estimate

$$\begin{aligned}\max_{k=1,2} V_{4,k} \lesssim & \left[\left(\int_0^t \|\hat{v}^m\|_{\dot{H}^{1-\frac{\gamma_2}{2}+\frac{\gamma_1}{2}}}^2 d\tau\right)^{\frac{1}{2}} + \left(\int_0^t \|\hat{v}\|_{\dot{H}^{1-\frac{\gamma_2}{2}+\frac{\gamma_1}{2}}}^2 d\tau\right)^{\frac{1}{2}}\right] \\ & \times \left(\int_0^t \|\hat{v}^m - \hat{v}\|_{\dot{H}^{1-\frac{\gamma_2}{2}}}^2 d\tau\right)^{\frac{1}{2}} \to 0 \end{aligned} \tag{4.63}$$

$\mathbb{P}$-a. s. as $m \to \infty$ thanks to (4.45) and (4.35). Thus we have shown that $g_m \to g$ $\mathbb{P}$-a. s. as $m \to \infty$. Similarly, we can prove the uniform integrability of the diffusive term and the convergence of $\int_0^t (\Lambda^{\gamma_1}\hat{v}^m(\tau) - \Lambda^{\gamma_1}\hat{v}(\tau), w) d\tau \to 0$ as $m \to \infty$ $\mathbb{P}$-a. s. Finally, we prove that $\mathbb{E}[h_m - h] \to 0$ as $m \to \infty$, where

$$h_m \triangleq \int_s^t (G(\hat{v}^m(\tau))^* P_m w, G(\hat{v}^m(\tau))^* P_m z) d\tau \varphi(\hat{v}^m|_{[0,s]}), \tag{4.64a}$$

$$h \triangleq \int_s^t (G(\hat{v}(\tau))^* w, G(\hat{v}(\tau))^* z) d\tau \varphi(\hat{v}|_{[0,s]}). \tag{4.64b}$$

For $r \in (1,\infty)$, using (4.30) and Lemma 2.4.11(1), we can compute

$$\begin{aligned}|h_m|^r &\lesssim \left|\int_s^t \|G(\hat{v}^m(\tau))^* P_m w\|_Z \|G(\hat{v}^m(\tau))^* P_m z\|_Z d\tau |\varphi(\hat{v}^m|_{[0,s]})|\right|^r \\ &\overset{(2.24)}{\lesssim} \left|\int_s^t \|G(\hat{v}^m(\tau))\|_{L_2(Z,H)}^2 d\tau\right|^r \overset{(4.27)(4.30)}{\lesssim} \left|\int_s^t 1 + \|\hat{v}^m(\tau)\|_{\dot{H}_\sigma^{1-\frac{\gamma_2}{2}}}^2 d\tau\right|^r 1.\end{aligned}$$

Next, to prove the pointwise convergence, writing

$$\begin{aligned}
&\mathbb{E}[h_m - h] \\
&= \mathbb{E}\Big[\int_s^t (G(\hat{v}^m(\tau))^* P_m w, G(\hat{v}^m(\tau))^* P_m z) d\tau [\varphi(\hat{v}^m|_{[0,s]}) - \varphi(\hat{v}|_{[0,s]})]\Big] \\
&\quad + \mathbb{E}\Big[\Big[\int_s^t (G(\hat{v}^m(\tau))^* P_m w, G(\hat{v}^m(\tau))^* P_m z) d\tau \\
&\quad - \int_s^t (G(\hat{v}(\tau))^* w, G(\hat{v}(\tau))^* z) d\tau\Big] \varphi(\hat{v}|_{[0,s]})\Big],
\end{aligned}$$

due to (4.59), it is clear that it suffices to show that

$$\int_s^t (G(\hat{v}^m(\tau))^* P_m w, G(\hat{v}^m(\tau))^* P_m z) d\tau - \int_s^t (G(\hat{v}(\tau))^* w, G(\hat{v}(\tau))^* z) d\tau \to 0$$

as $m \to \infty$ $\mathbb{P}$-a. s. To this end, we compute

$$\begin{aligned}
&\int_s^t (G(\hat{v}^m(\tau))^* P_m w, G(\hat{v}^m(\tau))^* P_m z) d\tau - \int_s^t (G(\hat{v}(\tau))^* w, G(\hat{v}(\tau))^* z) d\tau \\
&\le \|G(\hat{v}^m(\tau))^* P_m w - G(\hat{v}(\tau))^* w\|_{L^2(s,t;Z)} \|G(\hat{v}^m(\tau))^* P_m z\|_{L^2(s,t;Z)} \\
&\quad + \|G(\hat{v}(\tau))^* w\|_{L^2(s,t;Z)} \|G(\hat{v}^m(\tau))^* P_m z - G(\hat{v}(\tau))^* z\|_{L^2(s,t;Z)} \\
&\lesssim \|(P_m - \mathrm{Id}) w\|_{H^\alpha} + \|\hat{v}^m - \hat{v}\|_{L^2(s,t;\dot{H}^{1-\frac{\gamma_2}{2}})} + \|(P_m - \mathrm{Id}) z\|_{H^\alpha} \to 0
\end{aligned}$$

as $m \to \infty$. Taking the limit as $m \nearrow +\infty$ in (4.54), we obtain

$$\begin{aligned}
\mathbb{E}\Big[\Big(&(\hat{M}(t), w)(\hat{M}(t), z) - (\hat{M}(s), w)(\hat{M}(s), z) \\
&- \int_s^t (G(\hat{v}(\tau))^* w, G(\hat{v}(\tau))^* z) d\tau\Big) \varphi(\hat{v}|_{[0,s]})\Big] = 0,
\end{aligned}$$

where

$$\hat{M}(t) = \hat{v}(t) - \hat{v}(0) + \int_0^t [B(\hat{v}(\tau)) + \Lambda^{\gamma_1} \hat{v}(\tau)] d\tau$$

by (4.47b). Similarly, taking the limit as $m \nearrow \infty$ in (4.51) gives us

$$\mathbb{E}[(\hat{M}(t) - \hat{M}(s), w) \varphi(\hat{v}|_{[0,s]})] = 0. \tag{4.65}$$

This implies that $\hat{M}(t)$ is an $\dot{H}_\sigma^{-\alpha}$-valued martingale w. r. t. $\mathcal{F}(\{\hat{v}(\tau)\}_{\tau \in [0,T]})$ with quadratic variation

$$\langle\langle \hat{M} \rangle\rangle_t = \int_0^t G(\hat{v}(\tau)) G(\hat{v}(\tau))^* d\tau. \tag{4.66}$$

For all $p \in [2, \infty)$, relying on (4.50), (4.48), (4.19), (4.27), and (4.30) gives us

$$\sup_{m\in\mathbb{N}} \mathbb{E}[\|\hat{M}_m\|_H^p] \leq \sup_{m\in\mathbb{N}} \overline{\mathbb{E}}\left[\sup_{t\in[0,T]}\left\|\int_0^t P_m G(v^m(\tau))d\overline{W}(\tau)\right\|_H^p\right]$$

$$\lesssim \sup_{m\in\mathbb{N}}\left(\overline{\mathbb{E}}\left[\int_0^t \|P_m G(v^m(\tau))\|_{L_2(Z,H)}^2 d\tau\right]\right)^{\frac{p}{2}} \lesssim \sup_{m\in\mathbb{N}}\left(\overline{\mathbb{E}}\left[\int_0^t \|v^m\|_H^2 d\tau\right]\right)^{\frac{p}{2}} \lesssim 1.$$

Thus $\{\hat{M}_m\}_{m\in\mathbb{N}}$ is uniformly integrable, and $\lim_{m\to\infty} \mathbb{E}[\|\hat{M}_m(t)\|_H^2] = \mathbb{E}[\|\hat{M}(t)\|_H^2]$. Therefore $\hat{M}(t)$ is an $\dot{H}_\sigma^{-a}$-valued martingale w. r. t. $\mathcal{F}(\{\hat{v}(\tau)\}_{\tau\in[0,t]})$ and lies in $L^2(\Omega)$. If we define

$$\hat{N}_m(t) \triangleq \Lambda^{-a}\hat{M}_m(t), \tag{4.67}$$

then, as we did for $\hat{M}_m$, we can show that $\hat{N}_m$ is a square-integrable martingale on $(\Omega, \mathcal{F}, \{\mathcal{F}_t\}_{t\in[0,T]}, \mathbb{P})$ such that

$$\langle\langle \hat{N}_m(t)\rangle\rangle = \int_0^t \Lambda^{-a} P_m G(\hat{v}^m(\tau))G(\hat{v}^m(\tau))^* P_m \Lambda^{-a} d\tau.$$

We can also show that

$$\tilde{N}(t) \triangleq \Lambda^{-a}\tilde{v}(t) - \Lambda^{-a}\tilde{v}(0) + \int_0^t \Lambda^{-a}B(\tilde{v})(\tau) + \Lambda^{\gamma_1 - a}\tilde{v}(\tau)d\tau$$

is a square-integrable martingale w. r. t. $\{\tilde{\mathcal{F}}_t\} = \mathcal{F}(\{\tilde{v}(s) : s \leq t\})$ and

$$\langle\langle \tilde{N}(t)\rangle\rangle = \int_0^t \Lambda^{-a} N(\tilde{v}(\tau))N(\tilde{v}(\tau))^* \Lambda^{-a} d\tau.$$

By the martingale representation theorem (Lemma 4.4.9) we conclude that there exist $(\tilde{\Omega}, \tilde{\mathcal{F}}, \{\tilde{\mathcal{F}}_t\}_{t\in[0,T]}, \tilde{\mathbb{P}})$, a Wiener process $\tilde{W}$, and a predictable continuous process $\tilde{v}$ such that

$$\Lambda^{-a}\tilde{v}(t) - \Lambda^{-a}\tilde{v}(0) + \Lambda^{-a}\int_0^t [B(\tilde{v}) + \Lambda^{\gamma_1}\tilde{v}](\tau)d\tau = \Lambda^{-a}\int_0^t G(\tilde{v}(\tau))d\tilde{W}(\tau).$$

Applying Λ^a completes the proof of Theorem 4.5.5.

4.5.2 Martingale problem

There is a notion of martingale problem [310] used in [144] (also [153, 188, 327]). We refer the interested readers to [209, Chapter 5.4, Section B] concerning its relationship with the martingale solution that we investigated in Section 4.5.1. For clarity, we define this for the 3D Navier–Stokes equations (4.16) forced by a white-in-time additive noise on $\mathbb{T}^3$ following [144, Definition 3.3].

Definition 4.5.9. Define $\Omega_{\mathrm{NS}} \triangleq C([0,\infty);(H_\sigma^2)')$ and the corresponding Borel σ-algebra $\mathcal{B}_{\Omega_{\mathrm{NS}}}$. We define $\mathcal{B}_t^{\mathrm{NS}} \triangleq \mathcal{F}(\{\omega|_{[0,t]} : \omega \in \Omega_{\mathrm{N}}\})$ and say that ϑ is an a. s. $\mathcal{B}_t^{\mathrm{NS}}$-supermartingale under $\mathbb{P}$ if ϑ is $\mathbb{P}$-integrable and $\mathbb{E}^{\mathbb{P}}[\vartheta_t 1_A] \le \mathbb{E}^{\mathbb{P}}[\vartheta_s 1_A]$ for a. e. $s \ge 0$ and all $t \ge s$ and $A \in \mathcal{B}_s^{\mathrm{NS}}$.

Definition 4.5.10. We consider (4.16) with additive white-in-time noise, specifically $G : L_\sigma^2 \to L_\sigma^2$, $\{(\lambda_k, e_k)\}_{k=1}^\infty$ as pairs of eigenvectors and eigenvectors (recall (4.6)) such that $\{e_k\}_{k=1}^\infty$ forms a complete o. n. s. of L_σ^2, a family $\{\beta_k\}_{k=1}^\infty$ of 1D i. i. d. Brownian motions, and then $GdW = \sum_{k=1}^\infty \lambda_k e_k d\beta_k$.

Then given a probability measure μ_0 on L_σ^2, a probability measure $\mathbb{P}$ on $(\Omega_{\mathrm{NS}}, \mathcal{B}_{\Omega_{\mathrm{NS}}})$ is a solution to the martingale problem associated to (4.16) starting from μ_0 if the following hold:

(M1) $\mathbb{P}(L_{\mathrm{loc}}^\infty([0,\infty);L_\sigma^2) \cap L_{\mathrm{loc}}^2([0,\infty);\dot{H}_\sigma^1)) = 1$.

(M2) For all $\phi \in C^\infty(\mathbb{T}^3)$ that are mean-zero and divergence-free, the process M_t^ϕ defined $\mathbb{P}$-a. s. on $(\Omega_{\mathrm{NS}}, \mathcal{B}_{\Omega_{\mathrm{NS}}})$ by

$$M_t^\phi \triangleq \langle \zeta_t - \zeta(0), \phi\rangle - v\int_0^t \langle \zeta_s, \Delta\phi\rangle ds + \int_0^t \langle \mathcal{P}_L((\zeta_s\cdot\nabla\phi)), \zeta_s\rangle ds \tag{4.68}$$

is a square-integrable continuous $\mathcal{B}_t^{\mathrm{NS}}$-martingale under $\mathbb{P}$ with quadratic variation $\langle\langle M^\phi\rangle\rangle_t = t\|G^{\frac{1}{2}}\phi\|_{L^2}^2$.

(M3) The process E_t^1 defined $\mathbb{P}$-a. s. on $(\Omega_{\mathrm{NS}}, \mathcal{B}_{\Omega_{\mathrm{NS}}})$ as

$$E_t^1 \triangleq \|\zeta_t\|_{L^2}^2 + 2v\|\zeta\|_{L^2(0,t;\dot{H}^1(\mathbb{T}^3))}^2 - \|\zeta(0)\|_{L^2}^2 - t\sum_{k=1}^\infty \lambda_k$$

is a $\mathbb{P}$-integrable $\mathcal{B}_t^{\mathrm{NS}}$-supermartingale under $\mathbb{P}$.

(M4) For each $n \ge 2$, the process E_t^n, defined $\mathbb{P}$-a. s. on $(\Omega_{\mathrm{NS}}, \mathcal{B}_{\Omega_{\mathrm{NS}}})$ as

$$E_t^n \triangleq \|\zeta_t\|_{L^2}^{2n} + v2n\int_0^t \|\zeta_s\|_{L^2}^{2n-2}\|\zeta_s\|_{\dot{H}^1}^2 ds - \|\zeta(0)\|_{L^2}^{2n} - n(2n-1)\sum_{k=1}^\infty \lambda_k \|\zeta\|_{L^{2n-2}(0,t;L^2)}^{2n-2}$$

is a $\mathbb{P}$-integrable $\mathcal{B}_t^{\mathrm{NS}}$-supermartingale under $\mathbb{P}$.

(M5) μ_0 is the marginal of $\mathbb{P}$ at time $t = 0$.

4.5.3 Da Prato–Debussche trick and random PDEs

Let us now make important comments on the probabilistic Galerkin approximation that we elaborated in Section 4.5.1. First, the probabilistic Galerkin approximation will most certainly result in a probabilistically weak solution, because obtaining uniform estimates in the face of multiplicative noise requires taking mathematical expectation; once this step is taken, applications of the Prokhorov and Skorokhod theorems to deduce a completely new filtered probability space seems inevitable. Second, in case (4.20) consisted of an additive noise, i. e., $G(v) = G$ (recall Definition 4.2.11), we can subtract out the corresponding stochastic Stokes equation

$$dz + (\nabla\pi_z + \Lambda^{\gamma_1}v)dt = GdW, \quad \nabla\cdot z = 0. \tag{4.69}$$

This is a linear SPDE, and the solution theory for such equations is well known (e. g., [118, Chapter 5]). Therefore, when looking for a solution to the original (4.20), we only have to find a solution $X \triangleq v - z$ and $\tilde{\pi} \triangleq \pi - \pi_z$ that satisfies

$$\begin{aligned} &\partial_t X + (\Lambda^{2-\gamma_2}(X+z)\cdot\nabla)(X+z) \\ &\quad - \nabla(X+z)^T\cdot\Lambda^{2-\gamma_2}(X+z) + \nabla\tilde{\pi} + \Lambda^{\gamma_1}X = 0, \quad \nabla\cdot X = 0. \end{aligned} \tag{4.70}$$

A PDE of such a form is called a random PDE due to the implicit presence of the random force. Such a trick is called the Da Prato–Debussche trick and became famous in the works by Da Prato, Debussche, and Temam [116] on the 1D Burgers' equation forced by STWN and by Flandoli [139] on the 2D Navier–Stokes equations forced by noise that is white only in time. The reason why [116] was able to consider STWN is due to the low dimension and will be described in detail in Example 13. This trick will be elaborated furthermore in Sections 7.1 and 8.4. We can analyze (4.70) ω-wise, i. e., apply deterministic techniques for fixed $\omega \in \Omega$.

We can apply the same trick to the 3D Navier–Stokes equations forced by white-in-time additive noise (4.16); e. g., Flandoli and Schmalfuß [145, Proposition 2.2] proved the global-in-time existence of a probabilistically and analytically weak solution. The same trick also led Flandoli and Romito [144] to construct a global-in-time probabilistically weak, the Leray–Hopf weak solution, to the 3D Navier–Stokes equations forced by white-in-time additive noise (4.16) on $\mathbb{T}^3$; we call it the Leray–Hopf weak solution because its solution satisfies the appropriate energy inequality (cf. (3.137) from Definition 3.10.5). Additionally, Flandoli and Romito [144, Theorem 3.7] proved the existence of measurable selection (recall Definition 4.4.11); i. e., for every initial data $x \in L^2_\sigma$, the mapping $x \mapsto \mathbb{P}_x$ is measurable, where $\mathbb{P}_x$ is a probability measure on the path space $\Omega_{\text{NS}} = C([0,\infty);(H^2_\sigma)')$ (recall Definition 4.5.10). In fact, the selection was a Markov selection and thus measurable (see [144, Definition 2.4] and also [144, 153, 188, 327]). Unfortunately, because this approach relies on compactness argument, the converging subsequence depends on ω, and thus the measurable selection only applies to a given σ-algebra, and not a filtration. We quote Flandoli [140, p. 84] in 2008:

> existence of a progressively measurable selection remains an open problem.

The existence of a progressively measurable selection by Definition 4.3.1(b) would basically imply the existence of a probabilistically strong solution. As its consequence, we quote Flandoli [140, p. 84] in 2008:

> unfortunately ... there is no result of existence of [probabilistically] strong solutions for the 3D stochastic Navier–Stokes equations.

By Cherny's theorem (Lemma 4.3.3) the existence of probabilistically strong solution is half of the hypothesis needed to verify the pathwise uniqueness. The other half is the uniqueness in law, which is definitely weaker than the pathwise uniqueness according to the Yamada–Watanabe theorem (Lemma 4.3.2(a)) and Tanaka's example (Example 5). We quote Da Prato and Debussche [115, pp. 878–879] in 2003:

> The problem of path-wise uniqueness of such a solution seems to be as difficult as in the deterministic case. The uniqueness of the law might be less difficult.

Yet, the uniqueness in law of the solution to the 3D Navier–Stokes equations forced by additive white-in-time noise also remained open until the convex integration shed light; we will discuss this issue further in Chapter 7.

4.6 Further preliminaries

Before ending this chapter, we list a few more valuable preliminaries on stochastic analysis.

4.6.1 Preliminaries for deriving regularity of stochastic heat equation

The following Kolmogorov test and factorization formula have many applications, e. g., for proving temporal regularity of solutions to linear SPDEs.

Lemma 4.6.1 (Kolmogorov's test, [118, Theorem 3.3]). *Let $\{X_t\}_{t\in[0,T]}$ be a stochastic process on $(\Omega, \mathcal{F}, \mathbb{P})$ with values in a complete metric space (E, ρ) such that for some constants $C > 0$, $\epsilon > 0$, and $\delta > 1$,*

$$\mathbb{E}[(\rho(X_t, X_s))^\delta] \leq C|t-s|^{1+\epsilon}$$

for all $t, s \in [0, T]$. Then there exists a version of X with paths (recall Definition 4.1.1) that are $\mathbb{P}$-a. s. Hölder continuous with any exponent smaller than $\frac{\epsilon}{\delta}$.

Lemma 4.6.2 (Factorization formula, [118, p. 131]). *Let $t \in (0, \infty)$, $r \in [0, t]$, and $\alpha \in (0, 1)$. Then*

$$\int_r^t (t-s)^{\alpha-1}(s-r)^{-\alpha} ds = \frac{\pi}{\sin(\alpha\pi)}. \tag{4.71}$$

4.6.2 Wiener chaos decomposition

Here we follow [274, Chapter 1] to introduce a minimum amount of notations on Wiener chaos decomposition and Gaussian hypercontractivity theorem.

Definition 4.6.3. Let H be a real separable Hilbert space with scalar product $\langle\cdot,\cdot\rangle_H$ and norm $\|\cdot\|_H$. Fix a complete probability space $(\Omega,\mathcal{F},\mathbb{P})$.
(a) A stochastic process $W = \{W(h) : h \in H\}$ is an isonormal Gaussian process (or a Gaussian process on H) if $W(h)$ is Gaussian and centered (recall Definition 4.1.1(d)).
(b) Define the Hermite polynomials by

$$H_0(x) \triangleq 1 \quad\text{and}\quad H_n(x) \triangleq \frac{(-1)^n}{n!} e^{\frac{x^2}{2}} \frac{d^n}{(dx)^n} e^{-\frac{x^2}{2}}, \quad n \in \mathbb{N}. \tag{4.72}$$

(c) For each $n \in \mathbb{N}$, we denote by $\mathcal{H}_n$ the closed linear subspace of $L^2(\Omega,\mathcal{F},\mathbb{P})$ generated by the random variables $\{H_n(W(h)) : h \in H, \|h\|_H = 1\}$; $\mathcal{H}_0$ is the set of constants, and $\mathcal{H}_1$ is the set of random variables $\{W(h) : h \in H\}$. By [274, Theorem 1.1.1], $L^2(\Omega,\mathcal{F},\mathbb{P}) = \oplus_{n=0}^{\infty}\mathcal{H}_n$. The space $\mathcal{H}_n$ is called the homogeneous Wiener chaos of order n, and the inhomogeneous Wiener chaos of order l is $\otimes_{n=0}^{l}\mathcal{H}_n$.

We will come back to discuss the Wiener chaos in Section 8.3.2 to describe contraction of indices. For now, we limit ourselves to the following important result, which will be used in the proof of Lemma 7.1.1 and more thereafter, e. g., in Section 8.5.2.

Lemma 4.6.4 (Gaussian hypercontractivity theorem,. [167, Lemma 10.5], [201, Theorem 3.50]; see also [301, Theorem I.22]). *Let $X \in L^2(\Omega,\mathcal{F},\mathbb{P})$ be a random variable in the inhomogeneous Wiener chaos of order l. Then, for every $p \geq 1$, there exists a universal constant $C_{l,p}$ such that $\mathbb{E}[|X|^{2p}] \leq C_{l,p}(\mathbb{E}[|X|^2])^p$.*

For convenience, to define the Wick products in Section 8.3.1, we recall the following definition.

Definition 4.6.5. A Gaussian linear vector space H is a real linear vector space of random variables defined on some probability space $(\Omega,\mathcal{F},\mathbb{P})$ such that each variable in the space is centered Gaussian. It follows that it is a linear subspace of $L^2(\Omega,\mathcal{F},\mathbb{P})$, and we use the norm and inner product of $L^2(\Omega)$ on H. A Gaussian Hilbert space is a Gaussian linear vector space that is complete w. r. t. the $L^2(\Omega)$-norm.

4.6.3 Invariant measure and ergodicity

Because the notion of invariant measure is of importance, especially in stochastic quantization (recall Section 1.2.3), we briefly recall some definitions.

Definition 4.6.6. Let U and H be Hilbert spaces, and consider the equation

$$dX = (AX + F(X))dt + B(X)dW(t), \quad X(0) = \zeta, \tag{4.73}$$

where A informally represents diffusion, $F : \Omega \times H \to H$ is measurable from $(H, \mathcal{B}_H)$ into $(H, \mathcal{B}_H)$, and $B : \Omega \times H \to L_2^0$ is measurable from $(\Omega \times H, \mathcal{B}_H)$ into $(L_2^0, \mathcal{B}(L_2^0))$, where $L_2^0 \triangleq L^2(U_0, H)$, W is a Q-Wiener process on $U_1 \supset U$, and $U_0 \triangleq Q^{\frac{1}{2}} U$ (recall Definition 4.2.7).

(a) We define the corresponding transition semigroup and transition function for $t \geq 0$ and $\zeta \in H$ by

$$P_t \varphi(\zeta) \triangleq \mathbb{E}[\varphi(X(t,\zeta))] \quad \text{if } \varphi \in B_b(H),$$
$$P(t,\zeta,\Gamma) \triangleq P_t 1_\Gamma(\zeta) = \mathcal{L}(X(t,\zeta))(\Gamma) \quad \text{if } \Gamma \in \mathcal{B}_H,$$

where $B_b(H)$ is the set of all bounded Borel functions from H to $\mathbb{R}$.

(b) Let $\mathcal{M}(H)$ be the space of bounded measures on $(H, \mathcal{B}_H)$, and let

$$P_t^* \mu(\Gamma) \triangleq \int_H P(t,\zeta,\Gamma)\mu(d\zeta) \quad \forall \Gamma \in \mathcal{B}_H.$$

(i) We say that $\{P(t,\zeta,\Gamma)\}_{t\geq 0, \zeta \in H}$ is irreducible in H if $P(t,\zeta,\Gamma) > 0$ for all $t > 0$, $\zeta \in H$, and nonempty open sets $\Gamma \subset H$.

(ii) The associated transition semigroup $\{P_t\}_{t\geq 0}$ is Feller if $P_t(C_b(H)) \subset C_b(H)$ for all $t \geq 0$ and strong Feller if $P_t(B_b(H)) \subset C_b(H)$ for all $t \geq 0$, where $C_b(H)$ is the set of all continuous bounded functions from H to $\mathbb{R}$.

(c) A probability measure $\mu \in \mathcal{M}(H)$ is said to be invariant (stationary) measure for (4.73) if $P_t^* \mu = \mu$ for all $t > 0$ or, equivalently,

$$\int_H P(t,\zeta,\gamma)\mu(d\zeta) = \mu(\Gamma) \quad \forall \Gamma \in \mathcal{B}_H.$$

Recalling the definition of a stationary process (Definition 4.1.1(f)), we see that if a PDE admits a stationary solution, then its law is an invariant measure.

In this monograph, we focus on singular SPDEs and uniqueness/nonuniqueness issue, and not so much on ergodicity. We refer the interested readers to [118, Section 11.2] on the Krylov–Bogoliubov theorem to prove the existence of an invariant measure and [117, Proposition 4.1.1 and Theorem 4.2.1] on the Doob theorem, which states that irreducibility combined with strong Feller property implies the uniqueness of an invariant measure. Additionally, we refer to [115, 174] and [175, 352, 376] for ergodicity results on SPDEs in the case of white-in-time noise and STWN, respectively.

5 Global regularity of the 2D MHD system

This chapter is devoted to the global regularity issues of the deterministic systems related to MHD that have been open for some time. Section 5.1 is devoted to the global regularity of the 2D inviscid but resistive MHD system that caught much attention more than a decade ago but has remained open. In Section 5.2, we discuss the global regularity of logarithmically supercritical PDEs relevant to our discussions in Section 9.1 on the global-in-time solution theory of singular SPDEs. We conclude in Section 5.3 with review of recent developments and open problems.

5.1 2D inviscid resistive MHD system

First, let us use the fractional Laplacian from (3.10) to write down, on $\mathbb{R}^d$ for clarity, the generalized Navier–Stokes equations, originally introduced by Lions [241, eq. (8.7), p. 263], and the generalized MHD system:

$$\partial_t u + (u \cdot \nabla)u + \nabla \pi + \upsilon \Lambda^m u = 0, \quad \nabla \cdot u = 0; \tag{5.1}$$

$$\partial_t u + (u \cdot \nabla)u + \nabla \pi + \upsilon \Lambda^{m_1} u = (b \cdot \nabla)b, \quad \nabla \cdot u = 0, \tag{5.2a}$$

$$\partial_t b + (u \cdot \nabla)b + \eta \Lambda^{m_2} b = (b \cdot \nabla)u. \tag{5.2b}$$

Computations analogous to (2.29)–(2.30) show that $u \in L_T^\infty L_x^2 \cap L_T^2 \dot{H}_x^{\frac{m}{2}}$ for all smooth solutions starting from initial data in L_x^2. Analogous computations to (2.29)–(2.30) also show that

$$\int_{\mathbb{R}^d} (u \cdot \nabla)b \cdot b dx = 0, \qquad \int_{\mathbb{R}^d} (b \cdot \nabla)b \cdot u + (b \cdot \nabla)u \cdot b dx = 0,$$

leading to the energy conservation and dissipation, respectively when $\upsilon = \eta = 0$ and $\upsilon, \eta > 0$:

$$\mathcal{E}(t) \triangleq \frac{1}{2}\|(u,b)(t)\|_{L^2}^2, \tag{5.3a}$$

$$\mathcal{E}(t) + \int_0^t \upsilon\|u(s)\|_{\dot{H}^{\frac{m_1}{2}}}^2 + \eta\|b(s)\|_{\dot{H}^{\frac{m_2}{2}}}^2 ds = \mathcal{E}(0). \tag{5.3b}$$

The quantity $\mathcal{E}$ will be referred to as the total energy and discussed in Chapters 6–7. The notions of weak and Leray–Hopf weak solutions from Definition 3.10.5 can be modified analogously in both cases of (5.1)–(5.2).

Using the fractional chain rule from Proposition 3.1.14, it is readily verified that the solution (u, π) to (5.1) possesses the scaling property that if (u, π) is a solution, then so is

https://doi.org/10.1515/9783111349985-006

$$(u_\lambda, \pi_\lambda)(t,x) \triangleq (\lambda^{m-1}u, \lambda^{2m-2}\pi)(\lambda^m t, \lambda x) \tag{5.4}$$

for all $\lambda > 0$, and we can explicitly compute $\|u_\lambda\|^2_{L^\infty_T L^2_x} = \lambda^{2(m-1)-d}\|u\|^2_{L^\infty_{\lambda^m T} L^2_x}$, due to which we say that the generalized Navier–Stokes equations (5.1) are $L^2(\mathbb{R}^d)$-subcritical, critical, and supercritical if

$$m > \frac{2+d}{2}, \quad m = \frac{2+d}{2}, \quad m < \frac{2+d}{2}, \tag{5.5}$$

respectively (cf. notions of $L^\infty(\mathbb{T}^2)$-criticality discussed in Section 4.5.1). In general, the spaces in which

$$\|u_\lambda(t)\|_X = \|u(\lambda t)\|_X \tag{5.6}$$

are called critical spaces and examples in the case of the generalized Navier–Stokes equations (5.1) include $\dot{H}^{1-2m+\frac{d}{2}}(\mathbb{R}^d)$ and, more generally, $\dot{B}^{1-2m+\frac{d}{p}}_{p,q}(\mathbb{R}^d)$ for $p, q \in [1,\infty]$.

As far back as [242, Remark 6.11 on p. 96], Lions proved the uniqueness of the Leray–Hopf weak solution to (5.1) in the $L^2(\mathbb{R}^d)$-subcritical and critical cases for all $d \geq 2$. In the case of the MHD system, unless $m_1 = m_2$, there is no scaling property, and Wu [333, Theorem 2.3] proved the global regularity of the generalized MHD system in case $m_1 \geq 1 + \frac{d}{2}$ and $m_2 \geq 1 + \frac{d}{2}$.

Numerical analysis works such as [179, 281] suggested that the velocity plays a more crucial role in preserving the regularity of the solution pair (u, b). This can be heuristically seen by realizing that any L^p_x- and $W^{m,p}_x$-estimates of (u, b) that solve (5.2) will face nonlinear terms, all of which consist of u but not necessarily of b (see (A.8)), suggesting the importance of viscous diffusion, except the H^1_x-estimate in the 2D case, as we will shortly discuss. Taking this into account, it is reasonable to suggest that the statement of global regularity of (5.2) in the $L^2(\mathbb{R}^d)$-critical case should be the following, where we include all $d \geq 3$ but not $d = 2$, because the case $d = 2$ can be improved significantly, as we will shortly discuss.

Proposition 5.1.1. *Suppose that $d \geq 3$ and υ, η, m_1, m_2 in (5.2) satisfy*

$$\upsilon > 0, \quad m_1 \geq 1 + \frac{d}{2}, \quad \eta \geq 0, \quad m_2 \geq 0, \quad \text{such that } m_1 + m_2 = 2 + d, \tag{5.7}$$

where $\eta = 0$ in the case $m_2 = 0$. Then for all $(u^{\text{in}}, b^{\text{in}}) \in H^s(\mathbb{R}^d) \times H^s(\mathbb{R}^d)$ with $s > 1 + \frac{d}{2}$, there exists a unique solution pair $(u, b) \in H^s(\mathbb{R}^d) \times H^s(\mathbb{R}^d)$ such that $(u, b)|_{t=0} = (u^{\text{in}}, b^{\text{in}})$.

The $L^2(\mathbb{R}^d)$-subcritical case corresponds to (5.7), but $m_1 + m_2 > 2 + d$. Proposition 5.1.1 can be improved, as we will see in Section 5.2; however, because we could not locate its proof in the literature, as a good application of materials from Chapter 3, we sketch its proof in Appendix A.5.

We now turn to the 2D case, which admits a significantly better result than Proposition 5.1.1. As mentioned in (3.140), a smooth solution to the 2D Euler equations has $\omega \in L_T^\infty L_x^\infty$, and Yudovich [371] proved that it possesses a global-in-time smooth solution starting from sufficiently smooth initial data. Half a century later, many have investigated for possible improvements of Proposition 5.1.1 in the 2D case by taking advantage of the system of vorticity ω and current density $j \triangleq \nabla \times b$:

$$\partial_t \omega + (u \cdot \nabla)\omega + \upsilon \Lambda^{m_1} \omega = (b \cdot \nabla) j, \tag{5.8a}$$

$$\begin{aligned} \partial_t j + (u \cdot \nabla) j + \eta \Lambda^{m_2} j = (b \cdot \nabla)\omega \\ + 2[\partial_1 b_1(\partial_1 u_2 + \partial_2 u_1) - \partial_1 u_1(\partial_1 b_2 + \partial_2 b_1)]. \end{aligned} \tag{5.8b}$$

This already shows that the $L^\infty(\mathbb{R}^2)$-estimate of ω gives

$$\|\omega(t)\|_{L^\infty} \le \|\omega(0)\|_{L^\infty} + \int_0^t \|(b \cdot \nabla) j\|_{L^\infty} ds.$$

Therefore, if $\eta > 0$ and $m_2 > 6$, then even with $\upsilon = 0$, (5.3b) already allows this estimate to be closed since

$$\int_0^t \|(b(s) \cdot \nabla) j(s)\|_{L^\infty} ds \le \int_0^t \|b(s)\|_{L^\infty} \|\nabla j(s)\|_{L^\infty} ds \lesssim \int_0^t \|b(s)\|^2_{H^{\frac{m_2}{2}}} ds,$$

which promises potential to improve Proposition 5.1.1. In fact, we can benefit from (5.8) furthermore; the $L^2(\mathbb{R}^2)$-estimate of (5.8) leads not only to $\int_{\mathbb{R}^2} (u \cdot \nabla)\omega\omega dx = 0$, which was already observed in (3.138), but also to

$$\int_{\mathbb{R}^2} (u \cdot \nabla) j j dx = 0, \tag{5.9a}$$

$$\int_{\mathbb{R}^2} (b \cdot \nabla) j \omega + (b \cdot \nabla)\omega j dx = \int_{\mathbb{R}^2} (b \cdot \nabla)(j\omega) dx = 0, \tag{5.9b}$$

leaving the only nonlinear term to be dealt with

$$\int_{\mathbb{R}^2} [\partial_1 b_1(\partial_1 u_2 + \partial_2 u_1) - \partial_1 u_1(\partial_1 b_2 + \partial_2 b_1)] j dx.$$

We can estimate in a natural way

$$\begin{aligned} &\int_{\mathbb{R}^2} [\partial_1 b_1(\partial_1 u_2 + \partial_2 u_1) - \partial_1 u_1(\partial_1 b_2 + \partial_2 b_1)] j dx \lesssim \|\nabla u\|_{L^2} \|\nabla b\|_{L^4} \|j\|_{L^4} \\ &\lesssim \|\omega\|_{L^2} \|j\|_{L^2} \|\nabla j\|_{L^2} \le \frac{\eta}{2} \|\Delta b\|^2_{L^2} + C\|\omega\|^2_{L^2} \|j\|^2_{L^2} \end{aligned} \tag{5.10}$$

due to the Hölder's, Gagliardo–Nirenberg (Theorem 3.3.1), and Young's inequalities. This implies that for the inviscid resistive MHD system $(m_1, m_2) = (0, 2)$, we already have the $H^1(\mathbb{R}^2)$-bound

$$\omega \in L_T^\infty L_x^2 \quad \text{and} \quad j \in L_T^\infty L_x^2 \cap L_T^2 H_x^1. \tag{5.11}$$

Remark 5.1.2. For our subsequent discussion, we stress that in the first inequality of (5.10), we chose $\|\nabla u\|_{L^2}$ to play the role of "being estimated," $\|\nabla b\|_{L^4}$ and $\|j\|_{L^4}$ to share the role of "being a part of the diffusive term $\eta\|\nabla j\|_{L^2}^2$" and "to be integrable in time $\|j\|_{L^2}^2$."

Many papers have appeared in the past decade ([206, 320, 339, 341, 370]) to extend the $H^1(\mathbb{R}^2)$-bound (5.11) higher to attain global regularity results, and the best results to this day are the two cases
(a) $v = 0$ and $\eta\Lambda^{m_2} b$ for $m_2 > 2$ [57, 207],
(b) $v\Lambda^{m_1} u$ and $-\eta\Delta b$ for $m_1 > 0$ [134, 346].

We state the latter result and sketch its proof for completeness.

Theorem 5.1.3. *Let $v, \eta > 0$ and $m_1 > 0$. Then for every $(u^{\text{in}}, b^{\text{in}}) \in H^s(\mathbb{R}^2)$, $s \geq 3$, there exists a unique solution pair (u, b) to the generalized MHD system*

$$\partial_t u + (u \cdot \nabla)u + \nabla\pi + v\Lambda^{m_1} u = (b \cdot \nabla)b, \quad \nabla \cdot u = 0, \tag{5.12a}$$

$$\partial_t b + (u \cdot \nabla)b - \eta\Delta b = (b \cdot \nabla)u, \tag{5.12b}$$

such that $(u, b)|_{t=0} = (u^{\text{in}}, b^{\text{in}})$ and

$$u \in C([0, \infty); H^s(\mathbb{R}^2)) \cap L^2([0, \infty); H^{s+\frac{m_1}{2}}(\mathbb{R}^2)),$$
$$b \in C([0, \infty); H^s(\mathbb{R}^2)) \cap L^2([0, \infty); H^{s+1}(\mathbb{R}^2)).$$

Proof. We sketch the proof of Theorem 5.1.3 from [346]. Computations analogous to (5.3b), (5.9), (5.10), and (5.11) show that the smooth solution to (5.12) satisfies

$$\omega \in L_T^\infty L_x^2 \cap L_T^2 H_x^{\frac{m_1}{2}} \quad \text{and} \quad j \in L_T^\infty L_x^2 \cap L_T^2 H_x^1. \tag{5.13}$$

To ultimately prove the $H^s(\mathbb{R}^2)$-estimate, we take the $L^2(\mathbb{R}^2)$-inner products on (5.12) with $(\Lambda^{2s} u, \Lambda^{2s} b)$ to obtain

$$\frac{1}{2}\frac{d}{dt}(\|u\|_{\dot{H}^s}^2 + \|b\|_{\dot{H}^s}^2) + v\|u\|_{\dot{H}^{s+\frac{m_1}{2}}}^2 + \eta\|b\|_{\dot{H}^{s+1}}^2 = \sum_{k=1}^{4} \mathrm{VI}_k,$$

where

$$\mathrm{VI}_1 \triangleq -\int_{\mathbb{R}^2} \Lambda^s((u\cdot\nabla)u)\cdot\Lambda^s u dx, \quad \mathrm{VI}_2 \triangleq -\int_{\mathbb{R}^2} \Lambda^s((u\cdot\nabla)b)\cdot\Lambda^s b dx,$$
$$\mathrm{VI}_3 \triangleq \int_{\mathbb{R}^2} \Lambda^s((b\cdot\nabla)b)\cdot\Lambda^s u dx, \quad \mathrm{VI}_4 \triangleq \int_{\mathbb{R}^2} \Lambda^s((b\cdot\nabla)u)\cdot\Lambda^s b dx.$$

Similarly to the proofs of Theorem 3.9.1 and Proposition 5.1.1, we can use the Kato–Ponce commutator estimate (3.95) to estimate these terms. In our current case, the viscous diffusion is weak, and thus we focus on the most difficult term VI_1:

$$\begin{aligned}\mathrm{VI}_1 &= -\int_{\mathbb{R}^2} \Lambda^s((u\cdot\nabla)u)\cdot\Lambda^s u - (u\cdot\nabla)\Lambda^s u\cdot\Lambda^s u dx\\ &\lesssim \|\nabla u\|_{L^{\frac{4}{m_1}}}\|\Lambda^s u\|_{L^2}\|\Lambda^s u\|_{L^{\frac{4}{2-m_1}}} \le \frac{\nu}{8}\|\Lambda^{s+\frac{m_1}{2}}u\|_{L^2}^2 + C\|\nabla u\|_{L^{\frac{4}{m_1}}}^2\|\Lambda^s u\|_{L^2}^2.\end{aligned}$$

This convinces us that the proof will be complete if we bootstrap up to

$$\omega \in L_T^2 L_x^{\frac{4}{m_1}} \tag{5.14}$$

thanks to (3.73).

Now starting from (5.13), we perform the following steps: w. l. o. g. assuming that $m_1 \in (0, \frac{2}{3})$,

1. $j \in L_T^\infty H_x^{\frac{m_1}{2}} \cap L_T^2 H_x^{1+\frac{m_1}{2}}$. We can take the $L^2(\mathbb{R}^2)$-inner products on (5.12b) with $\Lambda^{m_1} j$ to deduce

$$\frac{d}{dt}\|\Lambda^{\frac{m_1}{2}} j\|_{L^2}^2 + \|\Lambda^{1+\frac{m_1}{2}} j\|_{L^2}^2 \lesssim (1 + \|\Lambda j\|_{L^2}^2 + \|\Lambda^{\frac{m_1}{2}}\omega\|_{L^2}^2)(1 + \|\Lambda^{\frac{m_1}{2}} j\|_{L^2}^2),$$

   and then use the fact that $\Lambda j, \Lambda^{\frac{m_1}{2}}\omega \in L_T^2 L_x^2$ by (5.13).

2. $\omega \in L_T^\infty L_x^{q_1} \cap L_T^{q_1} L_x^{\frac{2q_1}{2-m_1}}$ for $q_1 \in [2, \frac{2(2+m_1)}{2-m_1}]$. We multiply (5.8a) by $|\omega|^{q_1-2}\omega$ and integrate over $\mathbb{R}^2$ to estimate

$$\frac{1}{q_1}\frac{d}{dt}\|\omega\|_{L^{q_1}}^{q_1} + C(q_1)\|\omega\|_{L^{\frac{2q_1}{2-m_1}}}^{q_1} \lesssim \|b\|_{L^\infty}\|\nabla j\|_{L^{\frac{2q_1}{2+m_1}}}\|\omega\|_{L^{q_1}}^{q_1-2}\|\omega\|_{L^{\frac{2q_1}{2-m_1}}}.$$

   The upper bound of $q_1 \le \frac{2(2+m_1)}{2-m_1}$ is needed to deduce

$$\|\nabla j\|_{L^{\frac{2q_1}{2+m_1}}} \lesssim 1 + \|\Lambda^{1+\frac{m_1}{2}} j\|_{L^2} \in L_T^2$$

   due to Step 1.

3. $j \in L_T^\infty L_x^{q_2}$ for $q_2 \in [2, (\frac{2}{2-m_1})^2(2+m_1)]$. This is achieved by deducing

$$\frac{1}{q_2}\frac{d}{dt}\|j\|_{L^{q_2}}^{q_2} + \frac{C(q_2)}{2}\|\Lambda|j|^{\frac{q_2}{2}}\|_{L^2}^2 \lesssim (\|j\|_{L^{q_2}}^{q_2} + 1)(\|\omega\|_{L^{q_2}}^2 + \|\omega\|_{L^{q_2}}^{\frac{q_2}{q_2-1}})$$

and using Step 2 with $q_1 = \frac{2(2+m_1)}{2-m_1}$, so that $\|\omega\|_{L^{q_2}}^2$ becomes locally integrable in time.

4. $j \in L_T^2 B_{q_2,2}^{\delta}$ for $\frac{4}{2-m_1} < q_2 \le (\frac{2}{2-m_1})^2(2+m_1)$ and $\frac{4}{q_2(2-m_1)} < \delta < 1$. This is achieved using Besov space techniques. Applying Δ_k to the equation of j for $k \ge 3$, multiplying by $|\Delta_k j|^{q_2-2}\Delta_k j$, and integrating in space, we obtain

$$\begin{aligned}
&\frac{1}{q_2}\frac{d}{dt}\|\Delta_k j\|_{L^{q_2}}^{q_2} + \eta \int_{\mathbb{R}^2} |\Delta_k j|^{q_2-2}\Delta_k j \Lambda^2 \Delta_k j dx \\
&= \int_{\mathbb{R}^2} |\Delta_k j|^{q_2-2}\Delta_k j \Delta_k(-(u\cdot\nabla)j)dx + \int_{\mathbb{R}^2} |\Delta_k j|^{q_2-2}\Delta_k j \Delta_k((b\cdot\nabla)\omega)dx \\
&\quad + \int_{\mathbb{R}^2} |\Delta_k j|^{q_2-2}\Delta_k j \Delta_k(2\partial_1 b_1 \partial_1 u_2)dx + \int_{\mathbb{R}^2} |\Delta_k j|^{q_2-2}\Delta_k j \Delta_k(2\partial_1 b_1 \partial_2 u_1)dx \\
&\quad + \int_{\mathbb{R}^2} |\Delta_k j|^{q_2-2}\Delta_k j \Delta_k(-2\partial_1 u_1 \partial_1 b_2)dx + \int_{\mathbb{R}^2} |\Delta_k j|^{q_2-2}\Delta_k j \Delta_k(-2\partial_1 u_1 \partial_2 b_1)dx
\end{aligned}$$

and estimate each term using Step 3.

5. $\omega \in L_T^{\infty} L_x^{q_2} \cap L_T^{q_2} L_x^{\frac{2q_2}{2-m_1}}$ for $q_2 \in [2, (\frac{2}{2-m_1})^2(2+m_1)]$. This is achieved by obtaining

$$\begin{aligned}
&\frac{1}{q_2}\frac{d}{dt}\|\omega\|_{L^{q_2}}^{q_2} + C(q_2)\|\Lambda^{\alpha}|\omega|^{\frac{q_2}{2}}\|_{L^2}^2 \\
&\le \frac{C(q_2)}{2}\|\Lambda^{\alpha}|\omega|^{\frac{q_2}{2}}\|_{L^2}^2 + c(1 + \|\Lambda^{1-\alpha} j\|_{L^{q_2}}^2 + \|\Lambda^{1+\alpha} j\|_{L^2}^2)(\|\omega\|_{L^{q_2}}^{q_2} + 1),
\end{aligned}$$

where $\|\Lambda^{1-\alpha} j\|_{L^{q_2}}^2$ can be shown to be locally integrable in time thanks to Step 4, allowing us to close this estimate.

Comparing the range of q_1 in Step 2 and q_2 in Step 5, we see that these steps are improvements only because $m_1 > 0$. In particular, after k repetitions when we reach $\frac{4}{m_1} < (\frac{2}{2-m_1})^{k+1}(2+m_1)$, we are able to claim the desired $H^s(\mathbb{R}^2)$-bound according to (5.14). □

5.2 Logarithmically supercritical case

In general dimensions the global regularity of supercritical PDEs in fluid mechanics remains unknown, although Tao pointed out that the criticality threshold can be beaten logarithmically. This phenomenon is highly relevant to our discussion in Section 9.1, and thus we briefly elaborate on it.

First, Tao [313] considered the defocusing nonlinear wave equation

$$(\partial_t^2 - \Delta)u = -f(u) \tag{5.15}$$

solved by $u : I \times \mathbb{R}^3 \to \mathbb{R}$ for some $I \subset \mathbb{R}$, for which $f(u) = |u|^{p-1}u$ leads to the conservation of energy

$$E(u) \triangleq \int_{\mathbb{R}^3} \left(\frac{1}{2}|\partial_t u|^2 + \frac{1}{2}|\nabla u|^2 + \int_0^u f(v)dv \right) dx.$$

Consequently, since it has a scaling of $u_\lambda(t,x) = \lambda^{\frac{2}{p-1}} u(\lambda t, \lambda x)$, it is $E(u)$-subcritical, critical, and supercritical if $p \in (1,5)$, $p = 5$, and $p > 5$, respectively. Remarkably, Tao was able to prove that given sufficiently smooth initial data $(u^{\text{in}}, v^{\text{in}}) = (u, \partial_t u)|_{t=0}$, the logarithmically supercritical wave equation

$$(\partial_t^2 - \Delta)u = -u^5 \log(2 + u^2)$$

admits a global-in-time unique solution.

Tao [314] extended such a result to the Navier–Stokes equations for all $d \geq 3$, and we can explain the heuristic based on our previous computations (Sections 3.9 and A.5). Traditional a priori estimates would seek a bound of the form $\frac{d}{dt}\|u(t)\|_{H^s}^2 \leq a(t)\|u(t)\|_{H^s}^2$ for some $a \in L^1_{\text{loc}}(0,T)$. Even the amount of loss $\epsilon > 0$ is disallowed, i. e.,

$$\frac{d}{dt}\|u(t)\|_{H^s}^2 \leq a(t)\|u(t)\|_{H^s}^{2+\epsilon} \tag{5.16}$$

would only lead to a local-in-time bound, not global-in-time. However, logarithmic worsening is allowed, i. e.,

$$\frac{d}{dt}\|u(t)\|_{H^s}^2 \leq a(t)\|u(t)\|_{H^s}^2 \ln(e + \|u(t)\|_{H^s}^2) \tag{5.17}$$

suffices as it results in the finite bound of $\|u(t)\|_{H^s}^2 \leq \|u(0)\|_{H^s}^2 \exp(\exp(\int_0^T a(s)ds))$. Such a new affordability of logarithmically worse bound allows us to prove the global regularity of the following logarithmically supercritical Navier–Stokes equations for all $d \geq 3$:

$$\partial_t u + \operatorname{div}(u \otimes u) + \nabla\pi + \upsilon\mathcal{L}u = 0, \quad \nabla \cdot u = 0, \quad \widehat{\mathcal{L}u}(\xi) = \frac{|\xi|^{1+\frac{d}{2}}}{\sqrt{\ln(2 + |\xi|)}}\hat{u}(\xi). \tag{5.18}$$

We can obtain a logarithmically easier bound (5.17) starting from such a logarithmically worse diffusion by splitting the dyadic decomposition (3.43) of Littlewood–Paley theory by a time-dependent cutoff so that

$$\sum_{j\geq -1} = \sum_{j:2^j \leq e+\|u(t)\|_{H^s}^2} + \sum_{j:2^j > e+\|u(t)\|_{H^s}^2}, \tag{5.19}$$

of which the lower frequency part already gives

$$\sum_{j:2^j\le e+\|u(t)\|_{H^s}^2} 1 \approx \ln(e+\|u(t)\|_{H^s}^2),$$

indicating the heuristic of how the $\ln(e+\|u(t)\|_{H^s}^2)$ comes about. In fact, this technique is similar to the proofs of the Gagliardo–Nirenberg inequalities (Theorems 3.3.1 and 3.3.3), although the logarithmic function reminds us the Brezis–Wainger inequalities (Lemmas 3.4.7 and 3.4.8).

Wu [335] extended the result of Tao [314] to the MHD system (5.2) by proving the global regularity of the following system:

$$\partial_t u + (u\cdot\nabla)u + \nabla\pi + \upsilon\mathcal{L}_1 u = (b\cdot\nabla)b, \quad \nabla\cdot u = 0, \tag{5.20a}$$

$$\partial_t b + (u\cdot\nabla)b + \eta\mathcal{L}_2 b = (b\cdot\nabla)u, \tag{5.20b}$$

where

$$\widehat{\mathcal{L}_k f} = \frac{|\xi|^{m_k}}{\sqrt{\ln(2+|\xi|)}}, \quad \upsilon > 0,\ m_1 \ge 1+\frac{d}{2},\ \eta > 0,\ m_2 > 0,\ m_1+m_2 \ge 2+d. \tag{5.21}$$

It is clear that this result logarithmically improves Proposition 5.1.1 in all cases except the endpoint $\upsilon > 0$, $m_1 = 2+d$, $\eta = 0$. Inspecting the proof of Proposition 5.1.1, we see that the global regularity of this endpoint case already required the Brezis–Wainger inequality (see (A.12)) and thus suspects difficulty. However, the global regularity result of logarithmically supercritical MHD system in such endpoint case still holds and can be proven by tricks similar to those in Chapter 3.

Proposition 5.2.1 ([340], cf. [321]). *Suppose $d \ge 2$ and υ, η, m_1 in (5.20) satisfy*

$$\widehat{\mathcal{L}_1 f} = \frac{|\xi|^{m_1}}{\ln(2+\ln(2+|\xi|))}, \quad \upsilon > 0,\ m_1 = 2+d,\ \eta = 0.$$

Then, for every sufficiently smooth initial data $(u^{\mathrm{in}}, b^{\mathrm{in}})$, there exists a global-in-time smooth solution (u,b) to (5.20) such that $(u,b)|_{t=0} = (u^{\mathrm{in}}, b^{\mathrm{in}})$.

Proof. We sketch the proof. For generality, we assume that

$$\widehat{\mathcal{L}_1}(\xi) = \frac{|\xi|^{2+d}}{g(\xi)},$$

where $g : \mathbb{R}_{\ge 0} \to \mathbb{R}_{\ge 1}$ is a radially symmetric nondecreasing function. At the end of the proof, we will see that $g(\xi) = \ln(2+\ln(2+|\xi|))$ is a special case. First, we have the energy inequality

$$\frac{1}{2}\|(u,b)\|_{L_T^\infty L_x^2}^2 + \int_0^T \upsilon\|\mathcal{L}_1^{\frac{1}{2}} u\|_{L^2}^2 d\tau \le \frac{1}{2}\|(u^{\mathrm{in}}, b^{\mathrm{in}})\|_{L^2}^2. \tag{5.22}$$

Moreover, denoting $X(t) \triangleq \|(\nabla u, \nabla b)(t)\|_{L^2}^2$, for the $\dot{H}^1(\mathbb{R}^d)$-estimate, computations analogous to (A.8) show

$$\frac{1}{2}\frac{d}{dt}X + \nu\|\mathcal{L}_1^{\frac{1}{2}}\nabla u\|_{L^2}^2 \lesssim \|\nabla u\|_{L^\infty}X. \tag{5.23}$$

Similarly to the proofs of Theorems 3.3.1 and 3.3.3, we can estimate

$$\begin{aligned}\|\nabla u\|_{L^\infty} &\lesssim \sqrt{g(M)}\sum_{2^j\le M}\|\Delta_j\mathcal{L}_1^{\frac{1}{2}}u\|_{L^2} + \sum_{2^j>M}2^{-j}\sqrt{g(2^j)}\|\Delta_j\mathcal{L}_1^{\frac{1}{2}}\nabla u\|_{L^2}\\ &\lesssim \sqrt{g(M)}\sqrt{\ln(M)}\|\mathcal{L}_1^{\frac{1}{2}}u\|_{L^2} + M^{-\frac{1}{2}}\|\mathcal{L}_1^{\frac{1}{2}}\nabla u\|_{L^2}.\end{aligned} \tag{5.24}$$

This leads to, with a choice of $M(t) \approx e + X(t)$,

$$\partial_t X(t) + \nu\|\mathcal{L}_1^{\frac{1}{2}}\nabla u(t)\|_{L^2}^2 \lesssim [g(e+X(t))\ln(e+X(t))](e+\|\mathcal{L}_1^{\frac{1}{2}}u(t)\|_{L^2}^2)(e+X(t)). \tag{5.25}$$

This implies

$$\int_{e+X(0)}^{e+X(t)}\frac{dr}{g(r)\ln(r)r} \lesssim \int_0^t 1+\|\mathcal{L}_1^{\frac{1}{2}}u\|_{L^2}^2 dt \le \int_0^T 1+\|\mathcal{L}_1^{\frac{1}{2}}u\|_{L^2}^2 dt \overset{(5.22)}{\lesssim} 1. \tag{5.26}$$

Thus a special case of $g(\tau) = \ln(2+\ln(2+\tau))$ implies $\operatorname{esssup}_{t\in[0,T]} X(t) < \infty$. Higher regularity can be obtained from this $\dot{H}^1(\mathbb{R}^d)$-bound, and we refer the interested readers to [340] for details. □

We mention that such global regularity results of logarithmically supercritical PDEs were extended with various improvements to many other PDEs, e. g., the Boussinesq system [180], dyadic equation [14], Euler equations [65], MHD system [345], Navier–Stokes equations [15], SQG equations [64], and wave equations [292].

5.3 Recent developments and open problems

In this section, we list a few open problems related to discussions in this chapter. The first one was extensively discussed in Section 5.1.

Problem 1. Does the 2D MHD system that is inviscid but resistive with a full Laplacian admit a global-in-time unique smooth solution starting from any sufficiently smooth initial data or does there exist a smooth initial data from which the corresponding solution has a finite-time blowup? In relevance to this problem, we mention that the global regularity of the 2D MHD system that is viscous with a full Laplacian but nonresistive is also an open question to the best of the author's knowledge.

The next problem is related to discussions in Sections 5.1 and 5.2. As we saw in Section 5.1, it is now known that the 2D MHD system (5.2) with $m_2 = 2$ admits a globally unique smooth solution if $m_1 + m_2 > 2$ and $m_2 \geq 2$, thanks to many works, most of which appeared more than a decade ago [57, 206, 207, 134, 320, 339, 341, 346, 370] and they all relied on the crucial observation described in (5.10) that $m_2 = 2$ alone guarantees the $H^1(\mathbb{R}^2)$-bound of (u, b). Now let us consider the case $m_2 < 2$ and $2 \leq m_1 + m_2 < 4$. If $m_1 + m_2 \geq 4$, then (5.20)–(5.21) guarantee the global regularity. Let us recall the first inequality of (5.10):

$$\int_{\mathbb{R}^2} [\partial_1 b_1(\partial_1 u_2 + \partial_2 u_1) - \partial_1 u_1(\partial_1 b_2 + \partial_2 b_1)] j dx \lesssim \|\nabla u\|_{L^2} \|\nabla b\|_{L^4} \|j\|_{L^4}. \tag{5.27}$$

Recalling Remark 5.1.2, we interpolate $\|\nabla b\|_{L^4}$ by

$$\|\nabla b\|_{L^4} \lesssim \|\nabla b\|_{L^2}^{\frac{m_2-1}{m_2}} \|\Lambda^{\frac{m_2}{2}} \nabla b\|_{L^2}^{\frac{1}{m_2}}, \tag{5.28}$$

i. e., partially "being estimated" and "to be absorbed by the diffusive term $\eta \|\Lambda^{\frac{m_2}{2}} j\|_{L^2}^2$." However, this application of the Gagliardo–Nirenberg inequality requires $m_2 \geq 1$. On the other hand, since now $\|\Lambda^{\frac{m_2}{2}} u\|_{L^2}^2 \in L_T^1$ by the energy inequality, we interpolate $\|\nabla u\|_{L^4}$ from (5.27) between the "being estimated" and "to be square-integrable in time" as follows:

$$\|\nabla u\|_{L^4} \lesssim \|\nabla u\|_{L^2}^{\frac{m_1-3}{m_1-2}} \|\Lambda^{\frac{m_1}{2}} u\|_{L^2}^{\frac{1}{m_1-2}}.$$

However, this application of the Gagliardo–Nirenberg inequality requires $m_1 \geq 3$, and thus we already have the restriction $m_1 + m_2 \geq 4$. Therefore we conclude that there is a *large gap* in the literature for a technical reason: if $\eta > 0$ and $m_2 \geq 2$, then we only need $m_1 + m_2 > 2$, but once $m_2 < 2$, then the lower bound jumps to $m_1 + m_2 \geq 4$ modulo the logarithmic improvements described in Section 5.2 by [321, 335, 340]. The problem of filling this gap was proposed in [348], where partial progress was made; however, the author is not aware of any progress in this direction, and hence we state it again.

Problem 2. Consider the 2D MHD system with viscous and magnetic diffusion given by $\nu \Lambda^{m_1} u$ and $\eta \Lambda^{m_2} b$, respectively, where $\eta > 0$, $m_2 < 2$, and $m_1 + m_2 \in [2, 4)$. Does this system admit a global-in-time unique smooth solution starting from any sufficiently smooth initial data or does there exist a smooth initial data from which its corresponding solution has a finite-time blowup?

The next two problems are concerned with the Hall-MHD and electron MHD systems (1.9)–(1.10), about which we briefly review recent developments. Acheritogaray, Degond, Frouvelle, and Liu [1] observed, using (2.1b) upon an $L^2(\mathbb{T}^3)$-estimate of the solution (u, b) to the Hall-MHD system, that the Hall term vanishes, and thus identically to (5.3b), every smooth solution to the Hall-MHD system (1.9) with $\nu > 0$ and $\eta > 0$ satisfies $u, b \in L_T^\infty L_x^2 \cap L_T^2 \dot{H}_x^1$. This allowed them to construct a global-in-time weak solution in case $x \in \mathbb{T}^3$; later, this was extended to $\mathbb{R}^3$ in [66] and to stochastic case in [344]. Many other results were obtained in the past decade, e. g., a blowup and regularity criterion [67, 120, 285], temporal decay [68], blowup [70], and partial regularity [71, 72]. Recently, the ill-posedness results of (1.9)–(1.10) have attracted attention; e. g., [202, Theorem 1.1 (1)] states that in case of zero magnetic diffusion, for all $s > \frac{7}{2}$ and $\epsilon > 0$, there exists a divergence-free initial data $b^{\text{in}} \in H^s(\mathbb{R}^3)$ such that $\|b^{\text{in}}\|_{H^s} < \epsilon$, and there is no solution $b \in L^\infty(0, \delta; H^s(\mathbb{R}^3))$ to the electron MHD system that satisfies $b|_{t=0} = b^{\text{in}}$ for any $\delta > 0$; analogous statement for the Hall-MHD can be found therein too. We also refer to [368, 350] in the stochastic case. We now recall the notion of $2\frac{1}{2}$D Hall-MHD system from Section 1.1.6 and state two well-known open problems.

Problem 3. In either $2\frac{1}{2}$D or 3D case, for an arbitrary initial data in $H^s(U)$ with $U = \mathbb{R}^2$ or $\mathbb{T}^2$ and for $s > 0$ sufficiently large, does there exist $T > 0$ such that a unique strong solution to the Hall-MHD system (1.9) with $\nu = \eta = 0$ or the electron MHD system (1.10) with $\eta = 0$ exists on $[0, T)$?

We recall that such a result exists for the Euler equations or the ideal MHD system (recall Section 3.9) and mention that the best result toward this problem is by Chae, Wan, and Wu [69], who proved the local well-posedness in the inviscid case but with magnetic diffusion of $\eta\Lambda^{m_2} b$ for $m_2 > 1$.

Problem 4. With $\nu > 0$ and $\eta > 0$, given arbitrary initial data in $H^s(U)$ with $U = \mathbb{R}^2$ or $\mathbb{T}^2$ and $s > 0$ sufficiently large, does there exist a unique smooth solution for all $t > 0$ to the $2\frac{1}{2}$D Hall-MHD system (1.9) or the electron MHD system (1.10)? We quote Chae and Lee [67, Abstract] (2014):

> "Contrary to the usual MHD the global well-posedness in the $2\frac{1}{2}$-dimensional Hall-MHD is wide open."

Recently, the author, together with Rahman in [286, Theorem 2.3], was able to show that the $2\frac{1}{2}$D Hall-MHD system with $-\Delta u$ and $-\Delta b$ replaced by

$$\Lambda^a(u_1, u_2, 0) - \Delta(0, 0, u_3) \quad \text{and} \quad \Lambda^3(b_1, b_2, 0) + \Lambda^a(0, 0, b_3) \quad \text{for } a > 1,$$

respectively, is globally well posed in $H^3(\mathbb{R}^2)$. Curiously, the required sum of such derivatives is $11 + \epsilon$ for all $\epsilon > 0$, whereas the analogous sum for the $2\frac{1}{2}$D Hall-MHD system with $-\Delta u$ and $-\Delta b$, for which the question of global regularity is open, is 12.

6 Convex integration on deterministic PDEs

This chapter is devoted to discussion on convex integration technique applied on deterministic PDEs in fluid mechanics and will provide a basis for discussion in Chapter 7 in the stochastic case and Section 9.2 on singular SPDEs. Section 6.1 starts with a general review, Section 6.2 gives mathematical preliminaries, and Section 6.3 gives an example of hyperbolic Navier–Stokes equations (1.11). Section 6.4 concludes with more recent developments and open problems.

6.1 General review

We start with a review of motivation from physics perspective, such as Kolmogorov's zeroth law of turbulence, Onsager's and Taylor's conjectures in Section 6.1.1, and how convex integration started with the work of Nash in geometry in Section 6.1.2. Section 6.1.3 focuses on the works of Scheffer [296], Shnirelman [300], and Christ [91, 92] with detailed discussions on that of Christ. Finally, Section 6.1.4 describes most recent developments of systematic adaptation of convex integration to PDEs in fluid mechanics.

6.1.1 Onsager's and Taylor's conjectures

The celebrated Kolmogorov's 1941 theory [223, 224] lays down a foundational statistical framework for high-Reynolds-number turbulence and has been confirmed in various experiments. To limit ourselves to a simple setup, we consider the deterministic Navier–Stokes equations with time-independent mean-zero force $f \in L^2_\sigma$:

$$\partial_t u^\upsilon + \operatorname{div}(u^\upsilon \otimes u^\upsilon) + \nabla \pi^\upsilon = \upsilon \Delta u^\upsilon + f, \quad \nabla \cdot u^\upsilon = 0, \quad x \in \mathbb{T}^d,\ t \in (0,T]. \tag{6.1}$$

For every $\upsilon > 0$, we know from Section 3.10 that initial data $u^{\text{in}} \in L^2_\sigma$ leads to the existence of a global-in-time Leray–Hopf weak solution u^υ. Kolmogorov's zeroth law of turbulence postulates the existence of $\epsilon > 0$ such that

$$\lim_{\upsilon \searrow 0} \upsilon \mathbb{E}^{\mu^\upsilon}[\|\nabla u^\upsilon\|^2_{L^2_x}] = \epsilon > 0, \tag{6.2}$$

where μ^υ is the law of stationary process solution of (6.1), i. e., $\mathcal{L}(u^\upsilon(t)) = \mu^\upsilon$ for all $t \geq 0$ when the initial data u^{in} has distribution μ^υ (see, e. g., [146, Theorem 3.1] on the existence of statistical solutions). The Kolmogorov zeroth law of turbulence is also known as anomalous dissipation, and various numerical analysis supported such phenomena, e. g., [208] by Kaneda, Ishihara, Yokokawa, Itakura, and Uno.

Additionally, MHD turbulence describes the chaotic regimes of magnetofluid flow for small kinematic viscosity and magnetic resistivity, which often occur in laboratory

https://doi.org/10.1515/9783111349985-007

settings such as fusion confinement devices (e. g., the reversed field pinch) and astrophysical systems (e. g., the solar corona). Numerical analysis of anomalous dissipation for the MHD system can be found, e. g., in [111, 264].

Independently, physicist/chemist Onsager in 1949 conjectured the Hölder regularity of $\frac{1}{3}$ to be the critical threshold concerning conservation of energy for the Euler equations. We quote Onsager [276, p. 286] (1949):

> "It is of some interest to note that in principle, turbulent dissipation as described could take place just as readily without the final assistance by viscosity. In the absence of viscosity, the standard proof of the conservation of energy does not apply, because the velocity field does not remain differentiable! In fact it is possible to show that the velocity field in such 'ideal' turbulence cannot obey any Lipschitz condition of the form
>
> $$\left|\vec{v}(\vec{r}\,' + \vec{r}) - \vec{v}(\vec{r}\,')\right| < (\text{const})r^n,$$
>
> for any order n greater than $\frac{1}{3}$; otherwise, the energy is conserved."

For clarity, we state the Onsager conjecture.

Conjecture 6.1.1 (Onsager's conjecture). *Let $d \geq 2$ and $\alpha \in (0, 1)$.*
(a) *If $\alpha > \frac{1}{3}$, then every weak solution $u \in C_t C_x^\alpha$ to the Euler equations* (1.1) *with $\nu = 0$ conserves energy; i. e., $\|u(t)\|_{L^2}^2 = \|u(0)\|_{L^2}^2$.*
(b) *If $\alpha < \frac{1}{3}$, then there exists a weak solution $u \in C_t C_x^\alpha$ to the Euler equations* (1.1) *with $\nu = 0$ that fails the energy conservation; i. e., $\|u(t)\|_{L^2}^2$ is not constant in time.*

There seems to be no clear consensus of what to expect in the case $\alpha = \frac{1}{3}$.

The heuristic of the threshold $\frac{1}{3}$ can be understood as follows. Returning to (2.29), which led to the energy identity (2.30), we see that $\int_{\mathbb{R}^d} (u \cdot \nabla) u \cdot u dx$ must be well-defined. Informally, assuming that ∇ can be distributed equally among u that appears three times, we see that the minimum condition should be $C_t C_x^{\frac{1}{3}}$. We refer to [315, Remark 6] for an alternative heuristic derivation of the threshold $\alpha = \frac{1}{3}$.

We can observe a similarity between the Navier–Stokes problem (Statement 1) stated in Section 2.7 and Onsager's conjecture (Conjecture 6.1.1).

(a) Positive direction: In the case of the Navier–Stokes problem, given that local well-posedness is known in $H^s(\mathbb{T}^3)$ for all $s > 1 + \frac{d}{2}$, it suffices to obtain *a priori bounds* and show that solutions are unique. In the case of the Onsager conjecture, it suffices to show that an arbitrary solution to the Euler equations with regularity $C_t C_x^\alpha$ for $\alpha > \frac{1}{3}$ conserves energy.
(b) Negative direction: In the case of the Navier–Stokes problem, we need to discover a smooth initial data such that the corresponding solution exists, is unique and smooth up to time $T > 0$, and then blows up. In the case of the Onsager conjecture, we also need to construct, or at least prove the existence of, a solution $u \in C_t C_x^\alpha$ for $\alpha < \frac{1}{3}$ that does not conserve energy.

Toward the positive direction of the Navier–Stokes problem, we recall that [242] (and Proposition 5.1.1 for more detail) has a partial result in case of strengthened diffusion. Analogously, the positive direction of the Onsager conjecture was settled in 1994 partially by Eyink [133] (partially because the result therein required a norm slightly stronger than the Hölder continuity) and then completely by Constantin, E, and Titi [99] using Littlewood–Paley theory techniques. The negative direction for both problems is more subtle, and we will shortly describe this direction in more detail.

Having discussed the Onsager conjecture, we now briefly describe the Taylor's conjecture on the MHD system. The ideal MHD system, (1.8) with $\nu = \eta = 0$, has various invariants in addition to the total energy $\mathcal{E}(t)$ from (5.3a); in fact, so does the Hall-MHD system. To explain clearly, let us define the magnetic potential A by

$$\nabla \times A = b \tag{6.3}$$

and rewrite the ideal Hall-MHD system as (cf. (1.9))

$$\partial_t u + (u \cdot \nabla)u + \nabla \pi + \nu(-\Delta)^{m_1} u = (b \cdot \nabla)b, \quad \nabla \cdot u = 0, \tag{6.4a}$$

$$\partial_t A + b \times u + h(j \times b) + \nabla \chi + \eta(-\Delta)^{m_2} A = 0, \tag{6.4b}$$

for some χ and $x \in \mathbb{T}^d$ for $d \in \{2, 3\}$. If we define the magnetic helicity and cross helicity, which measure the linkage and twist of magnetic field lines, and entanglement of vorticity with magnetic field, respectively, as

$$\mathcal{H}_b(t) \triangleq \int_{\mathbb{T}^d} A(t) \cdot b(t) dx \quad \text{and} \quad \mathcal{H}_u(t) \triangleq \int_{\mathbb{T}^d} u(t) \cdot b(t) dx, \tag{6.5}$$

then using (2.1b), we see that smooth solutions to both ideal MHD and ideal Hall-MHD systems satisfy

$$\partial_t \mathcal{H}_b(t) \overset{(6.5)}{=} \int_{\mathbb{T}^d} \partial_t A \cdot b(t) + A \cdot \partial_t b(t) dx \overset{(6.4)(1.9b)}{=} 0, \tag{6.6}$$

so that $\mathcal{H}_b$ is invariant. Additionally, the smooth solution to the ideal MHD system also satisfies

$$\partial_t \mathcal{H}_u(t) \overset{(6.5)}{=} \int_{\mathbb{T}^d} \partial_t u \cdot b(t) + u \cdot \partial_t b(t) dx \overset{(1.8a)(1.8b)}{=} 0, \tag{6.7}$$

so that $\mathcal{H}_u$ is conserved too, although it is not conserved for the Hall-MHD system because $\int_{\mathbb{T}^3} \nabla \times (j \times b) \cdot u dx \neq 0$ in general. The invariance of magnetic helicity $\mathcal{H}_b$ was first discovered by Woltjer [332], and Taylor [316, 317] conjectured that it is approximately conserved for large Reynolds numbers (see also Moffatt [267, Section 4]). We quote Taylor [316, p. 1140] (1974):

"main consequence of any small departures from perfect conductivity is that topological properties of the magnetic field are no longer preserved... However, changes in field topology are accompanied by only very small changes in the field itself and ... thus the integral $\int A \cdot b$ over the total volume of the system will still be a good invariant..."

Taylor's conjecture has been successful in explaining magnetic structures in laboratory plasmas (e. g., in the prediction of the relaxed state of reversed field pinch) and lies at the heart of the Taylor's relaxation theory.

6.1.2 The origin of convex integration in geometry

The origin of convex integration is accredited to the work of Nash [273] on the isometric embedding theorem in 1954.

Definition 6.1.1. Consider nD compact Riemannian manifold (M^n, g) with local coordinates $\{x^i\}$ in M^n for $i = 1, \ldots, n$. An isometric immersion of (M^n, g) into $\mathbb{R}^k$ is a map z such that $\sum_\alpha \frac{\partial z^\alpha}{\partial x^i} \frac{\partial z^\alpha}{\partial x^j} = g_{ij}$; it is called a short immersion of (M^n, g) into $\mathbb{R}^k$ if the matrix $(g_{ij})_{i,j} - (\frac{\partial z}{\partial x^i} \cdot \frac{\partial z}{\partial x^j})_{i,j}$ is positive definite.

Theorem 6.1.2 (Nash–Kuiper isometric embedding theorem, [226, 273]). *Under the hypothesis of Definition* 6.1.1, *suppose* $k \geq n + 1$. *Then, for all* $\epsilon > 0$ *and all short immersions of* (M, g) *into* $\mathbb{R}^k$ *and* $\sigma \in C^1$, *there exists an isometric immersion* $z \in C^1$ *of* (M, g) *into* $\mathbb{R}^k$ *such that* $\|\sigma - z\|_{C^0} < \epsilon$.

To prove Theorem 6.1.2, Nash constructed a sequence of short isometric embeddings such that the error vanishes in the limit. We quote Nash [273, p. 384] (1954):

"The correction process has an infinite sequence of 'stages'. The first stage of correction begins with the initial short immersion, the second begins with the immersion constructed by the first, etc. Each stage has the purpose of modifying the immersion so that the induced metric is closer to the intrinsic metric, but in such a way that the immersion is still short, although less so."

Gromov considered Nash's work as part of "homotopy-principle" ([156, p. 3]) and coined the term "convex integration" in [156, Part 2.4].

6.1.3 Works of Scheffer, Shnirelman, and Christ

Let us briefly describe the works of Scheffer [296], Shnirelman [300], and Christ [91, 92]. None of them acknowledged any connection with convex integration of Nash or Gromov; yet, in retrospect, they can be seen as the very first step to adapt the approach of Nash to construct wild solutions to PDEs in fluid mechanics.

6.1.3.1 [296] by Scheffer

First, Scheffer [296] in 1993 constructed a weak solution $u \in L^2(\mathbb{R} \times \mathbb{R}^2)$ to the 2D Euler equations that has compact support in space and time, e. g.,

$$u(t,x) \equiv 0 \quad \text{for } |x|^2 + |t|^2 > 1.$$

Thus, fluid can start identically zero at initial time $t < -1$, then become nonzero for $t \in (-1,1)$, and then disappear again at $t = 1$!

6.1.3.2 [300] by Shnirelman

Subsequently in 1997, Shnirelman [300] constructed a weak solution $u \in L^2(\mathbb{R}\times\mathbb{T}^2)$ to the Euler equations that vanishes whenever the temporal variable satisfies $|t| > C$ for some $C > 0$. Importantly, the main idea of Shnirelman's proof was to construct a sequence of solutions not to the Euler equations but to the Euler equations with an external force that is in the limit oscillating at an infinitely fast pace in space such that it becomes unnoticed in the sense of distributions and consequently the limiting solution indeed solves the Euler equations weakly. We emphasize that $L^2_{t,x}$ is essentially the minimum regularity needed for a weak solution to guarantee that the nonlinear term in the weak formulation is well-defined, analogously to $L^3_{t,x}$ in Definition 3.10.2.

6.1.3.3 [91, 92] by Christ

The works of Christ, specifically [91, 92], seem less known in comparison to [296, 300], and yet they are remarkable for multiple reasons and surprisingly related to our discussions on singular SPDEs in Chapters 8–9. Hence we now elaborate on [91, 92]. First, Christ [91] considered the 1D cubic nonlinear Schrödinger (NLS) equation solved by $u : \mathbb{R}_+ \times \mathbb{T} \to \mathbb{C}$,

$$i\partial_t u + \omega |u|^2 u + \partial_x^2 u = 0, \tag{6.8}$$

where $i = \sqrt{-1}$ and $\omega \in \mathbb{R}\setminus\{0\}$. The Cauchy problem of (6.8) is locally well posed in $H^s(\mathbb{T})$ for all $s \geq 0$; more specifically, there exists a Banach space

$$Y \subset C_t H_x^s \cap L^3_{t,x} \tag{6.9}$$

such that any initial data $u^{\text{in}} \in H^s(\mathbb{T})$ leads to a unique solution $u \in Y$.

The question Christ raised is the uniqueness of the solution in $C([0,T];H^s(\mathbb{T}))$ with no additional regularity. Although it sounds natural, such a question is not even well formulated if s is not sufficiently large to guarantee that the nonlinear term $|u|^2 u$ is well-defined. Indeed, the presence of "$\omega|u|^2 u$" in (6.8) suggests the necessity of $L^3_{t,x}$ to pass the limit; we recall $L^3_{t,x}$ in Definition 3.10.2 concerning the Cahn–Hilliard equation precisely for this same reason. Remarkably, Christ proved the nonuniqueness of solutions in the class $C([0,T];H^s(\mathbb{T}))$ in case $s < 0$. Interestingly, this is precisely the same difficulty of the nonlinear term becoming ill-defined due to the irregularity of the solution that arises in the theory of singular SPDEs, to be discussed in Chapters 8–9. In fact, Christ is forced to work on a modified NLS equation of the form

$$i\partial_t u + \omega \mathbf{N}(u) + \partial_x^2 u = 0, \quad \mathbf{N}(u) \triangleq \left(|u|^2 - 2 \fint_{\mathbb{T}} |u|^2 dx \right) u. \tag{6.10}$$

Remarkably, $2 \fint_{\mathbb{T}} |u|^2 dx u$, which was subtracted out in (6.10), essentially plays the same role as the renormalization constant or, more specifically, the Wick product, which will be discussed in Chapters 8–9.

Let us explain in more detail. First, let us point out that $\fint_{\mathbb{T}} |u|^2 dx$ would be independent of time t if it were well-defined. We can see this by multiplying (6.10) by $\overline{u}$, integrating over $\mathbb{T}$, and observing that the imaginary part satisfies $\frac{1}{2}\frac{d}{dt}\|u\|_{L^2}^2 = 0$. This is already of great interest because the renormalization constants, as we will see in Chapters 8–9, are independent of space x. Because $\fint_{\mathbb{T}} |u|^2 dx$ is independent of t, the solution to (6.10) is merely $\tilde{u} = e^{-i2t\omega \fint_{\mathbb{T}} |u^{\text{in}}|^2 dx} u$, where u solves (6.8) if it is well-defined. As we described already, the solution Christ constructed is in $C_T H_x^s$ for $s < 0$, and thus $2 \fint_{\mathbb{T}} |u|^2 dx$ is not even well-defined; informally, this is taken as the limit of the Fourier truncation $u_N(x) = \sum_{k \in \mathbb{Z}: |k| \leq N} \hat{u}(k) e^{ik \cdot x}$ as $N \to \infty$.

Second, let us describe the heuristic of the reason to subtract out $2 \fint_{\mathbb{T}} |u|^2 dx$. With the Fourier truncation for later justification in mind, let us observe that the nonlinear term can be written as

$$|u|^2 u = \sum_{\substack{n, n_1, n_2, n_3 \in \mathbb{Z}: \\ n = n_1 - n_2 + n_3}} \hat{u}(n_1) \overline{\hat{u}(n_2)} \hat{u}(n_3) e^{inx}. \tag{6.11}$$

As we discussed in (3.83a) (see Lemma 8.1.4), the singularity comes from resonant terms that correspond to $n_1 = n_2$, $n_2 = n_3$, or $n_1 = n_3$, which are special cases of $\sum_{\substack{n, n_1, n_2, n_3 \in \mathbb{Z}: \\ n = n_1 - n_2 + n_3}} \hat{u}(n_1) \overline{\hat{u}(n_2)} \hat{u}(n_3) e^{inx}$, specifically,

$$\|u\|_{L^2}^2 u, \quad \|u\|_{L^2}^2 u, \quad \text{and} \quad \sum_{\substack{n, n_1, n_2 \in \mathbb{Z}: \\ n = 2n_1 - n_2}} \hat{u}(n_1) \overline{\hat{u}(n_2)} \hat{u}(n_1) e^{inx},$$

respectively. This displays precisely the reason why Christ subtracted out $2 \fint_{\mathbb{T}} |u|^2 dx u$ in (6.10). We also observe that Christ did not have to subtract nondiagonal oscillatory terms when $n_1 = n_3$ but subtracted off the diagonal terms when $n_1 = n_2$ and $n_2 = n_3$; again, this is philosophically identical to the notion of Wick products to be discussed in Section 8.3. To the best of the author's knowledge, this is the only work that investigated the deterministic analogue of Wick products and renormalization constants.

Now let us describe the notion of solution to (6.10) constructed by Christ.

Definition 6.1.3.

(a) We call any family of linear operators $\{\mathcal{P}_N\}_{N\in\mathbb{N}}$ on $C_c^\infty(\mathbb{T})$ a family of Fourier cutoff operators if its corresponding Fourier symbols $m_N : \mathbb{Z} \to \mathbb{C}$ (recall (3.7)) have finite support, are uniformly bounded, and $\lim_{N\to\infty} m_N(k) = 1$ for all $k \in \mathbb{Z}$.

(b) Let $\mathcal{N}$ be any nonlinear functional acting on functions of (t,x) and $u \in (C_c^\infty((0,T)\times\mathbb{T}))'$. Then $\mathcal{N}(u)$ is said to exist and equal $v \in (C_c^\infty((0,T)\times\mathbb{T}))'$ if for every family $\{\mathcal{P}_N\}_{N\in\mathbb{N}}$ of Fourier cutoff operators, $\lim_{N\to\infty}\mathcal{N}(\mathcal{P}_N u) = v$ in the topology of $(C_c^\infty((0,T)\times\mathbb{T}))'$.

(c) Given initial data $u^{\text{in}} \in H^s(\mathbb{T})$, we say that $u \in C([0,T];H^s(\mathbb{T}))$ is a weak solution to (6.10) if $u(0,\cdot) = u^{\text{in}}(\cdot)$, $\mathbf{N}(u)$ defined in (6.10) exists in the sense of Definition 6.1.3(b), and u satisfies (6.10) in the distributional sense in $(0,T)\times\mathbb{T}$.

Finally, we formally state the result by Christ.

Theorem 6.1.4 ([91, Theorem 2.1]). *For all $s < 0$, given initial data $u^{in} \equiv 0$, there exists $u \in C([0,T];H^s(\mathbb{T}))$ that is not identically zero and solves* (6.10) *starting from u^{in} in the sense of Definition* 6.1.3(c).

We end this section with a few more remarks and refer the interested readers to [91] for details.

Remark 6.1.5.

(a) In the proof of Theorem 6.1.4, Christ approximates (6.10) by adding a force, similarly to [300], and Nash-type convex integration schemes, which we will describe in the rest of this chapter.

(b) In [91, Section 2.7], he states that generalization to higher dimensions or diffusion can be achieved. However, we emphasize that (6.8) consists of i on the time derivative and that is why ∂_x^2 can be replaced by $(-\partial_{xx})^m$ for any $m \geq 0$, and the proof goes through. Let us be more precise. Multiplying (6.8) by $-i$, we get $\partial_t u - i\partial_x^2 u = i\omega|u|^2 u$, and thus Christ transforms the Fourier coefficient of u by $y_n(t) = e^{\text{i}n^2 t}\hat{u}(t,n)$ in [91, eq. (3.3)]. The fact that such $e^{\text{i}n^2 t}$ is unitary is crucial in his proof, and it is unitary thanks to the presence of i.

(c) Christ describes in [91, Section 2.7] his interest to extend the regularity of the nonunique solution to the Euler equations constructed in [296, 300] from $L_{t,x}^2$ to $C_t L_x^2$. In fact, in [92], he proves the nonuniqueness of weak solutions but with regularity of $C_t H^s$ for $s < 0$; however, he achieves this for the Navier–Stokes equations, in contrast to the Euler equations treated in [296, 300].

6.1.4 Adaptation of convex integration to PDEs in fluid mechanics

Although the works of Scheffer [296] and Shnirelman [300] were not directly linked to convex integration (and recall that [91, 92] by Christ emerged in 2005), the potential of convex integration as a systematic approach to construct wild nonphysical solutions to a large class of PDEs started to be acknowledged in the late 1990s. We quote Gromov [157] (1999):

> "The class of infinitesimal laws subjugated by the homotopy principle is wide, but it does not include most PDEs of physics with a few exceptions in favour of this principle leading to unexpected solutions. In fact, the presence of the *h*-principle would invalidate the very idea of a physical law..."

It is possible that the following breakthrough work of Müller and V. Śverák was one of the exceptions that Gromov had in mind because its proof consisted of extension of convex integration.

Theorem 6.1.6 ([271, Theorem 1.1] and [272, Theorem 4.1]). *There exists a smooth strongly elliptic* 2×2 *system*

$$- \operatorname{div} \sigma(Dv) = 0, \quad v : \mathbb{R}^2 \to \mathbb{R}^2$$

that admits nontrivial Lipschitz solutions with compact support and Lipschitz solutions that are nowhere C^1.

Theorem 6.1.6 led De Lellis and Székelyhidi Jr. [121] to use convex integration and Baire category argument to construct a weak solution $u \in L^\infty(\mathbb{R} \times \mathbb{R}^d)$ to the dD Euler equations for all $d \geq 2$ with compact support in space-time, extending [296, 300] from $L^2_{t,x}$. Subsequently, De Lellis and Székelyhidi Jr. [123] used Beltrami flows and Reynolds stress to prove that for any smooth function $e : [0,1] \to \mathbb{R}_+$, there exists a weak solution $u \in C(\mathbb{R}_+ \times \mathbb{T}^3)$ to the Euler equations such that $e(t) = \|u(t)\|^2_{L^2(\mathbb{T}^3)}$ for all $t \in [0,1]$ (see also [122]). The work of [123] was the first to construct such a solution with prescribed energy and the first step toward the negative direction of the Onsager conjecture (Conjecture 6.1.1) with $\alpha = 0$. This approach of the iterative construction of systems forced by Reynolds stress is generally referred to as the Nash-type iteration; the other approach of [121] is called the Baire category argument in the literature. Using the Nash-type iterations, Buckmaster, De Lellis, Isett, and Székelyhidi Jr. [44] improved [123] from $\alpha = 0$ to $\alpha < \frac{1}{5}$. After further extensions and improvements (see [48] for a survey), with a new strategy of utilizing Mikado flows and a gluing scheme, Isett [197] proved the negative direction of Onsager's conjecture (Conjecture 6.1.1) in all dimensions $d \geq 3$. The case $d = 2$ was ruled out in [197] due to a lack of suitable replacement of Mikado flows (see [197, p. 877]). We comment that although Conjecture 6.1.1 consisted of a threshold of $C_t C_x^{\frac{1}{3}}$, the solution constructed in [197] was $C^{\frac{1}{3}}$ in space-time; specifically, there exists $C \geq 0$ such that $|f(t + \Delta t, x + \Delta x) - f(t,x)| \leq C(|\Delta t| + |\Delta x|)^\alpha$ uniformly in t, x, Δt, and Δx.

Later, using intermittency, Buckmaster and Vicol [47] proved the nonuniqueness of weak solutions to the 3D Navier–Stokes equations; specifically, they proved that there exists $\beta > 0$ such that for any nonnegative smooth function $e(t)$ on $[0,T]$, there exists a weak solution (recall Definition 3.10.5(a)) $u \in C([0,T]; H^\beta(\mathbb{T}^3))$ such that $e(t) = \|u(t)\|^2_{L^2_x}$

for all $t \in [0, T]$. We emphasize that the solution constructed therein was not necessarily in $L_T^2 \dot{H}_x^1$, did not satisfy the energy inequality, and consequently, was not a Leray–Hopf weak solution (recall Definition 3.10.5(b)). Later, [47] was extended to the Lions exponent (recall Section 5.1) in [43, 253] in the 3D case and in [251] in the 2D case.

Let us highlight two consequences, among many others, of the result by [47]. First, it solved the Serrin's conjecture; we quote Serrin [299, p. 88] (1963):

> "Another problem of interest is the question whether a nonconstant [weak] solution of the [3D Navier–Stokes equations] can ever come to rest in a finite time."

Another consequence of [47] is the following. The 3D Navier–Stokes equations admit a global existence of a weak solution and regularity criteria to verify its uniqueness (as discussed after Definition 3.10.5), but at a low regularity class, ill-posedness occurs. As it will be discussed in Section 7.2, one of the consequences of [186] by Hofmanová, Zhu, and Zhu was the SPDE analogue; specifically, that Cherny's theorem (Lemma 4.3.3) implies the pathwise uniqueness if a probabilistically strong solution is unique in law; however, at a low regularity class, ill-posedness occurs.

In relevance to our subsequent discussions in Section 7.3.1, let us elaborate on convex integration on the MHD system and Taylor's conjecture from Section 6.1.1. Faraco and Lindberg [135] proved by forming a sequence of Leray-Hopf weak solutions depending on $v, \eta > 0$, that the weak solutions of the ideal MHD system that are weak*-limits in $L_t^\infty L_x^2$ conserve magnetic helicity $\mathcal{H}_b$ (recall (6.5)). The same authors, together with Székelyhidi Jr., [136] constructed infinitely many bounded non-trivial solutions that are compactly supported in space-time such that they preserve the magnetic helicity $\mathcal{H}_b$ but not total energy $\mathcal{E}$ or cross helicity $\mathcal{H}_u$ (recall (5.3a) and (6.5)). Finally, Beekie, Buckmaster, and Vicol [19] proved that there exists a weak solution (u, b) in $C_t H_x^\beta$ for some $\beta > 0$ to the 3D ideal MHD system such that $\mathcal{H}_b(1) > 2|\mathcal{H}_b(0)|$ and the corresponding total energy $\mathcal{E}(t)$ and cross-helicity $\mathcal{H}_u(t)$ are nontrivial and nonconstant. We also refer to [83, 137, 239] for more works that employed the convex integration technique on the MHD system.

Many more remarkable works on the convex integration technique appeared in the past decade, e. g., [88] on solutions to the Navier–Stokes equations in $C_t L^p$ for $p < 2$, [151] on the Onsager conjecture in the 2D case, and [252] on the Boussinesq system. Because one of the key motivations for this monograph is to describe connections between the convex integration technique and singular SPDEs theory, we choose not to describe any further and refer the interested readers to [48, 49] for a survey.

6.2 Preliminaries on deterministic convex integration

In Section 6.3, we will describe details of convex integration on the hyperbolic Navier–Stokes equations (1.11) as an example, and the purpose of this section is to give a few preliminaries on convex integration, some of which will also be useful in Chapter 7.

We denote the trace-free tensor product by

$$f\mathring{\otimes}g \triangleq \begin{pmatrix} f_1g_1 - \frac{1}{2}f\cdot g & f_1g_2 \\ f_2g_1 & f_2g_2 - \frac{1}{2}f\cdot g \end{pmatrix} \tag{6.12}$$

for the 2D case, and in the higher-dimensional case, it is defined analogously.

The following is an example of geometric lemma, which plays a crucial role in canceling out a previous error.

Lemma 6.2.1 (Geometric lemma from [46, Lemma 4.2]). *Let $B_\epsilon(\mathrm{Id})$ denote the ball of symmetric 2×2 matrices, centered at* Id *with radius $\epsilon > 0$. Then there exists $\epsilon_\gamma > 0$ for which there exist disjoint finite subsets $\Lambda^+, \Lambda^- \subset \mathbb{S}^1 \cap \mathbb{Q}^2$ and smooth positive functions $\gamma_\zeta \in C^\infty(B_{\epsilon_\gamma}(\mathrm{Id}))$, $\zeta \in \Lambda^\pm$, such that*

(a) $5\Lambda^\pm \subset \mathbb{Z}^2$,

(b) *if $\zeta \in \Lambda^\pm$, then $-\zeta \in \Lambda^\pm$ and $\gamma_\zeta = \gamma_{-\zeta}$,*

(c)

$$R = \frac{1}{2}\sum_{\zeta\in\Lambda^\pm}(\gamma_\zeta(R))^2(\zeta^\perp\otimes\zeta^\perp) \quad \forall R \in B_{\epsilon_\gamma}(\mathrm{Id}), \tag{6.13}$$

(d) $|\zeta+\zeta'| \geq \frac{1}{2}$ *for all $\zeta,\zeta' \in \Lambda^\pm$ such that $\zeta+\zeta' \neq 0$.*

We define $\Lambda \triangleq \Lambda^+ \cup \Lambda^-$.

Next, we introduce the inverse divergence operator; there are many other variants, and we only introduce the 2D version used in [336].

Lemma 6.2.2 ([105, Definition 9, Lemma 10], also [251, Definition 7.1, Lemmas 7.2 and 7.3]). *For $f \in C(\mathbb{T}^2)$, set*

$$\mathcal{B}f \triangleq \nabla g + (\nabla g)^T - (\nabla\cdot g)Id, \tag{6.14}$$

where g satisfies $\Delta g = f - \fint_{\mathbb{T}^2} f dx$ and $\fint_{\mathbb{T}^2} g dx = 0$. Then for every $f \in C(\mathbb{T}^2)$ such that $\fint_{\mathbb{T}^2} f dx = 0$, $\mathcal{B}f(x)$ is a trace-free symmetric matrix for all $x \in \mathbb{T}^2$. Moreover, $\nabla\cdot\mathcal{B}f = f$ and $\fint_{\mathbb{T}^2}\mathcal{B}f(x)dx = 0$. When f is not mean-zero, we overload the notation and denote $\mathcal{B}f \triangleq \mathcal{B}(f - \fint_{\mathbb{T}^2} f dx)$. Finally, for all $p \in (1,\infty)$,

$$\|\mathcal{B}\|_{L_x^p\to W_x^{1,p}} \lesssim 1, \quad \|\mathcal{B}\|_{C_x\to C_x} \lesssim 1, \quad \text{and} \quad \|\mathcal{B}f\|_{L_x^p} \lesssim \|(-\Delta)^{-\frac{1}{2}}f\|_{L_x^p}.$$

Next, we recall some results concerning the 2D intermittent stationary flows from [105, 251]. For all $\zeta \in \Lambda$ and a frequency parameter $\lambda \in 5\mathbb{N}$, we define b_ζ and its potential ψ_ζ as

$$b_\zeta(x) \triangleq b_{\zeta,\lambda}(x) \triangleq i\zeta^\perp e^{i\lambda\zeta\cdot x}, \quad \psi_\zeta(x) \triangleq \psi_{\zeta,\lambda}(x) \triangleq \frac{1}{\lambda}e^{i\lambda\zeta\cdot x}. \tag{6.15}$$

For $r \in \mathbb{N}$, we consider the 2D Dirichlet kernel

$$D_r(x) \triangleq \frac{1}{2r+1} \sum_{k \in \Omega_r} e^{ik \cdot x} \quad \text{with } \Omega_r \triangleq \{k : k_i \in \mathbb{Z} \cap [-r, r] \text{ for } i = 1, 2\}. \tag{6.16}$$

The role of r is to parameterize the number of frequencies along the edges of the cube Ω_r. We introduce σ such that $\lambda\sigma \in 5\mathbb{N}$ to parameterize the spacing between frequencies or, equivalently, such that the resulting rescaled kernel is $(\mathbb{T}/\lambda\sigma)^2$-periodic. Lastly, we introduce μ that measures the amount of temporal oscillation in the building blocks. The parameters we introduced are chosen to satisfy

$$1 \ll r \ll \mu \ll \sigma^{-1} \ll \lambda, \quad r \in \mathbb{N}, \quad \text{and} \quad \lambda, \lambda\sigma \in 5\mathbb{N}. \tag{6.17}$$

Now we define the directed-rescaled Dirichlet kernel

$$\eta_\zeta(t,x) \triangleq \eta_{\zeta,\lambda,\sigma,r,\mu}(t,x) \triangleq \begin{cases} D_r(\lambda\sigma(\zeta \cdot x + \mu t), \lambda\sigma\zeta^\perp \cdot x) & \text{if } \zeta \in \Lambda^+, \\ \eta_{-\zeta,\lambda,\sigma,r,\mu}(t,x) & \text{if } \zeta \in \Lambda^-, \end{cases} \tag{6.18}$$

and the intermittent 2D stationary flow

$$W_\zeta(t,x) \triangleq W_{\zeta,\lambda,\sigma,r,\mu}(t,x) \triangleq \eta_{\zeta,\lambda,\sigma,r,\mu}(t,x) b_{\zeta,\lambda}(x). \tag{6.19}$$

The following result has been used in many works on convex integration upon estimating the perturbation that always consists of composition with the Reynolds stress (see (6.40)).

Lemma 6.2.3 (E. g., [44, Proposition C.1]). *For $\Omega \subset \mathbb{R}^d$, let $\Psi : \Omega \to \mathbb{R}$ and $u : \mathbb{R}^d \to \Omega$ be smooth functions. Then, for all $m \in \mathbb{N}$, there exists a constant $C = C(m, d) \geq 0$ such that*

$$[\Psi \circ u]_{\dot{C}^m} \leq C([\Psi]_{\dot{C}^1}[u]_{\dot{C}^m} + \|D\Psi\|_{C^{m-1}}\|u\|_{C^0}^{m-1}[u]_{\dot{C}^m}),$$
$$[\Psi \circ u]_{\dot{C}^m} \leq C([\Psi]_{\dot{C}^1}[u]_{\dot{C}^m} + \|D\Psi\|_{C^{m-1}}[u]_{\dot{C}^1}^m).$$

The following estimates are often useful to treat Nash errors in convex integration for Euler equations, and stochastic analogue will be discussed in Section 7.3.4.

Lemma 6.2.4 (E. g., [44, Proposition D.1]). *For a smooth vector field u and smooth functions f_0 and g, consider a smooth solution f to*

$$\partial_t f + (u \cdot \nabla) f = g \quad \text{for } t > 0, \quad f(t_0, x) = f_0(x). \tag{6.20}$$

Let Φ be the inverse of the flux X of u starting at time t_0 as the identity (i. e., $\partial_t X = u(X, t)$, $X(t_0, x) = x$), so that $f(t, x) = f_0(\Phi(t, x))$ in case $g \equiv 0$. Then for all $t > t_0$,

$$\|f(t)\|_{C_x} \leq \|f_0\|_{C_x} + \int_{t_0}^t \|g(\tau)\|_{C_x} d\tau,$$

$$\|Df(t)\|_{C_x} \leq \|Df_0\|_{C_x} e^{(t-t_0)\|Du\|_{C_{t,x}}} + \int_{t_0}^{t} e^{(t-\tau)\|Du\|_{C_{t,x}}} \|Dg(\tau)\|_{C_x} d\tau$$

(recall D^k for $k \in \mathbb{N}$ from (1b)). More generally, for each $N \in \mathbb{N}\setminus\{1\}$, there exists a constant $C = C(N)$ such that for all $t > t_0$,

$$\begin{aligned}\|D^N f(t)\|_{C_x} \leq & (\|D^N f_0\|_{C_x} + C(t-t_0)\|D^N u\|_{C_{t,x}} \|Df_0\|_{C_x}) e^{C(t-t_0)\|Du\|_{C_{t,x}}} \\ & + \int_{t_0}^{t} e^{C(t-\tau)\|Du\|_{C_{t,x}}} (\|D^N g(\tau)\|_{C_x} + C(t-\tau)\|D^N u\|_{C_{t,x}} \|Dg(\tau)\|_{C_x}) d\tau.\end{aligned}$$

Additionally, for all $N \in \mathbb{N} \setminus \{1\}$ and $t > t_0$,

$$\begin{aligned}\|D\Phi(t) - \mathrm{Id}\|_{C_x} &\leq e^{(t-t_0)\|Du\|_{C_{t,x}}} - 1 \leq (t-t_0)\|Du\|_{C_{t,x}} e^{(t-t_0)\|Du\|_{C_{t,x}}}, \\ \|D^N \Phi(t)\|_{C_x} &\leq C(t-t_0)\|D^N u\|_{C_{t,x}} e^{C(t-t_0)\|Du\|_{C_{t,x}}}.\end{aligned}$$

6.3 Example of hyperbolic Navier–Stokes equations (1.11)

6.3.1 Review and statement of main result

As an example of deterministic convex integration, we describe [336] by Wu and the author concerning the hyperbolic Navier–Stokes equations (1.11). We choose [336] because it meets one of the goals of this monograph. Specifically, merely the existence of even local-in-time solutions is an open question for many singular SPDEs, and convex integration has the potential to construct global-in-time solutions to them. On the other hand, to the best of the author's knowledge, (1.11) represents the only deterministic PDE such that the global-in-time solution, even if not necessarily unique, was an open problem until it was solved via convex integration in [336].

Let us first describe the mathematical difficulty of (1.11). First, although introduced more than three quarters of a century ago, mathematical analysis of (1.11) is lagging behind due to technical reasons. Specifically, in contrast to the Navier–Stokes equations, among computations analogous to (2.29)–(2.30), we see that the energy identity fails due to the extra term

$$\gamma \int_0^t \int_{\mathbb{T}^d} \partial_s^2 u \cdot u dx ds.$$

Additionally, taking the $L^2(\mathbb{T}^d)$-inner products with $\partial_t u$, similarly to the wave equation, and summing the resulting equations create a different problem due to

$$\int_0^t \int_{\mathbb{T}^d} (u \cdot \nabla) u \cdot \partial_s u dx ds.$$

Consequently, there are no known bounded quantities for the solution to the hyperbolic Navier–Stokes equations (1.11). As we described in Section 3.10, a typical Galerkin approximation requires bounded quantities and compactness to take the limit in the nonlinear terms; thus a complete absence of bounded quantities in the case of (1.11) creates a formidable challenge to construct a global-in-time weak solution.

We remark that solutions to (1.11) can be constructed locally in time, and higher regularity estimates can allow us to extend such a local-in-time solution; particularly, Ji, Li, Tian, and Wu [203] proved that in the spatial domain $\mathbb{R}^d$ for $d \geq 2$, the hyperbolic Navier–Stokes equations (1.11) possess a unique global-in-time mild solution under constraints on the smallness of γ and initial data. Nevertheless, without any constraint on γ or initial data, it was not known if for any $T > 0$, solution can exist for (1.11), uniquely or not, until the following result from [336], although under constraints on dimension and diffusion strength.

Theorem 6.3.1 ([336, Theorem 2.1]). *Consider the following generalized hyperbolic Navier–Stokes equations:*

$$\gamma \partial_t^2 u + \partial_t u + \nu \Lambda^m u + (u \cdot \nabla) u + \nabla \pi = 0, \quad \nabla \cdot u = 0. \tag{6.21}$$

Fix $m \in (0, \frac{4}{3})$, arbitrary $T > 0$, and any

$$e \in C^1(\mathbb{R}; [\underline{e}, \infty)) \quad \text{such that } \|e\|_{C([-2,T])} \leq \bar{e} \text{ and } \|e'\|_{C([-2,T])} \leq \tilde{e}, \tag{6.22a}$$

$$\text{where } 4 \leq \underline{e} \leq \bar{e} < \infty, \tilde{e} \in [0, \infty). \tag{6.22b}$$

Then there exists a sufficiently small constant $\beta = \beta(m) \in (0,1)$ such that the following holds. There exists a mean-zero function

$$u \in C([0,T]; H^\beta(\mathbb{T}^2)) \cap C^\beta([0,T]; L^2(\mathbb{T}^2)) \tag{6.23}$$

such that

(a) *u solves (6.21) distributionally and is weakly divergence-free, i. e.,*

$$\int_0^T \int_{\mathbb{T}^2} u(s,x) \cdot (\gamma \partial_{ss}\phi - \partial_s \phi + \Lambda^m \phi - (u \cdot \nabla)\phi)(s,x) dx ds = 0,$$

$$\int_{\mathbb{T}^2} u(t) \cdot \nabla \psi = 0 \quad \forall t \in [0,T],$$

for every divergence-free $\phi \in C^2([0,T] \times \mathbb{T}^2)$ such that both ϕ and $\partial_t \phi$ vanish at $t = 0$ and $t = T$ and every $\psi \in C^1(\mathbb{T}^2)$;

(b)

$$\|u(t)\|_{L_x^2}^2 = e(t) \quad \forall t \in [0,T]. \tag{6.24}$$

Additionally, if two such energies e_1 *and* e_2 *obeying the same bounds* $\underline{e}$, $\overline{e}$, *and* $\tilde{e}$ *in* (6.22) *coincide on* $[0, t]$, *then there exist the corresponding solutions* u_1 *and* u_2 *that also coincide on* $[0, t \wedge T]$, *implying the nonuniqueness of weak solutions of* (6.21).

Let us make several remarks before starting the proof of Theorem 6.3.1.

Remark 6.3.2.

(a) Instead of prescribing energy $e(t)$ in (6.24), one may try to prescribe initial data and prove the existence of infinitely many solutions from it; we refer to [50] for such a result on the deterministic power-law model and [188, 364] on the stochastic Navier–Stokes equations and the stochastic MHD system, respectively. We also believe that using temporal intermittency may allow us to raise the upper bound on m, although potentially at the expense of decreasing the regularity or integrability of the solution u.

(b) During the workshop "Criticality and Stochasticity in Quasilinear Fluid Systems" at the American Institute of Mathematics in 2021, a question of whether we can apply the convex integration technique to dispersive or hyperbolic PDEs such as Korteweg–de Vries (KdV) or wave equation was raised. Theorem 6.3.1 was motivated out of interest to see how to handle ∂_t^2 in convex integration for the first time.

(c) Theorem 6.3.1 is interesting in that with less effort, we can construct wild solutions to (6.21) such that

$$\|u(T)\|_{L_x^2} > 2\|u(0)\|_{L_x^2}. \tag{6.25}$$

In the case of the Navier–Stokes equations, this will already imply nonuniqueness, as done in [48, Section 7] because we can construct a Galerkin approximation solution u^{G} from the same initial data $u(0)$ in (6.25) so that u^{G} satisfies the energy inequality, in contrast to (6.25). In contrast, this strategy is hopeless for (6.21) because the Galerkin approximation is not applicable. Interestingly, a stronger result of prescribing energy was able to prove nonuniqueness, though.

(d) Initial heuristic toward proving Theorem 6.3.1 was that linear terms such as diffusion only have to be treated as an error and made small in convex integration, and hence the hope was to simply estimate $\partial_t^2 u$ similarly to the diffusion. It turns out that this term is too singular.

6.3.2 Sketch of the proof of Theorem 6.3.1

For $q \in \mathbb{N}_0$, we set

$$\lambda_q \triangleq a^{b^q}, \quad \delta_q \triangleq \lambda_1^{2\beta} \lambda_q^{-2\beta} \tag{6.26}$$

for $a \in 10\mathbb{N}$, $a \geq a_0$, $b \in \mathbb{N}$ such that $b \geq 2$, and $\beta \in (0, 1)$ to be determined later. We define

$$t_q \triangleq -2 + \sum_{0 \leq \iota \leq q} \delta_\iota^{\frac{1}{2}} \leq -\frac{1}{2} \quad \text{for } q \in \mathbb{N}_0, \tag{6.27}$$

where the inequality can be justified by imposing $3 \leq a^{b\beta}$, so that $\sum_{0 \leq \iota \leq q} \delta_\iota^{\frac{1}{2}} \leq \frac{3}{2}$. Hereafter, including Chapter 8, we denote

$$C_{t,x,q} \triangleq C([t_q, t] \times \mathbb{T}^d), \tag{6.28a}$$

$$C_{t,q}L_x^p \triangleq C([t_q, t]; L^p(\mathbb{T}^d)), \tag{6.28b}$$

$$C_{t,q}H_x^s \triangleq C([t_q, t]; H^s(\mathbb{T}^d)). \tag{6.28c}$$

For $q \in \mathbb{N}_0$, on $[t_q, T]$, we consider

$$\partial_t^2 u_q + \partial_t u_q + \Lambda^m u_q + \operatorname{div}(u_q \otimes u_q) + \nabla \pi_q = \operatorname{div} \mathring{R}_q, \quad \nabla \cdot u_q = 0, \tag{6.29}$$

where $\mathring{R}_q$ is a trace-free symmetric matrix, and the divergence of a matrix was introduced in (3). Let us explain our inductive estimates.

Hypothesis 6.3.1 (Inductive hypothesis at level q). On $[t_q, T]$, the solution $(u_q, \mathring{R}_q)$ to (6.29) satisfies

$$\|u_q\|_{C_{t,q}L_x^2} \leq L\Big(1 + \sum_{1 \leq r \leq q} \delta_r^{\frac{1}{2}}\Big)\bar{e}^{\frac{1}{2}} \leq 3L\bar{e}^{\frac{1}{2}}, \tag{6.30a}$$

$$\|u_q\|_{C^1_{t,x,q}} \leq \lambda_q^3 \bar{e}^{\frac{1}{2}}, \tag{6.30b}$$

$$\|\mathring{R}_q\|_{C_{t,q}L_x^1} \leq \frac{\epsilon_\gamma}{36}\delta_{q+2} e(t), \tag{6.30c}$$

$$\frac{3}{4}\delta_{q+1} e(t) \leq e(t) - \|u_q(t)\|_{L_x^2}^2 \leq \frac{5}{4}\delta_{q+1} e(t) \tag{6.30d}$$

for a universal constant L (see [336, eq. (77)] for details).

Proposition 6.3.3 (Initial step $q = 0$). *Together with $\pi_0 \equiv 0$, the pair $(u_0, \mathring{R}_0) = (0, 0)$ solves* (6.29) *and satisfies Hypothesis* 6.3.1 *at level $q = 0$.*

The proof of Proposition 6.3.3 is immediate; we can use the fact that $\delta_1 = 1$ due to (6.26) to verify (6.30d).

The heart of the matter in the proof of Theorem 6.3.1 is the following proposition.

Proposition 6.3.4 (Step $q + 1$ assuming the step q). *Under the hypotheses of Theorem* 6.3.1, *there exists a choice of parameters a_0, β, and b such that for all $(u_q, \mathring{R}_q)$ that solves* (6.29) *and satisfies Hypothesis* 6.3.1, *there exists $(u_{q+1}, \mathring{R}_{q+1})$ that solves* (6.29) *and satisfies Hypothesis* 6.3.1 *at level $q + 1$ such that for all $t \in [t_{q+1}, T]$,*

$$\|u_{q+1} - u_q\|_{C_{t,q+1}L_x^2} \leq L\delta_{q+1}^{\frac{1}{2}}\bar{e}^{\frac{1}{2}}. \tag{6.31}$$

We sketch the proof of Theorem 6.3.1 assuming Proposition 6.3.4, the proof of which will be sketched later.

Proof of Theorem 6.3.1. A weak solution to (6.21) exists because $\{u_q\}_{q\in\mathbb{N}_0}$ is Cauchy in $C([-\frac{1}{2},T];H^{\beta'}(\mathbb{T}^2))\cap C^{\beta'}([-\frac{1}{2},T];L^2(\mathbb{T}^2))$ for $\beta'\in(0,\frac{\beta}{3+\beta})$, which can be verified by (6.31) and (6.30b). Taking the limit as $q\to\infty$ in (6.30d) also leads to (6.24).

The argument concerning nonuniqueness is as follows. From the proof of Proposition 6.3.4 it will become clear that if we start with two different energies e_1 and e_2 that satisfy (6.22) and $e_1 \equiv e_2$ on $[0,t\wedge T]$ for some $t>0$, then the perturbations $\{w^1_{q+1}\}_{q\in\mathbb{N}_0}$ and $\{w^2_{q+1}\}_{q\in\mathbb{N}_0}$ corresponding respectively to e_1 and e_2 are identical on $[0,t\wedge T]$ (see (6.39)). Thus we can start with identical initial choices $(u^1_0,\mathring{R}^1_0)=(u^2_0,\mathring{R}^2_0)=(0,0)$ according to Proposition 6.3.3 and see that $\{u^1_q\}_{q\in\mathbb{N}_0}$ and $\{u^2_q\}_{q\in\mathbb{N}_0}$ corresponding to e_1 and e_2 are also identical on $[0,t\wedge T]$. As a result, the constructed limiting solutions $u_1,u_2\in C([-\frac{1}{2},T];H^{\beta'}(\mathbb{T}^2))\cap C^{\beta'}([-\frac{1}{2},T];L^2(\mathbb{T}^2))$ corresponding respectively to e_1 and e_2 are identical on $[0,t\wedge T]$. This completes the proof of Theorem 6.3.1. □

Proof of Proposition 6.3.4. We sketch the proof of Proposition 6.3.4 with the main goal to clarify the cancellation of the previous error. In addition to a_0, b, and β from (6.26) and r, μ, and σ from (6.17), we introduce $a, l=l(a)\ll 1$, and $p^*\in(1,2)$ as follows:

$$r\triangleq\lambda_{q+1}^{\frac{11m-5}{7}},\quad \mu\triangleq\lambda_{q+1}^{\frac{8m-3}{7}},\quad \sigma^{-1}\triangleq\lambda_{q+1}^{\frac{3m}{2}}, \tag{6.32}$$

$$l\triangleq\lambda_{q+1}^{-\frac{2-3m}{112}}\lambda_q^{-\frac{3}{2}}, \tag{6.33}$$

$$b>\frac{(42)(56)}{2-3m}, \tag{6.34}$$

and

$$p^*\in\left(1,\frac{8(112)(11m-5)}{10795m-5106}\right). \tag{6.35}$$

These parameters are chosen carefully by optimization; we refer to [336, Remark 4.3]. The remaining a_0 and β will be taken respectively sufficiently large and sufficiently small as needed in the estimates.

Now we let $\phi_l(\cdot)\triangleq\frac{1}{l^2}\phi(\frac{\cdot}{l})$ and $\varphi_l(\cdot)\triangleq\frac{1}{l}\varphi(\frac{\cdot}{l})$ be families of standard mollifiers on $\mathbb{R}^2$ and $\mathbb{R}$, respectively, with mass one, where the latter has compact support in $(0,\delta_{q+1})$, and mollify $(u_q,\mathring{R}_q)$ in space-time to obtain over $[t_{q+1},T]$

$$u_l\triangleq(u_q *_x \phi_l)*_t\varphi_l,\quad \mathring{R}_l\triangleq(\mathring{R}_q *_x\phi_l)*_t\varphi_l. \tag{6.36}$$

It follows that $(u_l,\mathring{R}_l)$ satisfies over $[t_{q+1},T]$

$$\partial_t^2 u_l+\partial_t u_l+\Lambda^m u_l+\operatorname{div}(u_l\otimes u_l)+\nabla\pi_l=\operatorname{div}(\mathring{R}_l+R_{\text{com}}), \tag{6.37}$$

where

$$R_{\mathrm{com}} \triangleq u_l \mathring{\otimes} u_l - (u_l \mathring{\otimes} u_l) *_x \phi_l *_t \varphi_l, \tag{6.38a}$$

$$\pi_l \triangleq \pi_q *_x \phi_l *_t \varphi_l - \frac{1}{2}(|u_l|^2 - |u_l|^2 *_x \phi_l *_t \varphi_l). \tag{6.38b}$$

Now for $\epsilon_\gamma > 0$ from the Lemma 6.2.1, we define

$$\rho(t,x) \triangleq \epsilon_\gamma^{-1}\sqrt{l^2 + |\mathring{R}_l(t,x)|^2} + \gamma_l(t), \tag{6.39a}$$

$$\gamma_l(t) \triangleq (\gamma_q *_t \varphi_l)(t), \quad \gamma_q(t) \triangleq \frac{1}{2(2\pi)^2}[e(t)(1-\delta_{q+2}) - \|u_q(t)\|_{L_x^2}^2]; \tag{6.39b}$$

this is how the prescribed energy $e(t)$ is used within explicit construction of perturbation. Then we define the amplitude function

$$a_\zeta(t,x) \triangleq a_{\zeta,q+1}(t,x) \triangleq \frac{1}{2}\rho(t,x)^{\frac{1}{2}}\gamma_\zeta\left(\mathrm{Id} - \frac{\mathring{R}_l(t,x)}{\rho(t,x)}\right). \tag{6.40}$$

The following identities are crucial in canceling out the previous errors upon verifying (6.30c)–(6.30d):

$$\sum_{\zeta,\vartheta\in\Lambda} a_\zeta(t)a_\vartheta(t) \fint_{\mathbb{T}^2} W_\zeta \mathring{\otimes} W_\vartheta(t)dx = -\mathring{R}_l(t) + \rho(t)\left(1 - \frac{1}{8}\sum_{\zeta\in\Lambda}\gamma_\zeta\left(\mathrm{Id} - \frac{\mathring{R}_l(t)}{\rho(t)}\right)^2\right)\mathrm{Id}, \tag{6.41a}$$

$$\sum_{\zeta,\vartheta\in\Lambda} a_\zeta(t)a_\vartheta(t) \fint_{\mathbb{T}^2} W_\zeta \otimes W_\vartheta(t)dx = -\mathring{R}_l(t) + \rho(t)\,\mathrm{Id}. \tag{6.41b}$$

Let us sketch their derivations. First, by (6.40) and (6.19) we can prove

$$\sum_{\zeta,\vartheta\in\Lambda} a_\zeta(t,x)a_\vartheta(t,x) \fint_{\mathbb{T}^2} W_\zeta \otimes W_\vartheta(t,x)dx \tag{6.42}$$
$$= \sum_{\zeta,\vartheta\in\Lambda}\left[\frac{1}{2}\rho(t,x)^{\frac{1}{2}}\gamma_\zeta\left(\mathrm{Id} - \frac{\mathring{R}_l}{\rho}\right)\right]\left[\frac{1}{2}\rho(t,x)^{\frac{1}{2}}\gamma_\vartheta\left(\mathrm{Id} - \frac{\mathring{R}_l}{\rho}\right)\right]\fint_{\mathbb{T}^2}\eta_\zeta\eta_\vartheta b_\zeta \otimes b_\vartheta dx.$$

By (6.15), (6.18), and (6.16) we can show that if $\zeta + \zeta' \neq 0$, then $\fint_{\mathbb{T}^2}\eta_\zeta\eta_{\zeta'}b_\zeta \otimes b_{\zeta'}dx = 0$. Also, using the fact that $b_\zeta \otimes b_{-\zeta} = \zeta^\perp \otimes \zeta^\perp$ due to (6.15), (6.42) leads to

$$\sum_{\zeta,\vartheta\in\Lambda} a_\zeta(t,x)a_\vartheta(t,x) \fint_{\mathbb{T}^2} W_\zeta \otimes W_\vartheta(t,x)dx$$
$$= \sum_{\zeta\in\Lambda}\frac{1}{4}\rho(t,x)\gamma_\zeta\left(\mathrm{Id} - \frac{\mathring{R}_l}{\rho}\right)\gamma_{-\zeta}\left(\mathrm{Id} - \frac{\mathring{R}_l}{\rho}\right)\zeta^\perp \otimes \zeta^\perp \fint_{\mathbb{T}^2}\eta_\zeta\eta_{-\zeta}dx. \tag{6.43}$$

Next, η_ζ from (6.18) can be shown to satisfy $\fint_{\mathbb{T}^2}|\eta_\zeta|^2(t,x)dx = 1$, which, together with the fact that $\Lambda = \Lambda^+ \cup \Lambda^-$, $\Lambda^+ \cap \Lambda^- = \emptyset$, and $\gamma_\zeta = \gamma_{-\zeta}$ by Lemma 6.2.1, implies that by (6.43) we can continue to compute

$$\sum_{\zeta,\vartheta\in\Lambda} a_\zeta(t,x)a_\vartheta(t,x)\fint_{\mathbb{T}^2} W_\zeta\otimes W_\vartheta(t,x)dx$$
$$= \frac{1}{4}\rho(t,x)\sum_{\zeta\in\Lambda}\gamma_\zeta\left(\mathrm{Id}-\frac{\mathring{R}_l}{\rho}\right)\gamma_{-\zeta}\left(\mathrm{Id}-\frac{\mathring{R}_l}{\rho}\right)\zeta^\perp\otimes\zeta^\perp \overset{(6.13)}{=} -\mathring{R}_l+\rho\,\mathrm{Id},$$

which verifies (6.41b). Identity (6.41a) is verified similarly with the only difference that $b_\zeta\mathring{\otimes}b_{-\zeta}=\zeta^\perp\otimes\zeta^\perp-\frac{1}{2}\mathrm{Id}$ has to be applied instead of $b_\zeta\otimes b_{-\zeta}=\zeta^\perp\otimes\zeta^\perp$.

Next, with ψ_ζ, η_ζ, W_ζ, and μ from (6.15), (6.18), (6.19), and (6.17), respectively, we are ready to define the perturbation w_{q+1} and v_{q+1} as

$$w_{q+1}\triangleq w_{q+1}^{(p)}+w_{q+1}^{(c)}+w_{q+1}^{(t)} \quad\text{and}\quad v_{q+1}\triangleq v_l+w_{q+1}, \tag{6.44}$$

where the principal corrector, divergence corrector, and temporal corrector are defined as

$$w_{q+1}^{(p)}\triangleq\sum_{\zeta\in\Lambda}a_\zeta W_\zeta, \tag{6.45a}$$
$$w_{q+1}^{(c)}\triangleq\sum_{\zeta\in\Lambda}\nabla^\perp(a_\zeta\eta_\zeta)\psi_\zeta, \tag{6.45b}$$
$$w_{q+1}^{(t)}\triangleq\mu^{-1}\left(\sum_{\zeta\in\Lambda^+}-\sum_{\zeta\in\Lambda^-}\right)\mathcal{P}_L\mathbb{P}_{\neq0}\int_{t_{q+1}}^t(a_\zeta^2\mathbb{P}_{\neq0}\eta_\zeta^2\zeta)ds \tag{6.45c}$$

with

$$\mathbb{P}_{\neq0}f\triangleq f-\fint_{\mathbb{T}^2}fdx. \tag{6.46}$$

To verify (6.30c) at level $q+1$, relying on (6.29), (6.44), and (6.37) allows us to define $\mathring{R}_{q+1}$ by

$$\begin{aligned}\operatorname{div}\mathring{R}_{q+1}-\nabla\pi_{q+1}&\\ =\underbrace{\Lambda^m w_{q+1}+\partial_t(w_{q+1}^{(p)}+w_{q+1}^{(c)}+w_{q+1}^{(t)})+\operatorname{div}(u_l\otimes w_{q+1}+w_{q+1}\otimes u_l)}_{\operatorname{div}R_{\mathrm{lin1}}+\nabla\pi_{\mathrm{lin1}}}&\\ +\underbrace{\partial_{tt}(w_{q+1}^{(p)}+w_{q+1}^{(c)})}_{\operatorname{div}R_{\mathrm{lin2}}}+\underbrace{\operatorname{div}((w_{q+1}^{(c)}+w_{q+1}^{(t)})\otimes w_{q+1}+w_{q+1}^{(p)}\otimes(w_{q+1}^{(c)}+w_{q+1}^{(t)}))}_{\operatorname{div}R_{\mathrm{cor}}+\nabla\pi_{\mathrm{cor}}}&\\ +\underbrace{\operatorname{div}(w_{q+1}^{(p)}\otimes w_{q+1}^{(p)}+\mathring{R}_l)+\partial_{tt}w_{q+1}^{(t)}}_{\operatorname{div}R_{\mathrm{osc}}+\nabla\pi_{\mathrm{osc}}}+\operatorname{div}R_{\mathrm{com}}-\nabla\pi_l;&\end{aligned} \tag{6.47}$$

i. e.,

$$\operatorname{div}\mathring{R}_{q+1}-\nabla\pi_{q+1}=\operatorname{div}(R_{\mathrm{lin1}}+R_{\mathrm{lin2}}+R_{\mathrm{cor}}+R_{\mathrm{osc}}+R_{\mathrm{com}})-\nabla(\pi_{\mathrm{lin1}}+\pi_{\mathrm{cor}}+\pi_{\mathrm{osc}}-\pi_l),$$

where

$$R_{\text{lin1}} \triangleq \mathcal{B}\Lambda^m w_{q+1} + \mathcal{B}\partial_t(w_{q+1}^{(p)} + w_{q+1}^{(c)} + w_{q+1}^{(t)}) + u_l \mathring{\otimes} w_{q+1} + w_{q+1} \mathring{\otimes} u_l, \tag{6.48a}$$

$$\pi_{\text{lin1}} \triangleq u_l \cdot w_{q+1}, \tag{6.48b}$$

$$R_{\text{lin2}} \triangleq \mathcal{B}\partial_{\text{tt}}(w_{q+1}^{(p)} + w_{q+1}^{(c)}), \tag{6.48c}$$

$$R_{\text{cor}} \triangleq (w_{q+1}^{(c)} + w_{q+1}^{(t)}) \mathring{\otimes} w_{q+1} + w_{q+1}^{(p)} \mathring{\otimes} (w_{q+1}^{(c)} + w_{q+1}^{(t)}), \tag{6.48d}$$

$$\pi_{\text{cor}} \triangleq \frac{1}{2}[(w_{q+1}^{(c)} + w_{q+1}^{(t)}) \cdot w_{q+1} + w_{q+1}^{(p)} \cdot (w_{q+1}^{(c)} + w_{q+1}^{(t)})]. \tag{6.48e}$$

Concerning the most crucial R_{osc}, we use (6.41a) to write

$$\begin{aligned} & w_{q+1}^{(p)} \mathring{\otimes} w_{q+1}^{(p)} + \mathring{R}_l \\ &= w_{q+1}^{(p)} \mathring{\otimes} w_{q+1}^{(p)} - \sum_{\zeta,\vartheta\in\Lambda} a_\zeta a_\vartheta \fint_{\mathbb{T}^2} W_\zeta \mathring{\otimes} W_\vartheta dx + \rho\left(1 - \frac{1}{8}\sum_{\zeta\in\Lambda} \gamma_\zeta\left(\text{Id} - \frac{\mathring{R}_l}{\rho}\right)^2\right)\text{Id} \end{aligned}$$

and then rely on (6.45a) and the fact that $\mathbb{P}_{\geq \frac{\lambda\sigma}{2}}(W_\zeta \mathring{\otimes} W_\vartheta) = \mathbb{P}_{\neq 0}(W_\zeta \mathring{\otimes} W_\vartheta)$ from (6.19) to deduce

$$\begin{aligned} & w_{q+1}^{(p)} \mathring{\otimes} w_{q+1}^{(p)} + \mathring{R}_l \\ &= \sum_{\zeta,\vartheta\in\Lambda} a_\zeta a_\vartheta \mathbb{P}_{\geq \frac{\lambda_{q+1}\sigma}{2}}(W_\zeta \mathring{\otimes} W_\vartheta) + \rho\left(1 - \frac{1}{8}\sum_{\zeta\in\Lambda} \gamma_\zeta\left(\text{Id} - \frac{\mathring{R}_l}{\rho}\right)^2\right)\text{Id}. \end{aligned} \tag{6.49}$$

Upon taking divergence of (6.49), the second term can be part of gradient pressure, and there are various tools to handle the first term; see, e. g., [251, Lemma 7.4]. Of course, $\partial_{\text{tt}} w_{q+1}^{(t)}$ within R_{osc} plays a crucial role; for brevity, we refer the interested readers to [336] for details.

To compare with (3.21), let us show the estimate R_{com} in (6.38a):

$$\begin{aligned} \|R_{\text{com}}\|_{C_{t,q+1}L_x^1} &= \|u_l \mathring{\otimes} u_l - (u_q \mathring{\otimes} u_q) *_x \phi_l *_t \varphi_l\|_{C_{t,q+1}L_x^1} \\ &= \|(u_l - u_q) \mathring{\otimes} u_l + u_q \mathring{\otimes}(u_l - u_q) + u_q \mathring{\otimes} u_q - (u_q \mathring{\otimes} u_q) *_x \phi_l *_t \varphi_l\|_{C_{t,q+1}L_x^1} \\ &\lesssim l\|u_q\|_{C^1_{t,x,q+1}} \|u_q\|_{C_{t,q+1}L_x^2} \overset{(6.30a)(6.30b)}{\lesssim} lL\bar{e}\lambda_q^3 \ll \delta_{q+3} e(t). \end{aligned} \tag{6.50}$$

We refer to [239, eq. (6.36)] for an interesting Sobolev analogue of such an estimate and omit further details on the verification of (6.30c) at level $q+1$.

To verify (6.30d) at level $q+1$, we define

$$\delta E(t) \triangleq \left|e(t)(1 - \delta_{q+2}) - \|u_{q+1}(t)\|_{L_x^2}^2\right|, \tag{6.51}$$

so that (6.30d) at level $q+1$ holds once we show $\delta E(t) \le \frac{1}{4}\delta_{q+2}e(t)$. For this purpose, we first rewrite, using (6.39b),

$$\delta E = \left|\gamma_q 2(2\pi)^2 + \int_{\mathbb{T}^2} |u_q|^2 - |u_{q+1}|^2 dx\right|.$$

We further use (6.44) to continue by

$$\begin{aligned}\delta E = \Big|\gamma_q 2(2\pi)^2 + \int_{\mathbb{T}^2} &|u_q|^2 - |u_l|^2 - 2u_l \cdot w_{q+1}^{(p)} - 2u_l \cdot (w_{q+1}^{(c)} + w_{q+1}^{(t)}) \\ &- |w_{q+1}^{(p)}|^2 - 2w_{q+1}^{(p)} \cdot (w_{q+1}^{(c)} + w_{q+1}^{(t)}) - |(w_{q+1}^{(c)} + w_{q+1}^{(t)})|^2 dx\Big|.\end{aligned}$$

We focus on the most singular term, the product of $|w_{q+1}^{(p)}|^2$. To handle this, we use $\gamma_q(t)$ and write, using (6.45a),

$$|w_{q+1}^{(p)}(t)|^2 - \gamma_q(t)2 = \mathrm{Tr}\Big(\sum_{\zeta,\vartheta\in\Lambda} (a_\zeta a_\vartheta)(t)(W_\zeta \otimes W_\vartheta)(t)\Big) - \gamma_q(t)2. \tag{6.52}$$

Now we use (6.41b) to continue from (6.52) by

$$\begin{aligned}&|w_{q+1}^{(p)}(t)|^2 - \gamma_q(t)2 \\ &= \sum_{\zeta,\vartheta\in\Lambda} (a_\zeta a_\vartheta)(t)\,\mathrm{Tr}\,\mathbb{P}_{\neq 0}(W_\zeta \otimes W_\vartheta)(t) + \mathrm{Tr}(-\mathring{R}_l(t) + \rho(t)\,\mathrm{Id}) - \gamma_q(t)2 \\ &= \sum_{\zeta,\vartheta\in\Lambda} (a_\zeta a_\vartheta)(t)\,\mathrm{Tr}\,\mathbb{P}_{\neq 0}(W_\zeta \otimes W_\vartheta)(t) + 2(\rho(t) - \gamma_q(t)),\end{aligned}$$

where we used that $\mathrm{Tr}(\mathring{R}_l) = 0$. This can be made small by (6.39) and (6.30a)–(6.30c) at level q and the hypothesis of $\|e'\|_{C([-2,T])} \le \tilde{e}$ from (6.22). We omit further details in the verification of (6.30d) at level $q+1$ and conclude this proof. □

Let us emphasize how unorthodox the approach of convex integration is. In Galerkin approximation or any other classical approach to construct a solution to any given PDE, we would fix the initial value, construct a solution, and study its property such as energy. In the philosophy of convex integration scheme, we can fix energy and construct a solution based on the energy, and the initial value is just the solution constructed at time $t = 0$. Moreover, in general, even if two functions f and g have the same L^2-norm, it would not imply $f = g$. However, in the case of convex integration, because solution is constructed based on the prescribed energy, the same energy implies that the corresponding solutions are identical, as Theorem 6.3.1 showed.

6.4 Recent developments and open problems

This section is devoted to describe more recent developments and a few open problems of convex integration. Besides the Baire category argument or Nash-type iteration scheme, which we described in detail, another prominent approach to construct a nonunique solution was suggested by Jia and Šverák [204, 205], supported numerically in [163] by Guillod and Šverák. This approach uses spectral analysis and provides sufficient conditions for the nonuniqueness of Leray–Hopf weak solutions to the 3D Navier–Stokes equations.

As we discussed, the nonuniqueness of Leray–Hopf weak solutions to the 3D Navier–Stokes equations has been an open problem since [231], and the Baire category argument or Nash-type convex integration schemes seem to produce only solutions with low regularity. Some modifications can allow us to overcome this; particularly, if the diffusion $-\upsilon\Delta u$ is replaced by the fractional Laplacian $\upsilon\Lambda^m u$, then the authors of [96] in case $m < \frac{2}{5}$, and subsequently [124] in case $m < \frac{2}{3}$, were able to prove the nonuniqueness for the corresponding Leray–Hopf weak solutions. Inspired by Vishik [324, 325], Albritton, Brué, and Colombo [3] constructed a nonvanishing, divergence-free force $f \in L_t^1 L_x^2$ and initial data $u^{\text{in}} \in L_\sigma^2$ that produce infinitely many distinct Leray–Hopf weak solutions to the forced 3D Navier–Stokes equations.

Recently, Hou, Wang, and Yang [194] claimed to solve this major problem of proving the nonuniqueness of the Leray–Hopf weak solution to the unforced 3D Navier–Stokes equations. The construction is not by convex integration; instead, it follows the approach of Jia and Šverák. In short, they start from a forward self-similar solution and verify the sufficient spectral condition from [205] by computer-assisted proof.

Let us describe an open problem of interest, especially in the context of this monograph, where many PDEs from mathematical physics (Section 1.2) are discussed. To the best of the author's knowledge, almost all the PDEs to which convex integration was successfully applied have nonlinear terms of divergence form. The first exception is the isometric embedding by Nash [273]. Another exception is the NLS equation of [91], which we described in Section 6.1.3.3, although Christ himself did not call his approach convex integration. The third example that the author is aware of is [326] by Vo and Kim, in which the authors worked on deterministic conservation models that included the Burgers equation but rewrote it in the form of KPZ equation in its proof, and constructed an L^∞-solution that is nowhere continuous in the interior of its support in the absence of entropy condition. However, they relied on the approach of [272], not a Nash-type iteration, and were limited to the 1D, nondiffusive, and unforced case, and any extension seems unclear. The following question is intentionally not precise, and the aim is simply to encourage more extensions of convex integration to PDEs with various types of nonlinearity.

Problem 5. Can convex integration technique be extended to PDEs with nonlinearity in the nondivergence form in general?

An example of such a PDE with nonlinearity in the nondivergence form is the higher-dimensional Burgers' equation

$$\partial_t u + (u \cdot \nabla)u = \nu \Delta u. \tag{6.53}$$

A solution to such a PDE also does not satisfy the energy inequality because $\int_{\mathbb{T}^d} (u \cdot \nabla)u \cdot u dx \neq 0$ due to the absence of divergence-free property; only in the 1D case, we can get away with $\int_{\mathbb{T}} \theta \partial_x \theta \theta dx = 0$ for (6.53). Nevertheless, taking advantage of the maximum principle, e. g., Pooley and Robinson [283] showed the global well-posedness of (6.53) in $H^{\frac{1}{2}}(\mathbb{T}^3)$. Successful convex integration for (6.53) would be an interesting extension of [326] on the 1D inviscid Burgers' equation.

Of relevance, let us also briefly introduce the Kuramoto–Sivashinsky equation, which was originally proposed by Kuramoto and Tsuzuki [228, 229] in investigation of crystal growth and has found applications in various areas, such as the study of instabilities in laminar frame fronts [304] and reaction–diffusion systems. In the 1D case, it takes the form

$$\partial_t \theta + \frac{1}{2}|\partial_x \theta|^2 = -\lambda \partial_x^2 \theta - \partial_x^4 \theta, \quad \lambda > 0.$$

Analogously to the 1D Burgers equation, a global-in-time unique solution for a PDE with such a strong diffusion is well known ([312]). However, in the 2D case, it takes the form

$$\partial_t u + (u \cdot \nabla)u = \nu \Delta u - \Delta^2 u. \tag{6.54}$$

Not only does any solution u to (6.54) fail the energy conservation since $\int_{\mathbb{T}^2} (u \cdot \nabla)u \cdot u dx \neq 0$ due to the absence of divergence-free condition, it also fails the maximum principle due to $\Delta^2 u$, in contrast to the higher-dimensional Burgers' equation. Consequently, the global-in-time solution theory, unique or nonunique, is an open problem for the 2D Kuramoto–Sivashinsky equation despite some recent progress ([8, 109, 234, 235]).

Problem 6. Does the 2D deterministic Kuramoto–Sivashinsky equation admit a global-in-time unique smooth solution starting from an arbitrary smooth initial data? Can we construct a global-in-time weak solution to the 2D KSE, unique or nonunique?

Problem 6 allows for a nonunique case, and in light of [336], which constructed a global-in-time solution for the hyperbolic Navier–Stokes equations for which global-in-time solution theory was not known, it raises high hope that perhaps convex integration can lead to the resolution of this problem. Unfortunately, any progress toward the resolution of Problem 6 via convex integration must first make progress toward Problem 5, because (6.54) also has a nonlinear term in the nondivergence form.

7 Convex integration on SPDEs

This chapter is devoted to the convex integration technique applied on SPDEs with white-in-time noise; the case of singular SPDEs will be discussed in Section 9.2. Section 7.1 starts with a general review, Section 7.2 describes the main idea of how to adapt the convex integration technique to SPDEs, and Section 7.3 describes a limited amount of difficulties. Section 7.4 describes more recent developments. This is another area of rapid growth at the time of writing this monograph; we have no intention of claiming a thorough review and refer the interested readers to all the references we give.

7.1 General review

We first recall that adding an external random force often has a regularizing effect on the solution. It is well known that random perturbations can provide other interesting aspects that do not exist in the deterministic case. For example, a selection of solutions depending continuously on the initial condition in the deterministic case is not known; however, a probabilistic counterpart, namely Feller and even strong Feller properties under an assumption of sufficiently nondegenerate noise was obtained in [115, 144]. As the purpose of this monograph is to discuss convex integration technique applied on SPDEs, we limit ourselves to the following three examples of regularizing effects.

Example 6. The first example of regularization effect from random force is a damping effect of multiplicative noise, shown by Kim [215]. In Proposition 3.9.2, we saw that in the presence of damping, the Euler equations in any dimensions admit a global-in-time smooth solution starting from sufficiently small initial data. Its proof also clarified the important role played by the damping, convincing us why such a result does not exist for deterministic Euler equations without damping.

Remarkably, Kim showed that in the presence of certain multiplicative noise, for all $\epsilon \in (0,1)$, there exists $\mu = \mu(\epsilon) > 0$ such that $\mathbb{E}[\|u^{\text{in}}\|_{H^s}^4] \leq \mu$ implies $\mathbb{P}(\{T_{\max} = +\infty\})$, where $T_{\max}$ informally represents the maximal existence time of the local-in-time pathwise unique solution, and $s > 0$ is sufficiently large. We refer to [368, Theorem 2.3] for specific statement involving a stopping time.

Example 7. The second example of regularizing effect from random noise comes from [144], which we partially described in Section 4.5.2. Flandoli and Romito [144, Theorem 6.8] showed that informally stated, under certain assumptions on the covariance of the additive random force, if there exists (1) an initial data $x_0 \in \mathcal{W}$, where $\mathcal{W}$ is a certain Sobolev space, (2) deterministic time $t_0 > 0$, and (3) a solution P_{x_0} to the martingale problem (Definition 4.5.10) such that $P_{x_0}(C([0,t_0];\mathcal{W})) = 1$ starting from initial distribution $\mu_0 = \delta_x$, then, for every Markov selection $\{P_x\}_{x \in L^2_\sigma}$, $P_x(C([0,\infty);\mathcal{W})) = 1$, and the pathwise uniqueness holds for every $x \in \mathcal{W}$. On one hand, if such a result exists in the deterministic case, then we can consider the zero initial data, which leads to the

https://doi.org/10.1515/9783111349985-008

zero solution, that is unique for all time and conclude the global well-posedness of the 3D Navier–Stokes equations. On the other hand, a zero process is not a solution to the Navier–Stokes equations forced by additive noise and thus it remains unknown if the hypothesis of [144, Theorem 6.8] holds (see [144, Theorem 6.9] in case of uniqueness in law).

Example 8. Our last example of regularizing effect from random force comes from Flandoli, Gubinelli, and Priola [142] concerning the transport equation, a. k. a. the continuity equation, which has many applications not only in fluid mechanics but also in kinetic theory. For given $\rho^{\text{in}} : \mathbb{R}^d \to \mathbb{R}$ and $u : \mathbb{R}_+ \times \mathbb{R}^d \to \mathbb{R}^d$ such that $\nabla \cdot u = 0$, the transport equation solved by $\rho : \mathbb{R}_+ \times \mathbb{R}^d \to \mathbb{R}$ consists of

$$\partial_t \rho + u \cdot \nabla \rho = 0, \quad \rho(0,x) = \rho^{\text{in}}(x). \tag{7.1}$$

The Cauchy–Lipschitz theorem (e. g., [331, Theorems on p. 62 and 66]) informs us that if $u(t,x)$ is Lipschitz in x and continuous in t, then there exists a unique solution ρ to (7.1); in fact, $\rho(t, X(t,x)) = \rho(0,x)$ is a solution, where $\partial_t X(t,x) = u(t, X(t,x))$, and $X(0,x) = x$. The natural question is what happens if u is not Lipschitz in x or continuous in t. To address this issue, Di Perna and Lions [126, Corollary II.1] proved that for any initial data $\rho^{\text{in}} \in L^p(\mathbb{R}^d)$, $p \in [1,\infty]$, there exists a unique solution $\rho \in L_T^\infty L_x^p$ to (7.1) if $u \in L_T^1 W_x^{1,p'}$ and $\nabla \cdot u \in L_T^1 L_x^\infty$, where p' is the conjugate of p. Further extensions of [126] include [9] by Ambrosio and [31] by Bianchini and Bonicatto, and we refer the interested readers to [10] for a survey.

In relevance to our discussions, Flandoli, Gubinelli, and Priola [142] proved that transport noise in the Stratonovich interpretation (recall Remark 4.2.10 and see Example 11) can sufficiently regularize the solution to ensure the uniqueness. Specifically, they proved in [142, Theorem 20] that if $u \in L_T^\infty C_x^\alpha$ for some $\alpha \in (0,1)$ is bounded and $\nabla \cdot u \in L_{T,x}^p$ for some $p \in (2,\infty)$, then for any $\rho^{\text{in}} \in L^\infty(\mathbb{R}^d)$, there exists a unique process $\rho \in L^\infty(\Omega \times [0,T] \times \mathbb{R}^d)$ such that for all $\psi \in C_0^\infty(\mathbb{R}^d)$, the process $\int_{\mathbb{R}^d} \psi(x)\rho(t,x)dx$ has a continuous modification, which is a semimartingale and satisfies (7.1) forced by transport noise distributionally (see also [142, Theorem 21] for some variation).

The natural question is if convex integration can be applied to SPDEs, overcome the possible regularizing effects from noise, and prove the nonuniqueness in law or pathwise (recall Definition 4.3.1). The first idea that comes to mind would be to transform SPDEs to random PDEs (recall Section 4.5.3) so that the deterministic convex integration technique can be applied. Even this approach will create other difficulties, some of which will be described in Section 7.3; e. g. to deduce any desired property such as an energy inequality of a solution to SPDE by transforming the SPDE into a random PDE and applying convex integration to the random PDE, we still need to somehow come back to the original SPDE to deduce the desired property (see Section 7.3.1). Nevertheless, this is the initial approach that attracted much attention in the literature.

Let us describe three prototypes of random noise that allow for such transformations, certainly when hypotheses for such transformations hold; for simplicity, we use the transport equation (7.1) as a model, but our arguments readily apply to other SPDEs. Thus, for u such that $\nabla \cdot u = 0$ and ρ^{in}, we consider

$$d\rho + (\operatorname{div}(u\rho) - \Delta\rho)dt = G(\rho)dW, \quad \rho(0,x) = \rho^{\text{in}}(x). \tag{7.2}$$

Example 9 (Additive noise). Additive noise was defined in Definition 4.2.11 and briefly described in Section 4.5.3. This case boils down to $G(\rho)dW = dW$ where W is a GG^*-Wiener process for a Hilbert–Schmidt operator $G \in L_2(U_0, H)$ for some separable Hilbert spaces H, U, and $U_0 = GU$ (recall (4.9)). We can apply the Da Prato–Debussche trick introduced in (4.69)–(4.70), i. e.,

$$dz(t,x) = \Delta z(t,x)dt + dW(t,x), \quad z(0,x) \equiv 0, \tag{7.3}$$

and a random PDE

$$\partial_t\theta(t,x) + \operatorname{div}\big(u(t,x)\theta(t,x)\big) = \Delta\theta(t,x) - \operatorname{div}\big(u(t,x)z(t,x)\big), \quad \theta(0,x) = \rho^{\text{in}}(x), \tag{7.4}$$

so that $\rho = \theta + z$ recovers the original solution to (7.1). As we will discuss, convex integration on SPDEs with such a random force was first considered in [186].

Example 10 (Linear multiplicative noise). This case of linear multiplicative noise boils down to $G(\rho)dW = \rho dW$, where W is an $\mathbb{R}$-valued Wiener process. In this case, we may define $\theta(t,x) \triangleq \rho(t,x)e^{-W(t)}$ and see that an application of Itô's formula (Lemma 4.2.9) gives

$$de^{-W(t)} = -e^{-W(t)}dW(t) + \frac{1}{2}e^{-W(t)}dt.$$

Now Itô's product formula can be deduced from Lemma 4.2.9; in this simple case of $\mathbb{R}$-valued process and $\mathbb{R}$-valued Wiener process W, it implies

$$d(X_tY_t) = X_tdY_t + Y_tdX_t + dX_tdY_t$$

(e. g., [11, Theorem 4.4.13]). This leads us to

$$d\theta = e^{-W}[-[\operatorname{div}(u\rho) - \Delta\rho]dt + \rho dW] + \left[-e^{-W}dW + \frac{1}{2}e^{-W}dt\right]\rho + (-e^{-W}\rho)dt,$$

which, after simplification, leads to the random PDE

$$\partial_t\theta(t,x) + \operatorname{div}(u(t,x)\theta(t,x)) + \frac{1}{2}\theta(t,x) = \Delta\theta(t,x), \quad \theta(0,x) = \rho^{\text{in}}(x). \tag{7.5}$$

The first application of convex integration to SPDEs was accomplished in [33, 90] with this type of noise.

Example 11 (Transport noise in Stratonovich interpretation). This type of noise was mentioned upon discussing [142] and boils down to $G(\rho)dW = -\sum_{i=1}^{d}\frac{\partial}{\partial x_i}\rho \circ dW_i$ where $W = (W_1, \ldots, W_d)$ is a Brownian motion. In this case, we consider $\theta(t,x) \triangleq \rho(t, x + W(t))$, so that Itô's formula (e. g., [227, Theorem 3.3.2 on p. 93]) gives

$$d\theta(t,x) = d\rho|_{(t,g(t,x))} + \sum_{i=1}^{d}\frac{\partial \rho}{\partial x_i}(t, g(t,x)) \circ dg_i(t),$$

which leads to the random PDE

$$\partial_t\theta(t,x) + \operatorname{div}(u(t, x + W(t))\theta(t,x)) = \Delta\theta(t,x), \quad \theta(0,x) = \rho^{\text{in}}(x). \tag{7.6}$$

Convex integration on noise of this type was considered for the first time in [222].

Because most types of noise in singular SPDEs of our interest are additive, we focus on the additive type, except Section 7.3.2 in the case of linear multiplicative noise. With this in mind, whereas the Da Prato–Debussche trick separates the noise dW and allows us to focus on (7.4), the random PDE consists of z, and any analysis requires knowledge of its regularity. The following result is from [355, Proposition 4.4], which is originally from [186, Proposition 3.6] and [125, Proposition 34].

Lemma 7.1.1 ([355, Proposition 4.4]). *Consider*

$$dz + \Lambda^{\zeta} z dt = dW, \quad z(0,x) = 0, \tag{7.7}$$

where W is a GG^*-Wiener process, and $\zeta \in (0,2]$ for $x \in \mathbb{T}^d$, $d \in \{2,3\}$. Suppose that

$$\operatorname{Tr}(\Lambda^{d+2-\zeta+4\sigma}GG^*) < \infty \tag{7.8}$$

for some $\sigma > 0$. Then for all $\delta \in (0, \frac{1}{2})$, $T > 0$, and $l \in \mathbb{N}$,

$$\mathbb{E}[\|z\|^l_{C_T\dot{H}_x^{\frac{2+d+\sigma}{2}}} + \|z\|^l_{C_T^{\frac{1}{2}-\delta}\dot{H}_x^{\frac{d+\sigma}{2}}}] < \infty. \tag{7.9}$$

We may consider a stochastic Stokes equation instead of the heat equation, and the same regularity can be obtained. We give the proof in Appendix A.6 for readers' convenience. The proof makes it clear that various generalizations are possible; however, inherently from Remark 4.2.2, the temporal regularity has a strict upper bound of $\frac{1}{2} - \delta$ for some $\delta > 0$.

7.2 Adapting deterministic convex integration to SPDEs

In this section, we describe some of the first works that successfully adapted the convex integration to SPDEs. We certainly cannot review them all and will give an abundance of references in Section 7.4.

First, Breit, Feireisl, and Hofmanová [34] and Chiodaroli, Feireisl, and Flandoli [90] proved via convex integration the pathwise nonuniqueness of certain compressible Euler equations forced by linear multiplicative noise. In these pioneering works, the authors already realized a remarkable property of their solutions obtained via convex integration, specifically that they were probabilistically strong. This can be readily seen by recalling the proof of Theorem 6.3.1 and observing that the solution to approximation at every level q, a random PDE analogue of (6.29), is probabilistically strong, so that the limit of such a Cauchy sequence remains probabilistically strong.

Subsequently, Hofmanová, Zhu, and Zhu [186] considered the 3D Navier–Stokes equations forced by additive noise (Example 9) on $\mathbb{T}^3$, specifically

$$du + (\operatorname{div}(u \otimes u) + \nabla \pi - \nu \Delta u)dt = dW, \quad \nabla \cdot u = 0, \quad dW = \sum_{j=1}^{\infty} \sqrt{\lambda_j} e_j d\beta_j, \tag{7.10}$$

where W is a GG^*-Wiener process such that $GG^* e_j = \lambda_j e_j$ (cf. (4.6)). For a fixed $(\Omega, \mathcal{F}, \mathbb{P})$ and the Wiener process GdW, they proved that for any $T > 0$ and $\kappa \in (0,1)$, there exist $\gamma \in (0,1)$ and a $\mathbb{P}$-a. s. strictly positive stopping time $\mathfrak{t}$ satisfying $\mathbb{P}(\{\mathfrak{t} \geq T\}) > \kappa$ and probabilistically strong, analytically weak solution u that belongs to $C([0,\mathfrak{t}]; H^{\gamma}(\mathbb{T}^3))$ $\mathbb{P}$-a. s. starting from deterministic initial data u^{in} such that

$$\frac{1}{2}\|u(T)\|_{L^2}^2 > 2\left(\frac{1}{2}\|u^{\text{in}}\|_{L^2}^2 + \frac{T}{2}\operatorname{Tr}(GG^*)\right) \quad \text{on } \{\mathfrak{t} \geq T\}. \tag{7.11}$$

The types of noise considered in early preprint of [186] were additive, linear multiplicative, and nonlinear; the published version included nonlinear type, and we refer the interested readers to [186] for details. Neither energy nor initial data is prescribed here, and the initial data is merely taken as the constructed convex integration solution at time $t = 0$. We can take the initial data u^{in} that lies in L_σ^2 and construct another solution via probabilistic Galerkin approximation from Section 4.5 that would satisfy

$$\frac{1}{2}\mathbb{E}^{\mathbb{P}}[\|u(T)\|_{L^2}^2 ds] \leq \frac{1}{2}\|u^{\text{in}}\|_{L^2}^2 + \frac{T}{2}\operatorname{Tr}(GG^*). \tag{7.12}$$

Comparing (7.11) and (7.12) verifies the nonuniqueness in law of the 3D Navier–Stokes equations forced by additive noise and hence demonstrates the futility of proving pathwise uniqueness via Cherney's theorem (Lemma 4.3.3) at this low regularity level. The Yamada–Watanabe theorem (Lemma 4.3.2(a)) implies that nonuniqueness in law implies pathwise nonuniqueness, making this result more impressive. The deterministic analogue of (7.11) was proven earlier by Buckmaster and Vicol [48, Section 7]; we emphasize that it suffices for proving the nonuniqueness because Galerkin approximation is also available for the Navier–Stokes equations, deterministic or stochastic (recall Remark 6.3.2(c), where it was not available for the hyperbolic Navier–Stokes equations).

The main idea of [186] was to consider, as in Example 9,

$$dz + \nabla\pi^1 dt = \Delta z dt + dW, \quad \nabla \cdot z = 0 \text{ for } t > 0,\ z(0,x) \equiv 0, \tag{7.13}$$

and to apply convex integration to the random PDE

$$\partial_t v - \Delta v + \operatorname{div}((v+z)\otimes(v+z)) + \nabla\pi^2 = 0, \quad \nabla \cdot v = 0 \text{ for } t > 0. \tag{7.14}$$

Among many steps, they particularly
1. handled the stochastic error,
2. obtained an energy inequality for v analogous to (7.11),
3. transferred such an energy inequality for the solution u to (7.10).

We describe step 3 in Section 7.3.1 through the proof of Theorem 7.3.2 on the stochastic MHD system in detail and describe other difficulties in Sections 7.3.3–7.3.4.

7.3 Difficulties and ways to overcome them

In this section, by various examples we explain several difficulties of applying the convex integration technique to SPDEs.

7.3.1 Working on random PDE and coming back to SPDE

As we discussed, after applying the convex integration on any random PDE, there remains a task to come back to the original SPDE. This section is devoted to describing this by an example of the 3D MHD system for which an inequality that defies Taylor's conjecture (recall Sections 6.1.1 and 6.1.4) will be shown, similarly to (7.11). In Section 6.1.1 we described how in addition to the total $\mathcal{E}(t)$ from (5.3a), magnetic helicity $\mathcal{H}_b(t)$ and cross-helicity $\mathcal{H}_u(t)$ from (6.5) are conserved for the smooth solutions to the deterministic ideal MHD system. The stochastic MHD system has also attracted much attention, e. g., ergodicity [13], well-posedness [306], large deviation theory [93], tamed MHD system [297], Markov selection and global-in-time existence of probabilistically weak Leray–Hopf weak solution [349].

Remark 7.3.1. Before we proceed further, let us describe a quick way to apply convex integration on the deterministic MHD system. The following explanation will also be used in Section 9.2.1, where we discuss the idea of Hofmanová, Zhu, and Zhu [189], who applied convex integration to the 3D Navier–Stokes equations forced by STWN. For example, we may consider the following convex integration scheme for the MHD system (1.8):

$$\partial_t u_q + (u_q \cdot \nabla)u_q + \nabla \pi_q - \nu \Delta u_q = (b_q \cdot \nabla)b_q, \quad \nabla \cdot u_q = 0, \tag{7.15a}$$

$$\partial_t b_q + (u_q \cdot \nabla)b_q - \eta \Delta b_q = (b_q \cdot \nabla)u_q + \mathring{R}_q, \quad \nabla \cdot b_q = 0. \tag{7.15b}$$

We see that we may construct b_{q+1} by adding an appropriate perturbation w_{q+1} to b_q (recall (6.44)), substitute this b_{q+1} into (7.15a), and consider the resulting equation as the Navier–Stokes equation with smooth external force $(b_{q+1} \cdot \nabla)b_{q+1}$. Then a Galerkin approximation can readily construct a solution u_{q+1} (recall Remark 3.10.7). We can substitute such u_{q+1} into (7.15b), which now becomes linear, solve for $\mathring{R}_{q+1}$ using the inverse divergence operator (recall Lemma 6.2.2), and attempt to verify the inductive hypothesis (e. g., Hypothesis 6.3.1).

Among other issues, this approach is not favorable in the stochastic case because the solution obtained partially through Galerkin approximation would not be probabilistically strong (recall Section 4.5.3). This is the reason why in [360, 364], which we will describe next, the errors were placed on both equations of u_q and b_q.

To introduce our model, we fix a probability space $(\Omega, \mathcal{F}, \mathbf{P})$, choose for $k \in \{1, 2\}$ an additive force $G_k G_k^*$-Wiener process that is valued in some Hilbert space U_k, specifically

$$G_k dB_k = \sum_{j=1}^{\infty} G_{k,j} dB_{k,j} = \sum_{j=1}^{\infty} \sqrt{\lambda_{k,j}} e_{k,j} d\beta_{k,j}, \quad G_k G_k^* e_{k,j} = \lambda_{k,j} e_{k,j}, \tag{7.16}$$

where $\{\beta_{k,j}\}_{j=1}^{\infty}$ are mutually independent $\mathbb{R}$-valued Brownian motions on $(\Omega, \mathcal{F}, \mathbf{P})$, and $\{e_{k,j}\}_{j=1}^{\infty}$ is an o. n. b. of U_k. Then for $x \in \mathbb{T}^3 = [-\pi, \pi]^3$, we consider

$$du + (\operatorname{div}(u \otimes u - b \otimes b) + \nabla \pi)dt = G_1 dB_1, \quad \nabla \cdot u = 0, \tag{7.17a}$$

$$dA + (b \times u + \nabla \chi)dt = \nabla \times (-\Delta)^{-1} G_2 dB_2 \tag{7.17b}$$

(cf. (6.4)). Sufficiently smooth solutions to (7.17) starting from deterministic initial data satisfy

$$\mathbb{E}^{\mathbb{P}}[\mathcal{E}(t)] = \mathcal{E}(0) + \frac{t}{2} \sum_{k=1}^{2} \operatorname{Tr}(G_k G_k^*), \tag{7.18a}$$

$$\mathbb{E}^{\mathbb{P}}[\mathcal{H}_b(t)] = \mathcal{H}_b(0) + t C_{G_2}, \tag{7.18b}$$

$$\mathbb{E}^{\mathbb{P}}[\mathcal{H}_u(t)] = \mathcal{H}_u(0), \tag{7.18c}$$

$$C_{G_2} \triangleq \sum_{j=1}^{\infty} \langle \nabla \times (-\Delta)^{-1} G_2 G_2^* e_{2,j}, e_{2,j} \rangle. \tag{7.19}$$

The proof of (7.18) is similar to that of (7.12), (6.6), and (6.7). Taylor's argument in Section 6.1.1 can be extended to imply that even in the diffusive case, if $\nu, \eta > 0$ are arbitrarily small, then the magnetic helicity $\mathcal{H}_b(t)$ in mathematical expectation should grow linearly with the rate of C_{G_2}. The following result from [364] shows that Taylor's conjecture in the case of additive noise fails. It fails also in the case of linear multiplicative noise, and we refer to [364, Theorem 2.2]. Let B_k, $k = 1, 2$, be cylindrical Wiener processes on a Hilbert space U_k satisfying $B_k(0) = 0$, and let $\{\mathcal{F}_t\}_{t\geq 0}$ be the canonical filtration of (B_1, B_2) augmented by all the $\mathbb{P}$-negligible sets. We state the following result.

Theorem 7.3.2 ([364, Theorem 2.2]). *Consider* (7.17) *with* $\upsilon = \eta = 0$. *Suppose that for* $\delta \in (0, \frac{1}{12})$ *and* $k \in \{1, 2\}$, $G_k \in L_2(U_k, H_\sigma^{1-\delta})$. *Then for* $T > 0$ *and* $\kappa \in (0, 1)$, *there exist* $\gamma \in (0, 1)$ *and a* $\mathbb{P}$-*a. s. strictly positive stopping time* $\mathfrak{t}$ *such that* $\mathbb{P}(\{\mathfrak{t} \geq T\}) > \kappa$ *and the following is additionally satisfied: There exists a pair* (u, b) *of* $\{\mathcal{F}_t\}_{t\geq 0}$-*adapted processes that is a weak solution of* (7.17) *starting from a deterministic initial data* $(u^{\text{in}}, b^{\text{in}})$ *that satisfies*

$$\operatorname*{ess\,sup}_{\omega\in\Omega} \|u(\omega)\|_{C_t \dot{H}_x^\gamma} < \infty, \quad \operatorname*{ess\,sup}_{\omega\in\Omega} \|b(\omega)\|_{C_t \dot{H}_x^\gamma} < \infty,$$

and on $\{\mathfrak{t} \geq T\}$,

$$\mathcal{E}(T) > 2\left(\mathcal{E}(0) + \frac{T}{2}\sum_{k=1}^{2} \operatorname{Tr}(G_k G_k^*)\right), \quad \mathcal{H}_b(T) > 2(\mathcal{H}_b(0) + TC_{G_2}), \tag{7.20a}$$

$$\mathcal{H}_u(T) > 2\mathcal{H}_u(0). \tag{7.20b}$$

Remark 7.3.3. Theorem 7.3.2 does not lead to nonuniqueness outcome immediately, in contrast to the case of [186] we described in Section 7.2. Interestingly, this is because starting from the initial data $(u^{\text{in}}, b^{\text{in}})$ constructed in Theorem 7.3.2, probabilistic Galerkin approximation cannot yield a classical solution that satisfies the energy inequality due to the absence of diffusion (cf. the same reason in Remark 6.3.2(c)). Nevertheless, the nonuniqueness in law of the 3D MHD system forced by random noise was proven in other works, e. g., [56, 360] and [364, Theorem 2.3 and Corollary 2.4].

We provide only a sketch of the proof of Theorem 7.3.2 and refer the interested readers to [364] for details; our goal here is to highlight that we can apply convex integration on the corresponding random PDE of (7.17) but must come back to the original (7.17) to deduce (7.20). First, we apply the curl operator on (7.17b) and define

$$(v, \Xi) \triangleq (u - G_1 B_1, b - G_2 B_2) \tag{7.21}$$

to obtain

$$\partial_t v + \operatorname{div}((v + G_1 B_1) \otimes (v + G_1 B_1) - (\Xi + G_2 B_2) \otimes (\Xi + G_2 B_2)) + \nabla \pi = 0, \quad \nabla \cdot v = 0, \tag{7.22a}$$

$$\partial_t \Xi + \operatorname{div}((\Xi + G_2 B_2) \otimes (v + G_1 B_1) - (v + G_1 B_1) \otimes (\Xi + G_2 B_2)) = 0. \tag{7.22b}$$

For any $\sigma > 0$ arbitrarily small, $L > 1$ to be specified in (7.31), $\delta \in (0, \frac{1}{12})$, and a universal Sobolev constant $C_S = C_S(\sigma) \geq 0$ such that $\|f\|_{L^\infty(\mathbb{T}^3)} \leq C_S \|f\|_{\dot{H}^{\frac{3+\sigma}{2}}(\mathbb{T}^3)}$, we define the stopping time

$$\begin{aligned} T_L \triangleq \inf\Big\{t \geq 0 : C_S \max_{k\in\{1,2\}} \|G_k B_k(t)\|_{H^{1-\delta}} \geq L^{\frac{1}{4}}\Big\} \\ \wedge \inf\Big\{t \geq 0 : C_S \max_{k\in\{1,2\}} \|G_k B_k\|_{C_t^{\frac{1}{2}-2\delta} L^2} \geq L^{\frac{1}{2}}\Big\} \wedge L \end{aligned} \tag{7.23}$$

and observe that $T_L > 0$ and $\lim_{L\to\infty} T_L = +\infty$ $\mathbb{P}$-a. s. due to the hypothesis on the noise. For any $a \in 2\mathbb{N}$, $b \in \mathbb{N}$, and $\beta \in (0,1)$ to be specified, we define

$$\lambda_q \triangleq a^{b^q} \quad \text{for } q \in \mathbb{N}_0, \quad b \in \{\iota \in \mathbb{N} : \iota > 39\alpha^{-1}\}, \quad \delta_q \triangleq \lambda_q^{-2\beta} \quad \text{for } q \in \mathbb{N}_0, \tag{7.24}$$

$$M_0 \in C^\infty(\mathbb{R}) \quad \text{such that } M_0(t) = \begin{cases} L^4 & \text{if } t \leq 0, \\ L^4 e^{4Lt} & \text{if } t \geq T \wedge L, \end{cases} \tag{7.25a}$$

$$0 \leq M_0'(t) \leq 8LM_0(t), \quad M_0''(t) \leq 32L^2 M_0(t), \tag{7.25b}$$

and for $k = 1, 2$,

$$\alpha \triangleq \frac{1}{1600} \quad \text{and} \quad (G_k B_k)_q \triangleq \mathbb{P}_{\leq f(q)} G_k B_k, \quad \text{where } f(q) \triangleq \lambda_{q+1}^{\frac{3\alpha}{22}}. \tag{7.26}$$

We seek a solution $(v_q, \Xi_q, \mathring{R}_q^v, \mathring{R}_q^\Xi)$ to

$$\begin{aligned} &\partial_t v_q + \operatorname{div}((v_q + (G_1B_1)_q) \otimes (v_q + (G_1B_1)_q) \\ &\quad - (\Xi_q + (G_2B_2)_q) \otimes (\Xi_q + (G_2B_2)_q)) + \nabla \pi_q = \operatorname{div} \mathring{R}_q^v, \quad \nabla \cdot v_q = 0, \end{aligned} \tag{7.27a}$$

$$\begin{aligned} &\partial_t \Xi_q + \operatorname{div}((\Xi_q + (G_2B_2)_q) \otimes (v_q + (G_1B_1)_q) \\ &\quad - (v_q + (G_1B_1)_q) \otimes (\Xi_q + (G_2B_2)_q)) = \operatorname{div} \mathring{R}_q^\Xi, \quad \nabla \cdot \Xi_q = 0, \end{aligned} \tag{7.27b}$$

over $[t_q, T_L]$ for $q \in \mathbb{N}_0$, where $\mathring{R}_q^v$ and $\mathring{R}_q^\Xi$ are symmetric trace-free and skew-symmetric matrices, respectively, and $t_q \triangleq -1 + \sum_{1\leq \iota \leq q} \delta_\iota^{\frac{1}{2}}$. We extend B_k to $[-2, 0]$ by $B_k(t) \equiv B_k(0)$ for $k = 1, 2$ and all $t \in [-2, 0]$. For universal constants (see [359, eqs. (120) and (131)]), we state the inductive hypothesis appropriate for Theorem 7.3.2 (recall (6.28)).

Hypothesis 7.3.1. For universal constants $c_v, c_\Xi > 0$ from [360, eqs. (120) and (131)], the solution $(v_q, \Xi_q, \mathring{R}_q^v, \mathring{R}_q^\Xi)$ to (7.27) satisfies, for all $t \in [t_q, T_L]$,

$$\|v_q\|_{C_{t,q}L_x^2} \leq M_0(t)^{\frac{1}{2}} \Big(1 + \sum_{1\leq\iota\leq q} \delta_\iota^{\frac{1}{2}}\Big) \leq 2M_0(t)^{\frac{1}{2}}, \tag{7.28a}$$

$$\|\Xi_q\|_{C_{t,q}L_x^2} \leq M_0(t)^{\frac{1}{2}} \Big(1 + \sum_{1\leq\iota\leq q} \delta_\iota^{\frac{1}{2}}\Big) \leq 2M_0(t)^{\frac{1}{2}}, \tag{7.28b}$$

$$\|v_q\|_{C^1_{t,x,q}} \leq M_0(t)^{\frac{1}{2}} \lambda_q^4, \quad \|\Xi_q\|_{C^1_{t,x,q}} \leq M_0(t)^{\frac{1}{2}} \lambda_q^4, \tag{7.28c}$$

$$\|\mathring{R}_q^v\|_{C_{t,q}L_x^1} \leq c_v M_0(t) \delta_{q+1}, \quad \|\mathring{R}_q^\Xi\|_{C_{t,q}L_x^1} \leq c_\Xi M_0(t) \delta_{q+1}. \tag{7.28d}$$

The following two propositions are concerned with the initial step $q = 0$ and step $q + 1$, assuming that it holds at step q.

Proposition 7.3.4 ([364, Proposition 3.1]). *Let*

$$v_0(t,x) \triangleq \frac{M_0(t)^{\frac{1}{2}}}{(2\pi)^{\frac{3}{2}}}\begin{pmatrix}\sin(x_3)\\0\\0\end{pmatrix} \quad \text{and} \quad \Xi_0(t,x) \triangleq \frac{M_0(t)^{\frac{1}{2}}}{(2\pi)^{3}}\begin{pmatrix}\sin(x_3)\\\cos(x_3)\\0\end{pmatrix}. \tag{7.29}$$

Then, together with

$$\begin{aligned}\mathring{R}_0^v(t,x) \triangleq{}& \frac{\frac{d}{dt}M_0(t)^{\frac{1}{2}}}{(2\pi)^{\frac{3}{2}}}\begin{pmatrix}0 & 0 & -\cos(x_3)\\0 & 0 & 0\\-\cos(x_3) & 0 & 0\end{pmatrix}\\ &+ (v_0\mathring{\otimes}(G_1B_1)_0 + (G_1B_1)_0\mathring{\otimes}v_0 + (G_1B_1)_0\mathring{\otimes}(G_1B_1)_0\\ &- \Xi_0\mathring{\otimes}(G_2B_2)_0 - (G_2B_2)_0\mathring{\otimes}\Xi_0 - (G_2B_2)_0\mathring{\otimes}(G_2B_2)_0)(t,x),\\ \mathring{R}_0^{\Xi}(t,x) \triangleq{}& \frac{\frac{d}{dt}M_0(t)^{\frac{1}{2}}}{(2\pi)^{3}}\begin{pmatrix}0 & 0 & -\cos(x_3)\\0 & 0 & \sin(x_3)\\\cos(x_3) & -\sin(x_3) & 0\end{pmatrix}\\ &+ (\Xi_0 \otimes (G_1B_1)_0 + (G_2B_2)_0 \otimes v_0 + (G_2B_2)_0 \otimes (G_1B_1)_0\\ &- v_0 \otimes (G_2B_2)_0 - (G_1B_1)_0 \otimes \Xi_0 - (G_1B_1)_0 \otimes (G_2B_2)_0)(t,x),\end{aligned}$$

(v_0, Ξ_0) *satisfies Hypothesis* 7.3.1 *and* (7.27) *at level* $q = 0$ *over* $[t_0, T_L]$, *provided that* $(40)^{\frac{4}{3}} < L$,

$$\begin{aligned}&((2\pi)^3+1)^2 20\pi^{\frac{3}{2}}\max\{c_v^{-1}, c_\Xi^{-1}\}\\ &\quad < a^{2\beta b}20\pi^{\frac{3}{2}}\max\{c_v^{-1}, c_\Xi^{-1}\} \le L \le \frac{(2\pi)^{\frac{3}{2}}a^4 - 2}{4}.\end{aligned} \tag{7.30}$$

Finally, $v_0(t,x)$, $\Xi_0(t,x)$, $\mathring{R}_0^v(t,x)$, *and* $\mathring{R}_0^{\Xi}(t,x)$ *are all deterministic for all* $t \in [t_0, 0]$.

Proposition 7.3.5 ([364, Proposition 3.2]). *Let* L *satisfy*

$$\max\{(40)^{\frac{4}{3}}, ((2\pi)^3+1)^2 20\pi^{\frac{3}{2}}\max\{c_v^{-1}, c_\Xi^{-1}\}\} < L. \tag{7.31}$$

Then there exist parameters a, b, *and* β *such that* (7.30) *is fulfilled and the following holds. Suppose that* $(v_q, \Xi_q, \mathring{R}_q^v, \mathring{R}_q^{\Xi})$ *is a group of* $\{\mathcal{F}_t\}_{t\ge 0}$*-adapted processes that solves* (7.27) *and satisfies Hypothesis* 7.3.1. *Then there exists a group of* $\{\mathcal{F}_t\}_{t\ge 0}$*-adapted processes* $(v_{q+1}, \Xi_{q+1}, \mathring{R}_{q+1}^v, \mathring{R}_{q+1}^{\Xi})$ *that satisfies Hypothesis* 7.3.1 *and solves* (7.27) *at level* $q+1$ *and for all* $t \in [t_{q+1}, T_L]$,

$$\|v_{q+1}(t) - v_q(t)\|_{L_x^2} \le M_0(t)^{\frac{1}{2}}\delta_{q+1}^{\frac{1}{2}}, \quad \|\Xi_{q+1}(t) - \Xi_q(t)\|_{L_x^2} \le M_0(t)^{\frac{1}{2}}\delta_{q+1}^{\frac{1}{2}}. \tag{7.32}$$

Finally, if the solution $(v_q, \Xi_q, \mathring{R}_q^v, \mathring{R}_q^\Xi)(t,x)$ *is deterministic over* $[t_q, 0]$, *then so is* $(v_{q+1}, \Xi_{q+1}, \mathring{R}_{q+1}^v, \mathring{R}_{q+1}^\Xi)(t,x)$ *over* $[t_{q+1}, 0]$.

We only mention that the proof of Proposition 7.3.5 requires mollifying (v_q, Ξ_q) by mollifiers $\{\phi_l\}_{l>0}$ in space and $\{\varphi_l\}_{l>0}$ in time, constructing perturbations and thereby solutions at the step $q+1$ as

$$v_{q+1} \triangleq v_l + w_{q+1}, \quad v_l \triangleq v_q *_x \phi_l *_t \varphi_l, \tag{7.33a}$$

$$\Xi_{q+1} \triangleq \Xi_l + d_{q+1}, \quad \Xi_l \triangleq \Xi_q *_x \phi_l *_t \varphi_l, \tag{7.33b}$$

where

$$l \triangleq \lambda_{q+1}^{-\frac{3\alpha}{2}} \lambda_q^{-2}, \tag{7.34}$$

similarly to (6.44).

We are now ready to prove Theorem 7.3.2. The families $\{v_q\}_{q\in\mathbb{N}_0}$ and $\{\Xi_q\}_{q\in\mathbb{N}_0}$ can be verified to be Cauchy in $C([0,T_L]; \dot{H}^\gamma(\mathbb{T}^3))$ for $\gamma \in (0, \frac{\beta}{4+\beta})$, and the limits v and Ξ satisfy

$$\max\{\|\Xi(t) - \Xi_0(t)\|_{L_x^2}, \|v(t) - v_0(t)\|_{L_x^2}\} \le M_0(t)^{\frac{1}{2}} \left(\frac{a^{-b\beta}}{1 - a^{-b\beta}} \right). \tag{7.35}$$

We focus on highlighting the proof of the second inequality of (7.20a) concerning magnetic helicity while referring the readers to [364] for verification of other claims in (7.20). For fixed $T > 0$, we take $L > 0$ larger if necessary so that

$$\begin{aligned} &\frac{L^4}{(2\pi)^3} + \frac{3}{(2\pi)^{\frac{3}{2}} - 2} \left[2(2L^2 e^{2LT} + L^{\frac{1}{4}}) L^{\frac{1}{4}} + L^{\frac{1}{2}} + \left(\frac{(2\pi)^{\frac{3}{2}} - 2}{3} \right) T C_{G_2} \right] \\ &\le \frac{1}{(2\pi)^{\frac{9}{2}}} L^4 e^{4LT}. \end{aligned} \tag{7.36}$$

Because Ξ is deduced as a Cauchy limit of divergence-free and mean-zero smooth functions, Ξ retains such properties, and hence we can define its potential

$$\mathcal{F} \quad \text{such that } \nabla \times \mathcal{F} = \Xi. \tag{7.37}$$

Next, considering our choice of Ξ_0 at first iteration from (7.29), we can explicitly see its vector potential $\mathcal{F}_0$ defined by

$$\mathcal{F}_0(t,x) \triangleq \frac{M_0(t)^{\frac{1}{2}}}{(2\pi)^3} \begin{pmatrix} \sin(x_3) \\ \cos(x_3) \\ 0 \end{pmatrix}, \tag{7.38}$$

which can be verified to be identical to $\Xi_0(t,x)$ via direct computation. We define the random magnetic helicity corresponding to (7.22) as

$$\tilde{\mathcal{H}}_b(t) \triangleq \int_{\mathbb{T}^3} F(t,x)\cdot\Xi(t,x)dx \quad \text{and} \quad \tilde{\mathcal{H}}_{0,b}(t) \triangleq \int_{\mathbb{T}^3} F_0(t,x)\cdot\Xi_0(t,x)dx. \tag{7.39}$$

Due to the explicit expression of (7.29) and (7.38), we can directly compute

$$\|F_0(t)\|_{L^2} \overset{(7.38)}{=} \left\| \frac{M_0(t)^{\frac{1}{2}}}{(2\pi)^3} \begin{pmatrix} \sin(x_3) \\ \cos(x_3) \\ 0 \end{pmatrix} \right\|_{L^2} = \frac{M_0(t)^{\frac{1}{2}}}{(2\pi)^{\frac{3}{2}}}, \tag{7.40a}$$

$$\tilde{\mathcal{H}}_{0,b}(t) \overset{(7.39)(7.29)(7.38)}{=} \int_{\mathbb{T}^3} \left| \frac{M_0(t)^{\frac{1}{2}}}{(2\pi)^3} \begin{pmatrix} \sin(x_3) \\ \cos(x_3) \\ 0 \end{pmatrix} \right|^2 dx = \frac{M_0(t)}{(2\pi)^3}. \tag{7.40b}$$

Taking in (7.28b) the limit as $q \to \infty$, we see that

$$\|\Xi(t)\|_{L^2_x} \le 2M_0(t)^{\frac{1}{2}}. \tag{7.41}$$

Thus by Hölder's inequality we may first estimate

$$\begin{aligned} |\tilde{\mathcal{H}}_b(t) - \tilde{\mathcal{H}}_{0,b}(t)| &\overset{(7.39)}{=} \left| \int_{\mathbb{T}^3} (F(t)-F_0(t))\cdot\Xi(t) + F_0(t)\cdot(\Xi(t)-\Xi_0(t))dx \right| \\ &\le \|F(t)-F_0(t)\|_{L^2_x}\|\Xi(t)\|_{L^2_x} + \|F_0(t)\|_{L^2_x}\|\Xi(t)-\Xi_0(t)\|_{L^2_x} \end{aligned}$$

and then by (7.40) and (7.35) deduce

$$|\tilde{\mathcal{H}}_b(t) - \tilde{\mathcal{H}}_{0,b}(t)| \le \|F(t)-F_0(t)\|_{L^2_x} 2M_0(t)^{\frac{1}{2}} + \frac{M_0(t)}{(2\pi)^{\frac{3}{2}}} \frac{a^{-b\beta}}{1-a^{-b\beta}}. \tag{7.42}$$

We need to estimate $\|F(t)-F_0(t)\|_{L^2_x}$, for which we can write

$$\|F(t)-F_0(t)\|_{L^2_x} \overset{(7.37)}{\le} \sum_{q\ge0} \|\nabla\times(-\Delta)^{-1}(\Xi_{q+1}-\Xi_q)(t)\|_{L^2_x}. \tag{7.43}$$

We apply (7.33) to split (7.43) into two parts:

$$\|(F-F_0)(t)\|_{L^2_x} \le \mathrm{VII}_1(t) + \mathrm{VII}_2(t), \tag{7.44}$$

where

$$\text{VII}_1(t) \triangleq \sum_{q\geq 0} \|\nabla\times(-\Delta)^{-1}(\Xi_l - \Xi_q)(t)\|_{L_x^2}, \tag{7.45a}$$

$$\text{VII}_2(t) \triangleq \sum_{q\geq 0} \|\nabla\times(-\Delta)^{-1} d_{q+1}(t)\|_{L_x^2}. \tag{7.45b}$$

To work on VII_1, we use the mean-zero property of $\Xi_l - \Xi_q$ and mollifier estimate (3.19), which leads to

$$\begin{aligned}\|\nabla\times(-\Delta)^{-1}(\Xi_l - \Xi_q)(t)\|_{L_x^2} &\overset{(7.33b)}{\lesssim} \|\Xi_q *_x \phi_l *_t \varphi_l - \Xi_q *_x \phi_l\|_{C_{t,x}} + \|\Xi_q *_x \phi_l - \Xi_q\|_{C_{t,x}} \\ &\lesssim l\|\Xi_q\|_{C_{t,x}^1} \overset{(7.28c)(7.34)}{\lesssim} M_0(t)^{\frac{1}{2}} \lambda_{q+1}^{-\frac{3\alpha}{2}+\frac{2}{b}}.\end{aligned} \tag{7.46}$$

We apply (7.46) to VII_1 of (7.45) and use the fact that $b > 39\alpha^{-1}$ due to (7.24), so that $-\frac{3\alpha}{2} + \frac{2}{b} < 0$, and hence $a^{b^{q+1}(-\frac{3\alpha}{2}+\frac{2}{b})} \lesssim a^{b(q+1)(-\frac{3\alpha}{2}+\frac{2}{b})}$. Consequently,

$$\text{VII}_1(t) \lesssim M_0(t)^{\frac{1}{2}} \sum_{q\geq 0} \lambda_{q+1}^{-\frac{3\alpha}{2}+\frac{2}{b}} \lesssim M_0(t)^{\frac{1}{2}} \sum_{q\geq 0} a^{b(q+1)(-\frac{3\alpha}{2}+\frac{2}{b})} \ll M_0(t)^{\frac{1}{2}} \frac{a^{-1}}{1-a^{-1}}. \tag{7.47}$$

The estimate of VII_2 is much more delicate and requires thorough analysis of d_{q+1}, and we refer the interested readers to [364, eqs. (67)–(73)], which leads to

$$\text{VII}_2(t) \ll M_0(t)^{\frac{1}{2}} \frac{a^{-1}}{1-a^{-1}}. \tag{7.48}$$

Applying (7.47) and (7.48) to (7.44) gives us

$$\|(F - F_0)(t)\|_{L^2} \overset{(7.44)}{\leq} \text{VII}_1(t) + \text{VII}_2(t) \overset{(7.47)(7.48)}{\ll} M_0(t)^{\frac{1}{2}} \frac{a^{-1}}{1-a^{-1}}. \tag{7.49}$$

Now we use the fact that $(2\pi)^3 + 1 < a^{b\beta}$ for $a \gg 1$, which implies $\frac{a^{-b\beta}}{1-a^{b\beta}} \leq \frac{1}{(2\pi)^3}$ to deduce

$$\begin{aligned}|\tilde{\mathcal{H}}_b(t) - \tilde{\mathcal{H}}_{0,b}(t)| &\overset{(7.42)}{\leq} \|F(t) - F_0(t)\|_{L^2} 2M_0(t)^{\frac{1}{2}} + \frac{M_0(t)}{(2\pi)^{\frac{3}{2}}} \frac{a^{-b\beta}}{1-a^{-b\beta}} \\ &\overset{(7.49)}{\leq} M_0(t) \frac{a^{-1}}{1-a^{-1}} + \frac{M_0(t)}{(2\pi)^{\frac{3}{2}}} \frac{a^{-b\beta}}{1-a^{-b\beta}} \\ &= \frac{M_0(t)}{(2\pi)^3} \left[(2\pi)^3 \frac{a^{-1}}{1-a^{-1}} + (2\pi)^{\frac{3}{2}} \frac{a^{-b\beta}}{1-a^{-b\beta}}\right].\end{aligned}$$

Then (7.40) allows us to continue by

$$\begin{aligned}|\tilde{\mathcal{H}}_b(t) - \tilde{\mathcal{H}}_{0,b}(t)| &\overset{(7.40)}{\leq} \tilde{\mathcal{H}}_{0,b}(t) \left[(2\pi)^3 \frac{a^{-1}}{1-a^{-1}} + (2\pi)^{\frac{3}{2}} \frac{a^{-b\beta}}{1-a^{-b\beta}}\right] \\ &\leq \tilde{\mathcal{H}}_{0,b}(t) \left[\frac{(2\pi)^3 a^{-1}}{1-a^{-1}} + (2\pi)^{\frac{3}{2}} \frac{1}{(2\pi)^3}\right] \leq \frac{2\tilde{\mathcal{H}}_{0,b}(t)}{(2\pi)^{\frac{3}{2}}}\end{aligned} \tag{7.50}$$

for a sufficiently large. Consequently,

$$\tilde{\mathcal{H}}_{0,b}(t) - \tilde{\mathcal{H}}_b(t) \le |\tilde{\mathcal{H}}_b(t) - \tilde{\mathcal{H}}_{0,b}(t)| \overset{(7.50)}{\le} \frac{2\tilde{\mathcal{H}}_{0,b}(t)}{(2\pi)^{\frac{3}{2}}},$$

and thus along with (7.40) we can deduce

$$0 < \tilde{\mathcal{H}}_{0,b}(t)\left(1 - \frac{2}{(2\pi)^{\frac{3}{2}}}\right) \le \tilde{\mathcal{H}}_b(t). \tag{7.51}$$

We can compute

$$\begin{aligned}
|\tilde{\mathcal{H}}_b(0)| &\le |\tilde{\mathcal{H}}_b(0) - \tilde{\mathcal{H}}_{0,b}(0)| + |\tilde{\mathcal{H}}_{0,b}(0)| \\
&\overset{(7.50)(7.40)}{\le} \tilde{\mathcal{H}}_{0,b}(0)\frac{2}{(2\pi)^{\frac{3}{2}}} + \frac{M_0(0)}{(2\pi)^3} \overset{(7.25)}{=} \tilde{\mathcal{H}}_{0,b}(0)\frac{2}{(2\pi)^{\frac{3}{2}}} + \frac{L^4}{(2\pi)^3} \\
&\overset{(7.36)(7.40)}{\le} \tilde{\mathcal{H}}_{0,b}(0)\frac{2}{(2\pi)^{\frac{3}{2}}} + \frac{\tilde{\mathcal{H}}_{0,b}(T)}{(2\pi)^{\frac{3}{2}}} \\
&\quad - \frac{3}{(2\pi)^{\frac{3}{2}} - 2}\left[2(2M_0(T)^{\frac{1}{2}} + L^{\frac{1}{4}})L^{\frac{1}{4}} + L^{\frac{1}{2}} + \left(\frac{(2\pi)^{\frac{3}{2}} - 2}{3}\right)TC_{G_2}\right] \\
&\le \frac{3\tilde{\mathcal{H}}_{0,b}(T)}{(2\pi)^{\frac{3}{2}}} - \frac{3}{(2\pi)^{\frac{3}{2}} - 2}\left[2(2M_0(T)^{\frac{1}{2}} + L^{\frac{1}{4}})L^{\frac{1}{4}} + L^{\frac{1}{2}} + \left(\frac{(2\pi)^{\frac{3}{2}} - 2}{3}\right)TC_{G_2}\right],
\end{aligned}$$

where in the last inequality, we used the fact that $\tilde{\mathcal{H}}_{0,b}(t) = \frac{M_0(t)}{(2\pi)^3}$ from (7.40b) is nondecreasing. We use (7.51) to continue by

$$\begin{aligned}
|\tilde{\mathcal{H}}_b(0)| &\overset{(7.51)}{\le} \left(\frac{3}{(2\pi)^{\frac{3}{2}} - 2}\right)\tilde{\mathcal{H}}_b(T) \\
&\quad - \frac{3}{(2\pi)^{\frac{3}{2}} - 2}\left[2(2M_0(T)^{\frac{1}{2}} + L^{\frac{1}{4}})L^{\frac{1}{4}} + L^{\frac{1}{2}} + \left(\frac{(2\pi)^{\frac{3}{2}} - 2}{3}\right)TC_{G_2}\right].
\end{aligned} \tag{7.52}$$

Finally, we are ready to conclude that the magnetic helicity grows at least twice from the initial time on $\{t \ge T\}$ as follows. First, since $G_2B_2(0) = 0$, we can write

$$\begin{aligned}
\int_{\mathbb{T}^3} A(0) \cdot b(0)dx &\overset{(6.3)(7.21)}{=} \int_{\mathbb{T}^3} \nabla \times (-\Delta)^{-1}[\Xi + G_2B_2](0) \cdot [\Xi + G_2B_2](0)dx \\
&= \int_{\mathbb{T}^3} \nabla \times (-\Delta)^{-1}\Xi(0) \cdot \Xi(0)dx \overset{(7.37)}{=} \int_{\mathbb{T}^3} F(0) \cdot \Xi(0)dx \overset{(7.39)}{=} \tilde{\mathcal{H}}_b(0).
\end{aligned} \tag{7.53}$$

Then, using (7.53) and (7.52), we get

$$\int_{\mathbb{T}^3} A(0)\cdot b(0)dx \overset{(7.53)}{=} \tilde{\mathcal{H}}_b(0) \overset{(7.52)}{\leq} \left(\frac{3}{(2\pi)^{\frac{3}{2}}-2}\right)\tilde{\mathcal{H}}_b(T)$$

$$-\frac{3}{(2\pi)^{\frac{3}{2}}-2}\left[2(2M_0(T)^{\frac{1}{2}}+L^{\frac{1}{4}})L^{\frac{1}{4}}+L^{\frac{1}{2}}+\left(\frac{(2\pi)^{\frac{3}{2}}-2}{3}\right)TC_{G_2}\right],$$

from which we can continue on $\{t \geq T\}$ by

$$\begin{aligned}
&\int_{\mathbb{T}^3} A(0)\cdot b(0)dx \\
&\overset{(7.37)}{\leq} \left(\frac{3}{(2\pi)^{\frac{3}{2}}-2}\right)\left[\int_{\mathbb{T}^3} \nabla\times(-\Delta)^{-1}\Xi(T)\cdot\Xi(T)dx\right] \\
&\quad -\frac{3}{(2\pi)^{\frac{3}{2}}-2}\left[2(2M_0(T)^{\frac{1}{2}}+L^{\frac{1}{4}})L^{\frac{1}{4}}+L^{\frac{1}{2}}+\left(\frac{(2\pi)^{\frac{3}{2}}-2}{3}\right)TC_{G_2}\right] \\
&\overset{(7.21)(6.3)}{\leq} \left(\frac{3}{(2\pi)^{\frac{3}{2}}-2}\right)\left[\int_{\mathbb{T}^3} A(T)\cdot b(T)dx\right. \\
&\quad \left. +2(\|\Xi\|_{C_TL_x^2}+\|G_2B_2\|_{C_TL_x^2})\|G_2B_2\|_{C_TL_x^2}+\|G_2B_2\|_{C_TL_x^2}^2\right] \\
&\quad -\frac{3}{(2\pi)^{\frac{3}{2}}-2}\left[2(2M_0(T)^{\frac{1}{2}}+L^{\frac{1}{4}})L^{\frac{1}{4}}+L^{\frac{1}{2}}+\left(\frac{(2\pi)^{\frac{3}{2}}-2}{3}\right)TC_{G_2}\right] \\
&\overset{(7.28b)(7.23)}{\leq} \left(\frac{3}{(2\pi)^{\frac{3}{2}}-2}\right)\left[\int_{\mathbb{T}^3} A(T)\cdot b(T)dx+2(2M_0(T)^{\frac{1}{2}}+L^{\frac{1}{4}})L^{\frac{1}{4}}+L^{\frac{1}{2}}\right] \\
&\quad -\frac{3}{(2\pi)^{\frac{3}{2}}-2}\left[2(2M_0(T)^{\frac{1}{2}}+L^{\frac{1}{4}})L^{\frac{1}{4}}+L^{\frac{1}{2}}+\left(\frac{(2\pi)^{\frac{3}{2}}-2}{3}\right)TC_{G_2}\right] \\
&= \left(\frac{3}{(2\pi)^{\frac{3}{2}}-2}\right)\int_{\mathbb{T}^3} A(T)\cdot b(T)dx - TC_{G_2}.
\end{aligned} \tag{7.54}$$

Thus by the definition of $\mathcal{H}_b$ from (3.39b) we have shown that

$$\begin{aligned}
\frac{(2\pi)^{\frac{3}{2}}-2}{3}[\mathcal{H}_b(0)+TC_{G_2}] &\overset{(3.39b)}{=} \frac{(2\pi)^{\frac{3}{2}}-2}{3}\left[\int_{\mathbb{T}^3} A(0)\cdot b(0)dx+TC_{G_2}\right] \\
&\overset{(7.54)}{\leq} \int_{\mathbb{T}^3} A(T)\cdot b(T)dx \overset{(3.39b)}{=} \mathcal{H}_b(T),
\end{aligned}$$

which implies the second inequality in (7.20a) because $\frac{(2\pi)^{\frac{3}{2}}-2}{3} \approx 4.6$.

7.3.2 Deducing estimates in the linear multiplicative case

This section is devoted to describe a nice idea from [186], which essentially reduces the workload of many estimates of the oscillation error in the case of linear multiplicative noise to that of additive noise. Specifically, let us consider the Navier–Stokes equations forced by linear multiplicative noise

$$du + (-\Delta u + \operatorname{div}(u \otimes u) + \nabla \pi)dt = u dB, \quad \nabla \cdot u = 0, \quad t > 0,$$

where B is an $\mathbb{R}$-valued Wiener process on $(\Omega, \mathcal{F}, \mathbf{P})$. If we define $\Upsilon(t) \triangleq e^{B(t)}$, $v \triangleq \Upsilon^{-1} u$ and apply the transformation described in Example 10, then we deduce the corresponding random PDE

$$\partial_t v + \frac{1}{2} v - \Delta v + \Upsilon \operatorname{div}(v \otimes v) + \Upsilon^{-1} \nabla \pi = 0, \quad \nabla \cdot v = 0, \quad t > 0.$$

Now if $a_{(\xi)}(\omega, t, x)$ was an amplitude function (e. g. (6.40)) studied in the case of additive noise, let us mollify Υ in time to obtain $\Upsilon_l \triangleq \Upsilon *_t \varphi_l$ and define the modified amplitude function and modified principal corrector (cf. (6.45a)) respectively by

$$\overline{a_{(\xi)}}(\omega, t, x) \triangleq \overline{a}_{\xi, q+1}(\omega, t, x) \triangleq \Upsilon_l^{-\frac{1}{2}} a_{(\xi)}(\omega, t, x), \tag{7.55a}$$

$$w_{q+1}^{(p)} \triangleq \sum_{\xi \in \Lambda} \overline{a_{(\xi)}} W_{(\xi)}. \tag{7.55b}$$

As we stressed, the essence of every Nash-type convex integration scheme is to cancel out the previous error by the nonlinear term of the principal correctors, and in this case,

$$\begin{aligned} \Upsilon_l w_{q+1}^{(p)} \otimes w_{q+1}^{(p)} &\overset{(7.55b)}{=} \Upsilon_l \Big(\sum_{\xi \in \Lambda} \overline{a_{(\xi)}} W_{(\xi)} \Big) \otimes \Big(\sum_{\xi' \in \Lambda} \overline{a_{(\xi')}} W_{(\xi')} \Big) \\ &\overset{(7.55a)}{=} \Big(\sum_{\xi \in \Lambda} a_{(\xi)} W_{(\xi)} \Big) \otimes \Big(\sum_{\xi' \in \Lambda} a_{(\xi')} W_{(\xi')} \Big), \end{aligned}$$

which reduces to the oscillation error in the case of additive noise. This is a nice trick, which was used, e. g., in [363, eq. (91)].

7.3.3 Lack of temporal differentiability of solutions to SPDEs

This section is devoted to briefly describe convex integration on the SQG equations. In Section 4.5.1, we reviewed the literature on the well-posedness of the equation in both deterministic and stochastic cases, especially the global existence of weak solutions and global regularity issues. Concerning convex integration, in the deterministic case, Córdoba, Faraco, and Gancedo [107] proved the nonuniqueness of weak solutions to the incompressible porous media equation; later, Isett and Vicol [200] extended such a result to a more general class of active scalars. Interestingly, [200] was forced to exclude the SQG equations due to the technical reason arising from the odd Fourier multiplier of its nonlinear term, specifically the Riesz transform $\mathcal{R}^{\perp}$ in (1.4). This was overcome by Buckmaster, Shkoller, and Vicol [46] by introducing the momentum SQG equations (1.6) (see also [199]). Later, Cheng, Kwon, and Li [86] constructed nontrivial stationary weak solutions to the SQG equations (1.4) by considering the equation solved by the potential temperature $f \triangleq \Lambda^{-1}\theta$, in contrast to the potential velocity $v = \Lambda^{-1}u$. More recent advances include the Onsager conjecture on the SQG equations in [119, 198].

We will come back in Section 9.2.2 to discuss convex integration on singular SQG equations; the current section is limited to observing some difficulties of adapting the first convex integration scheme to deterministic momentum SQG equations (1.6) from [46] to the stochastic case. The authors of [46] considered, for all $q \in \mathbb{N}_0$,

$$\partial_t v_q + (u_q \cdot \nabla) v_q - (\nabla v_q)^T \cdot u_q + \nabla \pi_q + \Lambda^{\gamma} v_q = \operatorname{div} \mathring{R}_q, \tag{7.56a}$$

$$\nabla \cdot v_q = 0, \quad u_q = \Lambda v_q, \tag{7.56b}$$

where $\mathring{R}_q$ is a trace-free symmetric matrix. Starting from $(v_0, \mathring{R}_0) = (0,0)$, they constructed a sequence of solutions $(v_q, \mathring{R}_q)_{q\in\mathbb{N}}$ and took its limit as $q \to \infty$. Specifically, one of the inductive hypothesis of [46] on (7.56) was

$$\|\mathring{R}_q\|_{C_{t,x}} \leq \lambda_{q+1}\delta_{q+1}, \tag{7.57}$$

where

$$\lambda_q \triangleq \lambda_0^q, \delta_q \triangleq \lambda_0^2 \lambda_q^{-2\beta} \quad \text{for } 1 \ll \lambda_0,\ 0 < \beta \ll 1. \tag{7.58}$$

Denoting their perturbation as $w_{q+1} \triangleq v_{q+1} - v_q$, this leads us to

$$\begin{aligned} \operatorname{div} \mathring{R}_{q+1} \overset{(7.56)}{=} & \partial_t w_{q+1} + [-(u_q \cdot \nabla) v_q + (\nabla v_q)^T \cdot u_q - \nabla \pi_q - \Lambda^{\gamma} v_q + \operatorname{div} \mathring{R}_q] \\ & - (u_{q+1} \cdot \nabla) v_{q+1} - (\nabla v_{q+1})^T \cdot u_{q+1} + \nabla \pi_{q+1} + \Lambda^{\gamma} v_{q+1}, \end{aligned}$$

which, after some computations, produces the key oscillation error:

$$\operatorname{div} R_{\mathrm{osc}} \triangleq \operatorname{div} \mathring{R}_q + w_{q+1} \cdot \nabla w_{q+1} - (\nabla w_{q+1})^T \cdot \Lambda w_{q+1}, \tag{7.59}$$

where the crucial step is to cancel the previous error $\mathring{R}_q$. For this purpose, they considered χ to be a certain cutoff function such that

$$\sum_{j\in\mathbb{Z}} \chi^2(t-j) = 1 \quad \forall t \in \mathbb{R}, \quad \chi_j(t) \triangleq \chi(\tau_{q+1}^{-1} t - j), \tag{7.60}$$

and $\mathring{R}_{q,j}$ that solves

$$(\partial_t + u_q \cdot \nabla)\mathring{R}_{q,j} = 0, \quad \mathring{R}_{q,j}(\tau_{q+1}j, x) = \mathring{R}_q(\tau_{q+1}j, x). \tag{7.61}$$

This allowed them to use (7.60) and write

$$\mathring{R}_q(t) = \sum_{j\in\mathbb{Z}} \chi_j^2(t)\mathring{R}_q(t) = \sum_{j\in\mathbb{Z}} \chi_j^2(t)(\mathring{R}_q - \mathring{R}_{q,j})(t) + \sum_{j\in\mathbb{Z}} \chi_j^2(t)\mathring{R}_{q,j}(t), \tag{7.62}$$

where because $\mathring{R}_q - \mathring{R}_{q,j}$ satisfies

$$(\partial_t + u_q \cdot \nabla)(\mathring{R}_q - \mathring{R}_{q,j}) = (\partial_t + u_q \cdot \nabla)\mathring{R}_q, \quad (\mathring{R}_q - \mathring{R}_{q,j})(\tau_{q+1}j, x) = 0, \tag{7.63}$$

according to (7.61), they can estimate via Lemma 6.2.4

$$\|(\mathring{R}_q - \mathring{R}_{q,j})(t)\|_{C_x} \le \int_{\tau_{q+1}j}^{t} \|(\partial_s + u_q \cdot \nabla)\mathring{R}_q(s)\|_{C_x} ds. \tag{7.64}$$

Thus (7.64) motivated [46] to include another inductive hypothesis on

$$\|(\partial_t + u_q \cdot \nabla)\mathring{R}_q\|_{C_{t,x}} \quad \forall q \in \mathbb{N}_0. \tag{7.65}$$

However, this would be a huge problem in the stochastic case. Naively, similarly to (4.20), we want to work on

$$dv + ((u \cdot \nabla)v - (\nabla v)^T \cdot u + \nabla\pi + \Lambda^\gamma v)dt = dW, \quad u = \Lambda v, \quad \nabla \cdot v = 0, \tag{7.66}$$

let z solve

$$dz + (\nabla\pi_z + \Lambda^\gamma z)dt = dW, \tag{7.67}$$

so that $y \triangleq v - z$, together with $p \triangleq \pi - \pi_z$, solves

$$\partial_t y + (\Lambda(y+z) \cdot \nabla)(y+z) - (\nabla(y+z))^T \cdot \Lambda(y+z) + \nabla p + \Lambda^\gamma y = 0,$$

and attempt convex integration scheme on

$$\partial_t y_q + (\Lambda(y_q+z) \cdot \nabla)(y_q+z) - (\nabla(y_q+z))^T \cdot \Lambda(y_q+z) + \nabla p_q + \Lambda^\gamma y_q = \operatorname{div} \mathring{R}_q.$$

Unfortunately if we do this, then $\mathring{R}_q$ cannot possibly be differentiable because z is not differentiable in time (recall Lemma 7.1.1), and thus an inductive hypothesis such as (7.65) is impossible in the stochastic case.

In [362] this obstacle, along with other issues, was overcome by first mollifying $\mathring{R}_q$ to transform the random PDEs of $(v_q, \mathring{R}_q)$ to $(v_l, \mathring{R}_l)$ so that such a term can be readily handled by

$$\|(\partial_s + u_q \cdot \nabla)\mathring{R}_l(s)\|_{C_x} \leq \|\partial_s \mathring{R}_l\|_{C_{s,x}} + \|u_q\|_{C_{s,x}} \|\nabla \mathring{R}_l\|_{C_{s,x}}.$$

However, if we deviate from the approach of [46] this way, then we realize that the commutator error we saw in (6.38a) (and handled in (6.50)) arises again, and this time, it involves z within the products of nonlinear terms, which are not necessarily small. For this reason, the author in [362] developed a new idea to define $z_q \triangleq z *_x \tilde{\phi}_q$ where $\tilde{\phi}_q$ is a smooth cut-off function with frequency supported in $B(0, \frac{\lambda_q}{4})$, so that

$$\operatorname{supp} \hat{z}_q \subset B\left(0, \frac{\lambda_q}{4}\right), \tag{7.68}$$

and thus $z_q \to z$ as $q \to \infty$ in distribution. The same trick of Fourier-truncating z at every level q was independently used in [186], although its first version of preprint did not use this trick. Consequently,

$$\begin{aligned}
&\partial_t y_q + (\Lambda(y_q + z) \cdot \nabla)(y_q + z) - (\nabla(y_q + z))^T \cdot \Lambda(y_q + z) \\
&\quad + \nabla p_q + \Lambda^{\gamma_1} y_q = \operatorname{div} \mathring{R}_q, \quad \nabla \cdot y_q = 0, \quad t \in [t_q, T_L],
\end{aligned}$$

turns to

$$\begin{aligned}
&\partial_t y_l + (\Lambda(y_l + z_l) \cdot \nabla)(y_l + z_l) - (\nabla(y_l + z_l))^T \cdot \Lambda^{2-\gamma_2}(y_l + z_l) \\
&\quad + \nabla p_l + \Lambda^{\gamma_1} y_l = \operatorname{div}(\mathring{R}_l + R_{\text{Com1}}), \quad t \in [t_{q+1}, T_L]
\end{aligned} \tag{7.69a}$$

(recall t_q from (6.27)), where

$$\begin{aligned}
\operatorname{div} R_{\text{Com1}} \triangleq\ & \Lambda(y_l + z_l)^{\perp} \nabla^{\perp} \cdot (y_l + z_l) \\
& - [\Lambda^{2-\gamma_2}(y_q + z)^{\perp} \nabla^{\perp} \cdot (y_q + z)] *_x \phi_l *_t \varphi_l.
\end{aligned}$$

We refer to [362, eqs. (240)–(244)] on how to handle R_{Com1}, where an analogue of Lemma 7.1.1 is required. We also mention that convex integration on (7.69a) leads to an additional stochastic error of the form

$$\begin{aligned}
\operatorname{div} R_{\text{Com2}} \triangleq\ & \Lambda y_{q+1} \cdot \nabla(z_{q+1} - z) + \Lambda(z_{q+1} - z) \cdot \nabla y_{q+1} + (\nabla y_l)^T \cdot \Lambda(z - z_{q+1}) \\
& + (\nabla(z - z_{q+1}))^T \cdot \Lambda y_l + (\nabla \Lambda(z_{q+1} - z))^T \cdot w_{q+1} \\
& - \Lambda w_{q+1}^{\perp} \nabla^{\perp} \cdot (z_l - z) + (\nabla(z - z_{q+1}))^T \cdot \Lambda w_{q+1} \\
& + \Lambda y_l^{\perp} \nabla^{\perp} \cdot (z - z_l) + \Lambda(z - z_l) \cdot \nabla w_{q+1} + \Lambda(z - z_l)^{\perp} \nabla^{\perp} \cdot y_l
\end{aligned}$$

$$
\begin{aligned}
&+ \Lambda(z_{q+1} - z)^{\perp}\nabla^{\perp} \cdot z_{q+1} + \Lambda(z - z_l)^{\perp}\nabla^{\perp} \cdot z_{q+1} \\
&+ \Lambda z_l^{\perp}\nabla^{\perp} \cdot (z_{q+1} - z) + \Lambda z_l^{\perp}\nabla^{\perp} \cdot (z - z_l).
\end{aligned}
$$

We omit further details and refer the interested readers to [362] and [18, 327, 363].

Remark 7.3.6. The author is not aware of any convex integration on the generalized SQG equations, more specifically, at the level of momentum SQG equations with $v = \Lambda^{-1}u$ or $f = \Lambda^{-1}\theta$. We recall Remark 4.5.1, how the generalized Calderón commutator estimates (Proposition 3.8.4(b)) must play a crucial role. In fact, in [362], the author was able to generalize all the computations with $u = \Lambda^{2-\gamma_2}$, but only at the last step of taking the limit via Proposition 3.8.4(b), it became necessary to assume that $\gamma_2 = 1$. Specifically, if v has regularity only in $\dot{H}_x^{1-\frac{\gamma_2}{2}}$, then we need to estimate

$$
\begin{aligned}
\frac{1}{2}\langle \partial_i v_j, [\Lambda^{2-\gamma_2}, \phi_i] v_j \rangle &\lesssim \|\partial_i v^j\|_{\dot{H}_x^{-\frac{\gamma_2}{2}}} \|[\Lambda^{2-\gamma_2}, \phi_i] v_j\|_{\dot{H}_x^{\frac{\gamma_2}{2}}} \\
&\lesssim \|\partial_i v^j\|_{\dot{H}^{-\frac{\gamma_2}{2}}} (\|\phi\|_{C_x^{1,2-\gamma_2}} + \|\Lambda^{2-\gamma_2}\phi\|_{W_x^{1,\infty}}) \|v\|_{\dot{H}_x^{\frac{\gamma_2}{2}}},
\end{aligned}
$$

where the second inequality is due to Proposition 3.8.4(b). Then the problem is that we need $\frac{\gamma_2}{2} \leq 1 - \frac{\gamma_2}{2}$ for $\|v\|_{\dot{H}_x^{\frac{\gamma_2}{2}}}$ to be well-defined, and this is the case if and only if $\gamma_2 \leq 1$. Yet, Proposition 3.8.4(b) extended Proposition 3.8.4 (a) from $\gamma_2 = 1$ to $\gamma_2 \in (1,2)$, not to $\gamma_2 < 1$.

Interestingly, for the convex integration solution, the solution only has the regularity of $C([0,T]; \dot{H}^{1-\frac{\gamma_2}{2}})$. However, the solution we constructed via Galerkin approximation in Theorem 4.5.5 had the regularity of $L^2([0,T]; \dot{H}^{1-\frac{\gamma_2}{2}+\frac{\gamma_1}{2}}(\mathbb{T}^2))$ (see Definition 4.5.4 and (4.24)). This means that the estimate can be closed if $\frac{\gamma_2}{2} \leq 1 - \frac{\gamma_2}{2} + \frac{\gamma_1}{2}$, i. e., $\gamma_2 \leq 1 + \frac{\gamma_1}{2}$, which was precisely the hypothesis of Theorem 4.5.5. We refer to [362] for more detail and hope that convex integration becomes possible for generalized momentum SQG equations in the future.

7.3.4 Mismatch of transport error

In this section, we give another example of difficulty that arises when implementing deterministic convex integration technique to random PDEs, which basically is a stochastic analogue of Lemma 6.2.4.

In [358], convex integration technique was attempted on the following generalized Navier–Stokes equations forced by additive noise (cf. (5.1)):

$$
du + \nu\Lambda^m u dt + \operatorname{div}(u \otimes u) dt + \nabla\pi dt = dB, \quad \nabla \cdot u = 0, \ t > 0, \ x \in \mathbb{T}^3, \tag{7.70}
$$

for $m \in (0,1)$, where B is a GG^*-Wiener process under the hypothesis of $\operatorname{Tr}((-\Delta)^{\frac{5}{2}-2m+2\sigma} GG^*) < \infty$ for some $\sigma > 0$. This is a quasilinear case, and a convex integration scheme of [96] or [48, Section 5] on the Euler equations was found to be suitable. Thus we can consider

$$dz + \Lambda^m z dt + \nabla \pi_z dt = dB, \quad \nabla \cdot z = 0 \text{ for } t > 0, \; z(0,x) = 0,$$

and apply the convex integration on the resulting random PDEs, namely

$$\partial_t v_q + \upsilon \Lambda^m v_q + \operatorname{div}((v_q + z) \otimes (v_q + z)) + \nabla p_q = \operatorname{div} \mathring{R}_q, \quad \nabla \cdot v_q = 0$$

under the inductive hypothesis of

$$\|v_q\|_{C_{t,x,q}} \le M_0(t)^{\frac{1}{2}} \left(1 + \sum_{1 \le \iota \le q} \delta_\iota^{\frac{1}{2}}\right) \le 2M_0(t)^{\frac{1}{2}}, \tag{7.71a}$$

$$\|v_q\|_{C^1_{t,x,q}} \le M_0(t) \delta_q^{\frac{1}{2}} \lambda_q, \tag{7.71b}$$

$$\|\mathring{R}_q\|_{C_{t,x,q}} \le c_R M_0(t) \delta_{q+1}, \tag{7.71c}$$

where λ_q, δ_q, and $M_0(t)$ are identical to (7.24)–(7.25), and both L and c_R are universal constants from [358].

Given any $\zeta \in \mathbb{S}^2 \cap \mathbb{Q}^3$, let $A_\zeta \in \mathbb{S}^2$ satisfy $A_\zeta \cdot \zeta = 0$ and $A_{-\zeta} = A_\zeta$. Then we define $B_\zeta \triangleq 2^{-\frac{1}{2}}(A_\zeta + i\zeta \times A_\zeta) \in \mathbb{C}^3$. For any $\lambda \in \mathbb{Z}$ such that $\lambda\zeta \in \mathbb{Z}^3$, we define

$$W_{(\zeta)}(x) \triangleq W_{\zeta,\lambda}(x) \triangleq B_\zeta e^{i\lambda\zeta\cdot x}. \tag{7.72}$$

To attain acceptable estimates for transport and corrector errors at step $q+1$ on the time interval $[t_{q+1}, T]$, we split $[t_{q+1}, T]$ into an interval of size $l \triangleq \lambda_q^{-(\frac{1+b}{2})}$ and define $\Phi_j : [t_{q+1}, T] \times \mathbb{R}^3 \to \mathbb{R}^3$ for $j \in \{\lfloor t_{q+1} l^{-1} \rfloor, \ldots, \lceil l^{-1} T \rceil\}$ as a $\mathbb{T}^3$-periodic solution to

$$(\partial_t + (v_l + z_l) \cdot \nabla)\Phi_j = 0, \tag{7.73a}$$

$$\Phi_j(jl, x) = x \tag{7.73b}$$

(cf. (6.20)). This idea was independently suggested by Rehmeier and Schenke [288]. We now define the principal corrector as

$$w_{q+1}^{(p)}(t,x) \triangleq \sum_j \sum_{\zeta \in \Lambda_j} w_{(\zeta)}^{(p)}(t,x), \tag{7.74}$$

where Λ_j is an appropriate analogue of the set Λ from Lemma 6.2.1, and

$$w_{(\zeta)}^{(p)}(t,x) \triangleq w_{q+1,j,\zeta}^{(p)}(t,x) \triangleq a_{q+1,j,\zeta}(t,x) W_{\zeta,\lambda_{q+1}}(\Phi_j(t,x)), \tag{7.75}$$

where $a_{q+1,j,\zeta}$ is an appropriate amplitude function (see [358, eq. (80)] for details). We also define the incompressibility corrector as

$$w_{q+1}^{(c)}(t,x) \triangleq \sum_j \sum_{\zeta\in\Lambda_j} w_{(\zeta)}^{(c)}(t,x) \quad \text{and} \quad w_{q+1} \triangleq w_{q+1}^{(p)} + w_{q+1}^{(c)}, \tag{7.76}$$

where

$$w_{(\zeta)}^{(c)}(t,x) \triangleq (\lambda_{q+1}^{-1}\nabla a_{(\zeta)} + ia_{(\zeta)}\zeta \cdot (\nabla\Phi_j - \mathrm{Id}))(t,x) \times W_{(\zeta)}(\Phi_j(t,x)). \tag{7.77}$$

This leads to, in [358, eqs. (108)–(109)],

$$\begin{aligned}
&\operatorname{div} \mathring{R}_{q+1} - \nabla\pi_{q+1} \\
&= -\nabla\pi_l + \operatorname{div}\mathring{R}_l + \operatorname{div} R_{\mathrm{com1}} + \partial_t w_{q+1}^{(p)} + \partial_t w_{q+1}^{(c)} + \Lambda^m w_{q+1} \\
&\quad + \operatorname{div}(v_l \otimes w_{q+1}^{(p)} + v_l \otimes w_{q+1}^{(c)} + w_{q+1} \otimes v_l + w_{q+1}^{(p)} \otimes w_{q+1}^{(p)} \\
&\quad + w_{q+1}^{(p)} \otimes w_{q+1}^{(c)} + w_{q+1}^{(c)} \otimes w_{q+1} + v_{q+1} \otimes (z_{q+1} - z_l) + w_{q+1} \otimes z_l \\
&\quad + (z_{q+1} - z_l) \otimes v_{q+1} + z_l \otimes w_{q+1} + z_{q+1} \otimes (z_{q+1} - z_l) + (z_{q+1} - z_l) \otimes z_l).
\end{aligned} \tag{7.78}$$

A particularly difficult term is $\operatorname{div}(w_{q+1} \otimes z_l) = (z_l \cdot \nabla)w_{q+1}$, which is absent in the deterministic case. The derivative on w_{q+1} must be removed, and the first naive approach of writing

$$\operatorname{div}(z_l \otimes w_{q+1}) = \operatorname{div}(z_l \mathring{\otimes} w_{q+1}) + \nabla\left(\frac{1}{3} z_l \cdot w_{q+1}\right)$$

and estimating $\|z_l \mathring{\otimes} w_{q+1}\|_{C_{t,x,q+1}} \leq \|z_l\|_{C_{t,x,q+1}} \|w_{q+1}\|_{C_{t,x,q+1}}$ is not good enough because this cannot be made smaller than $c_R M_0(t)\delta_{q+2}$, which is needed to attain (7.71c) at level $q+1$.

The unique idea of [358] is to rely on (7.76) to rewrite

$$\operatorname{div}(z_l \otimes w_{q+1}) = (z_l \cdot \nabla)w_{q+1}^{(p)} + (z_l \cdot \nabla)w_{q+1}^{(c)}$$

and take the first term within the transport error and the second within the corrector error so that

$$\begin{aligned}
&\operatorname{div} \mathring{R}_{q+1} - \nabla\pi_{q+1} \\
&= \underbrace{\Lambda^m w_{q+1} + (w_{q+1} \cdot \nabla)z_l}_{\operatorname{div} R_{\mathrm{line}}} + \underbrace{(\partial_t + (v_l + z_l) \cdot \nabla)w_{q+1}^{(p)}}_{\operatorname{div} R_{\mathrm{tran}}} + \underbrace{\operatorname{div}(w_{q+1}^{(p)} \otimes w_{q+1}^{(p)} + \mathring{R}_l)}_{\operatorname{div} R_{\mathrm{osc}} + \nabla\pi_{\mathrm{osc}}} \\
&\quad + \underbrace{(w_{q+1} \cdot \nabla)v_l}_{\operatorname{div} R_{\mathrm{Nash}}} + \underbrace{(\partial_t + (v_l + z_l) \cdot \nabla)w_{q+1}^{(c)} + \operatorname{div}(w_{q+1}^{(c)} \otimes w_{q+1} + w_{q+1}^{(p)} \otimes w_{q+1}^{(c)})}_{\operatorname{div} R_{\mathrm{cor}} + \nabla\pi_{\mathrm{cor}}} \\
&\quad + \operatorname{div} R_{\mathrm{com1}} - \nabla\pi_l + \underbrace{\operatorname{div}(v_{q+1} \otimes (z_{q+1} - z_l) + (z_{q+1} - z_l) \otimes v_{q+1}}_{\operatorname{div} R_{\mathrm{com2}} + \nabla\pi_{\mathrm{com2}}} \\
&\quad \underbrace{+ z_{q+1} \otimes (z_{q+1} - z_l) + (z_{q+1} - z_l) \otimes z_l)}_{\operatorname{div} R_{\mathrm{com2}} + \nabla\pi_{\mathrm{com2}} \text{ continued}}.
\end{aligned} \tag{7.79}$$

The concern is that a gradient ∇ on

$$\nabla w_{q+1}^{(p)} \overset{(7.75)}{=} \nabla(a_{q+1,j,\zeta} W_{(\zeta)}(\Phi_j(t,x))$$

may fall on $W_{(\zeta)}(\Phi_j(t,x)) \overset{(7.72)}{=} B_\zeta e^{i\lambda_{q+1}\zeta\cdot\Phi_j(t,x)}$ and a chain rule may bring a singular term λ_{q+1}. However, our (7.73a) is designed so that the derivative falls only on the amplitude function in R_{tran} as follows:

$$\begin{aligned}
&(\partial_t + (v_l + z_l)\cdot\nabla) w_{q+1}^{(p)}(t,x) \\
\overset{(7.75)(7.74)}{=}\ & (\partial_t + (v_l + z_l)\cdot\nabla)\sum_j \sum_{\zeta\in\Lambda_j} a_{(\zeta)}(t,x) W_{(\zeta)}(\Phi_j(t,x)) \\
=\ & \sum_j \sum_{\zeta\in\Lambda_j} [\partial_t a_{(\zeta)} + (v_l + z_l)\cdot\nabla a_{(\zeta)}] W_{(\zeta)}(\Phi_j(t,x)) \\
& + a_{(\zeta)}(t,x)\nabla W_{(\zeta)}(\Phi_j(t,x))\cdot[\partial_t\Phi_j + (v_l + z_l)\cdot\nabla\Phi_j] \\
\overset{(7.73a)}{=}\ & \sum_j \sum_{\zeta\in\Lambda_j} [\partial_t a_{(\zeta)} + (v_l + z_l)\cdot\nabla a_{(\zeta)}] W_{(\zeta)}\circ\Phi_j.
\end{aligned}$$

Similarly, for $(\partial_t + (v_l + z_l)\cdot\nabla) w_{q+1}^{(c)}$,

$$\partial_t W_{(\zeta)}(\Phi_j(t,x)) = \partial_t(B_\zeta e^{i\lambda_{q+1}\zeta\cdot\Phi_j(t,x)}) = i\lambda_{q+1}\zeta\cdot W_{(\zeta)}|_{\Phi_j(t,x)}\partial_t\Phi_j(t,x)$$

due to (7.72) and

$$\begin{aligned}
(v_l + z_l)\cdot\nabla W_{(\zeta)}(\Phi_j(t,x)) &= (v_l + z_l)\cdot\nabla(B_\zeta e^{i\lambda_{q+1}\zeta\cdot\Phi_j(t,x)}) \\
&= (v_l + z_l)\cdot\nabla\Phi_j(t,x)\cdot i\lambda_{q+1}\zeta W_{(\zeta)}|_{\Phi_j(t,x)},
\end{aligned}$$

also due to (7.72), so that

$$\begin{aligned}
&(\partial_t + (v_l + z_l)\cdot\nabla) w_{q+1}^{(c)} \\
\overset{(7.76)}{=}\ & (\partial_t + (v_l + z_l)\cdot\nabla)\sum_j \sum_{\zeta\in\Lambda_j} w_{(\zeta)}^{(c)} \\
\overset{(7.77)}{=}\ & (\partial_t + (v_l + z_l)\cdot\nabla)\sum_j \sum_{\zeta\in\Lambda_j} [(\lambda_{q+1}^{-1}\nabla a_{(\zeta)} + i a_{(\zeta)}\zeta\cdot(\nabla\Phi_j - \mathrm{Id}))\times W_{(\zeta)}(\Phi_j)] \\
=\ & \sum_j \sum_{\zeta\in\Lambda_j} \partial_t(\lambda_{q+1}^{-1}\nabla a_{(\zeta)} + i a_{(\zeta)}\zeta\cdot(\nabla\Phi_j - \mathrm{Id}))\times W_{(\zeta)}(\Phi_j) \\
& + (v_l + z_l)\cdot\nabla(\lambda_{q+1}^{-1}\nabla a_{(\zeta)} + i a_{(\zeta)}\zeta\cdot(\nabla\Phi_j - \mathrm{Id}))\times W_{(\zeta)}(\Phi_j) \\
& + (\lambda_{q+1}^{-1}\nabla a_{(\zeta)} + i a_{(\zeta)}\zeta\cdot(\nabla\Phi_j - \mathrm{Id}))\times i\lambda_{q+1}\zeta\cdot W_{(\zeta)}|_{\Phi_j}[\partial_t\Phi_j + (v_l + z_l)\cdot\nabla\Phi_j] \\
\overset{(7.73a)}{=}\ & \sum_j \sum_{\zeta\in\Lambda_j} (\partial_t + (v_l + z_l)\cdot\nabla)(\lambda_{q+1}^{-1}\nabla a_{(\zeta)} + i a_{(\zeta)}\zeta\cdot(\nabla\Phi_j - \mathrm{Id}))\times W_{(\zeta)}(\Phi_j).
\end{aligned}$$

On the other hand, having modified (7.73) from the deterministic analogues, we would worry if they can give sufficient estimates. It turns out that they do, as the following proposition shows in comparison with Lemma 6.2.4.

Proposition 7.3.7 ([358, Proposition 4.9]). *For all $j \in \{\lfloor t_{q+1}l^{-1}\rfloor, \ldots, \lceil l^{-1}T\rceil\}$ and $t \in [l(j-1), l(j+1)]$ with appropriate modification in case $j = \lfloor t_{q+1}l^{-1}\rfloor$ and $\lceil l^{-1}T\rceil$,*

$$\begin{aligned}
\|\nabla\Phi_j(t) - \mathrm{Id}\|_{C_x} &\lesssim l\delta_q^{\frac{1}{2}}\lambda_q M_0(t) \ll 1,\\
\frac{1}{2} &\leq |\nabla\Phi_j(t,x)| \leq 2 \quad \forall x \in \mathbb{T}^3 \quad \text{and} \quad \|\Phi_j(t)\|_{C_x^1} \lesssim 1,\\
\|\partial_t\Phi_j(t)\|_{C_x} &\lesssim M_0(t)^{\frac{1}{2}},\\
\|\nabla\Phi_j(t)\|_{C_x^N} &\lesssim l^{-N+1}M_0(t)\delta_q^{\frac{1}{2}}\lambda_q \quad \forall N \in \mathbb{N},\\
\|\partial_t\nabla\Phi_j(t)\|_{C_x^N} &\lesssim l^{-N}M_0(t)^{\frac{3}{2}}\delta_q^{\frac{1}{2}}\lambda_q \quad \forall N \in \mathbb{N}_0.
\end{aligned}$$

We note that the linear multiplicative noise analogue of (7.73),

$$\begin{aligned}
(\partial_t + (\Upsilon_l v_l)\cdot\nabla)\Phi_j &= 0,\\
\Phi_j(jl, x) &= x,
\end{aligned}$$

appeared in [358, eq. (195)] and found further applications later, e. g., in [363, eq. (43)].

7.4 Recent developments

There have been many rapid developments with many novel ideas on the convex integration on SPDEs with white-in-time noise in the past decade. Particularly, we mention that Hofmanová, Zhu, and Zhu [188] applied convex integration on the 3D Navier–Stokes equations with prescribed energy or prescribed initial data and were able to extend the solution to be global-in-time by stopping-time argument at a regularity weaker than a classical Leray–Hopf weak solution. Although in the classical approach, to extend a local-in-time solution to a global-in-time one, the additional time interval of existence depends on the terminal value of the solution, the case of convex integration depends only on stopping time and hence is easy to extend. Additionally, we discussed in Section 4.5.3 the work of Flandoli and Romito [144] concerning the existence of a Markov selection for the 3D Navier–Stokes equations forced by additive white-in-time noise. In contrast, Hofmanová, Zhu, and Zhu [188] were able to prove the nonuniqueness of Markov selections via convex integration, although at a lower regularity level (see also [327] in the case of momentum SQG equations). Among additional works, we mention [354] on the 2D Navier–Stokes equations with $(-\Delta)^m$ for $m < 1$, [187] on the 3D Euler equations with additive noise (also [250]), [25] on the 3D Navier–Stokes equations with linear multiplicative noise and prescribed energy, and [192, 39] on the nonuniqueness of a Leray–Hopf weak solution in the case of 3D stochastic Navier–Stokes equations with an additional nonzero force. We cannot describe them all here, so we refer the interested readers to [26, 56, 58, 81, 82, 89, 143, 182, 184, 185, 191, 216, 222, 232, 243–246, 249, 266, 278].

8 Preliminaries on singular SPDEs

This chapter is devoted to preliminaries on singular SPDEs. As we mentioned in Preface, we intentionally split this from Chapter 4 to not spend too much before initiating research discussions starting in Chapter 5. Section 8.1 starts with discussions of ill-defined products considering Bony's paraproducts, followed by Section 8.2 with Young's integral and rough path theory. Section 8.3 introduces Wick products, whereas Section 8.4 describes renormalization constants and Da Prato–Debussche tricks (recall Example 9) introduced in Section 4.5.3, now applied on singular SPDEs. Section 8.5 describes the theories of regularity structures (RS) and paracontrolled distributions on the MHD system as an example. Finally, Section 8.6 concludes with more recent developments.

8.1 Ill-defined products

As we described in Section 4.2, [233] in 1957 pioneered the theory of hydrodynamic fluctuations, which already incorporated STWN. Many other models with STWN appeared in the literature, e. g., ferromagnet model [257], KPZ equation [210], MHD system [54], Navier–Stokes equations [148, 337], Φ^4 model [279, 152], and the Rayleigh–Bénard equations [2, 155, 181, 311, 372]. As we will discuss in detail, solution theories to these singular SPDEs are not well understood due to the roughness of the force.

Let us start preparation for the definition of parabolically scaled Besov–Hölder spaces of space-time distributions to be stated in Definition 8.1.1. For clarity, we assume a parabolic scaling of $\partial_t - \Delta$, but the argument can be readily generalized to $\partial_t + \Lambda^m$ for $m \geq 0$ ([24] and Section 9.2.2). We first define the corresponding scaling (see [167, p. 295])

$$\mathfrak{s} \triangleq (\mathfrak{s}_0, \mathfrak{s}_1, \dots, \mathfrak{s}_d) = (2, \underbrace{1, \dots, 1}_{d \text{ times}}). \tag{8.1}$$

We often write in this section $z = (t, x)$ and $\bar{z} = (\bar{t}, \bar{x})$. We associate to it a metric and scaling dimension defined respectively by

$$d_{\mathfrak{s}}((t,x),(s,y)) \triangleq |t-s|^{\frac{1}{\mathfrak{s}_0}} + \sum_{i=1}^{d} |x_i - y_i|^{\frac{1}{\mathfrak{s}_i}} \quad \text{and} \quad |\mathfrak{s}| \triangleq 2 + d \tag{8.2}$$

(see [167, p. 298]). We define $\|\cdot\|_{\mathfrak{s}}$ by

$$\|(t,x) - (s,y)\|_{\mathfrak{s}} \triangleq d_{\mathfrak{s}}((t,x),(s,y)), \tag{8.3}$$

although it is not a norm in general (see [167, p. 298] and [24, p. 336]). For a multiindex $k = (k_0, \dots, k_d)$, we define the scaled degree as

$$|k|_{\mathfrak{s}} \triangleq \sum_{i=0}^{d} \mathfrak{s}_i k_i. \tag{8.4}$$

https://doi.org/10.1515/9783111349985-009

We denote by $B_{\mathfrak{s}}(0,1)$ the ball of radius 1 in the distance $d_{\mathfrak{s}}$ centered at the origin. Additionally, we define the scaling maps (see [167, p. 300]) $\mathcal{S}_{\mathfrak{s}}^{\delta} : \mathbb{R}^{d+1} \to \mathbb{R}^{d+1}$ for $\delta > 0$ by

$$(\mathcal{S}_{\mathfrak{s},\bar{z}}^{\delta}\varphi)(z) \triangleq \delta^{-|\mathfrak{s}|}\varphi(\mathcal{S}_{\mathfrak{s}}^{\delta}(z-\bar{z})), \tag{8.5a}$$

$$\mathcal{S}_{\mathfrak{s}}^{\delta}((t,x)) \triangleq (\delta^{-\mathfrak{s}_0}t, \delta^{-\mathfrak{s}_1}x_1, \ldots, \delta^{-\mathfrak{s}_d}x_d). \tag{8.5b}$$

Now we recall the definitions of Hölder spaces C^r (Definition 2.3.2); in particular, C_c^r refers to an additional assumption of compact support. In the study of singular SPDEs, we will need to measure the regularity of distributions and rely on Besov–Hölder spaces $\mathcal{C}^\alpha$ from (3.50) hereafter.

Definition 8.1.1 ([167, Definition 3.7]). Let $\alpha < 0$ and $r = -\lceil \alpha \rceil > 0$. We say that a tempered distribution ξ belongs to $\mathcal{C}_{\mathfrak{s}}^\alpha$ if it belongs to the dual of $C_c^r(\mathbb{R}\times\mathbb{T}^d)$ and for every compact set $K \subset \mathbb{R}\times\mathbb{T}^d$, there exists a constant C such that

$$\langle \xi, \mathcal{S}_{\mathfrak{s},\bar{z}}^{\delta}\eta\rangle \leq C\delta^\alpha \quad \forall \eta \in \{\psi \in C^r : \|\psi\|_{C^r} \leq 1, \operatorname{supp}\psi \subset B_{\mathfrak{s}}(0,1)\}$$

for all $\delta \in (0,1]$ and $\bar{z} \in K$.

Remark 8.1.2.

(a) As stated in [167, Remark 3.8], $\mathcal{C}_{\mathfrak{s}}^\alpha$ is essentially $B_{\infty,\infty}^\alpha$ except that the definition is more local than global, and it allows for non-Euclidean scalings. In more detail, because $\|(t,x)\|_{\mathfrak{s}} \approx |t|^{\frac{1}{2}} + |x|$ due to (8.3) and (8.2), $f \in \mathcal{C}_{\mathfrak{s}}^\alpha$ if and only if $\|\Delta_j^{\mathfrak{s}} f\|_{L_{t,x}^\infty} \lesssim 2^{-j\alpha}$ for all $j \geq -1$, where $\Delta_j^{\mathfrak{s}}$ are parabolic dyadic projections onto $\{(\tau,\xi) : |\tau|^{\frac{1}{2}} + |\xi| \approx 2^j\}$.

(b) For our purpose, we only defined $\mathcal{C}_{\mathfrak{s}}^\alpha$ for $\alpha < 0$ in Definition 8.1.1. The definition in case $\alpha > 0$ can be found in [167, Definition 2.14]. Moreover, as stated in [167, Remark 1.13], when $\alpha > 0$, the elements in $\mathcal{C}_{\mathfrak{s}}^\alpha$ are α-Hölder continuous in space and $\frac{\alpha}{2}$-Hölder continuous in time, which can be readily understood as follows. We recall from Proposition 3.2.8 that if $\alpha \in \mathbb{R}_+ \setminus \mathbb{N}$, then $B_{\infty,\infty}^\alpha = \mathcal{C}^{\lfloor\alpha\rfloor, \alpha-\lfloor\alpha\rfloor}$. Thus $f \in \mathcal{C}_{\mathfrak{s}}^\alpha$ if and only if by Remark 8.1.2(a), for $z = (t,x)$ and $\bar{z} = (\bar{t},\bar{x})$,

$$|f(z) - f(\bar{z})| \lesssim \|z-\bar{z}\|_{\mathfrak{s}}^\alpha \overset{(8.2)(8.3)}{\approx} \left(|t-\bar{t}|^{\frac{1}{2}} + \sum_{i=1}^d |x_i - \bar{x}_i|\right)^\alpha \approx |t-\bar{t}|^{\frac{\alpha}{2}} + |x-\bar{x}|^\alpha.$$

This implies that we can take $t = \bar{t}$ to deduce

$$|f(t,x) - f(t,\bar{x})| \lesssim |x-\bar{x}|^\alpha$$

so that $f(t,\cdot) \in \mathcal{C}^\alpha(\mathbb{T}^d)$, and we can take $x = \bar{x}$ to deduce

$$|f(t,x) - f(\bar{t},x)| \lesssim |t-\bar{t}|^{\frac{\alpha}{2}},$$

so that $f(\cdot, x) \in \mathcal{C}^{\frac{\alpha}{2}}(\mathbb{R})$. Evidently, in the case $\alpha < 0$, $\mathcal{C}^{\alpha}_{\mathfrak{s}}$ consists of distributions with norms defined only through parabolic dyadic blocks, and thus pointwise computations above do not apply.

Lemma 8.1.3 ([167, Lemma 10.2 and Proposition 3.20] and [24, Lemma 4.1]). *Under the scaling of* (8.1), *the STWN* ξ *satisfies*

$$\xi \in \mathcal{C}^{\alpha}_{\mathfrak{s}}(\mathbb{R} \times \mathbb{T}^d) \quad \textit{for } \alpha < -\frac{|\mathfrak{s}|}{2} \quad \mathbb{P}\textit{-a. s.} \tag{8.6}$$

Let us now state further consequences of Bony's paraproducts introduced in Section 3.6.

Lemma 8.1.4 (Bony's estimates).
(a) ([269, *Proposition* A.7]) *Let* $\alpha, \beta \in \mathbb{R}$ *and* $p, p_1, p_2, q \in [1, \infty]$, *where* $\frac{1}{p} = \frac{1}{p_1} + \frac{1}{p_2}$. *Then*

$$\begin{aligned}
\|f \prec g\|_{B^{\alpha+\beta}_{p,q}} &\lesssim_{\alpha,\beta} \|f\|_{B^{\alpha}_{p_1,q}} \|g\|_{B^{\beta}_{p_2,q}} \quad \forall f \in B^{\alpha}_{p_1,q},\ g \in B^{\beta}_{p_2,q} \ \textit{if } \alpha < 0, \\
\|f \prec g\|_{B^{\beta}_{p,q}} &\lesssim_{\beta} \|f\|_{L^{p_1}} \|g\|_{B^{\beta}_{p_2,q}} \quad \forall f \in L^{p_1},\ g \in B^{\beta}_{p_2,q}, \\
\|f \circ g\|_{B^{\alpha+\beta}_{p,q}} &\lesssim_{\alpha,\beta} \|f\|_{B^{\alpha}_{p_1,q}} \|g\|_{B^{\beta}_{p_2,q}} \quad \forall f \in B^{\alpha}_{p_1,q},\ g \in B^{\beta}_{p_2,q} \ \textit{if } \alpha + \beta > 0.
\end{aligned}$$

(b) ([7, *Proposition* 3.1]) *Let* $\alpha, \beta \in \mathbb{R}$. *Then*

$$\|f \prec g\|_{H^{\alpha+\beta}} \lesssim_{\alpha,\beta} \|f\|_{H^{\alpha}} \|g\|_{\mathcal{C}^{\beta}} \quad \forall f \in H^{\alpha},\ g \in \mathcal{C}^{\beta} \ \textit{if } \alpha < 0, \tag{8.7a}$$

$$\|f \succ g\|_{H^{\alpha+\beta}} \lesssim_{\alpha,\beta} \|f\|_{H^{\alpha}} \|g\|_{\mathcal{C}^{\beta}} \quad \forall f \in H^{\alpha},\ g \in \mathcal{C}^{\beta} \ \textit{if } \beta < 0, \tag{8.7b}$$

$$\|f \circ g\|_{H^{\alpha+\beta}} \lesssim_{\alpha,\beta} \|f\|_{H^{\alpha}} \|g\|_{\mathcal{C}^{\beta}} \quad \forall f \in H^{\alpha},\ g \in \mathcal{C}^{\beta} \ \textit{if } \alpha + \beta > 0. \tag{8.7c}$$

The following is an immediate corollary, and its detailed proof can be found in [351, Corollary 1.2].

Corollary 8.1.5. *The product* fg *is well-defined for* $f \in \mathcal{C}^{\alpha}$ *and* $g \in \mathcal{C}^{\beta}$ *if*

$$\alpha + \beta > 0;$$

more precisely, if $\alpha + \beta > 0$, *then* $\|fg\|_{\mathcal{C}^{\min\{\alpha,\beta\}}} \lesssim \|f\|_{\mathcal{C}^{\alpha}} \|g\|_{\mathcal{C}^{\beta}}$.

We also refer to Colombeau [95] concerning multiplication of distributions.

Singular SPDEs are informally the SPDEs such that the roughness of the random force directly limits the regularity of the solution so that the products within the nonlinear term become ill-defined according to Corollary 8.1.5. For generality, let us consider the generalized singular Navier–Stokes equations (1.1) with $\upsilon = 1$ on $\mathbb{T}^d$ as

$$\partial_t u + \operatorname{div}(u \otimes u) + \nabla \pi + \Lambda^{m_1} u = \Lambda^{m_2} \mathcal{P}_L \mathbb{P}_{\neq 0} \xi, \quad \nabla \cdot u = 0, \tag{8.8}$$

starting from u^{in} at time $t = 0$, where ξ is an STWN, and $\mathbb{P}_{\neq 0}$ was defined in (6.46). For convenience of discussions, let us recall two preliminary lemmas concerning the Besov–Hölder spaces $\mathcal{C}^\alpha$.

Lemma 8.1.6 ([374, Lemma 3.6]). *Let $\mathcal{P}_L$ be the Leray projection from* (3.77). *Then for all $f \in \mathcal{C}^\alpha(\mathbb{T}^d)$, $\alpha \in \mathbb{R}$,*

$$\|\mathcal{P}_L f\|_{\mathcal{C}^\alpha(\mathbb{T}^d)} \lesssim \|f\|_{\mathcal{C}^\alpha(\mathbb{T}^d)}.$$

Lemma 8.1.7 ([159, Lemma A.7] and [374, Lemma 3.5]). *Let*

$$P_t \triangleq e^{t\Delta} \tag{8.9}$$

be the heat semigroup on $\mathbb{T}^d$. Then for all $f \in \mathcal{C}^\alpha(\mathbb{T}^d)$, $\alpha \in \mathbb{R}$, and $\delta \geq 0$,

$$\|P_t f\|_{\mathcal{C}^{\alpha+\delta}(\mathbb{T}^d)} \lesssim t^{-\frac{\delta}{2}} \|f\|_{\mathcal{C}^\alpha(\mathbb{T}^d)}.$$

Such a result can be readily generalized to heat semigroups with fractional Laplacians. We also refer to [167, *Proposition* 6.16] *in the case of $\mathcal{C}^\alpha_{\mathfrak{s}}$ setting.*

Now, analogously to (8.1), we define the scaling and scaling dimension corresponding to (8.8) as

$$\mathfrak{s} \triangleq (m_1, \underbrace{1, \ldots, 1}_{d \text{ times}}), \quad |\mathfrak{s}| = d + m_1. \tag{8.10}$$

Then we see that $\xi \in \mathcal{C}^\alpha_{\mathfrak{s}}$ for $\alpha < -\frac{d+m_1}{2}$ due to (8.6), leading to the regularity of $\Lambda^{m_2}\mathcal{P}_L \mathbb{P}_{\neq 0}\xi$ to be bounded by $\alpha < -\frac{d+m_1}{2} - m_2$ $\mathbb{P}$-a. s. The most common type of solutions to singular SPDEs is a mild solution, especially because a weak formulation offers no merits for some equations such as the KPZ equation (1.13) and stochastic Yang-Mills equation (1.18). A mild solution of (8.8) takes the form

$$\begin{aligned} u(t) = e^{-t\Lambda^{m_1}} u^{\text{in}} &- \int_0^t e^{-(t-s)\Lambda^{m_1}} \mathcal{P}_L \operatorname{div}(u \otimes u)(s)ds \\ &+ \int_0^t e^{-(t-s)\Lambda^{m_1}} \Lambda^{m_2} \mathcal{P}_L \mathbb{P}_{\neq 0}\xi(s)ds. \end{aligned} \tag{8.11}$$

Consequently, upon taking any norm, the regularity of the solution $u(t)$ is upper bounded by $\int_0^t e^{-(t-s)\Lambda^{m_1}} \Lambda^{m_2} \mathcal{P}_L \mathbb{P}_{\neq 0}\xi(s)ds$, which in turn is bounded by $\frac{m_1}{2} - m_2 - \frac{d}{2}$, and, consequently,

$$u \in C_t \mathcal{C}^\alpha_x \quad \text{for } \alpha < \frac{m_1}{2} - m_2 - \frac{d}{2} \quad \mathbb{P}\text{-a. s.} \tag{8.12}$$

Then Corollary 8.1.5 implies that the product $u \otimes u$ in (8.11) is well-defined only if

$$\frac{m_1}{2} - m_2 - \frac{d}{2} > 0. \tag{8.13}$$

We will come back to discuss this observation.

8.2 Young's integral and rough path theory

After describing the roughness of solutions to singular SPDEs in Section 8.1, we turn to the history of developments starting with Young's integral [369] and rough path theory (Lyons [255]). In Remark 4.1.8, we introduced a stochastic process of p-variation for which $\langle\langle X\rangle\rangle_t^{(p)} = \lim_{\Delta t_k \to 0} \sum_{t_k \le t} |X_{t_{k+1}}(\omega) - X_{t_k}(\omega)|^p < \infty$ (see (4.5)); taking X as a constant in $\omega \in \Omega$ gives us a deterministic analogue for our current discussions.

Remark 8.2.1 (Hölder regularity and p-variation). We mention that in connection with Hölder spaces from Definition 2.3.2, every $f \in C^{\frac{1}{p}}([0,t])$ for $p \in (0,\infty)$ and $t > 0$ is also continuous and of finite p-variation over $[0,t]$. It follows (see, e. g., [150, Proposition 5.3]) that if f is continuous over $[0,t]$ and $1 \le p \le q < \infty$, then $\langle\langle f\rangle\rangle_t^{(q)} \le \langle\langle f\rangle\rangle_t^{(p)}$. It is immediate that $f \in C^{\frac{1}{p}}$ for $p \in (0,\infty)$ is continuous and of finite p-variation. However, a function of finite p-variation need not be continuous; e. g., we can consider a step function (see [150, p. 78] for this discussion). The most relevant example for us is the Wiener process that has finite p-variation for every $p > 2$ but not for $p = 2$, considering Remark 4.2.2 (e. g., [255, p. 219]).

Now the well-known sufficient condition of the Lebesgue–Stieltjes integral $\int g df = \int g(x) df(x)$ requires that f and g are of bounded variation and that one of them is continuous (see [147, Theorem 3.30] for a specific statement). Young [369] in 1936 proved that the Lebesgue–Stieltjes integral can be defined as long as f is of p-variation, g is of q-variation with $p, q > 0$ such that

$$\frac{1}{p} + \frac{1}{q} > 1, \tag{8.14}$$

and they have no common discontinuities, e. g., one of them is continuous, whereas the other is not (see [369, p. 266], [150]). Such an improvement gives hope that perhaps $\int_0^t e^{-(t-s)\Lambda^{m_1}} \mathcal{P}_L \operatorname{div}(u \otimes u)(s) ds$ in (8.11) can be somehow defined through Young's integral. Unfortunately, Young's integral seems to fail in case m_1 in (8.13) is not sufficiently large w. r. t. m_2 or d.

At this point, we recall the stochastic integral constructed in Chapter 4, which can be further generalized to continuous semimartingales as discussed in Remark 4.2.13, and hope that it may extend Young's integral. Unfortunately, the limitations of stochastic integrals have been noticed over decades too. From Chapter 4 we saw that Itô's integral requires filtration, adapted integrands, a probability measure, and the integrand must have the L^2-orthogonal increments (see discussions on [159, p. 6] and [164, p. 1548]).

To complement the theory of Itô's integrals, Lyons [255, 256] developed a theory of rough paths; additionally, he also proved the existence of a unique solution to a differential equation driven by paths of finite p-variation with $p < 2$ in [254]. A rough path informally is a continuous path on which a sequence of iterated path integrals may be constructed. To highlight the power of rough path theory, let us discuss an example of fractional Brownian motion.

Example 12. Briefly stated, a fractional Brownian motion β^H with Hurst parameter $H \in (0,1)$ is a zero-mean Gaussian process with covariance

$$\mathbb{E}[\beta^H(s)\beta^H(t)] = \frac{1}{2}(t^{2H} + s^{2H} - |t-s|^{2H})$$

(e. g., [150, p. 405]). In comparison, recall from Remark 4.2.2 that the covariance of Wiener process is $\min\{s,t\}$, and thus β^H at $H = \frac{1}{2}$ reduces to the Wiener process. It is known that β^H for $H \neq \frac{1}{2}$ is not a semimartingale (e. g., [30, pp. 12–13]), so that Itô's integral cannot be extended to β^H (recall Remark 4.2.13). Nevertheless, using rough path theory, we may still construct a pathwise integral w. r. t. β^H (see [150, Proposition 15.5]).

We limit ourselves to only giving heuristics of rough path theory following [256, pp. 8–9] and refer the interested readers to [255, 256, 150] for details. If we consider

$$dY_t = f(t, Y_t)dX_t,$$

then in case the path $t \mapsto X_t$ is rough, dX_t cannot be understood in the traditional sense. A natural idea is to consider the incremental process $\{X^1_{s,t} \triangleq X_t - X_s : 0 \le s \le t\}$, approximate, and take the limit as the mesh size tends to zero. In the case where X is not of bounded variation but of p-variation for some $p > 1$, we consider higher-order integrals; specifically, X is the collection of all iterated path integrals

$$\left\{X^k_{s,t} = \int_{s<t_1<\dots<t_k<t} dX_{t_1} \otimes \cdots \otimes dX_{t_k} : 0 \le s \le t\right\}.$$

We have $X^1_{s,t} = X_t - X_s$ by definition, and for $k \ge 2$, we can compute $X^k_{s,t}$ explicitly:

$$X^2_{s,t} = \int_s^t \int_s^{t_2} dX_{t_1} dX_{t_2} = \int_s^t (X_{t_2} - X_s) dX_{t_2} = \frac{1}{2}(X_t - X_s)^2,$$

and $X^3_{s,t}, \dots, X^k_{s,t}$ may be computed similarly. Thus, if X is of bounded variation, then $\{X^1_{s,t} : 0 \le s \le t\}$ determines all higher-order iterated path integrals, and the fact that X is of bounded variation implies that

$$\sup_D \sum_l |X^k_{t_{l-1},t_l}|^{\frac{1}{k}} < \infty \quad \forall k = 1, 2, \dots,$$

where the supremum is taken over all finite partitions D of $[s,t]$. This motivates us to set $X_{s,t} \triangleq (1, X^1_{s,t}, \dots, X^k_{s,t}, \dots)$, and Chen's identity

$$X_{s,u} = X_{s,t} \otimes X_{t,u} \quad 0 \le s \le t \le u, \tag{8.15}$$

allows us to refine partitions ([256, p. 30]). However, X may not satisfy the finiteness of $\sup_D \sum_l |X^k_{t_{l-1},t_l}|^{\frac{1}{k}}$ for all $k = 1, 2, \dots$; e. g., a Brownian motion satisfies $\sup_D \sum_l |X^1_{t_{l-1},t_l}|^p < \infty$ only for all $p > 2$. One of the main contributions by Lyons is the following. Suppose that there exists a control function ω that is continuous and superadditive with values on $[0, \infty)$ such that $\omega(t,t) = 0$ and $|X^i_{s,t}| \le \omega(s,t)^{\frac{i}{p}}$ for all $i \in \{1, \dots, n\}$, so that

$$\sup_D \sum_l |X^k_{t_{l-1},t_l}|^{\frac{p}{k}} \le \sup_D \sum_l \omega(t_{l-1}, t_l) \le \sup_D \omega(s,t) = \omega(s,t) < \infty;$$

specifically, by a superadditive control function ω we mean that it satisfies $\omega(s,u) \ge \omega(s,t) + \omega(t,u)$ for all $s \le t \le u$ (recall Definition 2.3.7). In case X satisfies Chen's identity (8.15) and additionally is of p-variation in a way that such ω exists, then [255, Theorem 2.2.1] and [256, Theorem 3.2.1] show that even if $\sup_D \sum_l |X^k_{t_{l-1},t_l}|^{\frac{1}{k}}$ is infinite for some $k \in \mathbb{N}$, X^k for all $k \ge \lfloor p \rfloor + 1$ are uniquely determined by $X^1, \dots, X^{\lfloor p \rfloor}$. This implies that the rougher the path, say of p-variation for $p > 1$, the more the higher-order iterated path integrals become necessary, up to $\lfloor p \rfloor$ to be specific. Later, Gubinelli [158] pioneered the controlled rough path theory; we refer to [149, 150, 161] for further details.

Example 13. Let us describe the difficulty of the 1D Burgers' equation (1.3) forced by STWN through Young's integral. This case represents (8.8) with $d = 1$ and $(m_1, m_2) = (2, 0)$, so that (8.12) ensures that the solution u has regularity of $C^\alpha(\mathbb{T})$ for $\alpha < \frac{1}{2}$ $\mathbb{P}$-a. s. Thus (8.13) shows that the product u^2 is well-defined. Thus such 1D Burgers' equation with full Laplacian forced by STWN was proven to be globally well-posed by Bertini, Cancrini, and Jona-Lasinio [27, Theorem 2.2] via the stochastic Cole–Hopf transform and by Da Prato, Debussche, and Temam [116, Theorem 3.1] using the Da Prato–Debussche trick from Example 9, to be revisited in Section 8.4. However, inaccuracies in numerical approximations caused by the STWN were also pointed out by Hairer and Voss [177], which will be discussed more in Section 8.5.5. Hairer [164] considered the generalized Burgers' equation with nonlinear term of the form $g(u)\partial_x u$, where $g \neq DG$ for some function $G : \mathbb{R} \to \mathbb{R}$, specifically

$$\partial_t u + g(u)\partial_x u = \partial_x^2 u + \xi.$$

Upon considering its weak solution formulation, we are faced with

$$\int_{\mathbb{T}} \psi(x)g(u)\partial_x u dx, \tag{8.16}$$

where $\psi \in C^\infty(\mathbb{T})$, and we realize that integration by parts is not an option because $g(u)\partial_x u = \partial_x DG(u)$ only if $g = DG$. As $u \in \mathcal{C}^\alpha(\mathbb{T})$ for $\alpha < \frac{1}{2}$, Remark 8.2.1 ensures that u is of p-variation only for $p > 2$ $\mathbb{P}$-a. s. This implies that rewriting (8.16) as

$$\int_{\mathbb{T}} \psi(x)g(u)du$$

is also of no use since (8.14) is not satisfied. Hairer [164] overcomes this difficulty using the rough path theory and proved its global solution theory in [180, Theorem 3.6] for g that is sufficiently smooth (also [178] for extension to multiplicative STWN).

8.3 Wick products and contraction of indices

8.3.1 Wick products

The simplest example of a renormalization constant to be discussed later is the Wick product (8.18a), which is also useful for other aspects of analysis on singular SPDEs. To give a heuristic first, for simplicity, let $\{W_t\}_{t\geq 0}$ be a Wiener process, and let f be a deterministic function. Let us consider the product $(\int_0^T f(s)dW_s)^2$, which is a second-order Wiener chaos (recall Definition 4.6.3). If this were a deterministic equation, then we would write

$$\int_0^T \int_0^T f(s_1)f(s_2)dW(s_1)dW(s_2) \tag{8.17}$$

under suitable integrability condition. However, this is not a well-defined object as an Itô's integral, because upon writing (8.17), we recall that $\int_0^T f(s_1)f(s_2)dW(s_1)$ must be $\mathcal{F}_{s_2}$-measurable. Informally stated, in this case the Wick product is

$$: \int_0^T f(s)dW_s \int_0^T g(s)dW_s := \int_0^T \int_0^{s_1} \frac{f(s_1)g(s_2) + f(s_2)g(s_1)}{2} dW(s_2)dW(s_1).$$

It subtracts out the diagonal terms when $s_1 = s_2$ and symmetrizes the rest. We refer the readers to Definition A.7.4 for rigorous definitions; for practical purpose, we record the following examples computed in detail in [351] using Feynman diagrams and sketch ideas in Appendix A.7 for readers' convenience. We recall the Gaussian Hilbert space from Definition 4.6.5.

Example 14 (Wick product, [351, Example 2.1]). Let H be a Gaussian Hilbert space, and let $\xi, \xi_1, \xi_2, \ldots \in H$. Then

$$
\begin{aligned}
:\xi_1\xi_2: &= \xi_1\xi_2 - \mathbb{E}[\xi_1\xi_2], && \text{(8.18a)}\\
:\xi_1\xi_2\xi_3: &= \xi_1\xi_2\xi_3 - \mathbb{E}[\xi_2\xi_3]\xi_1 - \mathbb{E}[\xi_1\xi_3]\xi_2 - \mathbb{E}[\xi_1\xi_2]\xi_3, && \text{(8.18b)}\\
:\xi_1\xi_2\xi_3\xi_4: &= \xi_1\xi_2\xi_3\xi_4 - \mathbb{E}[\xi_1\xi_2]\xi_3\xi_4 - \mathbb{E}[\xi_1\xi_3]\xi_2\xi_4 - \mathbb{E}[\xi_1\xi_4]\xi_2\xi_3\\
&\quad - \mathbb{E}[\xi_2\xi_3]\xi_1\xi_4 - \mathbb{E}[\xi_2\xi_4]\xi_1\xi_3 - \mathbb{E}[\xi_3\xi_4]\xi_1\xi_2\\
&\quad + \mathbb{E}[\xi_1\xi_2]\mathbb{E}[\xi_3\xi_4] + \mathbb{E}[\xi_1\xi_3]\mathbb{E}[\xi_2\xi_4] + \mathbb{E}[\xi_1\xi_4]\mathbb{E}[\xi_2\xi_3]. && \text{(8.18c)}
\end{aligned}
$$

For example, (8.18c) becomes crucial upon estimating z  that satisfies (8.37), where a quadruple product of STWN can be seen, and also in (8.56).

By Lemma A.7.6, $\mathbb{E}[:\xi_1\xi_2:] = \mathbb{E}[:\xi_1\xi_2\xi_3:] = \mathbb{E}[:\xi_1\xi_2\xi_3\xi_4:] = 0$, so that

$$
\begin{aligned}
\mathbb{E}[\xi_1\xi_2\xi_3] &= 0, && \text{(8.19a)}\\
\mathbb{E}[\xi_1\xi_2\xi_3\xi_4] &= \mathbb{E}[\xi_2\xi_3]\mathbb{E}[\xi_1\xi_4] + \mathbb{E}[\xi_2\xi_4]\mathbb{E}[\xi_1\xi_3] + \mathbb{E}[\xi_3\xi_4]\mathbb{E}[\xi_1\xi_2]. && \text{(8.19b)}
\end{aligned}
$$

For some singular SPDEs, we also need to compute the mathematical expectation of the products of Wick products such as

$$
\mathbb{E}[:\xi_{11}\xi_{12}::\xi_{21}\xi_{22}:],\ \mathbb{E}[:\xi_{11}\xi_{12}\xi_{13}::\xi_{21}\xi_{22}\xi_{23}:],\ \mathbb{E}[:\xi_{11}\xi_{12}\xi_{13}\xi_{14}::\xi_{21}\xi_{22}\xi_{23}\xi_{24}:].
$$

In such a case, using Lemma A.7.6, we can compute the following:

Example 15 ([351, Example 2.2]). Let H be a Gaussian Hilbert space, and let $\xi, \xi_1, \xi_2, \ldots \in H$. Then

$$
\begin{aligned}
&\mathbb{E}[:\xi_{11}\xi_{12}::\xi_{21}\xi_{22}:] = \mathbb{E}[\xi_{11}\xi_{21}]\mathbb{E}[\xi_{12}\xi_{22}] + \mathbb{E}[\xi_{11}\xi_{22}]\mathbb{E}[\xi_{12}\xi_{21}],\\
&\mathbb{E}[:\xi_{11}\xi_{12}\xi_{13}::\xi_{21}\xi_{22}\xi_{23}:]\\
&= \mathbb{E}[\xi_{11}\xi_{21}]\mathbb{E}[\xi_{12}\xi_{22}]\mathbb{E}[\xi_{13}\xi_{23}] + \mathbb{E}[\xi_{11}\xi_{21}]\mathbb{E}[\xi_{12}\xi_{23}]\mathbb{E}[\xi_{13}\xi_{22}]\\
&\quad + \mathbb{E}[\xi_{11}\xi_{22}]\mathbb{E}[\xi_{12}\xi_{21}]\mathbb{E}[\xi_{13}\xi_{23}] + \mathbb{E}[\xi_{11}\xi_{22}]\mathbb{E}[\xi_{12}\xi_{23}]\mathbb{E}[\xi_{13}\xi_{21}]\\
&\quad + \mathbb{E}[\xi_{11}\xi_{23}]\mathbb{E}[\xi_{12}\xi_{21}]\mathbb{E}[\xi_{13}\xi_{22}] + \mathbb{E}[\xi_{11}\xi_{23}]\mathbb{E}[\xi_{12}\xi_{22}]\mathbb{E}[\xi_{13}\xi_{21}],\\
&\mathbb{E}[:\xi_{11}\xi_{12}\xi_{13}\xi_{14}::\xi_{21}\xi_{22}\xi_{23}\xi_{24}:]\\
&= \mathbb{E}[\xi_{11}\xi_{21}]\mathbb{E}[\xi_{12}\xi_{22}]\mathbb{E}[\xi_{13}\xi_{23}]\mathbb{E}[\xi_{14}\xi_{24}] + \mathbb{E}[\xi_{11}\xi_{21}]\mathbb{E}[\xi_{12}\xi_{22}]\mathbb{E}[\xi_{13}\xi_{24}]\mathbb{E}[\xi_{14}\xi_{23}]\\
&\quad + \mathbb{E}[\xi_{11}\xi_{21}]\mathbb{E}[\xi_{12}\xi_{23}]\mathbb{E}[\xi_{13}\xi_{22}]\mathbb{E}[\xi_{14}\xi_{24}] + \mathbb{E}[\xi_{11}\xi_{21}]\mathbb{E}[\xi_{12}\xi_{23}]\mathbb{E}[\xi_{13}\xi_{24}]\mathbb{E}[\xi_{14}\xi_{22}]\\
&\quad + \mathbb{E}[\xi_{11}\xi_{21}]\mathbb{E}[\xi_{12}\xi_{24}]\mathbb{E}[\xi_{13}\xi_{22}]\mathbb{E}[\xi_{14}\xi_{23}] + \mathbb{E}[\xi_{11}\xi_{21}]\mathbb{E}[\xi_{12}\xi_{24}]\mathbb{E}[\xi_{13}\xi_{23}]\mathbb{E}[\xi_{14}\xi_{22}]\\
&\quad + \mathbb{E}[\xi_{11}\xi_{22}]\mathbb{E}[\xi_{12}\xi_{21}]\mathbb{E}[\xi_{13}\xi_{23}]\mathbb{E}[\xi_{14}\xi_{24}] + \mathbb{E}[\xi_{11}\xi_{22}]\mathbb{E}[\xi_{12}\xi_{21}]\mathbb{E}[\xi_{13}\xi_{24}]\mathbb{E}[\xi_{14}\xi_{23}]\\
&\quad + \mathbb{E}[\xi_{11}\xi_{22}]\mathbb{E}[\xi_{12}\xi_{23}]\mathbb{E}[\xi_{13}\xi_{21}]\mathbb{E}[\xi_{14}\xi_{24}] + \mathbb{E}[\xi_{11}\xi_{22}]\mathbb{E}[\xi_{12}\xi_{23}]\mathbb{E}[\xi_{13}\xi_{24}]\mathbb{E}[\xi_{14}\xi_{21}]\\
&\quad + \mathbb{E}[\xi_{11}\xi_{22}]\mathbb{E}[\xi_{12}\xi_{24}]\mathbb{E}[\xi_{13}\xi_{21}]\mathbb{E}[\xi_{14}\xi_{23}] + \mathbb{E}[\xi_{11}\xi_{22}]\mathbb{E}[\xi_{12}\xi_{24}]\mathbb{E}[\xi_{13}\xi_{23}]\mathbb{E}[\xi_{14}\xi_{21}]\\
&\quad + \mathbb{E}[\xi_{11}\xi_{23}]\mathbb{E}[\xi_{12}\xi_{21}]\mathbb{E}[\xi_{13}\xi_{22}]\mathbb{E}[\xi_{14}\xi_{24}] + \mathbb{E}[\xi_{11}\xi_{23}]\mathbb{E}[\xi_{12}\xi_{21}]\mathbb{E}[\xi_{13}\xi_{24}]\mathbb{E}[\xi_{14}\xi_{22}]
\end{aligned}
$$

$$+ \mathbb{E}[\xi_{11}\xi_{23}]\mathbb{E}[\xi_{12}\xi_{22}]\mathbb{E}[\xi_{13}\xi_{21}]\mathbb{E}[\xi_{14}\xi_{24}] + \mathbb{E}[\xi_{11}\xi_{23}]\mathbb{E}[\xi_{12}\xi_{22}]\mathbb{E}[\xi_{13}\xi_{24}]\mathbb{E}[\xi_{14}\xi_{21}]$$
$$+ \mathbb{E}[\xi_{11}\xi_{23}]\mathbb{E}[\xi_{12}\xi_{24}]\mathbb{E}[\xi_{13}\xi_{21}]\mathbb{E}[\xi_{14}\xi_{22}] + \mathbb{E}[\xi_{11}\xi_{23}]\mathbb{E}[\xi_{12}\xi_{24}]\mathbb{E}[\xi_{13}\xi_{22}]\mathbb{E}[\xi_{14}\xi_{21}]$$
$$+ \mathbb{E}[\xi_{11}\xi_{24}]\mathbb{E}[\xi_{12}\xi_{21}]\mathbb{E}[\xi_{13}\xi_{22}]\mathbb{E}[\xi_{14}\xi_{23}] + \mathbb{E}[\xi_{11}\xi_{24}]\mathbb{E}[\xi_{12}\xi_{21}]\mathbb{E}[\xi_{13}\xi_{23}]\mathbb{E}[\xi_{14}\xi_{22}]$$
$$+ \mathbb{E}[\xi_{11}\xi_{24}]\mathbb{E}[\xi_{12}\xi_{22}]\mathbb{E}[\xi_{13}\xi_{21}]\mathbb{E}[\xi_{14}\xi_{23}] + \mathbb{E}[\xi_{11}\xi_{24}]\mathbb{E}[\xi_{12}\xi_{22}]\mathbb{E}[\xi_{13}\xi_{23}]\mathbb{E}[\xi_{14}\xi_{21}]$$
$$+ \mathbb{E}[\xi_{11}\xi_{24}]\mathbb{E}[\xi_{12}\xi_{23}]\mathbb{E}[\xi_{13}\xi_{21}]\mathbb{E}[\xi_{14}\xi_{22}] + \mathbb{E}[\xi_{11}\xi_{24}]\mathbb{E}[\xi_{12}\xi_{23}]\mathbb{E}[\xi_{13}\xi_{22}]\mathbb{E}[\xi_{14}\xi_{21}].$$

8.3.2 Contraction of indices and Wiener integrals

Let us discuss the notions of contraction of indices; here we follow [274, p. 8]. We also recall the Wiener chaos $\mathcal{H}_n$ of order n, isonormal Gaussian process $W = \{W(h) : h \in H\}$ from Definition 4.6.3, and the notion of atoms in Definition 2.2.3. Let $(X, \mathcal{B})$ be any measurable space, and let μ be a σ-finite measure without atoms. We define the separable Hilbert space $H \triangleq L^2(X, \mathcal{B}, \mu)$; e. g., X can be $X = [0,1]$ in the case of a Wiener process, whereas $X = [0,1] \times \mathbb{R}^d$ if STWN. In such a case, $W(h)$ can be regarded as $\int_X h dW$; hereafter we write $W(A) \triangleq W(1_A)$.

Now we fix $m \in \mathbb{N}$ and $\mathcal{B}_0 \triangleq \{A \in \mathcal{B} : \mu(A) < \infty\}$. We want to define the m-multiple stochastic integrals $I_m(f)$ for $f \in L^2(X^m, \mathcal{B}^m, \mu^m)$, a. k. a. mth-order Wiener integral. To do this, similarly to (4.7), we denote by $\mathcal{E}_m$ the set of elementary processes of the form

$$f(t_1, \dots, t_m) = \sum_{i_1,\dots,i_m=1}^{n} a_{i_1,\dots,i_m} 1_{A_{i_1} \times \dots \times A_{i_m}}(t_1, \dots, t_m)$$

for $n \in \mathbb{N}$, where $A_1, \dots, A_n$ are pairwise disjoint sets belonging to $\mathcal{B}_0$, and $a_{i_1,\dots,i_m}$ are real coefficients that are zero whenever any two of the indices $i_1, \dots, i_m$ match. For such an elementary process, it is natural to define

$$I_m(f) \triangleq \sum_{i_1,\dots,i_m=1}^{n} a_{i_1,\dots,i_m} W(A_{i_1}) \dots W(A_{i_m}), \tag{8.20}$$

and [274, p. 9] shows the following typical properties for such I_m:

(a) I_m is linear.

(b) $I_m(f) = I_m(\tilde{f})$, where $\tilde{f}$ denotes the symmetrization of f defined by

$$\tilde{f}(t_1, \dots, t_m) \triangleq \frac{1}{m!} \sum_{\sigma} f(t_{\sigma(1)}, \dots, t_{\sigma(m)}), \tag{8.21}$$

where the summation is over all the permutations of $\{1, \dots, m\}$.

(c)

$$\mathbb{E}[I_m(f) I_q(g)] = \begin{cases} m! \langle \tilde{f}, \tilde{g} \rangle_{L^2(X^m)} & \text{if } m = q, \\ 0 & \text{otherwise.} \end{cases}$$

In [274, pp. 9–10], such I_m are extended to $L^2(X^m, \mathcal{B}^m, \mu^m)$ by density argument, where the nonexistence of atoms is used. Thus we can now consider

$$I_m(f) = \int_{X^m} f(t_1, \ldots, t_m) W(dt_1) \ldots W(dt_m)$$

and note that I_0 is understood hereafter as the identity map.

Definition 8.3.1 ([274, p. 10]). If $f \in L^2(X^p)$ and $g \in L^2(X^q)$ are both symmetric functions, then for any $1 \le r \le \min\{p, q\}$, the contraction of r indices of f and g, denoted $f \otimes_r g$, is defined by

$$(f \otimes_r g)(t_1, \ldots, t_{p+q-2r}) \triangleq \int_{X^r} f(t_1, \ldots, t_{p-r}, s) g(t_{p+1}, \ldots, t_{p+q-r}, s) \mu^r(ds). \tag{8.22}$$

We extend $\otimes_r$ for $r \in \mathbb{N}$ to $\mathbb{N}_0$ by $\otimes_0 \triangleq \otimes$. The special case $p = q = r = 1$ gives

$$(f \otimes_1 g) = \int_X f(s) g(s) \mu(ds) \in \mathbb{R}. \tag{8.23}$$

Finally, we denote the symmetrizations of $f \otimes g$ and $f \otimes_r g$ by $f \tilde{\otimes} g$ and $f \tilde{\otimes}_r g$, respectively.

We end this section with a useful result from [274].

Lemma 8.3.2 ([274, Proposition 1.1.2]).
(a) *Let $f \in L^2(X^p)$ be a symmetric function, and let $g \in L^2(X)$. Then*

$$I_p(f) I_1(g) = I_{p+1}(f \otimes g) + p I_{p-1}(f \otimes_1 g). \tag{8.24}$$

(b) *Let $f \in L^2(X^p)$ and $g \in L^2(X^q)$ be symmetric functions. Then*

$$I_p(f) I_q(g) = \sum_{r=0}^{p \wedge q} r! \binom{p}{r} \binom{q}{r} I_{p+q-2r}(f \otimes_r g). \tag{8.25}$$

Remark 8.3.3. We record here a variation of (8.25) from [167, Lemma 10.3] that already includes combinatorial prefactors. Specifically, let $S(r)$ be the set of all permutations of r elements, and let $S(r, m) \subset S(m)$ be the set of all "shuffles" of r and $m - r$ elements, namely the set of permutations of m elements that preserves the order of the first r and of the last $m - r$ elements. For $x \in X^m$ and $\Sigma \in S(m)$, write $\Sigma(x) \in X^m$ as a shorthand for $\Sigma(x)_i = x_{\Sigma(i)}$. For $x \in X^r$ and $y \in X^{m-r}$, we denote by $x \sqcup y$ the element of X^m given by $(x_1, \ldots, x_r, y_1, \ldots, y_{m-r})$. Then [167, Lemma 10.3] states the following. Let $f \in L^2(X^p)$ and $g \in L^2(X^q)$. Then we have

$$I_p(f) I_q(g) = \sum_{r=0}^{p \wedge q} I_{p+q-2r}(f *_r g), \tag{8.26}$$

where for all $z \in X^{p-r}$ and $\widetilde{z} \in X^{q-r}$,

$$(f *_r g)(z \sqcup \widetilde{z}) = \sum_{\Sigma \in S(r,p), \widetilde{\Sigma} \in S(r,q)} \sum_{\sigma \in S(r)} \int_{X^r} f(\Sigma(x \sqcup z)) g(\widetilde{\Sigma}(x \sqcup \sigma(\widetilde{z}))) dx.$$

8.4 Renormalization constants and Da Prato–Debussche trick applied on singular SPDEs

The issue of ill-defined products we described in Section 8.1 suggests the need to subtract out renormalization constant, which is informally a constant to be subtracted out upon approximation that diverges to infinity as the approximation parameter approaches its limit. The following example in the case of the KPZ equation (1.13) is classical and particularly mentioned on [166, p. 562]; we give the details for readers' convenience.

Example 16. First, the scaling (8.1) for the dD KPZ would be $\mathfrak{s} = (2, 1, \ldots, 1)$, so that $|\mathfrak{s}| = 2 + d$ (recall (8.1) and (8.2)). Consequently, (8.6) implies $\xi \in \mathcal{C}^{\alpha}_{\mathfrak{s}}$ for $\alpha < -\frac{d+2}{2}$ $\mathbb{P}$-a. s. Then the same considerations as in (8.11)–(8.12) suggest that h has the regularity of $1 - \frac{d}{2}$ and, consequently, ∇h that of $-\frac{d}{2}$, so that the nonlinear term $|\nabla h|^2$ is ill-defined according to Corollary 8.1.5 even in the case $d = 1$.

Thus we focus on the case $d = 1$ and, for simplicity, $x \in \mathbb{R}$. We consider $\rho \in C^{\infty}(\mathbb{R})$ such that $\rho(0) = 1$ and $\|\rho\|_{L^2(\mathbb{R})} = 1$ and define $\rho_\epsilon(x) \triangleq \epsilon^{-1}\rho(x\epsilon^{-1})$, so that $\widehat{\rho_\epsilon}(\xi) = \widehat{\rho}(\epsilon\xi)$ according to (3.9c). Given $\xi = \frac{\partial W}{\partial t}$, where W is a Wiener process, we convolute W spatially with ρ_ϵ, define $W_\epsilon \triangleq W *_x \rho_\epsilon$, and consider the heat equation with linear multiplicative noise

$$dz^\epsilon = \Delta z^\epsilon dt + \lambda z^\epsilon dW_\epsilon, \quad \lambda > 0. \tag{8.27}$$

Here our intention is to indicate the mollification by subscript ϵ whereas dependence on ϵ by superscript ϵ. Since we convoluted W only in space, we still have W as a distribution in time; however, z^ϵ is continuous in time and smooth in space (cf. Lemma 7.1.1), justifying the products. We may define $h^\epsilon \triangleq \lambda^{-1}\ln(z^\epsilon)$, which is allowed because $z^\epsilon > 0$ $\mathbb{P}$-a. s. due to Mueller [270], so that Itô's formula (4.14) gives us

$$dh^\epsilon = [\Delta h^\epsilon + \lambda|\nabla h^\epsilon|^2]dt + dW_\epsilon - \frac{\lambda}{2} d\langle\langle W_\epsilon\rangle\rangle_t. \tag{8.28}$$

This transformation is known as the Cole–Hopf transform, which has been used by many researchers (e. g. [28]). Finally, we can use the assumption that $\|\rho\|_{L^2} = 1$ and directly compute $d\langle\langle W_\epsilon\rangle\rangle_t = \epsilon^{-1}dt$. Applying this to (8.28) gives us

$$\partial_t h^\epsilon = \Delta h^\epsilon + \lambda|\nabla h^\epsilon|^2 + \xi_\epsilon - \frac{\lambda}{2}\epsilon^{-1},$$

where $\xi_\epsilon \triangleq \frac{\partial W_\epsilon}{\partial t}$, and we observe that $\frac{\lambda}{2}\epsilon^{-1}$ diverges to ∞ as $\epsilon \searrow 0$.

8.5 Regularity structures and paracontrolled distributions

This section is devoted to describe the breakthrough works of the theories of RS by Hairer [167] and paracontrolled distributions by Gubinelli, Imkeller, and Perkowski. Specifically, we start with a review of other results on singular SPDEs and an important definition of local subcriticality in Definition 8.5.2 and verify that the Navier–Stokes equations forced by STWN are locally subcritical only if $d \in \{2, 3\}$. In Sections 8.5.1–8.5.2, we provide, from a more practical point of view, two different ways to see that the 4D is the critical dimension for the Navier–Stokes equations forced by STWN. In Section 8.5.3, we describe the theory of RS through an example of the 3D MHD system forced by STWN; in Section 8.5.4, we describe the theory of paracontrolled distributions very briefly, taking into account that it will be discussed in detail in Section 9.1. Finally, Section 8.5.5 describes other results obtained by applying the theories of RS and paracontrolled distributions.

Besides [116], which we described in Example 13, let us note that using their trick, Da Prato and Debussche [113] were able to study the well-posedness of the 2D Navier–Stokes equations forced by STWN. This is particularly surprising considering (8.13) with $m_1 = 2$, $m_2 = 0$, and $d = 2$ and realizing that the products are ill-defined. In Section 8.5.1, we will describe that in higher dimensions, we must repeat the Da Prato–Debussche trick multiple times, if possible, but in the case of the 2D Navier–Stokes equations forced by STWN, it suffices to apply the trick only once, and in this case the renormalization constant is only the most elementary Wick product, specifically (8.18a). This allowed [113] to use Besov space techniques to prove local-in-time solution theory. Additionally, the authors observed that based on (3.139), they can take advantage of the explicit knowledge of an invariant measure "$\mu = \times_{k \in \mathbb{Z}_0^2} \mathcal{N}(0, 1/2|k|^2)$" on [113, p. 185], particularly its Gaussianity, and deduce the existence of a pathwise unique solution globally in time starting from μ-a. e. initial data u^{in} (see [113, Theorem 5.1]). Analogous result of a global-in-time solution theory for a. e. initial data w. r. t. its invariant measure was obtained for the 2D Φ^4 model by Da Prato and Debussche [114, Theorem 4.2]. Using rough path theory, Hairer [166] was able to construct a local-in-time solution to the 1D KPZ equation (recall Example 16), of which the product is *not* almost well-defined!

The powerful theories of RS due to Hairer [167] and paracontrolled distributions [159] allow us to construct a local-in-time solution theory for singular SPDEs in which products are not just *almost* well-defined. For example, they can be applied to the 1D KPZ equation to recover the result from [166] and the 3D Φ^4 model (1.19) that we state here informally as an example. A more rigorous statement can be found in Theorem 8.5.5 on the 3D MHD system forced by STWN.

Example 17 ([167, Theorem 1.15]). Let $\rho : \mathbb{R}^4 \to \mathbb{R}$ be a smooth function with compact support such that its mass equals one and it is symmetric w. r. t. $x \in \mathbb{T}^3$. We define the space-time mollifier $\rho_\epsilon(t, x) \triangleq \epsilon^{-5}\rho(\frac{t}{\epsilon^2}, \frac{x}{\epsilon})$, $\xi_\epsilon \triangleq \xi *_{t,x} \rho_\epsilon$, and the renormalized equation

$$\partial_t \Phi_\epsilon = \Delta \Phi_\epsilon + C_\epsilon \Phi_\epsilon - \Phi_\epsilon^3 + \xi_\epsilon.$$

Let $\eta \in (-\frac{2}{3}, -\frac{1}{2})$. Then there exist constants C_ϵ such that for every initial data in $\mathcal{C}^\eta(\mathbb{T}^3)$, the sequence of solutions $\{\Phi_\epsilon\}_{\epsilon>0}$ converges to a limit Φ. Additionally, if $\{C_\epsilon\}_{\epsilon>0}$ is chosen suitably, then this limit is independent of the choice of mollifier ρ.

Remark 8.5.1. There is an informal heuristic why the regularity of $\mathcal{C}^\eta(\mathbb{T}^3)$, $\eta > -\frac{2}{3}$, is needed in Example 17, akin to how $H^s(\mathbb{T}^d)$ for $s > 1 + \frac{d}{2}$ is needed for deterministic Euler equations that we described in Section 3.9. In the mild formulation of the Φ^4 model (1.19), the nonlinear term is well-defined if

$$\int_0^t \| e^{(t-s)\Delta} \Phi^3(s) \|_{L_x^\infty} ds$$

is finite. Suppose that $\Phi \in C_t \mathcal{C}^\eta(\mathbb{T}^3)$ for some $\eta < 0$. Then

$$\int_0^t \| e^{(t-s)\Delta} \Phi^3(s) \|_{L^\infty} ds \lesssim \int_0^t ((t-s)^{\frac{\eta}{2}} \|\Phi(s)\|_{\mathcal{C}^\eta})^3 ds \lesssim 1$$

requires $1 + \frac{3\eta}{2} > 0$, which reduces to $\eta > -\frac{2}{3}$, as stated in Example 17. In Theorem 8.5.5, we will see that the initial data for the local solution theory of the 3D MHD system forced by STWN requires $\mathcal{C}^\eta(\mathbb{T}^3)$ for $\eta > -1$, which can also be verified analogously heuristically. We recall that $\mathcal{C}^{-1}(\mathbb{T}^3) = B^{-1}_{\infty,\infty}(\mathbb{T}^3)$ from (3.50) and $B^{-1}_{\infty,\infty}(\mathbb{R}^d)$ is actually one of the critical scaling spaces for the solution to the deterministic Navier–Stokes equations according to (5.4) and (5.6).

The theories of RS and paracontrolled distributions can be applied to singular SPDEs in general as long as they satisfy the condition of local subcriticality from [167, eq. (8.1) and Assumption 8.3], which we state here for completeness.

Definition 8.5.2 (Local subcriticality). Consider a fixed point problem of the form

$$u = K * F(u, \nabla u, \xi) + u^{\text{in}}, \tag{8.29}$$

where K is a singular integral operator that is β-regularizing on $\mathbb{R}^d$ w. r. t. some fixed scaling $\mathfrak{s}$, F is a smooth function, ξ denotes a rough force of regularity $\alpha < 0$, and u^{in} is a given initial data. Then we replace ξ by a dummy variable Θ. For any $i \in \{1, \ldots, d\}$, if $\beta + \alpha \leq \mathfrak{s}_i$, then replace $\partial_i u$ by a dummy variable P_i. Finally, if $\beta + \alpha \leq 0$, then we replace u by a dummy variable U. The PDE (8.29) is said to be locally subcritical under the following conditions. First, the resulting expression should be a polynomial of dummy variables. Second, we postulate that Θ has homogeneity α, U has homogeneity

$\beta + \alpha$, and P_i has homogeneity $\beta + \alpha - \mathfrak{s}_i$. Finally, we assume that the terms containing Θ do not contain the dummy variables and the remaining monomials all have homogeneity strictly greater than α. Under such conditions, we say that (8.29) is locally subcritical.

Informally stated, the local subcriticality boils down to the homogeneity of the nonlinear term strictly higher than the homogeneity of the force, where the homogeneity of a product is computed as the sum of homogeneities. Returning to (8.12) with $m_1 = 2$ and $m_2 = 0$, we can see that the 4D Navier–Stokes equations (1.1) forced by STWN are locally critical, and similarly so is the 4D stochastic Yang–Mills equation (1.18) (recall Remark 1.2.1). For another approach via rescaling to determine the local subcriticality threshold, see (9.72) and (9.73).

Via the theory of RS, Chandra, Chevyrev, Hairer, and Shen [74, 75] made significant progress on the analysis of the stochastic Yang–Mills equation (1.18) in both 2D and 3D cases; however, the locally critical 4D case remains completely open. In the following two sections, we describe two ways to see that the locally critical case faces significant obstacles, not just by Definition 8.5.2 but from a practical point of view.

8.5.1 First way to see the criticality of 4D

The first way to see the criticality of the 4D Navier–Stokes equations forced by STWN is what experts refer to as the "number of trees become infinite" and was described in detail in [129, Example 2.1]. For concreteness, let us consider the Navier–Stokes equations forced by STWN, specifically (8.8) with $m_1 = 2$ and $m_2 = 0$:

$$\partial_t u + \mathcal{P}_L \operatorname{div}(u \otimes u) = \Delta u + \mathcal{P}_L \mathbb{P}_{\neq 0} \xi, \quad \nabla \cdot u = 0. \tag{8.30}$$

Because we know that the 4D case is locally critical, we focus only on $d \geq 3$. We denote the STWN ξ by a black circle $\bullet$, convolution against a heat kernel $P_t = e^{t\Delta}$ by a straight black line $|$, and convolution against P_t after $\mathcal{P}_L$ div by a zigzag line $\wr$. Finally, multiplication of symbols is indicated by joining them at roots. We will come back to use such tree diagrams more carefully considering components in Section 8.5.3. We employ a series of the Da Prato–Debussche trick (i)–(iv) as follows, with explanation to follow afterward.

(i) First, we consider the following stochastic Stokes equation:

$$\partial_t z^{\stackrel{\bullet}{|}} - \Delta z^{\stackrel{\bullet}{|}} = \mathcal{P}_L \mathbb{P}_{\neq 0} \xi, \quad z^{\stackrel{\bullet}{|}}|_{t=0} = 0. \tag{8.31}$$

We define $u_1 \triangleq u - z^{\stackrel{\bullet}{|}}$, which satisfies

$$\partial_t u_1 - \Delta u_1 = -\mathcal{P}_L \operatorname{div}(u_1 + z^{\stackrel{\bullet}{|}})^{\otimes 2}, \tag{8.32}$$

where $X^{\otimes 2} \triangleq X \otimes X$.

(ii) We now consider

$$\partial_t z^{\mathbf{V}} - \Delta z^{\mathbf{V}} = -\mathscr{P}_L \operatorname{div}(z^{\mathbf{I}})^{\otimes 2}, \quad z^{\mathbf{V}}|_{t=0} = 0. \tag{8.33}$$

We then define $u_2 \triangleq u_1 - z^{\mathbf{V}}$, so that u_2 satisfies

$$\begin{aligned}&\partial_t u_2 - \Delta u_2\\ &= -\mathscr{P}_L \operatorname{div}((u_2 + z^{\mathbf{V}})^{\otimes 2} + (u_2 + z^{\mathbf{V}}) \otimes z^{\mathbf{I}} + z^{\mathbf{I}} \otimes (u_2 + z^{\mathbf{V}})).\end{aligned} \tag{8.34}$$

(iii) Next, we consider

$$\partial_t z^{\mathbf{Y}} - \Delta z^{\mathbf{Y}} = -\mathscr{P}_L \operatorname{div}(z^{\mathbf{V}} \otimes z^{\mathbf{I}} + z^{\mathbf{I}} \otimes z^{\mathbf{V}}), \quad z^{\mathbf{Y}}|_{t=0} = 0, \tag{8.35}$$

so that $u_3 \triangleq u_2 - z^{\mathbf{Y}}$ satisfies

$$\begin{aligned}&\partial_t u_3 - \Delta u_3\\ &= -\mathscr{P}_L \operatorname{div}((u_3 + z^{\mathbf{V}} + z^{\mathbf{Y}})^{\otimes 2} + (u_3 + z^{\mathbf{Y}}) \otimes z^{\mathbf{I}} + z^{\mathbf{I}} \otimes (u_3 + z^{\mathbf{Y}})).\end{aligned} \tag{8.36}$$

(iv) Lastly, we consider

$$\begin{aligned}&\partial_t z^{\mathbf{W}} - \Delta z^{\mathbf{W}} = -\mathscr{P}_L \operatorname{div}((z^{\mathbf{V}})^{\otimes 2} + z^{\mathbf{Y}} \otimes z^{\mathbf{I}} + z^{\mathbf{I}} \otimes z^{\mathbf{Y}}),\\ &z^{\mathbf{W}}|_{t=0} = 0,\end{aligned} \tag{8.37}$$

so that $u_4 \triangleq u_3 - z^{\mathbf{W}}$ satisfies

$$\begin{aligned}&\partial_t u_4 - \Delta u_4\\ &= -\mathscr{P}_L((u_4 + z^{\mathbf{W}} + z^{\mathbf{Y}})^{\otimes 2} + (u_4 + z^{\mathbf{W}} + z^{\mathbf{Y}}) \otimes z^{\mathbf{V}}\\ &\quad + z^{\mathbf{V}} \otimes (u_4 + z^{\mathbf{W}} + z^{\mathbf{Y}}) + (u_4 + z^{\mathbf{W}}) \otimes z^{\mathbf{I}} + z^{\mathbf{I}} \otimes (u_4 + z^{\mathbf{W}})).\end{aligned} \tag{8.38}$$

Now let us observe the regularity of $z^{\mathbf{I}}$, $z^{\mathbf{V}}$, $z^{\mathbf{Y}}$, and $z^{\mathbf{W}}$, understanding that the process can stop when all the products are *almost* well-defined, which informally requires the regularity of $\beta \geq 0$ according to Corollary 8.1.5. Since we just count the homogeneities, in the following informal computations, the homogeneity of products of objects with regularity α_1 and α_2 is $\min\{\alpha_1, \alpha_2, \alpha_1 + \alpha_2\}$. First, recalling that $\mathfrak{s} = (2, 1, \dots, 1)$ and $|\mathfrak{s}| = 2 + d$ by (8.10), it follows that

$$z^{\mathsf{I}} \text{ has a regularity of } \beta \text{ for } \beta < 1 - \frac{d}{2}, \tag{8.39}$$

yielding that $\operatorname{div}(z^{\mathsf{I}})^{\otimes 2}$ has a regularity of $1 - d$ $\mathbb{P}$-a. s. Considering (8.32) and (8.33), we see that

$$u_1, z^{\mathsf{Y}} \text{ has a regularity of } \beta < 3 - d. \tag{8.40}$$

The most singular term in (8.34) is $\operatorname{div}(z^{\mathsf{Y}} \otimes z^{\mathsf{I}})$ of regularity of $\beta < 3 - \frac{3d}{2}$, which allows us to conclude from (8.34) and (8.35) that

$$u_2, z^{\mathsf{Y}'} \text{ has a regularity of } \beta < 5 - \frac{3d}{2}. \tag{8.41}$$

The most singular products of (8.36) are $\operatorname{div}(z^{\mathsf{Y}})^{\otimes 2}$ and $\operatorname{div}(z^{\mathsf{Y}'} \otimes z^{\mathsf{I}})$, both of which have a regularity of $\beta < 5 - 2d$. Consequently,

$$u_3, z^{\mathsf{W}} \text{ has a regularity of } \beta < 7 - 2d. \tag{8.42}$$

Let us collect these results in the following table, denoting for simplicity

$$z_1 \triangleq z^{\mathsf{I}}, \quad z_2 \triangleq z^{\mathsf{Y}}, \quad z_3 \triangleq z^{\mathsf{Y}'}, \quad \text{and} \quad z_4 \triangleq z^{\mathsf{W}}.$$

Table 8.1: Singular structure and regularity of z_k.

k	The most singular term in z_k-equation, its regularity		Regularity of z_k
1	ξ,	$\beta < -\frac{d+2}{2}$	$\beta < 1 - \frac{d}{2}$
2	$\operatorname{div}(z_1)^{\otimes 2}$,	$\beta < 1 - d$	$\beta < 3 - d$
3	$\operatorname{div}(z_2 \otimes z_1 + z_1 \otimes z_2)$,	$\beta < 3 - \frac{3d}{2}$	$\beta < 5 - \frac{3d}{2}$
4	$\operatorname{div}(z_2^{\otimes 2} + z_3 \otimes z_1 + z_1 \otimes z_3)$,	$\beta < 5 - 2d$	$\beta < 7 - 2d$

We notice that the regularity of the most singular term in z_k-equation and consequently of z_k both improve by $2 - \frac{d}{2}$ at every step, which reveals that in the case $d = 4$, the regularity of z_k does not increase. This is why, informally stated, the locally critical case requires infinitely many repetitions of Da Prato–Debussche tricks, which, of course, is infeasible.

8.5.2 Second way to see the criticality of 4D

We describe another way to observe the criticality of the 4D case for the Navier–Stokes equations forced by STWN, namely that informally stated, the first Wick product $\mathbb{E}[\dot{z}_i \dot{z}_j]$, $1 \leq i, j \leq d$ (recall (8.18a)), already blows up. We explain by simply computing the most elementary renormalization constant given by (8.18a) in general dimension $\mathbb{T}^d, d \geq 2$. The author came to this observation while discussing with Prof. Matthew Novack. We note that the computation is standard; e. g., [374, p. 4486] states "$u_1^{\epsilon,i} \diamond u_1^{\epsilon,j} \to v_2^{ij}$" on [374, p. 4486], although details are completely omitted therein. For convenience, we first recall the following estimate from [374, Lemma 3.10].

Lemma 8.5.3. *Suppose that* $0 < l, m < d$, *and* $l + m > d$. *Then*

$$\sum_{\substack{k_1, k_2 \in \mathbb{Z}^d \setminus \{0\}: \\ k_{12} = k}} \frac{1}{|k_1|^l |k_2|^m} \lesssim \frac{1}{|k|^{l+m-d}}.$$

Let $\mathfrak{h} : [0, \infty) \to [0, \infty)$ be a smooth function such that

$$\mathfrak{h}(r) \triangleq \begin{cases} 1 & \text{if } r \geq 1, \\ 0 & \text{if } r \leq \frac{1}{2}, \end{cases} \qquad \mathfrak{l} \triangleq 1 - \mathfrak{h},$$

and consider, for $\lambda > 0$,

$$\check{\mathfrak{h}}_\lambda(x) \triangleq \mathcal{F}^{-1}\left(\mathfrak{h}\left(\frac{|\cdot|}{\lambda}\right)\right)(x), \quad \check{\mathfrak{l}}_\lambda(x) \triangleq \mathcal{F}^{-1}\left(\mathfrak{l}\left(\frac{|\cdot|}{\lambda}\right)\right)(x). \tag{8.43}$$

Then define the projections onto such higher and lower frequencies respectively by

$$\mathcal{H}_\lambda : \mathcal{S}'(\mathbb{R}^d) \to \mathcal{S}'(\mathbb{R}^d) \quad \text{by } \mathcal{H}_\lambda f \triangleq \check{\mathfrak{h}}_\lambda * f, \tag{8.44a}$$

$$\mathcal{L}_\lambda : \mathcal{S}'(\mathbb{R}^d) \to \mathcal{S}(\mathbb{R}^d) \quad \text{by } \mathcal{L}_\lambda f \triangleq f - \mathcal{H}_\lambda f = \check{\mathfrak{l}}_\lambda * f. \tag{8.44b}$$

We cut off the noise in lower frequencies so that $\xi_\lambda = \sum_{k \in \mathbb{Z}^d} \mathfrak{l}(\lambda^{-1}k)\hat{\xi}(k)$. Then we can solve the corresponding (8.31) with ξ replaced by such ξ_λ to see

$$\dot{z}_\lambda(t) = \int_0^t e^{(t-s)\Delta} \mathcal{P}_L \mathbb{P}_{\neq 0} \xi_\lambda(s) ds,$$

or, in coordinates,

$$\dot{z}_{\lambda,t}^{i} \triangleq \dot{z}_\lambda^{i}(t) = \int_0^t e^{(t-s)\Delta} \sum_{i_1=1}^d \mathcal{P}_L^{ii_1} \sum_{k \neq 0} \mathfrak{l}(\lambda^{-1}k)\hat{\xi}^{i_1}(k, s) ds e^{i2\pi k \cdot x}. \tag{8.45}$$

Denoting

$$\partial_t \zeta^{i_1}(s) = \hat{\xi}^{i_1}(s) \quad \text{and} \quad F_{i_1}(t,k) \triangleq \int_0^t e^{-(t-s)|k|^2} d\zeta^{i_1}(s,k) \tag{8.46}$$

and using (8.45) and (8.46), we can deduce

$$\mathcal{F}(z^{\dagger,i}_{\lambda,t})(k) = 1_{k\neq 0} \sum_{i_1=1}^{d} \widehat{\mathcal{P}_L}^{ii_1}(k)\iota(\lambda^{-1}k)F_{i_1}(t,k). \tag{8.47}$$

Denoting

$$k_{ij} \triangleq k_i + k_j \quad \text{and} \quad \psi_0(k_1,k_2) \triangleq \sum_{|c-d|\leq 1} \rho_c(k_1)\rho_d(k_2), \tag{8.48}$$

we recall the definition of resonance from (3.83a) and compute

$$\begin{aligned}\mathbb{E}[z^{\dagger,i}_{\lambda,t} \circ z^{\dagger,j}_{\lambda,t}] = &\sum_k \sum_{\substack{k_1,k_2:\\ k_{12}=k}} \sum_{c,d:|c-d|\leq 1} \phi_c(k_1)\phi_d(k_2)\\ &\times \mathbb{E}[\mathcal{F}(z^{\dagger,i}_{\lambda,t})(k_1)\mathcal{F}(z^{\dagger,j}_{\lambda,t})(k_2)]e^{i2\pi k\cdot x},\end{aligned}$$

so that applying the contractions of $k_1 + k_2 = 0$, $s = s'$, and $i_1 = j_1$ leads us to

$$\mathbb{E}[z^{\dagger,i}_{\lambda,t} \circ z^{\dagger,j}_{\lambda,t}] = C^{ij}(t), \tag{8.49a}$$

$$\text{where } C^{ij}(t) \triangleq \sum_{k\neq 0} \sum_{i_1=1}^{d} \frac{1-e^{-2t|k|^2}}{2|k|^2} \iota(\lambda^{-1}k)^2 \widehat{\mathcal{P}_L}^{ii_1}(k)\widehat{\mathcal{P}_L}^{ji_1}(k). \tag{8.49b}$$

Now analogous computations using the definition of $f \prec g$ from (3.83a) lead to

$$\begin{aligned}\mathbb{E}[z^{\dagger,i}_{\lambda,t} \prec z^{\dagger,j}_{\lambda,t}] = &\sum_k \sum_{\substack{k_1,k_2:\\ k_{12}=k}} \sum_{l\geq -1} \sum_{m\leq l-2} \phi_m(k_1)\phi_l(k_2)\\ &\times \mathbb{E}\Bigg[\int_0^t e^{-(t-s)|k_1|^2} \sum_{i_1=1}^{d} \widehat{\mathcal{P}_L}^{ii_1}(k_1)\iota(\lambda^{-1}k_1)\hat{\xi}^{i_1}(s,k_1,)ds\\ &\times \int_0^t e^{-(t-s')|k_2|^2} \sum_{j_1=1}^{d} \widehat{\mathcal{P}_L}^{jj_1}(k_2)\iota(\lambda^{-1}k_2)\hat{\xi}^{j_1}(s',k_2)ds'\Bigg],\end{aligned}$$

where the contraction of $k_1 + k_2 = 0$ will force $\phi_m(k_1)\phi_l(k_2) = 0$ due to (3.40), and hence

$$\mathbb{E}[z^{\dagger,i}_{\lambda,t} \prec z^{\dagger,j}_{\lambda,t}] = 0. \tag{8.50}$$

Then

$$
\begin{aligned}
:z_{\lambda,t}^{\text{¡},i} z_{\lambda,t}^{\text{¡},j}: &\overset{(8.18a)}{=} z_{\lambda,t}^{\text{¡},i} z_{\lambda,t}^{\text{¡},j} - \mathbb{E}\big[z_{\lambda,t}^{\text{¡},i} z_{\lambda,t}^{\text{¡},j}\big] \\
&\overset{(3.83b)}{=} z_{\lambda,t}^{\text{¡},i} z_{\lambda,t}^{\text{¡},j} - \mathbb{E}\big[z_{\lambda,t}^{\text{¡},i} \prec z_{\lambda,t}^{\text{¡},j} + z_{\lambda,t}^{\text{¡},i} \succ z_{\lambda,t}^{\text{¡},j} + z_{\lambda,t}^{\text{¡},i} \circ z_{\lambda,t}^{\text{¡},j}\big] \\
&\overset{(8.49a)(8.50)}{=} z_{\lambda,t}^{\text{¡},i} z_{\lambda,t}^{\text{¡},j} - C^{ij}(t).
\end{aligned}
\tag{8.51}
$$

The necessity to subtract out the renormalization constant arises from the resonant term, and thus our goal is to prove that for all $t \geq 0$, there exists a distribution $z_{\lambda,t}^{\text{¡},i} \diamond z_{\lambda,t}^{\text{¡},j} \in \mathcal{C}^{2-d-\kappa}(\mathbb{T}^d)$ such that

$$
z_{\lambda,t}^{\text{¡},i} \circ z_{\lambda,t}^{\text{¡},j} - C^{ij}(t) \to z_{\lambda,t}^{\text{¡},i} \diamond z_{\lambda,t}^{\text{¡},j} \quad \text{as } q \to \infty \tag{8.52}
$$

in $L^p(\Omega; C_{\mathrm{loc}}(\mathbb{R}_+; \mathcal{C}^{2-d-\kappa}(\mathbb{T}^d)))$; typically, we can also prove the convergence in $C_{\mathrm{loc}}(\mathbb{R}_+; \mathcal{C}^{2-d-\kappa}(\mathbb{T}^d))$ $\mathbb{P}$-a. s., but we limit ourselves to the former space to stay focused on our task of showing the criticality of the 4D case.

We compute the second chaos as

$$
\mathbb{E}\big[\big|\Delta_m\big(z_{\lambda,t}^{\text{¡},i} \circ z_{\lambda,t}^{\text{¡},j} - C^{ij}(t)\big)\big|^2\big] = \mathbb{E}\big[\big|\Delta_m\big(z_{\lambda,t}^{\text{¡},i} \circ z_{\lambda,t}^{\text{¡},j}\big)\big|^2\big] - \Delta_{-1} C^{ij}(t)^2. \tag{8.53}
$$

We focus on the first term, and for the ease of notations, w. l. o. g. we assume in the rest of these computations that $i = j = 1$; the other indices can be computed analogously. Since $z_{\lambda,t}^{\text{¡},i}$ is spatially mean-zero, by (8.47)–(8.48) we compute

$$
\begin{aligned}
\mathbb{E}\big[\big|\Delta_m\big(z_{\lambda,t}^{\text{¡},1} \circ z_{\lambda,t}^{\text{¡},1}\big)(t)\big|^2\big] = \mathbb{E}\Bigg[\Bigg| &\sum_{\substack{k,k' \in \mathbb{Z}^d: \\ k' \neq 0, k \neq k'}} e^{i2\pi k \cdot x} \phi_m(k) \psi_0(k-k',k') \\
&\times \Bigg(\sum_{i_1=1}^{d} \widehat{\mathcal{P}_L}^{1 i_1}(k-k') \iota(\lambda^{-1}(k-k')) F_{i_1}(t,k-k')\Bigg) \\
&\times \Bigg(\sum_{j_1=1}^{d} \widehat{\mathcal{P}_L}^{1 j_1}(k') \iota(\lambda^{-1}(k')) F_{j_1}(t,k')\Bigg)\Bigg|^2\Bigg].
\end{aligned}
$$

Then using (8.46), we can further simplify:

$$
\begin{aligned}
&\mathbb{E}\big[\big|\Delta_m\big(z_{\lambda,t}^{\text{¡},1} \circ z_{\lambda,t}^{\text{¡},1}\big)(t)\big|^2\big] \\
&= \sum_{k,k',k'',k''' \in \mathbb{Z}^d \setminus \{0\}} e^{i2\pi(k+k') \cdot x} \overline{e^{i2\pi(k''+k''') \cdot x}} \phi_m(k+k') \phi_m(k''+k''') \\
&\quad \times \psi_0(k,k') \psi_0(k'',k''') \sum_{i_1,j_1,i_1',j_1'=1}^{d} \widehat{\mathcal{P}_L}^{1 i_1}(k) \widehat{\mathcal{P}_L}^{1 j_1}(k') \widehat{\mathcal{P}_L}^{1 i_1'}(k'') \widehat{\mathcal{P}_L}^{1 j_1'}(k''') \\
&\quad \times \iota(\lambda^{-1}k) \iota(\lambda^{-1}k') \iota(\lambda^{-1}k'') \iota(\lambda^{-1}k''')
\end{aligned}
\tag{8.54}
$$

$$\times \mathbb{E}\left[\int_0^t e^{-(t-s)|k|^2} d\zeta^{i_1}(s,k) \int_0^t e^{-(t-s')|k'|^2} d\zeta^{j_1}(s',k') \right.$$
$$\left. \times \overline{\int_0^t e^{-(t-s'')|k''|^2} d\zeta^{i_1'}(s'',k'') \int_0^t e^{-(t-s''')|k'''|^2} d\zeta^{j_1'}(s''',k'''))}\right].$$

Applying (8.54) to (8.53) gives us

$$\begin{aligned}
&\mathbb{E}[|\Delta_m(z^{\dot{1},1}_{\lambda,t} \circ z^{\dot{1},1}_{\lambda,t} - C^{11}(t))|^2] \qquad (8.55)\\
&= \sum_{k,k',k'',k'''\in\mathbb{Z}^d\setminus\{0\}} e^{i2\pi(k+k')\cdot x} \overline{e^{i2\pi(k''+k''')\cdot x}} \phi_m(k+k')\phi_m(k''+k''')\\
&\times \psi_0(k,k')\psi_0(k'',k''') \sum_{i_1,j_1,i_1',j_1'=1}^d \widehat{\mathscr{P}_L}^{1i_1}(k)\widehat{\mathscr{P}_L}^{1j_1}(k')\widehat{\mathscr{P}_L}^{1i_1'}(k'')\widehat{\mathscr{P}_L}^{1j_1'}(k''')\\
&\times \mathfrak{l}(\lambda^{-1}k)\mathfrak{l}(\lambda^{-1}k')\mathfrak{l}(\lambda^{-1}k'')\mathfrak{l}(\lambda^{-1}k''')\\
&\times \mathbb{E}\left[\int_0^t e^{-(t-s)|k|^2} d\zeta^{i_1}(s,k) \int_0^t e^{-(t-s')|k'|^2} d\zeta^{j_1}(s',k')\right.\\
&\left.\times \overline{\int_0^t e^{-(t-s'')|k''|^2} d\zeta^{i_1'}(s'',k'') \int_0^t e^{-(t-s''')|k'''|^2} d\zeta^{j_1'}(s''',k'''))}\right] - \Delta_{-1}C^{11}(t)^2.
\end{aligned}$$

To handle the quadruple product of the STWN, we relying on (8.19b), we have

$$\begin{aligned}
&\mathbb{E}[|\Delta_m(z^{\dot{1},1}_{\lambda,t} \circ z^{\dot{1},1}_{\lambda,t} - C^{11}(t))|^2]\\
&= \sum_{k,k'\in\mathbb{Z}^d\setminus\{0\}} \left[e^{i4\pi(k+k')\cdot x} \phi_m(k+k')^2 \psi_0(k,k')^2 \right.\\
&\times \sum_{i_1,j_1=1}^d \widehat{\mathscr{P}_L}^{1i_1}(k)^2 \widehat{\mathscr{P}_L}^{1j_1}(k')^2 \mathfrak{l}(\lambda^{-1}k)^2 \mathfrak{l}(\lambda^{-1}k')^2\\
&\left.\times \int_0^t \int_0^t e^{-2(t-s)|k|^2} e^{-2(t-s')|k'|^2} ds ds'\right]\\
&+ \sum_{k,k'\in\mathbb{Z}^d\setminus\{0\}} \left[e^{i4\pi(k+k')\cdot x} \phi_m(k+k')^2 \psi_0(k,k')^2 \right.\\
&\times \sum_{i_1,j_1=1}^d \widehat{\mathscr{P}_L}^{1i_1}(k)^2 \widehat{\mathscr{P}_L}^{1j_1}(k')^2 \mathfrak{l}(\lambda^{-1}k)^2 \mathfrak{l}(\lambda^{-1}k')^2\\
&\left.\times \int_0^t \int_0^t e^{-2(t-s)|k|^2} e^{-2(t-s')|k'|^2} ds ds'\right]
\end{aligned}$$

$$
\begin{aligned}
&+ \sum_{k,k''\in\mathbb{Z}^d\setminus\{0\}} \Bigg[\phi_m(0)^2\psi_0(k,-k)\psi_0(k'',-k'') \\
&\times \sum_{i_1,i_1'=1}^{d} \widehat{\mathcal{P}_L}^{1i_1}(k)^2\widehat{\mathcal{P}_L}^{1i_1'}(k'')^2\iota(\lambda^{-1}k)^2\iota(\lambda^{-1}k'')^2 \\
&\times \int_0^t\int_0^t e^{-2(t-s)|k|^2}e^{-2(t-s'')|k''|^2}\,ds\,ds'' - \Delta_{-1}C^{11}(t)^2\Bigg].
\end{aligned}
$$

The last two terms cancel out leaving us with

$$
\begin{aligned}
&\mathbb{E}[|\Delta_m(z_{\lambda,t}^{\dot{|},1}\circ z_{\lambda,t}^{\dot{|},1} - C^{11}(t))|^2] \\
&= \frac{1}{2}\sum_{k,k'\in\mathbb{Z}^d\setminus\{0\}} e^{i4\pi(k+k')\cdot x}\phi_m(k+k')^2\psi_0(k,k')^2 \\
&\times \sum_{i_1,j_1=1}^{d} \widehat{\mathcal{P}_L}^{1i_1}(k)^2\widehat{\mathcal{P}_L}^{1j_1}(k')^2\iota(\lambda^{-1}k)^2\iota(\lambda^{-1}k')^2\left(\frac{1-e^{-2t|k|^2}}{|k|^2}\right)\left(\frac{1-e^{-2t|k'|^2}}{|k'|^2}\right).
\end{aligned}
\tag{8.56}
$$

Then we first estimate $\psi_0(k,k')$ as follows. We have $2^m \approx |k+k'|$, $2^c \approx |k|$, $2^d \approx |k'|$, so that $2^m \approx |k+k'| \lesssim |k| + |k'| \approx 2^c + 2^d \approx 2^c$, and therefore $m \lesssim c$, $|k'| \approx 2^d \approx 2^c \gtrsim 2^m$. From this, by (8.48) for all $\delta > 0$, we get the estimate

$$
(\psi_0(k,k'))^2 \lesssim |k'|^{\delta}2^{-m\delta}. \tag{8.57}
$$

Then for any $\delta \in (0,1)$, we are ready to estimate by (8.56) and Lemma 8.5.3, which is justified only for $d \in \{2,3\}$:

$$
\begin{aligned}
&\mathbb{E}[|\Delta_m(z_{\lambda,t}^{\dot{|},1}\circ z_{\lambda,t}^{\dot{|},1} - C^{11}(t))|^2] \\
&\lesssim 2^{-m\delta}\sum_{\substack{k,k'\in\mathbb{Z}^d\setminus\{0\}:\\ |k+k'|\approx 2^m}} \frac{1}{|k|^2}\frac{1}{|k'|^{2-\delta}} \lesssim 2^{m(2d-4)}.
\end{aligned}
\tag{8.58}
$$

We emphasize that the hypothesis of Lemma 8.5.3 requires "$l+m > d$", so that $2+(2-\delta) > d$, implying that only $d \in \{2,3\}$ is allowed. Naturally, we can retry to estimate without relying on Lemma 8.5.3 to hope for extension to $d \geq 4$. For example, first,

$$
\frac{1}{|k|^2|k'|^2} \lesssim \frac{1}{|k|^4} + \frac{1}{|k'|^4},
$$

so that by the symmetry of k and k' we reduce our task to estimate by (8.56), for any $\delta > 0$,

$$
\mathbb{E}[|\Delta_m(z_{\lambda,t}^{\dot{|},1}\circ z_{\lambda,t}^{\dot{|},1} - C^{11}(t))|^2] \lesssim \sum_{k,k'\in\mathbb{Z}^d\setminus\{0\}} \phi_m^2(k+k')(\psi_0(k,k'))^2\frac{1}{|k'|^4}
$$

$$\approx 2^{m(d-\delta)} \sum_{\substack{k' \in \mathbb{Z}^d \setminus \{0\}: \\ |k'| \gtrsim 2^m}} \frac{1}{|k'|^{4-\delta}} \approx 2^{2m(d-2)},$$

where the convergence requires $d \in \{2,3\}$. In other words, we only have $\frac{1}{|k'|^4}$ from the square of Fourier symbols of $(-\Delta)^{-1}$ and $|k'| \gtrsim 2^m$. To eliminate $(\sum_{|c-d| \le 1} \phi_c(k)\phi_d(k'))^2$, we need to give up a positive power of $|k'|$ (or $|k-k'|$), namely $\delta > 0$. Thus $\frac{1}{|k'|^{4-\delta}}$, and $\frac{1}{|k'|^{4-\delta}}|k'|^{d-1} = \frac{1}{|k'|^{5-d-\delta}}$ is no longer summable over $\{k' \in \mathbb{Z}^d : |k'| \gtrsim 2^m\}$ in $d \ge 4$. Thus we see the criticality of dimension 4 once again.

Let us complete the computations even under the compromise to $d \in \{2,3\}$ anyway. For all $\kappa > 0$ and $p \in (2 \vee (\frac{2d}{\kappa}), \infty)$, so that $B_{p,\infty}^{2-d-\frac{\kappa}{2}}(\mathbb{T}^d) \hookrightarrow B_{\infty,\infty}^{2-d-\kappa}(\mathbb{T}^d)$ due to (3.54b), using Gaussian hypercontractivity theorem (Lemma 4.6.4), we have

$$\begin{aligned}
&\mathbb{E}[\|z_{\lambda,t}^{\dot{\vert},1} \circ z_{\lambda,t}^{\dot{\vert},1} - C^{11}(t)\|_{\mathcal{C}^{2-d-\kappa}}^p] \lesssim \mathbb{E}[\|z_{\lambda,t}^{\dot{\vert},1} \circ z_{\lambda,t}^{11} - C^{11}(t)\|_{B_{p,p}^{2-d-\frac{\kappa}{2}}}^p] \\
&\approx \sum_{m \ge -1} 2^{(2-d-\frac{\kappa}{2})mp} \int_{\mathbb{T}^d} \|\Delta_m(z_{\lambda,t}^{\dot{\vert},1} \circ z_{\lambda,t}^{\dot{\vert},1} - C^{11}(t))\|_{L_\omega^p}^p \, dx \\
&\lesssim \sum_{m \ge -1} 2^{(2-d-\frac{\kappa}{2})mp} \int_{\mathbb{T}^d} \|\Delta_m(z_{\lambda,t}^{\dot{\vert},1} \circ z_{\lambda,t}^{\dot{\vert},1} - C^{11}(t))\|_{L_\omega^2}^p \, dx \overset{(8.58)}{\lesssim} \sum_{m \ge -1} 2^{-\frac{\kappa m p}{2}} \lesssim 1.
\end{aligned}$$

This will lead us to a conclusion that for any $t \ge 0$, there exists a distribution $z_{\lambda,t}^{\dot{\vert},i} \diamond z_{\lambda,t}^{\dot{\vert},j} \in \mathcal{C}^{2-d-\kappa}(\mathbb{T}^d)$ such that $z_{\lambda,t}^{\dot{\vert},i} \circ z_{\lambda,t}^{\dot{\vert},j} - C^{ij} \to z_{\lambda,t}^{\dot{\vert},i} \diamond z_{\lambda,t}^{\dot{\vert},j}$ as $\lambda \to \infty$ in $L^p(\Omega; C_{\text{loc}}(\mathbb{R}_+; \mathcal{C}^{2-d-\kappa}(\mathbb{T}^d)))$, $d \in \{2,3\}$. We refer to [359, Appendix A.5] for more details.

Sections 8.5.3, 8.5.4, and 8.5.5 will describe the applications of the theories of RS and paracontrolled distributions but limited to the locally subcritical case.

8.5.3 Example of RS on the 3D MHD system

Here we describe the main ideas of the theory of regularity structures (RS) [167] by an example and refer the interested readers to [167–169] for more detail; we will focus on one part of the proof and give detailed computations in the case of the MHD system [352].

The key intuition in the theory of RS is that a function is considered smooth when it can be approximated by polynomials. In contrast, the behavior of the solution to an SPDE forced by STWN is dominated by the STWN convoluted by a Green function. As a remedy, we take the following steps.

1. The first step is to construct a "regularity structure" endowed with all the calculus operators (multiply, differentiate, integrate) so that an abstract Taylor expansion for distributions becomes possible.
2. The second step is to find a local-in-time solution to the abstract formulation of the original SPDE via Picard iteration fixed point theory in RS.

3. The next step is to rely on the reconstruction theorem (see [167, Theorem 3.10] and also its extension to Besov spaces in [170, Theorem 3.1]); however, the reconstructed solution does not satisfy the original SPDE in a naive sense because the classical multiplication of distributions is ill-defined, and thus we rely on the renormalization group to identify the renormalized solutions.
4. The last step consists of verifying the convergence of the renormalized models to a limiting model, which will be the main focus of the following discussion.

The rest of this Section 8.5.3 is devoted to give a sketch of such steps in the case of the 3D MHD system forced by STWN:

$$\partial_t u + (u \cdot \nabla)u + \nabla\pi = \Delta u + (b \cdot \nabla)b + \xi_u, \quad \nabla \cdot u = 0, \tag{8.59a}$$

$$\partial_t b + (u \cdot \nabla)b = \Delta b + (b \cdot \nabla)u + \xi_b, \tag{8.59b}$$

where we assume particularly that the noises $\xi_u = (\xi_u^1, \xi_u^2, \xi_u^3)$ and $\xi_b = (\xi_b^1, \xi_b^2, \xi_b^3)$ are both STWN such that for all $i, j \in \{1, 2, 3\}$,

$$\mathbb{E}[\xi_u^i(x,t)\xi_u^j(y,s)] = 1_{i=j}\delta(x-y)\delta(t-s), \tag{8.60a}$$

$$\mathbb{E}[\xi_b^i(x,t)\xi_b^j(y,s)] = 1_{i=j}\delta(x-y)\delta(t-s), \tag{8.60b}$$

$$\mathbb{E}[\xi_u^i(x,t), \xi_b^j(y,s)] = 0, \tag{8.60c}$$

in accordance with those in the physics literature (e. g., [54, p. 1200]). Hereafter, for brevity, we may write $y \triangleq (u, b)$ and $\xi \triangleq (\xi_u, \xi_b)$.

Remark 8.5.4. Analogously to the singular Navier–Stokes equations, the MHD system forced by STWN becomes locally critical when $d = 4$ (recall Definition 8.5.2). We note that the Hall-MHD system forced by STWN, even in the $2\frac{1}{2}$D case, is locally supercritical, and no solution theory, even locally in time, is known. This was pointed out in [351, Section 3], and yet no progress is known.

For completeness, we state the precise type of local solution theory that can be obtained via the theory of RS for the 3D MHD system. In the statement and the proof of Theorem 8.5.5, a convolution $*_{t,x}$ w. r. t. both space and time will be simply denoted as $*$.

Theorem 8.5.5 ([352, Theorem 1.1]). *Let $u^{\text{in}}, b^{\text{in}} \in \mathcal{C}^\eta(\mathbb{T}^3)$ for $\eta \in (-1, a+2]$, where $a \in (-\frac{13}{5}, -\frac{5}{2})$. Let $\xi = (\xi_u, \xi_b)$, with each ξ_u^i and ξ_b^i being independent STWN on $\mathbb{R} \times \mathbb{T}^3$, which satisfy (8.60) and are extended periodically to $\mathbb{R}^4$. Define the space-time mollifier $\rho_\epsilon(t,x)$ identically to Example 17, $\xi_{u,\epsilon}^i \triangleq \rho_\epsilon * \xi_u^i$, and $\xi_{b,\epsilon}^i \triangleq \rho_\epsilon * \xi_b^i$. Suppose that (u_ϵ, b_ϵ) is the maximal solution to*

$$\partial_t u_\epsilon^i = \Delta u_\epsilon^i + \sum_{i_1=1}^{3} P^{ii_1}\xi_{u,\epsilon}^{i_1} - \frac{1}{2}\sum_{i_1=1}^{3} P^{ii_1}\left(\sum_{j=1}^{3} D_j(u_\epsilon^{i_1}u_\epsilon^j)\right)$$

$$+\frac{1}{2}\sum_{i_1=1}^{3} P^{ii_1}\left(\sum_{j=1}^{3} D_j(b_\epsilon^{i_1} b_\epsilon^j)\right), \quad u_\epsilon(0) = Pu^{\text{in}}, \tag{8.61a}$$

$$\partial_t b_\epsilon^i = \Delta b_\epsilon^i + \sum_{i_1=1}^{3} P^{ii_1}\xi_{b,\epsilon}^{i_1} - \frac{1}{2}\sum_{i_1=1}^{3} P^{ii_1}\left(\sum_{j=1}^{3} D_j(b_\epsilon^{i_1} u_\epsilon^j)\right)$$

$$+\frac{1}{2}\sum_{i_1=1}^{3} P^{ii_1}\left(\sum_{j=1}^{3} D_j(u_\epsilon^{i_1} b_\epsilon^j)\right), \quad b_\epsilon(0) = Pb^{\text{in}}, \tag{8.61b}$$

$i \in \{1,2,3\}$. *Then there exist* $u, b \in C([0,\tau); \mathcal{C}^\eta(\mathbb{T}^3))$ *and a sequence of random time* τ_L *converging to the explosion time* τ *such that for every* $\delta > 0$,

$$\lim_{\epsilon \to 0} \mathbb{P}\left(\left\{\omega : \sup_{t\in[0,\tau_L]} \|(u^\epsilon, b^\epsilon)(\omega) - (u,b)(\omega)\|_{\mathcal{C}^\eta} \geq \delta\right\}\right) = 0.$$

We remark that the global-in-time unique solution theory for the 2D MHD system forced by STWN was recently proven in [353], which will be descried in Chapter 9; its extension to the 3D case is completely open.

Remark 8.5.6. We make a quick remark here that in Section 8.5.2, we mollified the STWN only in space. In contrast, we mollified the STWN in both space and time in Theorem 8.5.5. This is typical. In application of the theory of RS, to construct an RS for a fixed SPDE, it is important to work directly with the heat kernel convoluted with STWN that has to be then mollified in space-time. In applications of the theory of paracontrolled distributions, one often works with the process z that solves (8.31) that is continuous in time. For example, Lemma 7.1.1 treated the case where the noise is white only in time, but analogous temporal regularity can be extended to the case of STWN; however, the spatial regularity will be much worse, and thus the need to mollify only in space (see, e. g., [183, Proposition 4.1]).

We start to give a sketch of the proof of Theorem 8.5.5. We emphasize again that the full details will take significant amount of notations and preliminaries, and instead we refer the interested readers to [352, 167–169] for details.

First, following [173, p. 1656], for any $l > 0$, let $\mathcal{B}_0^l(\mathbb{R}^d)$ be the space of $\mathcal{C}^l(\mathbb{R}^d)$-functions with compact support on $B_1(0)$ such that their $\mathcal{C}^l(\mathbb{R}^d)$-norms are bounded by one. For $\phi \in \mathcal{B}_0^l(\mathbb{R}^d)$, $\lambda > 0$, and $x, y \in \mathbb{R}^d$, let

$$\phi_x^\lambda(y) \triangleq \frac{1}{\lambda^d}\phi\left(\frac{y-x}{\lambda}\right) \quad \text{and} \quad \phi^\lambda \triangleq \phi_0^\lambda. \tag{8.62}$$

Let us give a complete definition of RS.

Definition 8.5.7 (Cf. [167, Definition 2.1]). A regularity structure is $\mathcal{J} = (\mathcal{A}, \mathcal{T}, \mathcal{G})$, where:

1. $\mathcal{A} \subset \mathbb{R}$ is an index set that is locally finite and bounded from below, and $0 \in \mathcal{A}$.

2. A model space $\mathcal{T} = \oplus_{a\in\mathcal{A}}\mathcal{T}_a$ is a graded normed vector space, where $\mathcal{T}_a$ is a Banach space for every $a \in \mathcal{A}$. Because $0 \in \mathcal{A}$, we set $\mathcal{T}_0$ to be one-dimensional with a basis vector of 1. For $a \in \mathcal{T}$, we denote by $\mathcal{Q}_a a$ the component of a in $\mathcal{T}_a$ and by $\|a\|_a$ the norm of the component (see [167, p. 292] and [173, Remark 2.3]).
3. $\mathcal{G}$ is a structure group of linear operators acting on $\mathcal{T}$ such that for all $\Gamma \in \mathcal{G}$, $a \in \mathcal{A}$, and $a \in \mathcal{T}_a$, we have $\Gamma a - a \in \mathcal{T}_{<a} \triangleq \oplus_{\beta<a}\mathcal{T}_\beta$. Moreover, $\Gamma 1 = 1$ for all $\Gamma \in \mathcal{G}$.

Example 18 (Cf. [173, Remark 2.2] and [167, p. 295]). An important example of an RS, through which an inhomogeneous model will be defined in Definition 8.5.8, is that generated by an abstract polynomial in $(1 + d)$ indeterminates $X_0, \ldots, X_d$ of scaled degree r. Given a multiindex $k = (k_0, \ldots, k_d)$, we denote $X^k \triangleq X_0^{k_0} \ldots X_d^{k_d}$ with $X^0 \triangleq 1$ in the case of an empty monoid. In this case, $\mathcal{A} = \{a \in \mathbb{N} : a \le r\}$, and each $\mathcal{T}_a$ for $a \in \mathcal{A}$ contains all monomials in X_i that are homogeneous of scaled degree a. The structure group $\mathcal{G}_{\text{poly}}$ is a group of translations in $\mathbb{R}^{1+d}$ acting on X^k by $h \mapsto (X - h)^k$, i. e., $\Gamma_h X^k \triangleq (X - h)^k$ for all $h \in \mathbb{R}^{1+d}$. We fix $r > 0$ sufficiently large, denote by $\mathcal{T}_{\text{poly}}$ the space of such polynomials with scaled degree r, and define $\mathcal{F}_{\text{poly}} \triangleq \{X^k : |k|_{\mathfrak{s}} \le r\}$.

To consider a heat kernel composed with Leray projection $\mathcal{P}_L$, we rely on the following inhomogeneous model from [173].

Definition 8.5.8 (Cf. [173, Definition 2.4 and Remark 2.5]). Given an RS $\mathcal{J} = (\mathcal{A}, \mathcal{T}, \mathcal{G})$, an inhomogeneous model $\Upsilon \triangleq (\Pi, \Gamma, \Sigma)$ consists of the following elements.
1. Γ is a collection of groups $\Gamma^t : \mathbb{R}^d \times \mathbb{R}^d \to \mathcal{G}$ parameterized by $t \in \mathbb{R}$ and such that

$$\Gamma_{xx}^t = 1, \quad \Gamma_{xy}^t \Gamma_{yz}^t = \Gamma_{xz}^t \tag{8.63}$$

for all $x, y, z \in \mathbb{R}^d$ and $t \in \mathbb{R}$, and the action of Γ_{xy}^t on polynomials is given as in Example 18 with $h = (0, y - x)$.
2. Σ is a collection of maps $\Sigma_x : \mathbb{R} \times \mathbb{R} \to \mathcal{G}$ parameterized by $x \in \mathbb{R}^d$ and such that

$$\Sigma_x^{tt} = 1, \quad \Sigma_x^{sr}\Sigma_x^{rt} = \Sigma_x^{st}, \quad \Sigma_x^{st}\Gamma_{xy}^t = \Gamma_{xy}^s \Sigma_y^{st} \tag{8.64}$$

for all $x \in \mathbb{R}^d$ and $s, r, t \in \mathbb{R}$, and the action of Σ_x^{st} on polynomials is given as in Example 18 with $h = (t - s, 0)$.
3. Π is a collection of linear maps $\Pi_x^t : \mathcal{T} \to \mathcal{S}'(\mathbb{R}^d)$ such that

$$\Pi_y^t = \Pi_x^t \Gamma_{xy}^t, \quad (\Pi_x^t X^{(0,\bar{k})})(y) = (y - x)^{\bar{k}}, \quad (\Pi_x^t X^{(k_0,\bar{k})})(y) = 0 \tag{8.65}$$

for all $x, y \in \mathbb{R}^d$, $t \in \mathbb{R}$, $\bar{k} \in \mathbb{N}^d$, and $k_0 \in \mathbb{N}$ such that $k_0 > 0$.

Finally, for all $\gamma > 0$ and $T > 0$, we assume that there exists a constant c with which the following analytic bounds hold uniformly over all $\tau \in \mathcal{T}_l$ for $l, m \in \mathcal{A}$ such that

$m < l < \gamma$, all $\lambda \in (0,1]$, all $\phi \in \mathscr{B}_0^r(\mathbb{R}^d)$ with $r > -\lfloor \min \mathscr{A} \rfloor$, all $t, s \in [-T, T]$ such that $|t-s| \le 1$, and all $x, y \in \mathbb{R}^d$ such that $|x-y| \le 1$:

$$|\langle \Pi_x^t \tau, \phi_x^\lambda \rangle| \le c\|\tau\|_l \lambda^l, \tag{8.66a}$$

$$\|\Gamma_{xy}^t \tau\|_m \le c\|\tau\|_l |x-y|^{l-m}, \tag{8.66b}$$

$$\|\Sigma_x^{st} \tau\|_m \le c\|\tau\|_l |t-s|^{\frac{l-m}{\mathfrak{s}_0}}. \tag{8.66c}$$

We formally define $\|\Pi\|_{\gamma;T}$, $\|\Gamma\|_{\gamma;T}$, $\|\Sigma\|_{\gamma;T}$ as the minimum of the constants for which (8.66) hold, respectively. Hereafter, we refer to any inhomogeneous model as simply a model. Finally, for models $\Upsilon = (\Pi, \Gamma, \Sigma)$ and $\overline{\Upsilon} = (\overline{\Pi}, \overline{\Gamma}, \overline{\Sigma})$, we define

$$|||\Upsilon; \overline{\Upsilon}|||_{\gamma;T} \triangleq \|\Pi - \overline{\Pi}\|_{\gamma;T} + \|\Gamma - \overline{\Gamma}\|_{\gamma;T} + \|\Sigma - \overline{\Sigma}\|_{\gamma;T}, \tag{8.67}$$

$$\text{where } |||\Upsilon|||_{\gamma;T} \triangleq \|\Pi\|_{\gamma;T} + \|\Gamma\|_{\gamma;T} + \|\Sigma\|_{\gamma;T}. \tag{8.68}$$

Definition 8.5.9 (Cf. [173, p. 1660], [375, Definition 2.4]). Denote $|t|_0 \triangleq |t|^{\frac{1}{\mathfrak{s}_0}} \wedge 1$ and $|t,s|_0 \triangleq |t|_0 \wedge |s|_0$. Given an RS $\mathscr{G} = (\mathscr{A}, \mathscr{T}, \mathscr{G})$ with model $\Upsilon = (\Pi, \Gamma, \Sigma)$, $\gamma > 0$, $\eta \in \mathbb{R}$, and final time $T > 0$, $\mathscr{D}_T^{\gamma,\eta} = \mathscr{D}_T^{\gamma,\eta}(\Upsilon)$ consists of all functions $H : (0,T] \times \mathbb{R}^d \to \mathscr{T}_{<\gamma} = \oplus_{\beta<\gamma} \mathscr{T}_\beta$ such that

$$|||H|||_{\gamma,\eta;T} \triangleq \|H\|_{\gamma,\eta;T} + \sup_{\substack{s,t\in(0,T]:s\neq t \\ |t-s|\le|t,s|_0^{\mathfrak{s}_0}}} \sup_{x\in\mathbb{R}^d} \sup_{l<\gamma} \frac{\|H_t(x) - \Sigma_x^{ts} H_s(x)\|_l}{|t,s|_0^{\eta-\gamma} |t-s|^{\frac{\gamma-l}{\mathfrak{s}_0}}} < \infty, \tag{8.69}$$

where

$$\|H\|_{\gamma,\eta;T} \triangleq \sup_{t\in(0,T]} \sup_{x\in\mathbb{R}^d} \sup_{l<\gamma} |t|_0^{(l-\eta)\vee 0} \|H_t(x)\|_l + \sup_{t\in(0,T]} \sup_{\substack{x,y\in\mathbb{R}^d: \\ x\neq y, |x-y|\le 1}} \sup_{l<\gamma} \frac{\|H_t(x) - \Gamma_{xy}^t H_t(y)\|_l}{|x-y|^{\gamma-l} |t|_0^{\eta-\gamma}}, \tag{8.70}$$

with $H_t(x) \triangleq H(t,x)$. Such H is called an inhomogeneous modeled distribution. The norm $|||H|||_{\gamma,\eta;T}$ depends not only on γ, η, and T but also on Γ and Σ, while not on Π. With this in mind, for $H \in \mathscr{D}_T^{\gamma,\eta}(\Upsilon)$ and $\overline{H} \in \mathscr{D}_T^{\gamma,\eta}(\overline{\Upsilon})$, where $\overline{\Upsilon} = (\overline{\Pi}, \overline{\Gamma}, \overline{\Sigma})$ is another model, we define

$$|||H; \overline{H}|||_{\gamma,\eta;T} \triangleq \|H; \overline{H}\|_{\gamma,\eta;T} + \sup_{\substack{s,t\in(0,T]:s\neq t \\ |t-s|\le|t,s|_0}} \sup_{x\in\mathbb{R}^d} \sup_{l<\gamma} \frac{\|H_t(x) - \Sigma_x^{ts} H_s(x) - \overline{H}_t(x) + \overline{\Sigma}_x^{ts} \overline{H}_s(x)\|_l}{|t,s|_0^{\eta-\gamma} |t-s|^{\frac{\gamma-l}{\mathfrak{s}_0}}}, \tag{8.71}$$

where

$$\|H;\overline{H}\|_{\gamma,\eta;T} \triangleq \sup_{t\in(0,T]} \sup_{x\in\mathbb{R}^d} \sup_{l<\gamma} |t|_0^{(l-\eta)\vee 0} \|H_t(x) - \overline{H}_t(x)\|_l$$
$$+ \sup_{t\in(0,T]} \sup_{\substack{x,y\in\mathbb{R}^d:x\neq y\\|x-y|\leq 1}} \sup_{l<\gamma} \frac{\|H_t(x) - \Gamma^t_{xy}H_t(y) - \overline{H}_t(x) + \overline{\Gamma}^t_{xy}\overline{H}_t(y)\|_l}{|x-y|^{\gamma-l}|t|_0^{\eta-\gamma}}. \tag{8.72}$$

Definition 8.5.10 (Cf. [173, Definition 2.14]). Given an RS $\mathcal{G} = (\mathcal{A},\mathcal{T},\mathcal{G})$, a linear map $\mathcal{I} : \mathcal{T} \to \mathcal{T}$ is an abstract integration map of order $\beta > 0$ if it satisfies the following three properties.

1. $\mathcal{I} : \mathcal{T}_m \to \mathcal{T}_{m+\beta}$ for all $m \in \mathcal{A}$ such that $m + \beta \in \mathcal{A}$,
2. $\mathcal{I}\tau = 0$ for all $\tau \in \mathcal{T}_{\text{poly}}$ (recall Example 18),
3. $\mathcal{I}\Gamma\tau - \Gamma\mathcal{I}\tau \in \mathcal{T}_{\text{poly}}$ for all $\tau \in \mathcal{T}$ and $\Gamma \in \mathcal{G}$.

Definition 8.5.11 (Cf. [173, Definition 2.16] and [167, p. 351–352]). A function $K : \mathbb{R}^{1+d} \setminus \{0\} \to \mathbb{R}$ is regularizing of order $\beta > 0$ if there exists $r > 0$ such that

1. $K = \sum_{n\geq 0} K_n$, where each K_n is smooth and compactly supported in $\{z \in \mathbb{R}^{1+d} : \|z\|_{\mathfrak{s}} \leq c2^{-n}\}$ for some $c > 0$,
2. for all multiindices k such that $|k|_{\mathfrak{s}} \leq r$, $\sup_z |D^k K_n(z)| \lesssim 2^{(1+d-\beta+|k|_{\mathfrak{s}})n}$ uniformly in n,
3. K annihilates every polynomial of scaled degree r, i. e., $\int_{\mathbb{R}^{1+d}} z^k K_n(z)dz = 0$ for all $k \in \mathbb{N}^{1+d}$ such that $|k|_{\mathfrak{s}} \leq r$ (recall (8.4)).

Definition 8.5.12 (Cf. [173, Definition 2.17] and [167, Definition 5.9]). Let $\mathcal{I}$ be an abstract integration map of order $\beta > 0$ for an RS $\mathcal{G} = (\mathcal{A},\mathcal{T},\mathcal{G})$ with model $\Upsilon = (\Pi,\Gamma,\Sigma)$, and let K be regularizing of order β with $r > -\lfloor \min \mathcal{A} \rfloor$. We say that Υ realizes K for $\mathcal{I}$ if the following two conditions are satisfied.

1. For all $\alpha \in \mathcal{A}$, $\tau \in \mathcal{T}_\alpha$, and $x \in \mathbb{R}^d$,
$$\Pi^t_x(\mathcal{I}\tau + J_{t,x}\tau)(y) = \int_{\mathbb{R}} \langle \Pi^s_x \Sigma^{st}_x \tau, K_{t-s}(y - \cdot)\rangle ds, \tag{8.73}$$
where
$$J_{t,x}\tau \triangleq \sum_{|k|_{\mathfrak{s}}<\alpha+\beta} \frac{X^k}{k!} \int_{\mathbb{R}} \langle \Pi^s_x \Sigma^{st}_x \tau, D^k K_{t-s}(x - \cdot)\rangle ds, \tag{8.74}$$
$k \in \mathbb{N}^{1+d}$, and D^k is a differentiation operator in time-space.
2. We have the identities
$$\Gamma^t_{xy}(\mathcal{I} + J_{t,y}) = (\mathcal{I} + J_{t,x})\Gamma^t_{xy}, \quad \Sigma^{st}_x(\mathcal{I} + J_{t,x}) = (\mathcal{I} + J_{s,x})\Sigma^{st}_x \tag{8.75}$$
for all $s,t \in \mathbb{R}$, and $x,y \in \mathbb{R}^d$.

Now as a first step, we rely on [167, Lemma 5.5] to split the heat kernel $e^{t\Delta}$ as $K(t,x)+R(t,x)$, where K is a 2-regularizing kernel, and $R(t,x) \in C^{\infty}(\mathbb{R}^4)$. We can apply [352, Lemma 2.5] to the (i,j)-entry of the Leray projection $\mathcal{P}_L^{ij}$ to analogously split it to $\overline{\mathcal{P}_L}^{ij} + R_0^{ij}$. Next, we define abstract integration maps (see [167, Definition 5.7], [173, Definition 2.14]) $\mathcal{I}$ as an abstract integration corresponding to K, $\mathcal{I}_0^{ij}$ corresponding to $\overline{\mathcal{P}_L}^{ij}$, $\mathcal{I}_l$ corresponding to $D^l K$, and $\mathcal{I}^{ij} = \mathcal{I}_0^{ij}\mathcal{I}$ and $\mathcal{I}_l^{ij} = \mathcal{I}_0^{ij}\mathcal{I}_l$. The key fact here is that the abstract integration map does not need to be defined for R or R_0^{ij} because they are smooth and thus contribute to symbols of strictly positive regularity; additionally, the integration maps cannot be directly defined on $e^{t\Delta}$ or $\mathcal{P}_L^{ij}$ because they do not satisfy [167, Assumptions 5.1 and 5.4]. We also denote by $\Theta_{u,i}$ and $\Theta_{b,i}$ to represent $\xi_{u,i}$ and $\xi_{b,i}$ on the RS.

We define a set $\mathcal{F}$ by postulating that $\{1\}\{\Theta_{u,i}, \Theta_{b,i}\}_i, \{X_j\} \subset \mathcal{F}$, as well as

$$\tau\overline{\tau} \in \mathcal{F}, \quad \mathcal{I}_k\tau \in \mathcal{F}, \quad \text{and} \quad \mathcal{I}_{0,k}^{ij}\tau \in \mathcal{F}$$

whenever $\tau, \overline{\tau} \in \mathcal{F}$. To each $\tau \in \mathcal{F}$ we associate a weight $|\tau|_{\mathfrak{s}}$ by setting

$$\begin{aligned} |1|_{\mathfrak{s}} &= 0, \quad |\tau\overline{\tau}|_{\mathfrak{s}} = |\tau|_{\mathfrak{s}} + |\overline{\tau}|_{\mathfrak{s}}, \quad |X_i|_{\mathfrak{s}} = \mathfrak{s}_i, \\ |\Theta_i|_{\mathfrak{s}} &= \alpha \quad \text{for all } i \in \{1,\dots,6\}, \text{ i.e., } |\Theta_{u,i}|_{\mathfrak{s}} = \alpha, \quad |\Theta_{b,i}|_{\mathfrak{s}} = \alpha, \\ |\mathcal{I}_k(\tau)|_{\mathfrak{s}} &= |\tau|_{\mathfrak{s}} + 2 - |k|_{\mathfrak{s}}, \quad |\mathcal{I}_{0,k}^{ij}(\tau)|_{\mathfrak{s}} = |\tau|_{\mathfrak{s}} - |k|_{\mathfrak{s}} \quad \text{for } i,j \in \{1.2\}. \end{aligned} \tag{8.76}$$

As we already did in Section 8.5.1, let us use tree diagrams to describe some notations. We must distinguish the notation of a black circle for Θ_u and Θ_b by labeling it u or b beside it. Following the previous notations in Section 8.5.1, $\mathcal{I}^{ij}$ is denoted by a downward line, $\mathcal{I}_{i_1}^{ij}$ by a downward dotted line, and, finally, $\mathcal{I}_{i_1}$ by a self-crossing curve. For example,

$\mathcal{I}^{ii_1}(\Theta_{u,i_1}) =$ , $\mathcal{I}^{ii_1}(\Theta_{u,i_1})\mathcal{I}^{jj_1}(\Theta_{u,j_1}) =$ 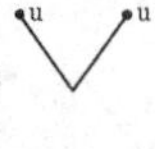,

$\mathcal{I}_j^{ii_1}(\mathcal{I}^{i_1i_2}(\Theta_{u,i_2})) =$ , $\mathcal{I}_j^{ii_1}(\mathcal{I}^{i_1i_2}(\Theta_{u,i_2})\mathcal{I}^{jj_1}(\Theta_{u,j_1})) =$ 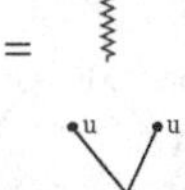,

$\mathcal{I}_k^{ii_1}(\mathcal{I}^{i_1i_2}(\Theta_{u,i_2})\mathcal{I}^{kk_1}(\Theta_{u,k_1}))\mathcal{I}_l^{jj_1}(\mathcal{I}^{j_1j_2}(\Theta_{u,j_2})\mathcal{I}^{ll_1}(\Theta_{u,l_1})) =$,

$\mathcal{I}_j(\mathcal{I}^{ii_1}(\Theta_{u,i_1})\mathcal{I}^{jj_1}(\Theta_{u,j_1})) =$  .

In contrast to the tree diagrams introduced in Section 8.5.1, is new. The reason for this necessity, which may seem surprising because the products in the MHD system always have Leray projection, is because we are now trying to employ the theory of RS to construct a solution theory, and such abstract integration $\mathcal{I}_{i_1}$ w. r. t. $D_{i_1}K$ arises on various occasions. We refer readers to $\mathcal{U}_{u,n}$, $\mathcal{U}_{b,n}$ in [352, eq. (39)] for details.

Remark 8.5.13. As described in Theorem 8.5.5, we work with mollified noise ξ_ϵ. For such a fixed mollified noise $\xi_\epsilon = (\xi_{u,\epsilon}, \xi_{b,\epsilon})$, we can build a canonical model $\Upsilon_\epsilon \triangleq (\Pi^{(\epsilon)}, \Gamma^{(\epsilon)}, \Sigma^{(\epsilon)})$ for the RS (see [167, p. 433]). For all $i, i_1 \in \{1,2,3\}$, we set

$$(\Pi_x^{(\epsilon,t)} \mathcal{I}^{ii_1}(\Theta_{u,i_1}))(y) = K^{ii_1} * \xi_{u,\epsilon}^{i_1}(t,y), \tag{8.77a}$$

$$(\Pi_x^{(\epsilon,t)} \mathcal{I}^{ii_1}(\Theta_{b,i_1}))(y) = K^{ii_1} * \xi_{b,\epsilon}^{i_1}(t,y), \tag{8.77b}$$

and, recursively,

$$(\Pi_x^{(\epsilon,t)} \tau\overline{\tau})(y) = (\Pi_x^{(\epsilon,t)} \tau)(y)(\Pi_x^{(\epsilon,t)} \overline{\tau})(y), \tag{8.78a}$$

$$(\Sigma_x^{(\epsilon,st)} \tau\overline{\tau}) = (\Sigma_x^{(\epsilon,st)} \tau)(\Sigma_x^{(\epsilon,st)} \overline{\tau}), \tag{8.78b}$$

$$(\Gamma_{xy}^{(\epsilon,t)} \tau\overline{\tau}) = (\Gamma_{xy}^{(\epsilon,t)} \tau)(\Gamma_{xy}^{(\epsilon,t)} \overline{\tau}) \tag{8.78c}$$

(see [173, eqs. (4.33a)–(4.33b)] and [375, eq. (2.6)]).

Next, we can define the following precise family of symbols needed to employ the fixed argument in RS:

$\mathcal{F}_0 \triangleq$

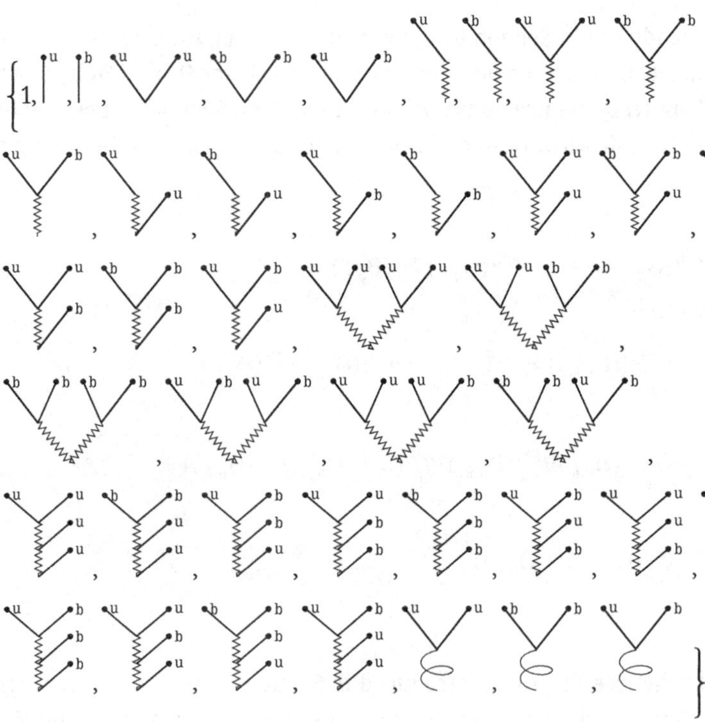

(8.79)

Description (8.79) is not fully detailed, as there are missing components; for completeness, we include them in Section A.8, where the elements of $\mathcal{F}_0$ are split into groups G_j, $j \in \{1, \ldots, 13\}$.

Next, we define $\mathcal{H}_0 \triangleq \langle \mathcal{F}_0 \rangle$, a set of all linear combinations of $\mathcal{F}_0$, and then renormalization linear maps $M\colon \mathcal{H}_0 \to \mathcal{H}_0$ by

$$M(\mathcal{G}^{ii_1}(\Theta_{u,i_1})\mathcal{G}^{jj_1}(\Theta_{u,j_1})) \triangleq \mathcal{G}^{ii_1}(\Theta_{u,i_1})\mathcal{G}^{jj_1}(\Theta_{u,j_1}) - C^{1,uu}_{ii_1jj_1}, \tag{8.80a}$$

$$M(\mathcal{G}^{ii_1}(\Theta_{b,i_1})\mathcal{G}^{jj_1}(\Theta_{b,j_1})) \triangleq \mathcal{G}^{ii_1}(\Theta_{b,i_1})\mathcal{G}^{jj_1}(\Theta_{b,j_1}) - C^{1,bb}_{ii_1jj_1}, \tag{8.80b}$$

$$\begin{aligned} &M(\mathcal{G}^{ii_1}_k(\mathcal{G}^{i_1i_2}(\Theta_{u,i_2})\mathcal{G}^{kk_1}(\Theta_{u,k_1}))\mathcal{G}^{jj_1}_l(\mathcal{G}^{j_1j_2}(\Theta_{u,j_2})\mathcal{G}^{ll_1}(\Theta_{b,l_1}))) \\ &\triangleq \mathcal{G}^{ii_1}_k(\mathcal{G}^{i_1i_2}(\Theta_{u,i_2})\mathcal{G}^{kk_1}(\Theta_{u,k_1}))\mathcal{G}^{jj_1}_l(\mathcal{G}^{j_1j_2}(\Theta_{u,j_2})\mathcal{G}^{ll_1}(\Theta_{b,l_1})) \\ &\quad - C^2_{ii_1i_2jj_1j_2kk_1ll_1}, \end{aligned} \tag{8.80c}$$

$$\begin{aligned} &M(\mathcal{G}^{ii_1}_l(\mathcal{G}^{i_1i_2}_k(\mathcal{G}^{i_2i_3}(\Theta_{u,i_3})\mathcal{G}^{kk_1}(\Theta_{u,k_1}))\mathcal{G}^{ll_1}(\Theta_{u,l_1}))\mathcal{G}^{jj_1}(\Theta_{u,j_1})) \\ &\triangleq \mathcal{G}^{ii_1}_l(\mathcal{G}^{i_1i_2}_k(\mathcal{G}^{i_2i_3}(\Theta_{u,i_3})\mathcal{G}^{kk_1}(\Theta_{u,k_1}))\mathcal{G}^{ll_1}(\Theta_{u,l_1}))\mathcal{G}^{jj_1}(\Theta_{u,j_1}) \\ &\quad - C^3_{ii_1i_2i_3ll_1kk_1jj_1}, \end{aligned} \tag{8.80d}$$

$$\begin{aligned} &M(\mathcal{G}^{ii_1}_l(\mathcal{G}^{i_1i_2}(\Theta_{u,i_2})\mathcal{G}^{ll_1}_k(\mathcal{G}^{l_1l_2}(\Theta_{u,l_2})\mathcal{G}^{kk_1}(\Theta_{u,k_1})))\mathcal{G}^{jj_1}(\Theta_{u,j_1})) \\ &\triangleq \mathcal{G}^{ii_1}_l(\mathcal{G}^{ll_1}_k(\mathcal{G}^{l_1l_2}(\Theta_{u,l_2})\mathcal{G}^{kk_1}(\Theta_{u,k_1}))\mathcal{G}^{i_1i_2}(\Theta_{u,i_2}))\mathcal{G}^{jj_1}(\Theta_{u,j_1}) \\ &\quad - C^4_{ii_1i_2ll_1l_2kk_1jj_1}, \end{aligned} \tag{8.80e}$$

$$M(\tau) \triangleq \tau \quad \text{for all other remaining elements of } \mathcal{F}_0, \tag{8.80f}$$

where $C^{1,uu}_{ii_1jj_1}$, $C^{1,bb}_{ii_1jj_1}$, $C^2_{ii_1i_2jj_1j_2kk_1ll_1}$, $C^3_{ii_1i_2i_3ll_1kk_1jj_1}$, and $C^4_{ii_1i_2ll_1l_2kk_1jj_1}$ are all renormalization constants. For precise forms of such renormalization constants, we refer to [352, eqs. (131), (180), (184), and (192)]. Denoting by $\mathcal{R}_0$ the set of all such maps, such as M, we can verify that it satisfies the definition of a renormalization group in [167, Definition 8.41]. Additionally, as described on [167, eq. (8.33)], we can build a new model $(\Pi^M, \Gamma^M, \Sigma^M)$ such that

$$\Pi^{M,t}_x \tau = \Pi^t_x M\tau \tag{8.81}$$

(see [352, eq. (63)]).

The rest of this section is devoted to give an idea on the proof of the convergence of the renormalized models to a limiting model, specifically [352, Theorem 3.7], which we summarize here.

Proposition 8.5.14 ([352, Theorem 3.7]). *Under the hypothesis of* $\Upsilon_\epsilon \triangleq (\Pi^{(\epsilon)}, \Gamma^{(\epsilon)}, \Sigma^{(\epsilon)})$ *from Remark* 8.5.13, *let* $\{M_\epsilon\}_{\epsilon>0}$ *be a family of renormalization linear maps corresponding to* $\epsilon > 0$, *and define*

$$\widehat{\Upsilon}_\epsilon \triangleq M_\epsilon \Upsilon_\epsilon \quad \textit{where } M_\epsilon \Upsilon_\epsilon \triangleq (\Pi^{M,(\epsilon)}, \Gamma^{M,(\epsilon)}, \Sigma^{M,(\epsilon)}). \tag{8.82}$$

There exist a random model $\widehat{\Upsilon}$ independent of the mollifier ρ and $M_\epsilon \in \mathscr{R}_0$ such that $M_\epsilon \Upsilon_\epsilon \to \widehat{\Upsilon}$ as $\epsilon \searrow 0$ in probability. More precisely, for all $\theta \in (0, -\frac{5}{2} - \alpha)$ (recall $\alpha < -\frac{5}{2}$ from the hypothesis of Theorem 8.5.5) and $\gamma < r$,

$$\mathbb{E}[\|\!|M_\epsilon \Upsilon_\epsilon; \widehat{\Upsilon}|\!\|_{\gamma;T}] \lesssim \epsilon^\theta$$

uniformly over $\epsilon \in (0,1]$.

To prove Proposition 8.5.14, it suffices to show that for all $\tau \in G_j$ from (A.17) of negative order, test functions $\phi \in \mathscr{B}_0^r(\mathbb{R}^3)$, $x \in \mathbb{R}^3$, and $t \in [0,T]$, there exist random variables $\widehat{\Pi}_x^t \tau(\phi)$ such that

$$\mathbb{E}[|(\widehat{\Pi}_x^t \tau)(\phi_x^\lambda)|^2] \lesssim \lambda^{2|\tau|_\mathfrak{s}+\kappa} \tag{8.83}$$

for some $\kappa > 0$ sufficiently small and

$$\mathbb{E}[|(\widehat{\Pi}_x^t \tau - \widehat{\Pi}_x^{(\epsilon,t)} \tau)(\phi_x^\lambda)|^2] \lesssim \epsilon^{2\theta} \lambda^{2|\tau|_\mathfrak{s}+\kappa} \tag{8.84}$$

for some $\theta \in (0, -\frac{5}{2} - \alpha)$. We can verify as follows that G_js with elements of negative orders are $j = 2, 3, 5, 7, 13$, where we recall that G_js are defined in Appendix A.8.

1. $G_1 = \{1\}$, and $|1|_\mathfrak{s} = 0$,
2. G_2 consists of $\mathcal{I}^{ii_1}(\Theta_{u,i_1})$ that satisfies $|\mathcal{I}^{ii_1}(\Theta_{u,i_1})|_\mathfrak{s} = 2 + \alpha < 0$,
3. G_3 consists of $\mathcal{I}^{ii_1}(\Theta_{u,i_1})\mathcal{I}^{jj_1}(\Theta_{u,j_1})$ that satisfies $|\mathcal{I}^{ii_1}(\Theta_{u,i_1})\mathcal{I}^{jj_1}(\Theta_{u,j_1})|_\mathfrak{s} = 2(2+\alpha) < 0$,
4. G_4 consists of $\mathcal{I}_j^{ii_1}(\mathcal{I}^{i_1i_2}(\Theta_{u,i_2}))$ that satisfies $|\mathcal{I}_j^{ii_1}(\mathcal{I}^{i_1i_2}(\Theta_{u,i_2}))|_\mathfrak{s} = 3 + \alpha > 0$,
5. G_5 consists of $\mathcal{I}_j^{ii_1}(\mathcal{I}^{i_1i_2}(\Theta_{u,i_2})\mathcal{I}^{jj_1}(\Theta_{u,j_1}))$ that satisfies
$$|\mathcal{I}_j^{ii_1}(\mathcal{I}^{i_1i_2}(\Theta_{u,i_2})\mathcal{I}^{jj_1}(\Theta_{u,j_1}))|_\mathfrak{s} = 5 + 2\alpha < 0,$$
6. G_6 is similar to G_4,
7. G_7 that consists of $\mathcal{I}_k^{ii_1}(\mathcal{I}^{i_1i_2}(\Theta_{u,i_2}))\mathcal{I}^{jj_1}(\Theta_{u,j_1})$ that satisfies
$$|\mathcal{I}_k^{ii_1}(\mathcal{I}^{i_1i_2}(\Theta_{u,i_2}))\mathcal{I}^{jj_1}(\Theta_{u,j_1})|_\mathfrak{s} = 5 + 2\alpha < 0,$$
8. G_8 is similar to G_7,
9. G_9 consists of $\mathcal{I}_k^{ii_1}(\mathcal{I}^{i_1i_2}(\Theta_{u,i_2})\mathcal{I}^{kk_1}(\Theta_{u,k_1}))\mathcal{I}^{jj_1}(\Theta_{u,j_1})$ that satisfies
$$|\mathcal{I}_k^{ii_1}(\mathcal{I}^{i_1i_2}(\Theta_{u,i_2})\mathcal{I}^{kk_1}(\Theta_{u,k_1}))\mathcal{I}^{jj_1}(\Theta_{u,j_1})|_\mathfrak{s} = 7 + 3\alpha < 0,$$
10. G_{10} that consists of $\mathcal{I}_k^{ii_1}(\mathcal{I}^{i_1i_2}(\Theta_{u,i_2})\mathcal{I}^{kk_1}(\Theta_{u,k_1}))\mathcal{I}_l^{jj_1}(\mathcal{I}^{j_1j_2}(\Theta_{u,j_2})\mathcal{I}^{ll_1}(\Theta_{u,l_1}))$ that satisfies $|\mathcal{I}_k^{ii_1}(\mathcal{I}^{i_1i_2}(\Theta_{u,i_2})\mathcal{I}^{kk_1}(\Theta_{u,k_1}))\mathcal{I}_l^{jj_1}(\mathcal{I}^{j_1j_2}(\Theta_{u,j_2})\mathcal{I}^{ll_1}(\Theta_{u,l_1}))|_\mathfrak{s} = 2(5+2\alpha) < 0$,
11. G_{11} consists of $\mathcal{I}_l^{ii_1}(\mathcal{I}_k^{i_1i_2}(\Theta_{u,i_3})\mathcal{I}^{kk_1}(\Theta_{u,k_1}))\mathcal{I}^{ll_1}(\Theta_{u,l_1})\mathcal{I}^{jj_1}(\Theta_{u,j_1})$ that satisfies $|\mathcal{I}_l^{ii_1}(\mathcal{I}_k^{i_1i_2}(\Theta_{u,i_3})\mathcal{I}^{kk_1}(\Theta_{u,k_1}))\mathcal{I}^{ll_1}(\Theta_{u,l_1})\mathcal{I}^{jj_1}(\Theta_{u,j_1})|_\mathfrak{s} = 10 + 4\alpha < 0$,

12. G_{12} is similar to G_{11},
13. G_{13} is similar to G_5.

We now fix $(\Omega, \mathcal{F}, \mathbb{P})$, a Hilbert space $H = L^2(\mathbb{R}\times\mathbb{T}^3)$, and a collection $W_h = (W_{u,h}, W_{b,h})$ of jointly Gaussian random variables indexed by $h \in H$ such that $\mathbb{E}[W_h W_{\overline{h}}] = \langle h, \overline{h}\rangle$. We consider a family of mth-order Wiener integrals (recall Section 8.3.2)

$$I_m : H^{\otimes m} \to L^2(\Omega, \mathcal{F}, \mathbb{P}).$$

Let $\{S_z\}_{z\in\mathbb{R}^4}$ be a group of unitary operators via which $\mathbb{R}^4$ acts onto H and ρ_ϵ such that $\xi_\epsilon(z) = I_1(S_z\rho_\epsilon)$. In our case of vector-valued ξ, we can consider

$$\xi^i(\psi) = W^i_\psi = (I_1(\psi))^i. \tag{8.85}$$

As in [167, p. 479], we can take $(S_z f)(\overline{z}) = f(\overline{z} - z)$ for $z = (t, x)$. By (8.82) we know that $\widehat{\Pi}^{(\epsilon,t)}_0 = M_\epsilon \Pi^{(\epsilon,t)}_0$. Because the mapping $\phi \mapsto (\widehat{\Pi}^{(\epsilon,t)}_x \tau)(\phi)$ is linear due to Definition 8.5.8(3), we can find distributions $\widehat{\mathcal{W}}^{(\epsilon;k)}\tau$ such that $(\widehat{\mathcal{W}}^{(\epsilon;k)}\tau)(z) \in L^2(\mathbb{R}\times\mathbb{T}^3)^{\otimes k}$ for $z = (t, x) \in \mathbb{R}^4$ and

$$\langle \widehat{\Pi}^{(\epsilon,t)}_0 \tau, \phi\rangle \triangleq (\widehat{\Pi}^{(\epsilon,t)}_0 \tau)(\phi) = \sum_{k\le\|\tau\|_n} I_k\Big(\int \phi(y)(\widehat{\mathcal{W}}^{(\epsilon;k)}\tau)(t,y)dy\Big), \tag{8.86}$$

where $\|\tau\|_n$ denotes the sum of the numbers of occurrences of Θ_u and Θ_b. Hence, to complete the proof, it suffices to find $\widehat{\mathcal{W}}^{(k)}\tau \in L^2(\mathbb{R}\times\mathbb{T}^3)^{\otimes k}$, define

$$\langle \widehat{\Pi}^t_x \tau, \phi\rangle \triangleq (\widehat{\Pi}^t_x \tau)(\phi) \triangleq \sum_{k\le\|\tau\|_n} I_k\Big(\int \phi(y) S^{\otimes k}_{(t,x)}(\widehat{\mathcal{W}}^{(k)}\tau)(t,y)dy\Big), \tag{8.87}$$

where we note that the law of the random variable is independent of x, and then estimate the terms

$$|\langle(\widehat{\mathcal{W}}^{(\epsilon;k)}\tau)(t,y), (\widehat{\mathcal{W}}^{(\epsilon;k)}\tau)(t,\overline{y})\rangle|, \tag{8.88a}$$

$$|\langle(\delta\widehat{\mathcal{W}}^{(\epsilon;k)}\tau)(t,y), (\delta\widehat{\mathcal{W}}^{(\epsilon;k)}\tau)(t,\overline{y})\rangle|, \tag{8.88b}$$

where

$$\delta\widehat{\mathcal{W}}^{(\epsilon;k)}\tau \triangleq \widehat{\mathcal{W}}^{(\epsilon;k)}\tau - \widehat{\mathcal{W}}^{(k)}\tau. \tag{8.89}$$

To this endeavor, we can start by explicitly computing $\langle \widehat{\Pi}^{(\epsilon,t)}_0 \tau, \phi\rangle$ using the definition of $\widehat{\Pi}^{(\epsilon,t)}_0$ and compare with (8.86) to determine $\widehat{\mathcal{W}}^{(\epsilon,k)}$.

Example 19. (G_5 in (A.17e)) As a first example, let us select

$$\tau = \mathcal{J}^{ii_1}_j(\mathcal{J}^{i_1i_2}(\Theta_{u,i_2}))\mathcal{J}^{jj_1}(\Theta_{b,j_1}) \tag{8.90}$$

from G_5 in (A.17e), which is informally written

$$(\mathcal{P}_L^{ii_1} * (D_j e^{-t\Delta} * (\mathcal{P}_L^{i_1 i_2} * (e^{-t\Delta} * \xi_{u,i_2}))))(\mathcal{P}_L^{jj_1} * (e^{-t\Delta} * \xi_{b,j_1})).$$

Denote by $z = (t, y)$ and $\overline{z} = (t, \overline{y})$. Observe that

$$\widehat{\Pi}_0^{(\epsilon,t)}\tau \overset{(8.82)}{=} \Pi_0^{M_\epsilon,(\epsilon,t)}\tau \overset{(8.81)}{=} \Pi_0^{(\epsilon,t)} M_\epsilon \tau \overset{(8.80f)}{=} \Pi_0^{(\epsilon,t)}\tau. \tag{8.91}$$

Thus

$$\langle \widehat{\Pi}_0^{(\epsilon,t)}\tau, \phi\rangle = \int \Pi_0^{(\epsilon,t)} \mathcal{G}_j^{ii_1}(\mathcal{G}^{i_1 i_2}(\Theta_{u,i_2})\mathcal{G}^{jj_1}(\Theta_{b,j_1}))(y)\phi(y)dy. \tag{8.92}$$

Then, because $|\mathcal{G}^{i_1 i_2}(\Theta_{u,i_2})\mathcal{G}^{jj_1}(\Theta_{b,j_1})|_{\mathfrak{s}} = 4 + 2\alpha$, (8.73)–(8.74) and (8.77)–(8.78) allow us to write

$$\begin{aligned}\langle \widehat{\Pi}_0^{(\epsilon,t)}\tau, \phi\rangle = &\iiint D_j K^{ii_1}(t - s_0, y - y_0)\\ &\times [K^{i_1 i_2} * \xi_{u,\epsilon}^{i_2}(s_0, y_0)][K^{jj_1} * \xi_{b,\epsilon}^{j_1}(s_0, y_0)]dy_0 ds_0 \phi(y)dy.\end{aligned} \tag{8.93}$$

We now apply (8.18a) to deduce

$$\begin{aligned}\langle \widehat{\Pi}_0^{(\epsilon,t)}\tau, \phi\rangle = &\iiint D_j K^{ii_1}(t - s_0, y - y_0) \iint K^{i_1 i_2}(z_0 - z_1) K^{jj_1}(z_0 - z_2)\\ &\times [: \xi_{u,\epsilon}^{i_2}(z_1)\xi_{b,\epsilon}^{j_1}(z_2) : + \mathbb{E}[\xi_{u,\epsilon}^{i_2}(z_1)\xi_{b,\epsilon}^{j_1}(z_2)]]dz_1 dz_2 dy_0 ds_0 \phi(y)dy.\end{aligned} \tag{8.94}$$

We can rely on (8.60c), so that $\mathbb{E}[\xi_{u,\epsilon}^{i_2}(z_1)\xi_{b,\epsilon}^{j_1}(z_2)]$ vanishes. Additionally, we recall (8.85) and identify $\xi_{u,\epsilon}^{i_2}(z_1)$ with $I_1^{i_2}(S_{z_1}\rho_\epsilon)$ and $\xi_{b,\epsilon}^{j_1}(z_2)$ with $I_1^{j_1+3}(S_{z_2}\rho_\epsilon)$; here the second case has the superscript $j_1 + 3$ because it corresponds to ξ_b and $\xi = (\xi_u, \xi_b)$. Additionally, we use (8.24) to realize that

$$I_1(f)I_1(g) = I_2(f \otimes g) + I_0(f \otimes_1 g)$$

and more precisely in our case, with $\{e_k\}_{k=1}^6$ as a standard basis of $\mathbb{R}^6$ temporarily,

$$I_1^{i_2}(f)I_1^{j_1+3}(g) = I_2((fe_{i_2}) \otimes (ge_{j_1+3})) + \langle fe_{i_2}, ge_{j_1+3}\rangle_{L^2(\mathbb{R}\times\mathbb{T}^3;\mathbb{R}^6)},$$

where $\langle\cdot,\cdot\rangle_{L^2(\mathbb{R}\times\mathbb{T}^3;\mathbb{R}^6)}$ is the inner product of $L^2(\mathbb{R} \times \mathbb{T}^3; \mathbb{R}^6)$, which vanishes since $i_2 \neq j_1 + 3$. This allows us to deduce

$$\begin{aligned}\langle \widehat{\Pi}_0^{(\epsilon,t)}\tau, \phi\rangle = &\iint I_2((S_{z_1} \otimes S_{z_2})(e_{i_2} \otimes e_{j_1+3}))\\ &\times \left[\int \phi(y)\left[\int D_j K^{ii_1}(z - z_0)K_\epsilon^{i_1 i_2}(z_0 - z_1)K_\epsilon^{jj_1}(z_0 - z_2)dz_0\right]dy\right]dz_1 dz_2.\end{aligned}$$

In comparison with (8.87), we see that τ is of second homogeneous Wiener chaos and

$$\widehat{\mathcal{W}}^{(2)}\tau(t,y;z_1,z_2) \triangleq \int D_j K^{ii_1}(z-z_0)K^{i_1i_2}(z_0-z_1)K^{jj_1}(z_0-z_2)dz_0. \tag{8.95}$$

Example 20. (G_7 in (A.17g)) In the previous example, we described how to derive the specific form of $\widehat{\mathcal{W}}^{(2)}\tau(t,y;z_1,z_2)$ in (8.95). In this example, let us denote

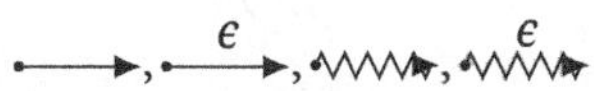

to represent $K, K_\epsilon, DK, DK_\epsilon$, where K_ϵ is K convoluted via mollifier. Moreover, if a vertex is replaced by ○, then it implies that the corresponding variable is integrated out. Without loss of generality, let us work on $\tau = \mathcal{G}_k^{ii_1}(\mathcal{G}^{i_1i_2}(\Theta_{b,i_2}))\mathcal{G}^{jj_1}(\Theta_{u,j_1})$ from G_7 in (A.17g), for which analogous computations in Example 19 show that

$$\begin{aligned}\widehat{\mathcal{W}}^{(\epsilon;2)}\tau(z;z_1,z_2,z_3) = \int & D_k K^{ii_1}(z-z_1)K_\epsilon^{i_1i_2}(z_1-z_2)K_\epsilon^{jj_1}(z-z_3) \\ & - D_k K^{ii_1}(t-s_1,-y_1)K_\epsilon^{i_1i_2}(z_1-z_2)K_\epsilon^{jj_1}(z-z_3)dz_1\end{aligned}$$

$=$ 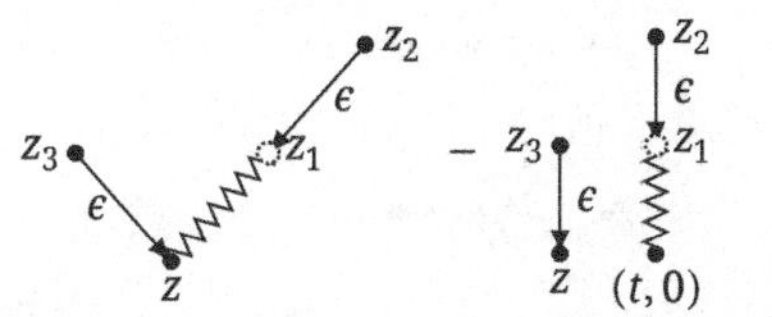 (8.96)

and give ideas of how to set up estimates toward (8.88a), specifically a bound on

$$|\langle \widehat{\mathcal{W}}^{(\epsilon;2)}\tau(z), \widehat{\mathcal{W}}^{(\epsilon;2)}\tau(\bar{z})\rangle|.$$

Within $\langle \widehat{\mathcal{W}}^{(\epsilon;2)}\tau(z), \widehat{\mathcal{W}}^{(\epsilon;2)}\tau(\bar{z})\rangle$, z and $\bar{z}$ are the fixed variables, and z_1 is integrated out; hence it is actually an L^2-inner product in z_2 and z_3. For the first term, pictorially we first flip 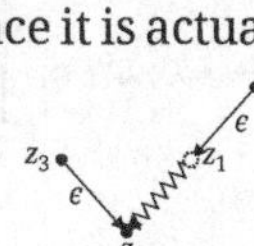 to 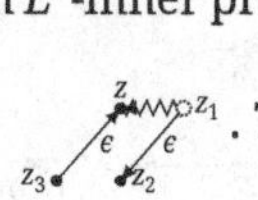. Then because we need to multiply with 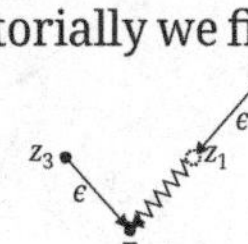, as z_1 is integrated in both, we change this dummy variable to z_0 to obtain 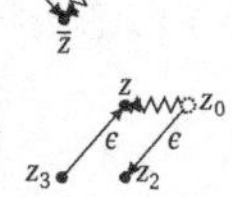, and hence matching them up leads to

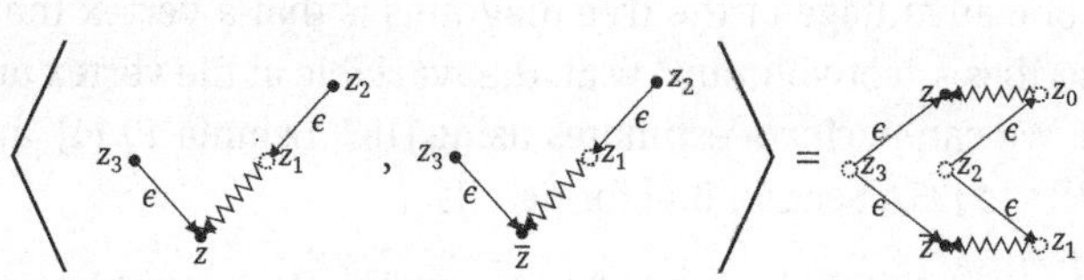

Similarly,

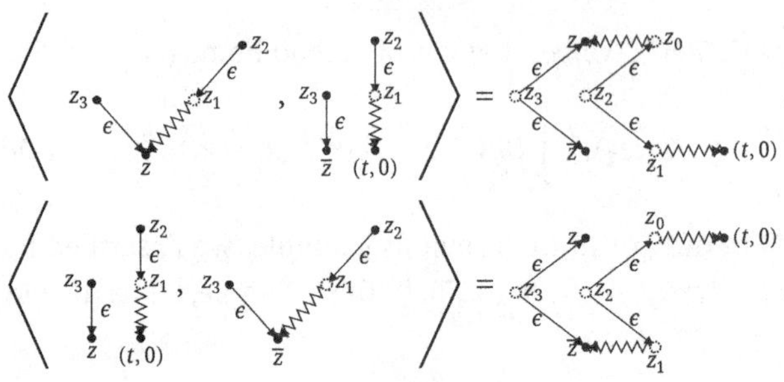

and

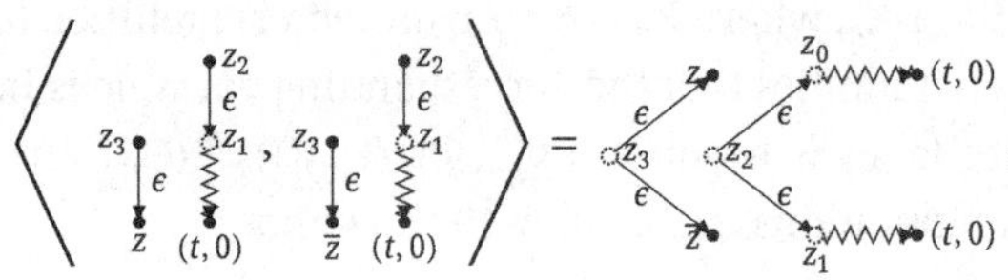

Then we see from (8.96) we see that

$$\langle \widehat{\mathcal{W}}^{(\epsilon;2)}\tau(z), \widehat{\mathcal{W}}^{(\epsilon;2)}\tau(\overline{z})\rangle$$

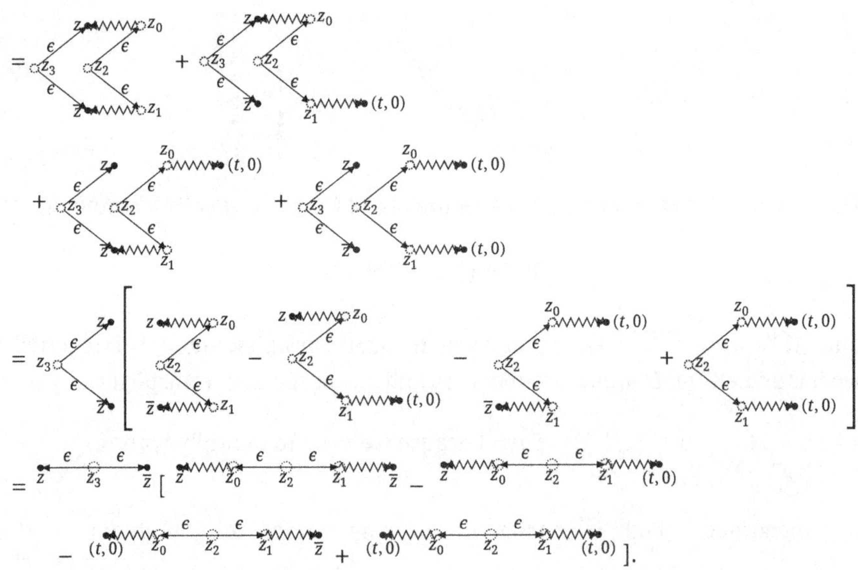

As we can see, one advantage of the tree diagrams is that a vertex that connects only two arrows describes a “convolution” w.r.t. that variable at the vertex and thus the easiest to estimate. We can perform estimates using [167, Lemma 10.14], and we refer the interested readers to [352, Section B.4] for details.

Example 21. (G_{10} in (A.17j)) In Example 20, we saw how to manipulate tree diagrams to ease the notations in computations. In this example, we work w. l. o. g. on

$$\tau = \mathcal{I}_k^{ii_1}(\mathcal{I}^{i_1i_2}(\Theta_{u,i_2})\mathcal{I}^{kk_1}(\Theta_{u,k_1}))\mathcal{I}_l^{jj_1}(\mathcal{I}^{j_1j_2}(\Theta_{u,j_2})\mathcal{I}^{ll_1}(\Theta_{b,l_1}))$$

from G_{10} in (A.17j) and describe how the renormalization constant pictorially appears in tree diagrams as the zeroth Wiener chaos. By computations similar to (8.91), but with (8.80c) instead of (8.80f), we get

$$\widehat{\Pi}_0^{(\epsilon,t)}\tau \overset{(8.82)(8.81)}{=} \Pi_0^{(\epsilon,t)}M_\epsilon\tau \overset{(8.80c)}{=} \Pi_0^{(\epsilon,t)}(\tau - C^2_{ii_1i_2jj_1j_2kk_1ll_1}),$$

and, consequently, similarly to (8.92),

$$\begin{aligned}\langle\widehat{\Pi}_0^{(\epsilon,t)}\tau,\phi\rangle = \int \Pi_0^{(\epsilon,t)}[&\mathcal{I}_k^{ii_1}(\mathcal{I}^{i_1i_2}(\Theta_{u,i_2})\mathcal{I}^{kk_1}(\Theta_{u,k_1}))\\ &\times \mathcal{I}_l^{jj_1}(\mathcal{I}^{j_1j_2}(\Theta_{u,j_2})\mathcal{I}^{ll_1}(\Theta_{b,l_1})) - C^2_{ii_1i_2jj_1j_2kk_1ll_1}]\phi(y)dy.\end{aligned} \tag{8.97}$$

Then, similarly to (8.93), because, e. g., $|\mathcal{I}^{i_1i_2}(\Theta_{u,i_2})\mathcal{I}^{kk_1}(\Theta_{u,k_1})|_{\mathfrak{s}} = 4 + 2\alpha$, (8.73)–(8.74) and (8.77)–(8.78) allow us to write

$$\begin{aligned}&\langle\widehat{\Pi}_0^{(\epsilon,t)}\tau,\phi\rangle = \int\!\!\int\!\!\int\!\!\int\!\!\int\!\!\int\!\!\int\\ &\times [D_kK^{ii_1}(t-s_0,y-y_0)K_\epsilon^{i_1i_2}(z_0-z_2)(I_1(S_{z_2}))^{i_2}K_\epsilon^{kk_1}(z_0-z_3)(I_1(S_{z_3}))^{k_1}\\ &\times D_lK^{jj_1}(t-s_1,y-y_1)K_\epsilon^{j_1j_2}(z_1-z_4)(I_1(S_{z_4}))^{j_2}K_\epsilon^{ll_1}(z_1-z_5)(I_1(S_{z_5}))^{l_1+3}]\\ &\times\phi(y)\prod_{j=0}^{5}dz_jdy - \int C^2_{ii_1i_2jj_1j_2kk_1ll_1}\phi(y)dy.\end{aligned} \tag{8.98}$$

Using (8.22), (8.23), and Remark 8.3.3, we compute

$$\begin{aligned}&I_1(S_{z_2})I_1(S_{z_3})I_1(S_{z_4})I_1(S_{z_5})\\ &= I_4([S_{z_2}\otimes S_{z_3}]\otimes[S_{z_4}\otimes S_{z_5}])\\ &\quad + I_2[\langle S_{z_2},S_{z_4}\rangle S_{z_3}\otimes S_{z_5} + \langle S_{z_2},S_{z_5}\rangle S_{z_3}\otimes S_{z_4} + \langle S_{z_3},S_{z_4}\rangle S_{z_2}\otimes S_{z_5}\\ &\quad + \langle S_{z_3},S_{z_5}\rangle S_{z_2}\otimes S_{z_4} + (S_{z_2}\otimes S_{z_3})\langle S_{z_4},S_{z_5}\rangle + \langle S_{z_2},S_{z_3}\rangle(S_{z_4}\otimes S_{z_5})]\\ &\quad + I_0[\langle S_{z_2},S_{z_4}\rangle\langle S_{z_3},S_{z_5}\rangle + \langle S_{z_2},S_{z_5}\rangle\langle S_{z_3},S_{z_4}\rangle + \langle S_{z_2},S_{z_3}\rangle\langle S_{z_4},S_{z_5}\rangle].\end{aligned} \tag{8.99}$$

Applying (8.99) to (8.98) shows that we have the fourth, second, and zeroth Wiener chaos respectively from

$$\begin{aligned}\langle\widehat{\Pi}_0^{(\epsilon,t)}\tau,\phi\rangle = \int\!\!\int\!\!\int\!\!\int\!\!\int\!\!\int\!\!\int &I_4([S_{z_2}\otimes S_{z_3}]\otimes[S_{z_4}\otimes S_{z_5}])\\ &\times [D_kK^{ii_1}(t-s_0,y-y_0)K_\epsilon^{i_1i_2}(z_0-z_2)K_\epsilon^{kk_1}(z_0-z_3)\\ &\times D_lK^{jj_1}(t-s_1,y-y_1)K_\epsilon^{j_1j_2}(z_1-z_4)K_\epsilon^{ll_1}(z_1-z_5)]\\ &\times\phi(y)\prod_{j=0}^{5}dz_jdy - \int C^2_{ii_1i_2jj_1j_2kk_1ll_1}\phi(y)dy,\end{aligned}$$

$$
\begin{aligned}
\langle \widehat{\Pi}_0^{(\epsilon,t)}\tau, \phi\rangle = \iiiint\!\!\iint I_2\big[&\langle S_{z_2}, S_{z_4}\rangle S_{z_3}\otimes S_{z_5} + \langle S_{z_2}, S_{z_5}\rangle S_{z_3}\otimes S_{z_4} + \langle S_{z_3}, S_{z_4}\rangle S_{z_2}\otimes S_{z_5}\\
&+ \langle S_{z_3}, S_{z_5}\rangle S_{z_2}\otimes S_{z_4} + (S_{z_2}\otimes S_{z_3})\langle S_{z_4}, S_{z_5}\rangle + \langle S_{z_2}, S_{z_3}\rangle (S_{z_4}\otimes S_{z_5})\big]\\
&\times \big[D_k K^{ii_1}(t-s_0, y-y_0) K_\epsilon^{i_1 i_2}(z_0 - z_2) K_\epsilon^{kk_1}(z_0 - z_3)\\
&\times D_l K^{jj_1}(t-s_1, y-y_1) K_\epsilon^{j_1 j_2}(z_1 - z_4) K_\epsilon^{ll_1}(z_1 - z_5)\big]\\
&\times \phi(y)\prod_{j=0}^{5} dz_j dy - \int C^2_{ii_1 i_2 jj_1 j_2 kk_1 ll_1}\phi(y)dy,
\end{aligned}
$$

and

$$
\begin{aligned}
\langle \widehat{\Pi}_0^{(\epsilon,t)}\tau, \phi\rangle = \iiiint\!\!\iint I_0\big[&\langle S_{z_2}, S_{z_4}\rangle\langle S_{z_3}, S_{z_5}\rangle + \langle S_{z_2}, S_{z_5}\rangle\langle S_{z_3}, S_{z_4}\rangle + \langle S_{z_2}, S_{z_3}\rangle\langle S_{z_4}, S_{z_5}\rangle\big]\\
&\times \big[D_k K^{ii_1}(t-s_0, y-y_0) K_\epsilon^{i_1 i_2}(z_0 - z_2) K_\epsilon^{kk_1}(z_0 - z_3)\\
&\times D_l K^{jj_1}(t-s_1, y-y_1) K_\epsilon^{j_1 j_2}(z_1 - z_4) K_\epsilon^{ll_1}(z_1 - z_5)\big]\\
&\times \phi(y)\prod_{j=0}^{5} dz_j dy - \int C^2_{ii_1 i_2 jj_1 j_2 kk_1 ll_1}\phi(y)dy.
\end{aligned}
$$

Specifically, they take the form

$$(\widehat{\mathscr{W}}^{(\epsilon;4)}\tau)(z) =$$ 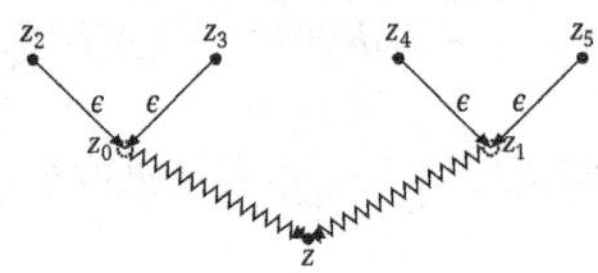 ,

$$(\widehat{\mathscr{W}}^{(\epsilon;2)}\tau)(z) = \sum_{i=1}^{4}(\widehat{\mathscr{W}}_i^{(\epsilon;2)}\tau)(z),$$

where

$$\sum_{i=1}^{4}(\widehat{\mathscr{W}}_i^{(\epsilon;2)}\tau)(z)$$

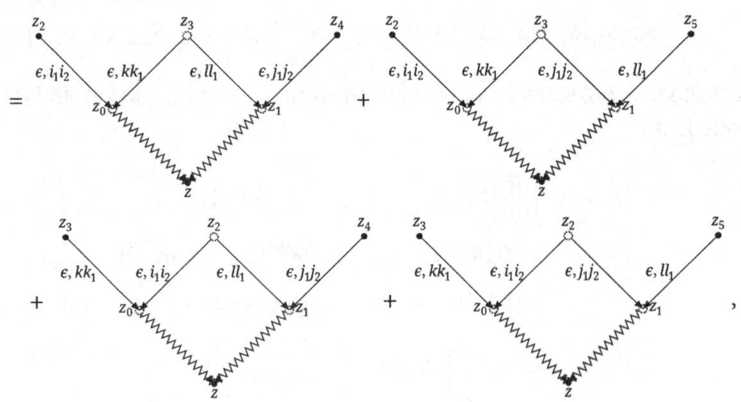

,

and

$$(\widehat{\mathcal{W}}^{(\epsilon;0)}\tau)(z) =$$ 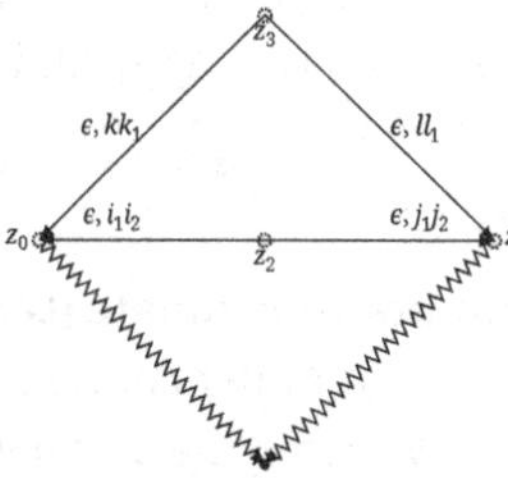

$+$ 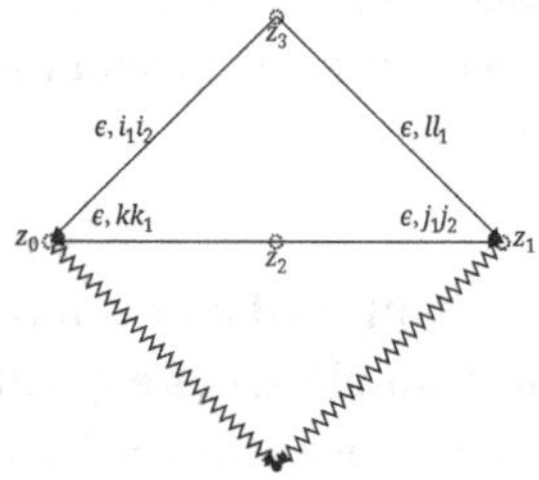

$$- C^2_{ii_1i_2jj_1j_2kk_1ll_1},$$

which leads to a natural choice of

$$C^2_{ii_1i_2jj_1j_2kk_1ll_1} \triangleq$$ 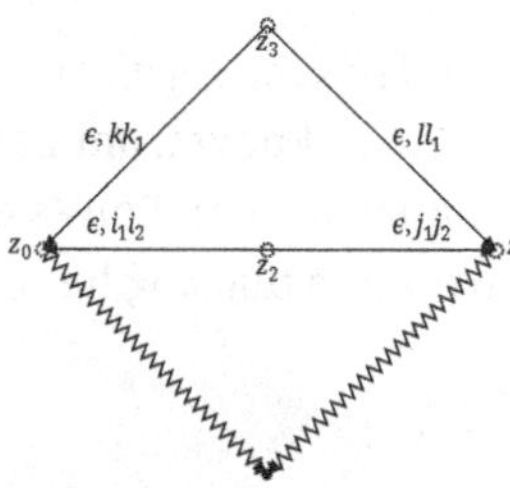

$+$ 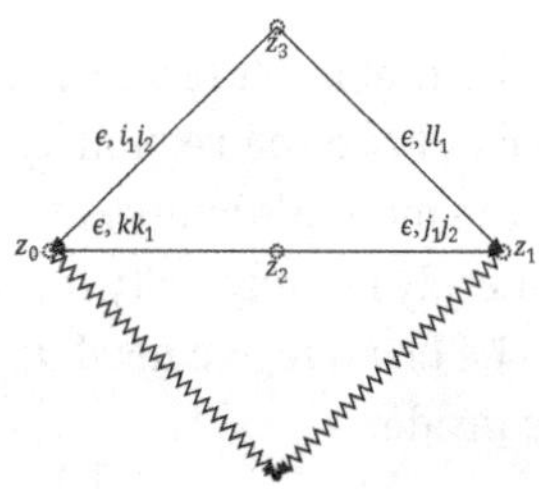

. (8.100)

The fact that such a renormalization constant is indeed independent of z can be seen from

$$\begin{aligned}
&\iiiint D_k K^{ii_1}(z-z_0)K_\epsilon^{i_1i_2}(z_0-z_2)K_\epsilon^{kk_1}(z_0-z_3)\\
&\quad\times D_l K^{jj_1}(z-z_1)K_\epsilon^{j_1j_2}(z_1-z_2)K_\epsilon^{ll_1}(z_1-z_3)dz_2dz_3dz_1dz_0\\
&= D_k K^{ii_1} * ([(K_\epsilon^{i_1i_2} * K_\epsilon^{j_1j_2})(K_\epsilon^{kk_1} * K_\epsilon^{ll_1})] * D_l K^{jj_1})(0)
\end{aligned}$$

and, similarly,

$$\begin{aligned}
&\iiiint D_k K^{ii_1}(z-z_0)K_\epsilon^{i_1i_2}(z_0-z_3)K_\epsilon^{kk_1}(z_0-z_2)\\
&\quad\times D_l K^{jj_1}(z-z_1)K_\epsilon^{j_1j_2}(z_1-z_2)K_\epsilon^{ll_1}(z_1-z_3)dz_2dz_3dz_1dz_0\\
&= D_k K^{ii_1} * ([(K_\epsilon^{i_1i_2} * K_\epsilon^{ll_1})(K_\epsilon^{kk_1} * K_\epsilon^{j_1j_2})] * D_l K^{jj_1})(0).
\end{aligned}$$

For the interested readers, we give complete details on G_{11} from (A.171), which was omitted in [352], in Appendix A.9.

8.5.4 Theory of paracontrolled distributions

In this section, we briefly describe the theory of paracontrolled distributions from [159], which will be discussed in Section 9.1 in more detail. In contrast to the theory of RS, this approach blends harmonic analysis and controlled rough path theory and makes use of paracontrolled ansatz (see (8.102)) and commutator estimate from Lemma 3.8.5. There are many great expository notes such as [159, 160], and hence here we only give a heuristic idea.

8.5.4.1 Paracontrolled ansatz and commutator estimate (Lemma 3.8.5)

As a simple example, let us explain the case of Anderson Hamiltonian studied in [7], to be discussed in more detail in Section 9.1. We recall the parabolic Anderson model (1.14) with $v = 1$ for simplicity:

$$\partial_t u = \Delta u + u\xi = (\Delta + \xi)u,$$

where ξ is a noise that is white only in space and has the regularity $\alpha < -\frac{d}{2}$. Heuristically, we see that u can have the regularity of $\alpha + 2$. We know from Lemma 8.1.4 and Corollary 8.1.5 that the main difficulty is the resonance term in Bony's estimate. Considering regularity as merely homogeneity for now, we agree that $u \circ \xi$ has the regularity of $2\alpha + 2$, and for this to be positive, we need $\alpha > -1$.

Now we consider

$$\partial_t Q = \Delta Q + \xi, \quad Q(0, x) = 0, \tag{8.101}$$

so that $Q(t, x) = \int_0^t e^{(t-s)\Delta} \xi(s) ds$, which has the regularity of essentially $\alpha + 2$. Then we postulate that the solution u takes the form of the following paracontrolled ansatz:

$$u = u \prec Q - u^\sharp \tag{8.102}$$

for some remainder $u^\sharp$; although it has to be formally derived, we can agree that $u^\sharp$ takes the regularity of the sum of u and Q and thus $2\alpha + 4$. At this point,

$$u \circ \xi = (u \prec Q - u^\sharp) \circ \xi, \tag{8.103}$$

and because the regularity of $u \prec Q$ is basically that of Q, for the resonance term $(u \prec Q) \circ \xi$ to be well-defined, we continue to have the same lower bound of $2\alpha + 2 > 0$, and thus $\alpha > -1$. However, using the definition of $\mathcal{C}$ from (3.103), we now rewrite

$$u \circ \xi = u(\xi \circ Q) + \mathcal{C}(u, Q, \xi) + u^\sharp \circ \xi. \tag{8.104}$$

The key result from [159] is that due to Lemma 3.8.5, $\mathcal{C}(u, Q, \xi)$ has the regularity of $3\alpha + 4$ if $3\alpha + 4 > 0$, which requires only a lower bound of $\alpha > -\frac{4}{3}$. We also see that $u^\sharp \circ \xi$ in (8.104) would be defined if $3\alpha + 4 > 0$ and thus the same improved lower bound of $\alpha > -\frac{4}{3}$. Now we have to still work on $\xi \circ Q$, but this is better than $\xi \succ Q$. In fact, this is when we subtract out a renormalization constant; e. g., in the 2D case with white-in-space noise,

$\alpha < -1$, so that the renormalization constant of the form (8.18a) suffices, and it grows only logarithmically.

Now, to make the system closed, along with ansatz (8.102), we want an explicit equation solved by $u^\sharp$. To this end, using (1.14) and (8.102), we can compute

$$\Delta u + u\xi = \partial_t u = \partial_t u \prec Q + u \prec \partial_t Q - \partial_t u^\sharp. \tag{8.105}$$

Here the choice of Q in (8.101) comes into play; we can directly see that after substituting $\partial_t Q$ by $\Delta Q + \xi$, $u \prec \xi$ will appear, but due to $u\xi$ on the left-hand side, it reduces to $u \succ \xi + u \circ \xi$, which is the desired situation where $u \prec \xi$ is absent. We may be concerned about the $u \prec \Delta Q$, where ΔQ is as bad as $u \prec \xi$ in terms of regularity; however, this will be absorbed as part of $\Delta u^\sharp$, as we will see in (8.107). Let us proceed in this manner to first substitute $\partial_t Q$ by $\Delta Q + \xi$ due to (8.101) and solve for the equation solved by $\partial_t u^\sharp$ as

$$\partial_t u^\sharp = \partial_t u \prec Q + u \prec \Delta Q - u \succ \xi - u \circ \xi - \Delta u. \tag{8.106}$$

Now we substitute $u \circ \xi$ by (8.104) to conclude

$$\begin{aligned} \partial_t u^\sharp - \Delta u^\sharp = {} & \partial_t u \prec Q - u \succ \xi - u(\xi \circ Q) \\ & - \mathcal{C}(u, Q, \xi) - u^\sharp \circ \xi - \Delta u \prec Q - 2\nabla u \prec \nabla Q. \end{aligned} \tag{8.107}$$

We refer to [7] for further details.

In the case of the 3D MHD system forced by STWN (8.59), the following paracontrolled ansatz appeared in [357, eqs. (2.12), (2.15), and (2.17)], although in different notations. First, we apply the Da Prato–Debussche trick repeatedly similarly to Section 8.5.1:

$$(\partial_t - \Delta)u_1^i = \sum_{i_1=1}^{3} \mathcal{P}_L^{ii_1} \xi_u^{i_1}; \quad (\partial_t - \Delta)b_1^i = \sum_{i_1=1}^{3} \mathcal{P}_L^{ii_1} \xi_b^{i_1},$$

$$(\partial_t - \Delta)u_2^i = -\frac{1}{2}\sum_{i_1,j=1}^{3} \mathcal{P}_L^{ii_1} \partial_{x_j}(u_1^{i_1} \diamond u_1^j - b_1^{i_1} \diamond b_1^j), \quad u_2(\cdot, 0) = 0,$$

$$(\partial_t - \Delta)b_2^i = -\frac{1}{2}\sum_{i_1,j=1}^{3} \mathcal{P}_L^{ii_1} \partial_{x_j}(b_1^{i_1} \diamond u_1^j - u_1^{i_1} \diamond b_1^j), \quad b_2(\cdot, 0) = 0,$$

$$(\partial_t - \Delta)u_3^i = -\frac{1}{2}\sum_{i_1,j=1}^{3} \mathcal{P}_L^{ii_1} \partial_{x_j}(u_1^{i_1} u_2^j + u_2^{i_1} u_1^j - b_1^{i_1} b_2^j - b_2^{i_1} b_1^j), \quad u_3(\cdot, 0) = 0,$$

$$(\partial_t - \Delta)b_3^i = -\frac{1}{2}\sum_{i_1,j=1}^{3} \mathcal{P}_L^{ii_1} \partial_{x_j}(b_1^{i_1} u_2^j + b_2^{i_1} u_1^j - u_1^{i_1} b_2^j - u_2^{i_1} b_1^j), \quad b_3(\cdot, 0) = 0,$$

and

$$\begin{aligned}
(\partial_t - \Delta)u_4^i &= -\frac{1}{2}\sum_{i_1,j=1}^{3}\mathcal{P}_L^{ii_1}\partial_{x_j}[u_1^{i_1} \diamond (u_3^j + u_4^j) + (u_3^{i_1} + u_4^{i_1}) \diamond u_1^j + u_2^{i_1} \diamond u_2^j \\
&\quad + u_2^{i_1}(u_3^j + u_4^j) + u_2^j(u_3^{i_1} + u_4^{i_1}) + (u_3^{i_1} + u_4^{i_1})(u_3^j + u_4^j) \\
&\quad - b_1^{i_1} \diamond (b_3^j + b_4^j) - (b_3^{i_1} + b_4^{i_1}) \diamond b_1^j - b_2^{i_1} \diamond b_2^j \\
&\quad - b_2^{i_1}(b_3^j + b_4^j) - b_2^j(b_3^{i_1} + b_4^{i_1}) - (b_3^{i_1} + b_4^{i_1})(b_3^j + b_4^j)], \\
u_4(\cdot, 0) &= \mathcal{P}_L u^{\text{in}}(\cdot) - u_1(\cdot, 0), \\
(\partial_t - \Delta)b_4^i &= -\frac{1}{2}\sum_{i_1,j=1}^{3}\mathcal{P}_L^{ii_1}\partial_{x_j}[b_1^{i_1} \diamond (u_3^j + u_4^j) + (b_3^{i_1} + b_4^{i_1}) \diamond u_1^j + b_2^{i_1} \diamond u_2^j \\
&\quad + b_2^{i_1}(u_3^j + u_4^j) + u_2^j(b_3^{i_1} + b_4^{i_1}) + (b_3^{i_1} + b_4^{i_1})(u_3^j + u_4^j) \\
&\quad - u_1^{i_1} \diamond (b_3^j + b_4^j) - (u_3^{i_1} + u_4^{i_1}) \diamond b_1^j - u_2^{i_1} \diamond b_2^j \\
&\quad - u_2^{i_1}(b_3^j + b_4^j) - b_2^j(u_3^{i_1} + u_4^{i_1}) - (u_3^{i_1} + u_4^{i_1})(b_3^j + b_4^j)], \\
b_4(\cdot, 0) &= \mathcal{P}_L b^{\text{in}}(\cdot) - b_1(\cdot, 0).
\end{aligned}$$

All the products involving $\diamond$ require extra care such as renormalizations, and we refer the interested readers to [357] for details. We limit ourselves to only mentioning that the key solution components u_4 and b_4 are postulated to satisfy paracontrolled ansatz of

$$\begin{aligned}
u_4^i &= -\frac{1}{2}\sum_{i_1,j_1=1}^{3}\mathcal{P}_L^{ii_1}\partial_{x^{j_1}}[\pi_<(u_3^{i_1} + u_4^{i_1}, Q_u^{j_1}) + \pi_<(u_3^{j_1} + u_4^{j_1}, Q_u^{i_1}) \\
&\quad - \pi_<(b_3^{i_1} + b_4^{i_1}, Q_b^{j_1}) - \pi_<(b_3^{j_1} + b_4^{j_1}, Q_b^{i_1})] + u^{\sharp,i}, \\
b_4^i &= -\frac{1}{2}\sum_{i_1,j_1=1}^{3}\mathcal{P}_L^{ii_1}\partial_{x^{j_1}}[-\pi_<(u_3^{i_1} + u_4^{i_1}, Q_b^{j_1}) + \pi_<(u_3^{j_1} + u_4^{j_1}, Q_b^{i_1}) \\
&\quad + \pi_<(b_3^{i_1} + b_4^{i_1}, Q_u^{j_1}) - \pi_<(b_3^{j_1} + b_4^{j_1}, Q_u^{i_1})] + b^{\sharp,i},
\end{aligned}$$

where

$$(\partial_t - \Delta)Q_u^i = u_1^i, \quad Q_u^i(0) = 0 \quad \text{and} \quad (\partial_t - \Delta)Q_b^i = b_1^i, \quad Q_b^i(0) = 0.$$

8.5.5 Further results using the theory of paracontrolled distributions

The theory of paracontrolled distributions is considered by some to be easier to adapt than the theory of RS. As we will describe in Section 9.2.1, the first application of the convex integration technique to singular SPDEs was accomplished by Hofmanová, Zhu, and Zhu [189] on the 3D Navier–Stokes equations forced by STWN via the theory of paracontrolled distributions. This seems natural, considering that harmonic analysis tools are already used extensively in the application of convex integration technique.

8.5.5.1 Local solution theory

Besides [7, 357], which we already mentioned, the theory of paracontrolled distributions was applied on the 3D Φ^4 model [61] and the 3D Navier–Stokes equations forced by STWN [374].

8.5.5.2 Approximation under spatial discretization

Here, as another example of application of the theory of paracontrolled distributions, we briefly describe a particular feature of the STWN that has led to intensive investigations, namely approximating solutions to singular SPDEs under spatial discretization, briefly previewed in Example 13. To explain this phenomena, we start with a simple example that represents a widely accepted folklore that if a PDE is well-posed, then a solution to any reasonable approximation via spatial discretization will converge in the limit to the unique solution of the original PDE. We leave the preliminaries used in Example 22 to Section B.1 for completeness.

Example 22. Let us consider the deterministic Burgers' equation (1.3), with $\upsilon = 1$ for simplicity,

$$\partial_t \theta + \frac{1}{2}\partial_x \theta^2 = \partial_x^2 \theta \tag{8.108}$$

on $\mathbb{T}$ with initial data θ^{in}. We define the spatial approximation of differentiation operators

$$D^{\epsilon} f(x) \triangleq \frac{f(x+\epsilon) - f(x)}{\epsilon}, \quad \Delta_{\epsilon} f(x) = \frac{f(x+\epsilon) - 2f(x) + f(x+\epsilon)}{\epsilon^2}, \tag{8.109}$$

and consider the following approximations of (8.108):

$$\partial_t \theta^{\epsilon} + \frac{1}{2} D^{\epsilon} (\theta^{\epsilon})^2 = \Delta_{\epsilon} \theta^{\epsilon}, \quad \theta^{\epsilon}(0) = \theta^{\text{in},\epsilon}, \tag{8.110}$$

where the superscript ϵ is intended to emphasize the dependence on $\epsilon > 0$. It is well known that both (8.108) and (8.110) have unique solutions $\theta, \theta^{\epsilon} \in C([0,T]; L^{\infty}(\mathbb{T}))$ starting from continuous initial data. Because we want to only investigate the convergence of θ^{ϵ} to θ in $C([0,T]; L^{\infty}(\mathbb{T}))$, let us assume that

$$\lim_{\epsilon \searrow 0} \|\theta^{\text{in}} - \theta^{\text{in},\epsilon}\|_{L^{\infty}(\mathbb{T})} = 0. \tag{8.111}$$

To show that $\lim_{\epsilon \to 0} \sup_{t \in [0,T]} \|(\theta - \theta^{\epsilon})(t)\|_{L^{\infty}} = 0$, we can write both solutions in a mild form as

$$\theta(t,x) = P_t\theta^{\text{in}}(x) - \frac{1}{2}\int_0^t P_{t-s}(\partial_x\theta^2)ds, \quad P_t \overset{(8.9)}{=} e^{t\Delta}, \tag{8.112a}$$

$$\theta^\epsilon(t,x) = P_t\theta^{\text{in},\epsilon}(x) - \frac{1}{2}\int_0^t P_{t-s}^\epsilon(D^\epsilon(\theta^\epsilon)^2)ds, \quad P_t^\epsilon \triangleq e^{t\Delta_\epsilon}, \tag{8.112b}$$

and estimate

$$\|(\theta - \theta^\epsilon)(t)\|_{L^\infty} \lesssim \|\theta^{\text{in}} - \theta^{\text{in},\epsilon}\|_{L^\infty} + \sum_{i=1}^3 \text{VIII}_i, \tag{8.113}$$

where

$$\text{VIII}_1 \triangleq \int_0^t \|(P_{t-s} - P_{t-s}^\epsilon)\partial_x\theta^2\|_{L^\infty}ds,$$
$$\text{VIII}_2 \triangleq \int_0^t \|P_{t-s}^\epsilon(\partial_x - D^\epsilon)\theta^2\|_{L^\infty}ds,$$
$$\text{VIII}_3 \triangleq \int_0^t \|P_{t-s}^\epsilon D^\epsilon(\theta^2(\theta^\epsilon)^2)\|_{L^\infty}ds.$$

First, we estimate

$$\text{VIII}_1 \lesssim \int_0^t \epsilon^{\frac{1}{4}}(t-s)^{-\frac{3}{4}}\|\theta^2\|_{C^0}ds \lesssim \epsilon^{\frac{1}{4}}t^{\frac{1}{4}} \sup_{t\in[0,T]}\|\theta(t)\|_{L^\infty}^2 \tag{8.114}$$

by Lemma B.1.2. Second, we estimate

$$\text{VIII}_2 \lesssim \int_0^t (t-s)^{-\frac{3}{4}}\epsilon^{\frac{1}{8}}\|\theta^2\|_{C^0}ds \lesssim \epsilon^{\frac{1}{8}}t^{\frac{1}{4}} \sup_{t\in[0,T]}\|\theta(t)\|_{L^\infty}^2 \tag{8.115}$$

by Lemmas B.1.1 and B.1.4. Finally, we see that there exists a uniform constant $C_0 > 0$ such that

$$\begin{aligned}\text{VIII}_3 &\leq C_0\int_0^t (t-s)^{-\frac{3}{4}}\|D^\epsilon(\theta^2 - (\theta^\epsilon)^2)\|_{C^{-\frac{5}{4}}}ds \\ &\leq C_0 t^{\frac{1}{4}}\Big(\sup_{s\in[0,T]}\|(\theta,\theta^\epsilon)(s)\|_{L^\infty}\Big)\sup_{s\in[0,t]}\|(\theta-\theta^\epsilon)(s)\|_{L^\infty}\end{aligned} \tag{8.116}$$

by Lemmas B.1.1 and B.1.3.

Applying estimates (8.114)–(8.116) to (8.113) leads to

$$\sup_{t\in[0,T]} \|(\theta-\theta^{\epsilon})(t)\|_{L^\infty} \lesssim \|\theta^{\mathrm{in}}-\theta^{\mathrm{in},\epsilon}\|_{L^\infty} + \epsilon^{\frac{1}{4}}T^{\frac{1}{4}} \sup_{t\in[0,T]} \|\theta(t)\|_{L^\infty}^{2} + +\epsilon^{\frac{1}{8}}T^{\frac{1}{4}} \sup_{t\in[0,T]} \|\theta(t)\|_{L^\infty}^{2} + \\ + T^{\frac{1}{4}}\Big(\sup_{t\in[0T]} \|(\theta,\theta^{\epsilon})(t)\|_{L^\infty}\Big) \sup_{t\in[0,T]} \|(\theta-\theta^{\epsilon})(t)\|_{L^\infty},$$

so that for T small, $\lim_{\epsilon\to 0}\sup_{t\in[0,T]}\|(\theta-\theta^{\epsilon})(t)\|_{L^\infty}=0$, as expected.

As this example showed, when an equation is well-posed, even locally in time, then we expect that a solution to a spatially discretized PDE depending on a parameter $\epsilon>0$ converges to the solution of the original equation as $\epsilon\searrow 0$. However, Hairer [165] showed that a certain family of solutions to a semilinear variation of a heat equation forced by STWN that depends on $\epsilon>0$ converges as $\epsilon\searrow 0$ to a solution of the same equation with $\epsilon=0$, but with an addition of a correction term, akin to that of Itô–Stratonovich correction mentioned in Remark 4.2.10. Moreover, Hairer and Voss [177] argued that for the Burgers' equation (1.3), different finite-difference schemes converge to different limiting processes as the mesh size tends to zero and provided supporting evidence via numerical analysis; subsequently, a rigorous proof of their argument was given by Hairer and Maas [171]. The results of [171] were extended by Hairer, Maas, and Weber [172] from the case where the nonlinearity is a gradient type and the noise is additive to the case where the nonlinearity is of nongradient type and the noise is multiplicative. Finally, extension of such results to the 3D Navier–Stokes equations and MHD system forced by STWN can be found respectively in [375, 361], and we refer the interested readers to references therein for more detail.

8.6 Recent developments and open problems

We conclude this chapter by mentioning further developments related to our discussions. Initially in [167], for every fixed locally subcritical SPDE, it was necessary to create an RS and verify issues such as the well-posedness of the abstract formulation of the system, identification of the renormalized solutions, and convergence of the renormalized models to a limiting model we mentioned. However, a series of works [41, 40, 76] unified these aspects and ultimately provided a more general theorem, which leads to local-in-time solution theory and stability for a wide class of singular SPDEs (e. g., "Metatheorem in [41, Theorem 2.18]").

As we already discussed, besides the issue of locally subcritical limitation, the solutions obtained by the theories of paracontrolled distributions or RS are only local-in-time without further work, which is the topic of the next chapter. We conclude this chapter by stating an open problem mentioned in Remark 8.5.4.

Problem 7. Can we construct any solution theory, even local-in-time, for the Hall-MHD system forced by STWN, even in the $2\frac{1}{2}$D case? Of course, this is a particular case of the greater problem of constructing any solution theory in locally critical and supercritical SPDEs.

9 Global solution theory of singular SPDEs

As we described, there has not been any general way to extend the local-in-time solution to a singular SPDE to a global-in-time solution. This chapter is devoted to a limited amount of global-in-time solution theory of singular SPDEs. Sections 9.1 and 9.2 describe the directions of uniqueness and nonuniqueness, the latter proven via convex integration. Section 9.3 describes the recent developments and open problems.

9.1 Global-in-time unique solution

First and foremost, we recall from Chapter 8 that even local-in-time solution theory is limited to the locally subcritical case in general, even if constructed via the theories of RS or paracontrolled distributions, which is an issue that originates from the regularity of the random force. Considering (8.6), we see that this already places possibilities to only low dimensions. Broadly stated, examples of cases where the global-in-time solution theory was proven, with an exception of convex integration solution to be described in Section 9.2, are limited to cases with favorable transformation, damping effect, and explicit knowledge of invariant measure.

1. First, Gubinelli and Perkowski [160] observed that the solution constructed by Hairer [166] to the 1D KPZ equation was actually global-in-time due to a favorable transformation (see [160, p. 170, and Corollary 7.5]).
2. As we already described in Section 8.5, Da Prato and Debussche [114] in $\mathbb{T}^2$ were able to provide global-in-time solution theory for the 2D Φ^4 model, for a. e. initial data w. r. t. its invariant measure. Mourrat and Weber [268] extended such a result to $\mathbb{R}^2$. Later, Hairer and Matetski [173, Corollary 1.3] considered the 3D Φ^4 model with the damping term Φ^3 replaced by $\lambda\Phi^3$ on $\mathbb{T}^3$ analogously to $\lambda > 0$ in (1.13), and proved the global-in-time solution theory for a. e. initial data w. r. t. an invariant measure when the coupling constant $\lambda > 0$ is sufficiently small. Finally, Mourrat and Weber [269] provided the global-in-time solution theory for the 3D Φ^4 model on $[-1,1]^3$ starting from every initial data. A heuristic idea behind this phenomenon is related to the fact that the nonlinear term in the Φ^4 model has a damping effect, which will be discussed in Section 9.3.1 as well.
3. As we already described in Section 8.5, Da Prato and Debussche [113] considered the 2D Navier–Stokes equations forced by STWN, split their solution u into two parts $u = X + v$,

$$\partial_t X = \upsilon \Delta X + \mathcal{P}_L \mathbb{P}_{\neq 0} \xi, \quad \partial_t v + \mathcal{P}_L \operatorname{div}(v + X)^{\otimes 2} = \upsilon \Delta v, \tag{9.1}$$

   and relied on Wick products and (3.139) to provide global-in-time solution theory for a. e. initial data w. r. t. its invariant measure.

https://doi.org/10.1515/9783111349985-010

The resolution in [113] on the 2D Navier–Stokes equations naturally leads to a question of possible extension to the 2D MHD system forced by STWN. However, as already pointed out in (5.8), (5.9), and (5.10), an analogue of (3.139) does not exist for the MHD system (see (9.11)), making it impossible to adapt the approach of [113]. Another curious question would be for the 2D stochastic Yang–Mills equation (1.18). Although its local-in-time solution theory was established in [74], the complex structure of the nonlinear terms disallows any good transformation, nonlinear terms are not just damping, and there is no explicit knowledge of invariant measure, in contrast to the KPZ equation, Φ^4 model, and the 2D Navier–Stokes equations; consequently, the global-in-time solution theory of the 2D stochastic Yang–Mills equation (1.18) is currently open.

As we will see in Theorem 9.1.2, the case of the 2D MHD system forced by STWN is now settled thanks to the new approach of Hairer and Rosati [176], which we will describe next; on the other hand, the case of the 2D Yang–Mills equation remains open, and we will state it as Problem 8.

9.1.1 The approach of Hairer and Rosati [176]

Hairer and Rosati [176] gave a second proof of the global-in-time solution theory of the 2D Navier–Stokes equations forced by STWN without relying on invariant measure. To emphasize their capability, they included an additional force $\zeta \in \mathcal{C}_{\mathfrak{s}}^{-2+3\kappa}(\mathbb{R} \times \mathbb{T}^2; \mathbb{R}^2)$, intentionally not in the Cameron–Martin space $L^2_{\text{loc}} L^2_x$ (e. g., [274, p. 32]), so that the law of the solution u to the 2D Navier–Stokes equations forced by STWN and ζ have no obvious link to the law of X in (9.1), and the approach via the explicit knowledge of invariant measure from [113] becomes inapplicable. Specifically, the 2D Navier–Stokes equations forced by STWN and such ζ considered in [176] read

$$\partial_t u + \operatorname{div}(u \otimes u) + \nabla \pi = \nu \Delta u + \mathbb{P}_{\neq 0}(\xi + \zeta), \quad \nabla \cdot u = 0. \tag{9.2}$$

Hairer and Rosati started with the Da Prato–Debussche trick, so that we may explain continuing from (9.1). To exclude the additional ill-defined products, Hairer and Rosati considered an equation of Y that solves

$$\partial_t Y = \nu \Delta Y - \mathcal{P}_L \operatorname{div}(2X \otimes_s Y + X^{\otimes 2}) + \mathcal{P}_L \mathbb{P}_{\neq 0} \zeta, \quad \nabla \cdot u = 0. \tag{9.3}$$

This leads to an equation of $w \triangleq v - Y$,

$$\partial_t w + \mathcal{P}_L (w \cdot \nabla) X = \nu \Delta w + \text{“l. s. t.”}, \tag{9.4}$$

where for simplicity we denoted, and hereafter will denote interchangeably, the less singular terms by “l. s. t.”. Since the 2D Navier–Stokes equations are $L^2(\mathbb{T}^2)$-critical (recall (5.5)), it is natural to focus on the $L^2(\mathbb{T}^2)$-estimate of w, although more careful decomposition to lower- and higher-frequency parts will be given in Section 9.1.2 in the case of the MHD system. Taking the $L^2(\mathbb{T}^2)$-inner products on (9.4) by $2w$, we obtain

$$\partial_t \|w\|_{L^2}^2 = 2 \int_{\mathbb{T}^2} w \cdot (\nu\Delta \operatorname{Id} - \nabla_{\text{sym}} X) w dx + \text{“l. s. t.”}, \tag{9.5}$$

where ∇_{sym} is the symmetric gradient, to be defined in (9.17). As X that solves (9.1) can be seen to possess the regularity of $C_t\mathcal{C}_x^\alpha$ for all $\alpha < 0$, (9.4) indicates that $w \in C_t\mathcal{C}_x^\alpha$ for all $\alpha < 1$, and as a result, the product $w \cdot \nabla_{\text{sym}} X$ in (9.5) is ill-defined according to Bony's estimates (recall Corollary 8.1.5). To compensate this issue, Hairer and Rosati defined the time-dependent operator

$$\mathcal{A}_t^\lambda \triangleq \nu\Delta \operatorname{Id} - \nabla_{\text{sym}} X - r_\lambda(t) \operatorname{Id}, \tag{9.6}$$

where $r_\lambda(t)$ is the appropriate renormalization constant (see (9.25b)), so that (9.5) transforms to

$$\frac{d}{dt}\|w\|_{L^2}^2 = 2 \int_{\mathbb{T}^2} w \cdot \mathcal{A}_t^\lambda w dx + 2r_\lambda(t)\|w\|_{L^2}^2 + \text{“l. s. t.”}. \tag{9.7}$$

This suggests that the rest of the proof of [176] consists of the following three key steps.

(a) Renormalization: Recalling the lesson concerning (5.17) due to Tao, we realize that closing the estimate (9.7) requires

$$r_\lambda(t) \lesssim \ln(\|w(t)\|_{L^2}^2). \tag{9.8}$$

(b) Anderson Hamiltonian: To close estimate (9.7), it suffices that

$$\left| \int_{\mathbb{T}^2} w \cdot \mathcal{A}_t^\lambda w dx \right| \lesssim \|w\|_{L^2}^2. \tag{9.9}$$

We recall the parabolic Anderson model (1.14), which can be written as

$$\partial_t u = (\Delta + \xi)u, \tag{9.10}$$

and thus the investigation of the eigenvalues of the operator $\Delta + \xi$ has attracted much attention, e. g., in [7, 230, 162], as discussed in Section 8.5.4.1. The authors of [176] were able to directly rely on such a result from [7] to accomplish (9.9).

(c) Estimates: There will be a difficult commutator term (9.27). As we will discuss in Section 9.1.3, this is significantly more difficult in the case of a fractional Laplacian.

Let us explain these steps in more detail by an example of the 2D MHD system forced by STWN.

9.1.2 Example of the global-in-time unique solution for the 2D MHD system

In this section, we sketch the results and main ideas of [359] on the 2D MHD system forced by STWN. As already stressed, because

$$\int_{\mathbb{T}^2} (u\cdot\nabla)u\cdot\Delta u - (b\cdot\nabla)b\cdot\Delta u dx + \int_{\mathbb{T}^2} (u\cdot\nabla)b\cdot\Delta b - (b\cdot\nabla)u\cdot\Delta b dx$$
$$= -2\int_{\mathbb{T}^2} [\partial_1 b_1(\partial_1 u_2 + \partial_2 u_1) - \partial_1 u_1(\partial_1 b_2 + \partial_2 b_1)]\cdot\Delta b dx \neq 0, \tag{9.11}$$

an analogue of (3.139) is absent, and the approach of [113] does not seem applicable.

To be precise, we fix $(\Omega, \mathcal{F}, \mathbb{P})$ and consider the 2D MHD system forced by STWN (ξ_u, ξ_b):

$$\partial_t u + \mathcal{P}_L \operatorname{div}(u\otimes u - b\otimes b) = \nu\Delta u + \mathcal{P}_L\mathbb{P}_{\neq 0}(\xi_u + \zeta_u), \tag{9.12a}$$
$$\partial_t b + \mathcal{P}_L \operatorname{div}(b\otimes u - u\otimes b) = \nu\Delta b + \mathcal{P}_L\mathbb{P}_{\neq 0}(\xi_b + \zeta_b), \tag{9.12b}$$

where we assume that for sufficiently small $\kappa > 0$, the additional force (ζ_u, ζ_b) satisfies the regularity of

$$\zeta_u, \zeta_b \in C_t\mathcal{C}_x^{-2+3\kappa} \cap \mathcal{C}_t^{\frac{-2+3\kappa}{2}}\mathcal{C}_x. \tag{9.13}$$

We note that we compromised on the case of common diffusivity constant in (9.12), not merely for simplicity but for technical difficulties (see [359, Remark 7]).

As already previewed, analogously to (9.1), we consider

$$\partial_t X_u = \nu\Delta X_u + \mathcal{P}_L\mathbb{P}_{\neq 0}\xi_u, \quad \text{and} \quad \partial_t X_b = \nu\Delta X_b + \mathcal{P}_L\mathbb{P}_{\neq 0}\xi_b,$$

both starting from zero data. It follows that both equations possess unique solutions $X_u, X_b \in C([0,\infty); \mathcal{C}^{-\kappa}(\mathbb{T}^2))$ for all $\kappa > 0$. Thus we look for (v_u, v_b) that satisfies

$$\partial_t v_u + \mathcal{P}_L \operatorname{div}((v_u + X_u)^{\otimes 2} - (v_b + X_b)^{\otimes 2}) = \nu\Delta v_u + \mathcal{P}_L\mathbb{P}_{\neq 0}\zeta_u, \tag{9.14a}$$
$$\partial_t v_b + \mathcal{P}_L \operatorname{div}((v_b + X_b)\otimes(v_u + X_u) - (v_u + X_u)\otimes(v_b + X_b)) = \nu\Delta v_b + \mathcal{P}_L\mathbb{P}_{\neq 0}\zeta_b, \tag{9.14b}$$
$$v_u(0,x) = u^{\text{in}}(x), \quad v_b(0,x) = b^{\text{in}}(x) \tag{9.14c}$$

for $(u^{\text{in}}, b^{\text{in}})$ that are both mean-zero and divergence-free. The following result is classical.

Proposition 9.1.1 ([359, Proposition 1]). *There exists a null set $\mathcal{N} \subset \Omega$ such that for all $\omega \in \Omega \setminus \mathcal{N}$ and $\kappa > 0$, we have the following. For any $(u^{\text{in}}, b^{\text{in}}) \in \mathcal{C}^{-1+\kappa} \times \mathcal{C}^{-1+\kappa}$ that are both divergence-free and mean-zero, there exist*

$$T^{\max}(\omega, u^{\text{in}}, b^{\text{in}}) \in (0, \infty]$$

and a unique maximal mild solution $(v_u, v_b)(\omega)$ *to* (9.14) *on* $[0, T^{\max}(\omega, u^{\text{in}}, b^{\text{in}}))$ *such that* $(v_u, v_b)(\omega, 0, x) = (u^{\text{in}}, b^{\text{in}})(x)$.

The following is the global-in-time result from [359]; the case $b \equiv 0$ reduces to [176, Theorem 2.5].

Theorem 9.1.2 ([359, Theorem 1]). *There exists a null set* $\mathcal{N}' \subset \Omega$ *such that for all* $\omega \in \Omega \setminus \mathcal{N}'$, *we have the following. For any* $\kappa > 0$ *and* $(u^{\text{in}}, b^{\text{in}}) \in \mathcal{C}^{-1+\kappa}(\mathbb{T}^2) \times \mathcal{C}^{-1+\kappa}(\mathbb{T}^2)$ *that are both divergence-free and mean-zero, there exists a unique maximal mild solution* $(v_u, v_b)(\omega)$ *to* (9.14) *on* $[0, T^{\max}(\omega, u^{\text{in}}, b^{\text{in}}))$ *such that* $(v_u, v_b)(\omega, 0, x) = (u^{\text{in}}, b^{\text{in}})(x)$ *and*

$$T^{\max}(\omega, u^{\text{in}}, b^{\text{in}}) = \infty \quad \forall \omega \in \Omega \setminus \mathcal{N}'.$$

The rest of this section is devoted to the proof of Theorem 9.1.2. First, analogously to how $\mathring{R}_q^{\Xi}$ in (9.15) on the convex integration on the MHD system was skew-symmetric, we must introduce tensors that respect the antisymmetric structure of the MHD system. We first recall the symmetric tensor product

$$u \otimes_s v \triangleq \frac{1}{2}(u \otimes v + v \otimes u) \tag{9.15}$$

introduced in [176] and introduce the antisymmetric tensor product

$$u \otimes_a v \triangleq \frac{1}{2}(u \otimes v - v \otimes u). \tag{9.16}$$

Similarly, in addition to the symmetric gradient

$$(\nabla_{\text{sym}}\phi)_{i,j} = \frac{1}{2}(\partial_i \phi_j + \partial_j \phi_i) \tag{9.17}$$

used in [176], we introduce the antisymmetric gradient

$$(\nabla_{\text{anti}}\phi)_{i,j} \triangleq \frac{1}{2}(\partial_i \phi_j - \partial_j \phi_i). \tag{9.18}$$

Moreover, we extend Bony's paraproducts from (3.83a) by

$$f \circledless g \triangleq \sum_{i \geq -1} S_{i-1}f \otimes \Delta_i g,$$
$$f \circledgtr g \triangleq \sum_{i \geq -1} \Delta_i f \otimes S_{i-1} g,$$
$$f \circledcirc g \triangleq \sum_{i \geq -1} \sum_{j : |j| \leq 1} \Delta_i f \otimes \Delta_{i+j} g,$$

so that $f \otimes g = f \olessthan g + f \ogreaterthan g + f \circledcirc g$. We further extend such Bony's decomposition to the symmetric case of

$$f \olessthan_s g \triangleq \sum_{i \geq -1} S_{i-1} f \otimes_s \Delta_i g,$$
$$f \ogreaterthan_s g \triangleq \sum_{i \geq -1} \Delta_i f \otimes_s S_{i-1} g,$$
$$f \circledcirc_s g \triangleq \sum_{i \geq -1} \sum_{j:|j| \leq 1} \Delta_i f \otimes_s \Delta_{i+j} g,$$

so that $f \otimes_s g = f \olessthan_s g + f \ogreaterthan_s g + f \circledcirc_s g$, as well as its antisymmetric counterpart

$$f \olessthan_a g \triangleq \sum_{i \geq -1} S_{i-1} f \otimes_a \Delta_i g, \tag{9.19a}$$
$$f \ogreaterthan_a g \triangleq \sum_{i \geq -1} \Delta_i f \otimes_a S_{i-1} g, \tag{9.19b}$$
$$f \circledcirc_a g \triangleq \sum_{i \geq -1} \sum_{j:|j| \leq 1} \Delta_i f \otimes_a \Delta_{i+j} g, \tag{9.19c}$$

so that $f \otimes_a g = f \olessthan_a g + f \ogreaterthan_a g + f \circledcirc_a g$.

Returning to (9.14) solved by (v_u, v_b), because $\zeta_u, \zeta_b \in C_t \mathcal{C}_x^{-2+3\kappa}$ are the roughest terms there, we introduce the following equations of Y_u, Y_b:

$$\partial_t Y_u = \nu \Delta Y_u - \mathcal{P}_L \operatorname{div}(2X_u \otimes_s Y_u + X_u^{\otimes 2} - 2X_b \otimes_s Y_b - X_b^{\otimes 2}) + \mathcal{P}_L \mathbb{P}_{\neq 0} \zeta_u, \tag{9.20a}$$
$$\partial_t Y_b = \nu \Delta Y_b - \mathcal{P}_L \operatorname{div}(2X_b \otimes_a Y_u + 2Y_b \otimes_a X_u + 2X_b \otimes_a X_u) + \mathcal{P}_L \mathbb{P}_{\neq 0} \zeta_b, \tag{9.20b}$$
$$Y_u(0, x) = 0, \quad Y_b(0, x) = 0. \tag{9.20c}$$

We do this so that if we define $D_u \triangleq 2(X_u + Y_u)$ and $D_b \triangleq 2(X_b + Y_b)$, then

$$(w_u, w_b) \triangleq (v_u - Y_u, v_b - Y_b) \tag{9.21}$$

satisfies

$$\begin{aligned} &\partial_t w_u + \mathcal{P}_L \operatorname{div}(w_u^{\otimes 2} + D_u \otimes_s w_u + Y_u^{\otimes 2} \\ &\quad - w_b^{\otimes 2} - D_b \otimes_s w_b - Y_b^{\otimes 2}) = \nu \Delta w_u, \quad w_u(0, x) = u^{\text{in}}(x), \end{aligned} \tag{9.22a}$$
$$\begin{aligned} &\partial_t w_b + \mathcal{P}_L \operatorname{div}(w_b \otimes w_u + w_b \otimes_a D_u + Y_b \otimes Y_u \\ &\quad - w_u \otimes w_b - w_u \otimes_a D_b - Y_u \otimes Y_b) = \nu \Delta w_b, \quad w_b(0, x) = b^{\text{in}}(x). \end{aligned} \tag{9.22b}$$

Because $\zeta_u, \zeta_b \in \mathcal{C}^{-2+3\kappa}(\mathbb{T}^2)$ $\mathbb{P}$-a.s., we can expect that $Y_u, Y_b \in \mathcal{C}^{2\kappa}(\mathbb{T}^2)$ $\mathbb{P}$-a.s. Thus, in $L^2(\mathbb{T}^2)$-estimates of (w_u, w_b), the ill-defined terms that require renormalization are

$$\langle w_u, \nu \Delta w_u + \mathcal{P}_L \operatorname{div}(2X_u \otimes_s w_u - 2X_b \otimes_s w_b) \rangle,$$
$$\langle w_b, \nu \Delta w_b + \mathcal{P}_L \operatorname{div}(2w_b \otimes_a X_u - 2w_u \otimes_a X_b) \rangle.$$

Let us formally write down the definition of a mild solution to (9.22).

Definition 9.1.3 ([359, Definition 2]). For $\gamma > 0$, $T > 0$, and $\beta \in \mathbb{R}$, we define

$$\mathcal{M}_T^\gamma \mathcal{C}_x^\beta \triangleq \{f : t \mapsto t^\gamma \|f(t)\|_{\mathcal{C}_x^\beta} \text{ is continuous over } [0,T], \|f\|_{\mathcal{M}_T^\gamma \mathcal{C}_x^\beta} < \infty\},$$

where $\|f\|_{\mathcal{M}_T^\gamma \mathcal{C}_x^\beta} \triangleq \|t^\gamma \|f(t)\|_{\mathcal{C}_x^\beta}\|_{C_T}$. Then a pair $(w_u, w_b) \in \mathcal{M}_T^\gamma \mathcal{C}_x^\beta \times \mathcal{M}_T^\gamma \mathcal{C}_x^\beta$ is a mild solution to (9.22a)–(9.22b) over $[0,T]$ if for all $t \in [0,T]$,

$$\begin{aligned} w_u(t) = P_t u^{\text{in}} - \int_0^t & P_{t-s}\mathcal{P}_L \operatorname{div}(w_u^{\otimes 2} + D_u \otimes_s w_u + Y_u^{\otimes 2} \\ & - w_b^{\otimes 2} - D_b \otimes_s w_b - Y_b^{\otimes 2})(s)ds, \\ w_b(t) = P_t b^{\text{in}} - \int_0^t & P_{t-s}\mathcal{P}_L \operatorname{div}(w_b \otimes w_u + w_b \otimes_a D_u + Y_b \otimes Y_u \\ & - w_u \otimes w_b - w_u \otimes_a D_b - Y_u \otimes Y_b)(s)ds. \end{aligned}$$

Now, considering the special structure of the MHD system, we define the special 4×4 matrix

$$\nabla_{\text{spec}}(X_u, X_b) \in \mathbb{M}^4 \quad \text{by} \quad \nabla_{\text{spec}}(X_u, X_b) \triangleq \begin{pmatrix} \nabla_{\text{sym}} X_u & \nabla_{\text{anti}} X_b \\ -\nabla_{\text{anti}} X_b & -\nabla_{\text{sym}} X_u \end{pmatrix}. \tag{9.23}$$

Next, for $\lambda \geq 1$ and $t \in [0,\infty)$, we define the *enhanced noise* by

$$t \mapsto (\nabla_{\text{spec}}\mathcal{L}_\lambda(X_u, X_b)(t), \nabla_{\text{spec}}\mathcal{L}_\lambda(X_u, X_b)(t) \circledcirc P^\lambda(t) - r_\lambda \operatorname{Id}), \tag{9.24}$$

where $\mathcal{L}_\lambda$ is a projection onto Fourier modes below λ defined in (8.44), Id is the identity matrix in $\mathbb{M}^4$, and

$$P^\lambda(t,x) \triangleq \left(-\frac{\nu\Delta}{2} + 1\right)^{-1} \nabla_{\text{spec}}\mathcal{L}_\lambda(X_u, X_b)(t,x), \tag{9.25a}$$

$$r_\lambda(t) \triangleq \sum_{k \in \mathbb{Z}^2 \setminus \{0\}} \frac{1}{4} \mathfrak{l}\left(\frac{|k|}{\lambda}\right)^2 \left(\frac{1 - e^{-2\nu|k|^2 t}}{\nu}\right)\left(\frac{\nu|k|^2}{2} + 1\right)^{-1}. \tag{9.25b}$$

With $\{\lambda^i\}_{i \in \mathbb{N}}$ to be defined in (9.35), now for all $t \in [0,\infty)$ and $\kappa > 0$, we define

$$L_t^\kappa \triangleq 1 + \sum_{j \in \{u,b\}} [\|X_j\|_{C_t \mathcal{C}_x^{-\kappa}} + \|Y_j\|_{C_t \mathcal{C}_x^{2\kappa}}], \tag{9.26a}$$

$$N_t^\kappa \triangleq L_t^\kappa + \sup_{i \in \mathbb{N}} \|(\nabla_{\text{spec}}\mathcal{L}_{\lambda^i}(X_u, X_b) \circledcirc P^{\lambda^i} - r_{\lambda^i}\operatorname{Id}\|_{C_t \mathcal{C}_x^{-\kappa}}. \tag{9.26b}$$

The following gives a local-in-time solution theory for (9.22) in mild formulation.

Proposition 9.1.4 ([359, Proposition 2]). *Fix any* $\kappa \in (0, \frac{1}{2})$ *and then* $\gamma \triangleq 1-\frac{\kappa}{2}$. *Suppose that* $D_u, D_b \in C([0,\infty); \mathcal{C}^{-\kappa}(\mathbb{T}^2))$ *and* $Y_u^{\otimes 2}$, $Y_b^{\otimes 2}$, $Y_u \otimes Y_b$, *and* $Y_b \otimes Y_u$ *belong to* $C([0,\infty); \mathcal{C}^{2\kappa}(\mathbb{T}^2))$. *Then, for all* $(u^{\text{in}}, b^{\text{in}}) \in \mathcal{C}^{-1+2\kappa}(\mathbb{T}^2) \times \mathcal{C}^{-1+2\kappa}(\mathbb{T}^2)$ *that are both divergence-free and mean-zero,* (9.22) *has a unique mild solution* $(w_u, w_b) \in \mathcal{M}^{\frac{\gamma}{2}}_{T^{\max}} \mathcal{C}^{\frac{3\kappa}{2}} \times \mathcal{M}^{\frac{\gamma}{2}}_{T^{\max}} \mathcal{C}^{\frac{3\kappa}{2}}$ *over* $[0, T^{\max})$, *where* $T^{\max} = T^{\max}(\{L_t^\kappa\}_{t\geq 0}, u^{\text{in}}, b^{\text{in}}) \in (0, \infty]$.

Toward estimates, we introduce the symmetric and antisymmetric commutators $C^{\otimes_s}$ and $C^{\otimes_a}$ respectively as

$$C^{\otimes_s}(f,g) \triangleq (\partial_t - \nu\Delta)(f \otimes_s g) - f \otimes_s (\partial_t - \nu\Delta)g, \tag{9.27a}$$

$$C^{\otimes_a}(f,g) \triangleq (\partial_t - \nu\Delta)(f \otimes_a g) - f \otimes_a (\partial_t - \nu\Delta)g. \tag{9.27b}$$

We define Q_u and Q_b to solve

$$(\partial_t - \nu\Delta)Q_u = 2X_u, \quad Q_u(0) = 0, \quad \text{and} \quad (\partial_t - \nu\Delta)Q_b = 2X_b, \quad Q_b(0) = 0 \tag{9.28}$$

(cf. (8.101)) and then postulate that the solution (w_u, w_b) takes the form

$$w_u = -\mathcal{P}_L \operatorname{div}(w_u \otimes_s Q_u - w_b \otimes_s Q_b) + w_u^\sharp, \tag{9.29a}$$

$$w_b = -\mathcal{P}_L \operatorname{div}(w_b \otimes_a Q_u - w_u \otimes_a Q_b) + w_b^\sharp. \tag{9.29b}$$

The reason for the strange constant 2 in (9.28) will be clear in the following derivations of the equations of $(w_u^\sharp, w_b^\sharp)$, of which we give details for $w_b^\sharp$ as follows. First, using (9.29b), we see

$$\begin{aligned}\partial_t w_b^\sharp &= \partial_t w_b + \mathcal{P}_L \operatorname{div}(2w_b \otimes_a X_u + \partial_t w_b \otimes_a Q_u + w_b \otimes_a (\partial_t Q_u - 2X_u) \\ &\quad - 2w_u \otimes_a X_b - \partial_t w_u \otimes_a Q_b - w_u \otimes_a (\partial_t Q_b - 2X_b)).\end{aligned} \tag{9.30}$$

We now substitute (9.28) into $\partial_t Q_b - 2X_b$ and rely on (9.22b), (9.27b), and (9.29b) to compute

$$\begin{aligned}\partial_t w_b^\sharp &= \partial_t w_b + \mathcal{P}_L \operatorname{div}(2w_b \otimes_a X_u + \partial_t w_b \otimes_a Q_u + \nu w_b \otimes_a \Delta Q_u \\ &\quad - 2w_u \otimes_a X_b - \partial_t w_u \otimes_a Q_b - \nu w_u \otimes_a \Delta Q_b) \\ &= \nu\Delta w_b^\sharp - \mathcal{P}_L \operatorname{div}(w_b \otimes w_u + w_b \otimes_a D_u - 2w_b \otimes_a X_u - C^{\otimes_a}(w_b, Q_u) + Y_b \otimes Y_u \\ &\quad - w_u \otimes w_b - w_u \otimes_a D_b + 2w_u \otimes_a X_b + C^{\otimes_a}(w_u, Q_b) - Y_u \otimes Y_b).\end{aligned} \tag{9.31}$$

The equation of $w_u^\sharp$ can be derived analogously; consequently,

$$\begin{aligned}&\partial_t w_u^\sharp - \nu\Delta w_u^\sharp \\ &\quad + \mathcal{P}_L \operatorname{div}(w_u^{\otimes 2} + D_u \otimes_s w_u - 2X_u \otimes_s w_u - C^{\otimes_s}(w_u, Q_u) + Y_u^{\otimes 2} \\ &\quad - w_b^{\otimes 2} - D_b \otimes_s w_b + 2X_b \otimes_s w_b + C^{\otimes_s}(w_b, Q_b) - Y_b^{\otimes 2}) = 0,\end{aligned} \tag{9.32a}$$

$$\partial_t w_b^\sharp - \nu\Delta w_b^\sharp \tag{9.32b}$$
$$+ \mathcal{P}_L \operatorname{div}(w_b \otimes w_u + w_b \otimes_a D_u - 2w_b \circledS_a X_u - C^{\circledS_a}(w_b, Q_u) + Y_b \otimes Y_u$$
$$- w_u \otimes w_b - w_u \otimes_a D_b + 2w_u \circledS_a X_b + C^{\circledS_a}(w_u, Q_b) - Y_u \otimes Y_b) = 0.$$

Definition 9.1.5 ([359, Definition 3]). We fix any $a > 0$ and define a family of stopping times $\{T_i\}_{i\in\mathbb{N}_0}$ by

$$T_0 \triangleq 0, \tag{9.33a}$$
$$T_{i+1}(\omega, u^{\text{in}}, b^{\text{in}}) \triangleq \inf\{t \geq T_i : \|w_u(t)\|_{L^2} + \|w_b(t)\|_{L^2} \geq i+1\}$$
$$\wedge T^{\max}(\omega, u^{\text{in}}, b^{\text{in}}), \tag{9.33b}$$

with $T^{\max}(\omega, u^{\text{in}}, b^{\text{in}})$ from Proposition 9.1.1. We set

$$i_0(u^{\text{in}}, b^{\text{in}}) \triangleq \max\{i \in \mathbb{N}_0 : i \leq \|u^{\text{in}}\|_{L^2} + \|b^{\text{in}}\|_{L^2}\}, \tag{9.34}$$

so that $T_i = 0$ if and only if $i \leq i_0(u^{\text{in}}, b^{\text{in}})$. We define

$$\lambda^i \triangleq (i+1)^a, \tag{9.35}$$

$$\lambda_t \triangleq \begin{cases} (1 + \lceil \|u^{\text{in}}\|_{L^2} + \|b^{\text{in}}\|_{L^2} \rceil)^a & \text{if } t = 0, \\ (1 + \|w_u(T_i)\|_{L^2} + \|w_b(T_i)\|_{L^2})^a & \text{if } t > 0 \text{ and } t \in [T_i, T_{i+1}). \end{cases} \tag{9.36}$$

We define their high-frequency components as

$$Q_u^{\mathcal{H}}(t) \triangleq \mathcal{H}_{\lambda_t} Q_u(t), \quad Q_b^{\mathcal{H}}(t) \triangleq \mathcal{H}_{\lambda_t} Q_b(t), \tag{9.37a}$$
$$w_u^{\mathcal{H}} \triangleq -\mathcal{P}_L \operatorname{div}(w_u \circledS_s Q_u^{\mathcal{H}} - w_b \circledS_s Q_b^{\mathcal{H}}), \quad w_u^{\mathcal{L}} \triangleq w_u - w_u^{\mathcal{H}}, \tag{9.37b}$$
$$w_b^{\mathcal{H}} \triangleq -\mathcal{P}_L \operatorname{div}(w_b \circledS_a Q_u^{\mathcal{H}} - w_u \circledS_a Q_b^{\mathcal{H}}), \quad w_b^{\mathcal{L}} \triangleq w_b - w_b^{\mathcal{H}}. \tag{9.37c}$$

The higher- and low-frequency parts of w_u, w_b can be estimated as follows; below we assume Proposition 9.1.8, which can be proven independently of Proposition 9.1.6.

Proposition 9.1.6 ([359, Proposition 4]). *Fix any $a > 0$ from Definition 9.1.5, $\mathcal{N}$ from Proposition 9.1.1, and $\mathcal{N}''$ from Proposition 9.1.8, and define N_t^κ by (9.26b). Then, for any $\delta > 0$ and $\omega \in \Omega \setminus (\mathcal{N} \cup \mathcal{N}'')$, there exists a constant $C(\delta) > 0$ such that $w_u^{\mathcal{H}}$ and $w_b^{\mathcal{H}}$ defined by (9.37) satisfy*

$$(\|w_u^{\mathcal{H}}\|_{H^{1-2\kappa-\delta}} + \|w_b^{\mathcal{H}}\|_{H^{1-2\kappa-\delta}})(\omega, t) \tag{9.38}$$
$$\leq C(\delta)(1 + \|w_u\|_{L^2} + \|w_b\|_{L^2})^{1-a\delta}(\omega, t) N_t^\kappa(\omega) \quad \forall t \in [0, T^{\max}(\omega, u^{\text{in}}, b^{\text{in}})).$$

We now focus on the low-frequency part $(w_u^{\mathcal{L}}, w_b^{\mathcal{L}})$. Upon $L^2(\mathbb{T}^2)$-estimate of $(w_u^{\mathcal{L}}, w_b^{\mathcal{L}})$, we obtain

$$\partial_t \|(w_u^{\mathcal{L}}, w_b^{\mathcal{L}})(t)\|_{L^2}^2 = \sum_{k=1}^{4} \mathrm{IX}_k,$$

where

$$\begin{aligned}
\mathrm{IX}_1 \triangleq{} & 2\langle w_u^{\mathcal{L}}, \upsilon\Delta w_u^{\mathcal{L}} - \mathrm{div}(2(\mathcal{L}_{\lambda_t} X_u) \otimes_s w_u^{\mathcal{L}} - 2(\mathcal{L}_{\lambda_t} X_b) \otimes_s w_b^{\mathcal{L}})\rangle(t) \\
& + 2\langle w_b^{\mathcal{L}}, \upsilon\Delta w_b^{\mathcal{L}} - \mathrm{div}(2w_b^{\mathcal{L}} \otimes_a (\mathcal{L}_{\lambda_t} X_u) - 2w_u^{\mathcal{L}} \otimes_a (\mathcal{L}_{\lambda_t} X_b))\rangle(t), \qquad (9.39a)\\
\mathrm{IX}_2 \triangleq{} & -2\langle w_u^{\mathcal{L}}, \mathrm{div}(2(\mathcal{H}_{\lambda_t} X_u) \otimes_s w_u^{\mathcal{L}} - 2(\mathcal{H}_{\lambda_t} X_u) \circledcirc_s w_u^{\mathcal{L}} \\
& - 2(\mathcal{H}_{\lambda_t} X_b) \otimes_s w_b^{\mathcal{L}} + 2(\mathcal{H}_{\lambda_t} X_b) \circledcirc_s w_b^{\mathcal{L}})\rangle(t) \\
& - 2\langle w_b^{\mathcal{L}}, \mathrm{div}(2w_b^{\mathcal{L}} \otimes_a (\mathcal{H}_{\lambda_t} X_u) - 2w_b^{\mathcal{L}} \circledcirc_a (\mathcal{H}_{\lambda_t} X_u) \\
& - 2w_u^{\mathcal{L}} \otimes_a (\mathcal{H}_{\lambda_t} X_b) + 2w_u^{\mathcal{L}} \circledcirc_a (\mathcal{H}_{\lambda_t} X_b))\rangle(t), \qquad (9.39b)\\
\mathrm{IX}_3 \triangleq{} & -2\langle w_u^{\mathcal{L}}, \mathrm{div}(2X_u \otimes_s w_u^{\mathcal{H}} - 2(\mathcal{H}_{\lambda_t} X_u) \circledcirc_s w_u^{\mathcal{H}} \\
& - 2X_b \otimes_s w_b^{\mathcal{H}} + 2(\mathcal{H}_{\lambda_t} X_b) \circledcirc_s w_b^{\mathcal{H}})\rangle(t) \\
& - 2\langle w_b^{\mathcal{L}}, \mathrm{div}(2w_b^{\mathcal{H}} \otimes_a X_u - 2w_b^{\mathcal{H}} \circledcirc_a (\mathcal{H}_{\lambda_t} X_u) \\
& - 2w_u^{\mathcal{H}} \otimes_a X_b + 2w_u^{\mathcal{H}} \circledcirc_a (\mathcal{H}_{\lambda_t} X_b))\rangle(t), \qquad (9.39c)\\
\mathrm{IX}_4 \triangleq{} & -2\langle w_u^{\mathcal{L}}, \mathrm{div}(w_u^{\otimes 2} + 2Y_u \otimes_s w_u - w_b^{\otimes 2} - 2Y_b \otimes_s w_b \\
& - C^{\circledcirc_s}(w_u, Q_u^{\mathcal{H}}) + Y_u^{\otimes 2} + C^{\circledcirc_s}(w_b, Q_b^{\mathcal{H}}) - Y_b^{\otimes 2})\rangle(t) \\
& - 2\langle w_b^{\mathcal{L}}, \mathrm{div}(w_b \otimes w_u + 2w_b \otimes_a Y_u - w_u \otimes w_b - 2w_u \otimes_a Y_b \\
& - C^{\circledcirc_a}(w_b, Q_u^{\mathcal{H}}) + Y_b \otimes Y_u + C^{\circledcirc_a}(w_u, Q_b^{\mathcal{H}}) - Y_u \otimes Y_b)\rangle(t). \qquad (9.39d)
\end{aligned}$$

We observe that $\frac{\upsilon}{2}\Delta \,\mathrm{Id} - \nabla_{\mathrm{spec}}\mathcal{L}_{\lambda_t}(X_u, X_b)$ is reminiscent of the parabolic Anderson operator $\Delta + \xi$ (recall (9.5) and (9.10)). In consideration of the renormalization constant, we define the time-dependent family of operators

$$\mathcal{A}_t \triangleq \frac{\upsilon}{2}\Delta \,\mathrm{Id} - \nabla_{\mathrm{spec}}(X_u, X_b)(t) - \infty \quad \forall t \geq 0 \qquad (9.40)$$

as the limit of

$$\mathcal{A}_t^{\lambda} \triangleq \frac{\upsilon}{2}\Delta \,\mathrm{Id} - \nabla_{\mathrm{spec}}\mathcal{L}_{\lambda}(X_u, X_b) - r_{\lambda}(t)\,\mathrm{Id} \quad \forall t \geq 0 \qquad (9.41)$$

as $\lambda \nearrow +\infty$ for r_λ from (9.25), allowing us to rewrite (9.39a) as

$$\begin{aligned}
\mathrm{IX}_1 ={} & -\upsilon\|(w_u^{\mathcal{L}}, w_b^{\mathcal{L}})(t)\|_{\dot{H}^1}^2 \\
& + 2\left\langle \begin{pmatrix} w_u^{\mathcal{L}} \\ w_b^{\mathcal{L}} \end{pmatrix}, \mathcal{A}_t^{\lambda_t} \begin{pmatrix} w_u^{\mathcal{L}} \\ w_b^{\mathcal{L}} \end{pmatrix} \right\rangle (t) + r_{\lambda_t}(t)\|(w_u^{\mathcal{L}}, w_b^{\mathcal{L}})(t)\|_{L^2}^2. \qquad (9.42)
\end{aligned}$$

Here we split the diffusion to halves: one in $\upsilon\|(w_u^{\mathcal{L}}, w_b^{\mathcal{L}})(t)\|_{\dot{H}^1}^2$ for energy estimates and another within $\mathcal{A}_t^{\lambda}$ (9.41) as an Anderson Hamiltonian operator; the choice of $\frac{1}{2}$ is arbitrary.

As we described in Section 9.1.1, there are now three key steps to complete the proof of Theorem 9.1.2: (a) Renormalization, (b) Anderson Hamiltonian, and (c) Estimates.

9.1.2.1 (a) Renormalization

Concerning the renormalization constant $r_\lambda(t)$, we can directly show from (9.25b) using Definition 2.1.1 that

$$r_\lambda(t) \lesssim \int_1^\lambda \frac{1}{\mu^2}\mu d\mu \approx \ln(\lambda). \tag{9.43}$$

Computations similar to those in Section 8.5.2 can verify the following proposition.

Proposition 9.1.7 ([359, Proposition 15]). *For all $\kappa > 0$ and $t \geq 0$, there exists a distribution $\nabla_{spec}(X_u, X_b)(t) \diamond P_t \in \mathcal{C}^{-\kappa}(\mathbb{T}^2; \mathbb{M}^4)$ such that*

$$\begin{aligned}&(\nabla_{spec}(\mathcal{L}_{\lambda^n}X_u, \mathcal{L}_{\lambda^n}X_b), (\nabla_{spec}(\mathcal{L}_{\lambda^n}X_u, \mathcal{L}_{\lambda^n}X_b)) \circledcirc P^{\lambda^n} - r_{\lambda^n} Id) \\ &\to (\nabla_{spec}(X_u, X_b), \nabla_{spec}(X_u, X_b) \diamond P)\end{aligned}$$

as $n \to \infty$ $\mathbb{P}$-a. s.

Proof. We give details here only concerning how precisely the renormalization constant $r_\lambda(t)$ appears. For brevity, we denote $X_{a,\lambda} \triangleq \mathcal{L}_\lambda X_a$ for $a \in \{u, b\}$ and using (9.25a) and (6.15), compute

$$\begin{aligned}
&4(\nabla_{\text{spec}}(\mathcal{L}_\lambda X_u, \mathcal{L}_\lambda X_b) \circledcirc P^\lambda) \qquad (9.44)\\
&= \left(2\nabla_{\text{spec}}(\mathcal{L}_\lambda X_u, \mathcal{L}_\lambda X_b) \circledcirc \left(-\frac{\upsilon\Delta}{2} + 1\right)^{-1} 2\nabla_{\text{spec}}(\mathcal{L}_\lambda X_u, \mathcal{L}_\lambda X_b)\right)\\
&= \begin{pmatrix} \begin{pmatrix} \partial_1 X^1_{u,\lambda} + \partial_1 X^1_{u,\lambda} & \partial_1 X^2_{u,\lambda} + \partial_2 X^1_{u,\lambda} \\ \partial_2 X^1_{u,\lambda} + \partial_1 X^2_{u,\lambda} & \partial_2 X^2_{u,\lambda} + \partial_2 X^2_{u,\lambda} \end{pmatrix} & \begin{pmatrix} 0 & \partial_1 X^2_{b,\lambda} - \partial_2 X^1_{b,\lambda} \\ \partial_2 X^1_{b,\lambda} - \partial_1 X^2_{b,\lambda} & 0 \end{pmatrix} \\ -\begin{pmatrix} 0 & \partial_1 X^2_{b,\lambda} - \partial_2 X^1_{b,\lambda} \\ \partial_2 X^1_{b,\lambda} - \partial_1 X^2_{b,\lambda} & 0 \end{pmatrix} & -\begin{pmatrix} \partial_1 X^1_{u,\lambda} + \partial_1 X^1_{u,\lambda} & \partial_1 X^2_{u,\lambda} + \partial_2 X^1_{u,\lambda} \\ \partial_2 X^1_{u,\lambda} + \partial_1 X^2_{u,\lambda} & \partial_2 X^2_{u,\lambda} + \partial_2 X^2_{u,\lambda} \end{pmatrix} \end{pmatrix}\\
&\circledcirc \left(-\frac{\upsilon\Delta}{2} + 1\right)^{-1}\\
&\times \begin{pmatrix} \begin{pmatrix} \partial_1 X^1_{u,\lambda} + \partial_1 X^1_{u,\lambda} & \partial_1 X^2_{u,\lambda} + \partial_2 X^1_{u,\lambda} \\ \partial_2 X^1_{u,\lambda} + \partial_1 X^2_{u,\lambda} & \partial_2 X^2_{u,\lambda} + \partial_2 X^2_{u,\lambda} \end{pmatrix} & \begin{pmatrix} 0 & \partial_1 X^2_{b,\lambda} - \partial_2 X^1_{b,\lambda} \\ \partial_2 X^1_{b,\lambda} - \partial_1 X^2_{b,\lambda} & 0 \end{pmatrix} \\ -\begin{pmatrix} 0 & \partial_1 X^2_{b,\lambda} - \partial_2 X^1_{b,\lambda} \\ \partial_2 X^1_{b,\lambda} - \partial_1 X^2_{b,\lambda} & 0 \end{pmatrix} & -\begin{pmatrix} \partial_1 X^1_{u,\lambda} + \partial_1 X^1_{u,\lambda} & \partial_1 X^2_{u,\lambda} + \partial_2 X^1_{u,\lambda} \\ \partial_2 X^1_{u,\lambda} + \partial_1 X^2_{u,\lambda} & \partial_2 X^2_{u,\lambda} + \partial_2 X^2_{u,\lambda} \end{pmatrix} \end{pmatrix},
\end{aligned}$$

in which we focus only on the (4, 4)-entry:

$$- (\partial_2 X^1_{b,\lambda} - \partial_1 X^2_{b,\lambda}) \circ \left(-\frac{\upsilon\Delta}{2}+1\right)^{-1} (\partial_1 X^2_{b,\lambda} - \partial_2 X^1_{b,\lambda}) \tag{9.45}$$
$$+ (\partial_2 X^1_{u,\lambda} + \partial_1 X^2_{u,\lambda}) \circ \left(-\frac{\upsilon\Delta}{2}+1\right)^{-1} (\partial_1 X^2_{u,\lambda} + \partial_2 X^1_{u,\lambda})$$
$$+ (\partial_2 X^2_{u,\lambda} + \partial_2 X^2_{u,\lambda}) \circ \left(-\frac{\upsilon\Delta}{2}+1\right)^{-1} (\partial_2 X^2_{u,\lambda} + \partial_2 X^2_{u,\lambda}).$$

Analogous analysis applies to other diagonal entries, and nondiagonal entries actually vanish. We define

$$c^{i,l}_{j,m}(k,k') \triangleq -k_i k_j^{\perp} k'_l (k')^{\perp}_m |k|^{-1}|k'|^{-1}, \tag{9.46}$$

so that for all $i,j,l,m \in \{1,2\}$ and $\alpha,\gamma \in \{u,b\}$, we can write

$$\partial_i X^j_{\alpha,\lambda} \circ \left(-\frac{\upsilon\Delta}{2}+1\right)^{-1} \partial_l X^m_{\gamma,\lambda}(x)$$
$$= \sum_{\substack{k,k' \in \mathbb{Z}^2: k' \neq 0, k \neq k' \\ |c-d| \leq 1}} e^{i2\pi k\cdot x} \rho_c(k-k')\rho_d(k')$$
$$\times i(k-k')_i \mathcal{F}(X^j_{\alpha,\lambda})(k-k') \left(\frac{\upsilon|k'|^2}{2}+1\right)^{-1} i k'_l \mathcal{F}(X^m_{\gamma,\lambda})(k'). \tag{9.47}$$

Now taking mathematical expectation $\mathbb{E}$ and using the fact that $c^{i,l}_{j,m}(-k',k') = c^{i,l}_{j,m}(k',k')$ by (9.46) give us

$$\mathbb{E}\left[\partial_i X^j_{\alpha,\lambda} \circ \left(-\frac{\upsilon\Delta}{2}+1\right)^{-1} \partial_l X^m_{\gamma,\lambda}\right](x)$$
$$= - \sum_{k' \in \mathbb{Z}^2 \setminus \{0\}, |c-d| \leq 1} \rho_c(-k')\rho_d(k')$$
$$\times \imath\left(\frac{|k'|}{\lambda}\right)^2 \frac{1-e^{-2\upsilon|k'|^2 t}}{2\upsilon|k'|^2} \left(\frac{\upsilon|k'|^2}{2}+1\right)^{-1} c^{i,l}_{j,m}(-k',k') 1_{\{\alpha=\gamma\}}$$
$$= \sum_{k \in \mathbb{Z}^2 \setminus \{0\}} \imath\left(\frac{|k|}{\lambda}\right)^2 \frac{1-e^{-2\upsilon|k|^2 t}}{2\upsilon|k|^4} \left(\frac{\upsilon|k|^2}{2}+1\right)^{-1} k_i k_j^{\perp} k_l k_m^{\perp} 1_{\{\alpha=\gamma\}}. \tag{9.48}$$

Finally, we apply (9.48) to the (4, 4)-entry in (9.45) to obtain

$$\mathbb{E}\Big[-\partial_2 X^1_{b,\lambda} \circ \left(-\frac{\upsilon\Delta}{2}+1\right)^{-1} \partial_1 X^2_{b,\lambda} + \partial_2 X^1_{b,\lambda} \circ \left(-\frac{\upsilon\Delta}{2}+1\right)^{-1} \partial_2 X^1_{b,\lambda}$$
$$+ \partial_1 X^2_{b,\lambda} \circ \left(-\frac{\upsilon\Delta}{2}+1\right)^{-1} \partial_1 X^2_{b,\lambda} - \partial_1 X^2_{b,\lambda} \circ \left(-\frac{\upsilon\Delta}{2}+1\right)^{-1} \partial_2 X^1_{b,\lambda}$$
$$+ \partial_2 X^1_{u,\lambda} \circ \left(-\frac{\upsilon\Delta}{2}+1\right)^{-1} \partial_1 X^2_{u,\lambda} + \partial_2 X^1_{u,\lambda} \circ \left(-\frac{\upsilon\Delta}{2}+1\right)^{-1} \partial_2 X^1_{u,\lambda}$$

$$
\begin{aligned}
&+ \partial_1 X_{u,\lambda}^2 \circ \left(-\frac{\upsilon\Delta}{2}+1\right)^{-1} \partial_1 X_{u,\lambda}^2 + \partial_1 X_{u,\lambda}^2 \circ \left(-\frac{\upsilon\Delta}{2}+1\right)^{-1} \partial_2 X_{u,\lambda}^1 \\
&+ 4\partial_2 X_{u,\lambda}^2 \circ \left(-\frac{\upsilon\Delta}{2}+1\right)^{-1} \partial_2 X_{u,\lambda}^2 \Big](t) \\
= &- \sum_{k\in\mathbb{Z}^2\setminus\{0\}} \iota\left(\frac{|k|}{\lambda}\right)^2 \frac{(1-e^{-2\upsilon|k|^2 t})}{2\upsilon|k|^4} \left(\frac{\upsilon|k|^2}{2}+1\right)^{-1} \\
&\times [k_2k_1^\perp k_1k_2^\perp - k_2k_1^\perp k_2k_1^\perp - k_1k_2^\perp k_1k_2^\perp + k_1k_2^\perp k_2k_1^\perp \\
&\quad - k_2k_1^\perp k_1k_2^\perp - k_2k_1^\perp k_2k_1^\perp - k_1k_2^\perp k_1k_2^\perp - k_1k_2^\perp k_2k_1^\perp - 4k_2k_2^\perp k_2k_2^\perp],
\end{aligned}
$$

so that due to (9.25b), further simplifications lead to

$$
\begin{aligned}
\mathbb{E}\Big[&-\partial_2 X_{b,\lambda}^1 \circ \left(-\frac{\upsilon\Delta}{2}+1\right)^{-1} \partial_1 X_{b,\lambda}^2 + \partial_2 X_{b,\lambda}^1 \circ \left(-\frac{\upsilon\Delta}{2}+1\right)^{-1} \partial_2 X_{b,\lambda}^1 \\
&+ \partial_1 X_{b,\lambda}^2 \circ \left(-\frac{\upsilon\Delta}{2}+1\right)^{-1} \partial_1 X_{b,\lambda}^2 - \partial_1 X_{b,\lambda}^2 \circ \left(-\frac{\upsilon\Delta}{2}+1\right)^{-1} \partial_2 X_{b,\lambda}^1 \\
&+ \partial_2 X_{u,\lambda}^1 \circ \left(-\frac{\upsilon\Delta}{2}+1\right)^{-1} \partial_1 X_{u,\lambda}^2 + \partial_2 X_{u,\lambda}^1 \circ \left(-\frac{\upsilon\Delta}{2}+1\right)^{-1} \partial_2 X_{u,\lambda}^1 \\
&+ \partial_1 X_{u,\lambda}^2 \circ \left(-\frac{\upsilon\Delta}{2}+1\right)^{-1} \partial_1 X_{u,\lambda}^2 + \partial_1 X_{u,\lambda}^2 \circ \left(-\frac{\upsilon\Delta}{2}+1\right)^{-1} \partial_2 X_{u,\lambda}^1 \\
&+ 4\partial_2 X_{u,\lambda}^2 \circ \left(-\frac{\upsilon\Delta}{2}+1\right)^{-1} \partial_2 X_{u,\lambda}^2 \Big](t) = 4r_\lambda(t).
\end{aligned}
$$

This is the reasoning behind the definition in (9.25b), where the factor $\frac{1}{4}$ is due to the multiplication by 4 in (9.44). □

Consequently, we have the following proposition that verifies that $N_t^\kappa = L_t^\kappa + \sup_{i\in\mathbb{N}} \|\nabla_{\text{spec}}\mathcal{L}_{\lambda^i}(X_u, X_b) \circledcirc P^{\lambda^i} - r_{\lambda^i}\text{Id}\|_{C_t\mathcal{C}_x^{-\kappa}}$ in (9.26b) is finite.

Proposition 9.1.8 ([359, Proposition 3]). *There exists a null set $\mathcal{N}'' \subset \Omega$ such that*

$$
N_t^\kappa(\omega) < \infty \quad \forall\, \omega \in \Omega \setminus \mathcal{N}'' \quad \forall\, t \geq 0, \quad \forall\, \kappa > 0.
$$

This justifies that quantity (9.26b).

9.1.2.2 (b) Anderson Hamiltonian

We return to $\mathcal{A}_t, \mathcal{A}_t^\lambda$, and the enhanced noise defined in (9.40), (9.41), and (9.24). We now define the space of enhanced noise $\Theta_\kappa \subset \mathcal{C}^{-1-\kappa}(\mathbb{T}^2; \mathbb{M}^4) \times \mathcal{C}^{-2\kappa}(\mathbb{T}^2; \mathbb{M}^4)$, $\kappa > 0$, by

$$
\Theta_\kappa \triangleq \overline{\left\{\left(X_1, X_1 \circledcirc \left(-\frac{\upsilon\Delta}{2}+1\right)^{-1} X_1 - c\right) : X_1 \in C^\infty(\mathbb{T}^2; \mathbb{M}^4), c \in \mathbb{R}\right\}}, \tag{9.49}
$$

where the closure is taken w. r. t. the product norm of $\mathcal{C}^{-1-\kappa}(\mathbb{T}^2;\mathbb{M}^4)\times\mathcal{C}^{-2\kappa}(\mathbb{T}^2;\mathbb{M}^4)$. To define

$$\mathcal{U}(X)\triangleq\frac{\upsilon\Delta}{2}\operatorname{Id}+X_1$$

for all $X=(X_1,X_2)\in\Theta_\kappa$ for some $\kappa>0$, we define the space of strongly paracontrolled distributions ([7, Definitions 4.1 and 4.17, Section 4.2])

$$P\triangleq\left(-\frac{\upsilon\Delta}{2}+1\right)^{-1}X_1, \tag{9.50a}$$

$$\mathcal{X}_\kappa(X)\triangleq\{\phi\in H^{1-\kappa}(\mathbb{T}^2):\phi=\phi\otimes P+\phi^\sharp,\phi^\sharp\in H^{2-2\kappa}(\mathbb{T}^2)\}, \tag{9.50b}$$

$$\|\phi\|_{\mathcal{X}_\kappa}\triangleq\|\phi\|_{H^{1-\kappa}}+\|\phi-\phi\otimes P\|_{H^{2-2\kappa}}. \tag{9.50c}$$

The following proposition can be obtained from [7] (e. g., see [7, Propositions 4.13 and 4.23, Lemma 4.15]).

Proposition 9.1.9 (Cf. [176, Proposition 7.1]). *Define $\Theta\triangleq\cup_{0<\kappa<\kappa_0}\Theta_\kappa$ for $\kappa_0>0$, with Θ_κ defined in* (9.49), *and C_{op}, the space of closed self-adjoint operators with the graph distance where the convergence in this distance is implied by the convergence in the resolvent sense. Then there exist $\kappa_0>0$ and a unique map $\mathcal{U}:\Theta\mapsto C_{op}$ such that we have the following.*

(a) *For any smooth $X=(X_1,X_2)\in[C^\infty(\mathbb{T}^2;\mathbb{M}^4)\times C^\infty(\mathbb{T}^2;\mathbb{M}^4)]\cap\Theta$ and $\phi\in H^2(\mathbb{T}^2)$, $\mathcal{U}$ satisfies*

$$\mathcal{U}(X)\phi=\frac{\upsilon\Delta}{2}\phi+X_1\otimes\phi+X_1\circledcirc\phi+X_1\odot\phi^\sharp+\phi\otimes X_2+C^{\odot}(\phi,P,X_1),$$

where P is defined in (9.50a), *and*

$$C^{\odot}(\phi,P,X_1)\triangleq X_1\odot(\phi\otimes P)-\phi\otimes(P\odot X_1).$$

If $X_2=P\odot X_1$, then $\mathcal{U}(X)\phi=\frac{\upsilon\Delta}{2}\phi+X_1\phi$.

(b) *For any $\{X^n\}_{n\in\mathbb{N}}\subset C^\infty(\mathbb{T}^2;\mathbb{M}^4)\times C^\infty(\mathbb{T}^2;\mathbb{M}^4)$ such that $X^n\to X$ in Θ_κ as $n\to\infty$ for some $\kappa\in(0,\kappa_0)$ and $X\in\Theta_\kappa$, $\mathcal{U}(X^n)$ converges to $\mathcal{U}(X)$ in resolvent sense. Moreover, for any $\kappa\in(0,\kappa_0)$, there exist two continuous maps $\mathbf{m},\mathbf{c}:\Theta_\kappa\mapsto\mathbb{R}_+$ such that*

$$[\mathbf{m}(X),\infty)\subset\rho(\mathcal{U}(X))\quad\forall X\in\Theta_\kappa,$$

where $\rho(\mathcal{U}(X))$ is the resolvent set of an operator $\mathcal{U}(X)$ with upper bound

$$\|(-\mathcal{U}(X)+m)^{-1}\phi\|_{\mathcal{X}_\kappa}\leq\mathbf{c}(X)\|\phi\|_{L^2}\quad\forall m\geq\mathbf{m}(X).$$

We apply Proposition 9.1.9 for our case and see that

$$\mathcal{U}(\mathbf{X})=\frac{\upsilon}{2}\Delta\operatorname{Id}-\nabla_{\mathrm{spec}}\mathcal{L}_\lambda(X_u,X_b)\overset{(9.41)}{=}\mathcal{A}_t^\lambda+r_\lambda\operatorname{Id}. \tag{9.51}$$

Then we see that $\langle -\mathcal{U}(\mathbf{X})\phi, \phi\rangle \geq -\sup\{\lambda \in \mathbb{R} \setminus \rho(\mathcal{U}(\mathbf{X}))\}\|\phi\|_{L^2}^2$, so that $-\mathcal{U}(\mathbf{X}) + \mathbf{m}(N_t^\kappa)$ is nonnegative. We can now estimate the key term within (9.42) as

$$\left\langle \begin{pmatrix} w_u^{\mathcal{L}} \\ w_b^{\mathcal{L}} \end{pmatrix}, \mathcal{A}_t^{\lambda_t} \begin{pmatrix} w_u^{\mathcal{L}} \\ w_b^{\mathcal{L}} \end{pmatrix} \right\rangle (t) \leq \left\langle \begin{pmatrix} w_u^{\mathcal{L}} \\ w_b^{\mathcal{L}} \end{pmatrix}, \mathcal{U}(\mathbf{X}) \begin{pmatrix} w_u^{\mathcal{L}} \\ w_b^{\mathcal{L}} \end{pmatrix} \right\rangle (t)$$
$$\leq \mathbf{m}(N_t^\kappa)\|(w_u^{\mathcal{L}}, w_b^{\mathcal{L}})(t)\|_{L^2}^2, \tag{9.52}$$

where in the first inequality, we used the fact that $r_\lambda(t) \geq 0$, as can be seen from (9.25b).

9.1.2.3 (c) Estimates

The estimate of $\mathbb{IX}_1$ was completed due to the estimates from (9.43) and (9.52):

$$\mathbb{IX}_1 \leq -\nu\|(w_u^{\mathcal{L}}, w_b^{\mathcal{L}})(t)\|_{\dot{H}^1}^2 + 2\mathbf{m}(N_t^\kappa)\|(w_u^{\mathcal{L}}, w_b^{\mathcal{L}})(t)\|_{L^2}^2 + C\ln(\lambda_t)\|(w_u^{\mathcal{L}}, w_b^{\mathcal{L}})(t)\|_{L^2}^2.$$

The estimates for $\mathbb{IX}_2$, $\mathbb{IX}_3$, and $\mathbb{IX}_4$ can be handled separately (see [359, Propositions 5–7] for details). We formally state a sufficient bound on $\mathbb{IX}_4$ because we wish to elaborate on the commutator term within $\mathbb{IX}_4$ for a later purpose in Section 9.1.3.

Proposition 9.1.10 ([359, Proposition 7]). *Let $t \in [T_i, T_{i+1})$ and fix λ_t from* (9.36) *with $\mathfrak{a} \in [\frac{11}{4}, \infty)$ and $\kappa_0 \in (0, \frac{3}{44}]$. Then there exists $\theta \in [\frac{3}{4} + 3\kappa_0, 1)$ such that for all $\kappa \in (0, \kappa_0]$, $\mathbb{IX}_4$ from* (9.39d) *satisfies*

$$\mathbb{IX}_4 \lesssim (\|w_u^{\mathcal{L}}\|_{H^{1-\frac{3\kappa}{2}}} + \|w_b^{\mathcal{L}}\|_{H^{1-\frac{3\kappa}{2}}})(t)(\|w_u^{\mathcal{L}}(t)\|_{H^{2\kappa}} + \|w_b^{\mathcal{L}}(t)\|_{H^{2\kappa}} + N_t^\kappa)N_t^\kappa$$
$$+ (\|w_u^{\mathcal{L}}\|_{H^\theta} + \|w_b^{\mathcal{L}}\|_{H^\theta})(t)(\|w_u^{\mathcal{L}}(t)\|_{H^\theta} + \|w_b^{\mathcal{L}}(t)\|_{H^\theta} + N_t^\kappa)(N_t^\kappa)^2. \tag{9.53}$$

Proof. We only sketch the proof of the commutator term within $\mathbb{IX}_4$ in (9.39d). Without loss of generality, we focus on $2\langle w_b^{\mathcal{L}}, \operatorname{div}(C^{\circledS_a}(w_b, Q_u^{\mathcal{H}}))\rangle$. It turns out that we can rewrite the commutator in a simpler form. First, directly from (9.27b) we can rewrite

$$C^{\circledS_a}(f, g) = \partial_t f \circledS_a g - \nu\Delta(f \circledS_a g) + \nu f \circledS_a \Delta g. \tag{9.54}$$

Now we focus on the second and third terms, which we write out carefully using (9.19) and (9.16):

$$-\Delta(f \circledS_a g) = -\sum_{k=1}^{2}\sum_{i\geq -1}\sum_{-1\leq j\leq i-2}\frac{1}{2}$$
$$\times \begin{pmatrix} 0 & \begin{matrix} \Delta_j\partial_k^2 f_1\Delta_i g_2 + 2\Delta_j\partial_k f_1\Delta_i\partial_k g_2 \\ + \Delta_j f_1\Delta_i\partial_k^2 g_2 - \Delta_i\partial_k^2 g_1\Delta_j f_2 \\ - 2\Delta_i\partial_k g_1\Delta_j\partial_k f_2 - \Delta_i g_1\Delta_j\partial_k^2 f_2 \end{matrix} \\ \begin{matrix} \Delta_j\partial_k^2 f_2\Delta_i g_1 + 2\Delta_j\partial_k f_2\Delta_i\partial_k g_1 \\ + \Delta_j f_2\Delta_i\partial_k^2 g_1 - \Delta_i\partial_k^2 g_2\Delta_j f_1 \\ - 2\Delta_i\partial_k g_2\Delta_j\partial_k f_1 - \Delta_i g_2\Delta_j\partial_k^2 f_1 \end{matrix} & 0 \end{pmatrix}$$

and

$$
\begin{aligned}
&\nu f \circledleq_a \Delta g \\
&= \sum_{k=1}^{2} \sum_{i \geq -1} \sum_{-1 \leq j \leq i-2} \frac{1}{2} \begin{pmatrix} 0 & \Delta_j f_1 \Delta_i \partial_k^2 g_2 - \Delta_i \partial_k^2 g_1 \Delta_j f_2 \\ \Delta_j f_2 \Delta_i \partial_k^2 g_1 - \Delta_i \partial_k^2 g_2 \Delta_j f_1 & 0 \end{pmatrix}.
\end{aligned}
$$

We can see the cancellations that lead to

$$
-\Delta(f \circledleq_a g) + f \circledleq_a \Delta g = -\Delta f \circledleq_a g - 2 \sum_{k=1}^{2} \partial_k f \circledleq_a \partial_k g. \tag{9.55}
$$

Putting (9.54) together with (9.55) gives us

$$
C^{\circledleq_a}(f, g) = ((\partial_t - \nu\Delta) f) \circledleq_a g - 2\nu \sum_{k=1}^{2} \partial_k f \circledleq_a \partial_k g. \tag{9.56}
$$

Analogously, we can rewrite (9.27a) as

$$
C^{\circledleq_s}(f, g) = ((\partial_t - \nu\Delta) f) \circledleq_s g - 2\nu \sum_{k=1}^{2} \partial_k f \circledleq_s \partial_k g.
$$

Heuristically, we just simply used the Leibniz rule

$$
\Delta(fg) - (\Delta f) g - f(\Delta g) = 2(\nabla f) \cdot (\nabla g). \tag{9.57}
$$

In the theory of nonlinear PDEs, it is well known that the extreme cases of distributions of derivatives on a product present the most difficulty since $(\Delta f) g$ and $f(\Delta g)$ require both f and g to be twice differentiable (recall Lemma 3.6.1). In this perspective, $(\nabla f) \cdot (\nabla g)$ in (9.57) is the most convenient scenario. Putting all these together, we must estimate $2\langle w_b^{\mathcal{L}}, \operatorname{div}(C^{\circledleq_a}(w_b, Q_u^{\mathcal{H}}))\rangle$, where

$$
C^{\circledleq_a}(w_b, Q_u^{\mathcal{H}}) = ((\partial_t - \nu\Delta) w_b) \circledleq_a Q_u^{\mathcal{H}} - 2\nu \sum_{k=1}^{2} \partial_k w_b \circledleq_a \partial_k Q_u^{\mathcal{H}}.
$$

Now we estimate using merely (8.7a), not even a commutator estimate, since $\frac{3\kappa}{2} - \theta < 0$ by the hypothesis of Proposition 9.1.10:

$$
\begin{aligned}
&2 \sum_{k=1}^{2} \langle w_b^{\mathcal{L}}, \operatorname{div}(\partial_k w_b \circledleq_a \partial_k Q_u^{\mathcal{H}})\rangle(t) \\
&\lesssim \sum_{k=1}^{2} \|w_b^{\mathcal{L}}(t)\|_{H^\theta} \|\partial_k w_b(t)\|_{H^{\frac{3\kappa}{2}-\theta}} \|\partial_k Q_u^{\mathcal{H}}(t)\|_{\mathcal{C}^{1-\frac{3\kappa}{2}}}.
\end{aligned} \tag{9.58}
$$

We continue by

$$
\begin{aligned}
&2\sum_{k=1}^{2}\langle w_b^{\mathcal{L}}, \operatorname{div}(\partial_k w_b \circledS_a \partial_k Q_u^{\mathcal{H}})\rangle(t) \\
&\overset{(9.37)}{\lesssim} \|w_b^{\mathcal{L}}(t)\|_{H^\theta}(\|w_b^{\mathcal{L}}\|_{H^\theta} + \|w_b^{\mathcal{H}}\|_{H^{1-2\kappa-\frac{1}{\mathfrak{a}}}})(t)\|\mathcal{H}_{\lambda_t}Q_u(t)\|_{\mathcal{C}^{2-\frac{3\kappa}{2}}} \\
&\overset{(9.38)}{\lesssim} \|w_b^{\mathcal{L}}(t)\|_{H^\theta}(\|w_b^{\mathcal{L}}(t)\|_{H^\theta} + [1+\|w_u\|_{L^2}+\|w_b\|_{L^2}]^{1-\mathfrak{a}(\frac{1}{\mathfrak{a}})}(t)N_t^\kappa)\|Q_u(t)\|_{\mathcal{C}^{2-\frac{3\kappa}{2}}} \\
&\overset{(9.28)}{\lesssim} \|w_b^{\mathcal{L}}(t)\|_{H^\theta}(\|w_b^{\mathcal{L}}(t)\|_{H^\theta} + N_t^\kappa)\left\|\int_0^t e^{\upsilon\Delta(t-s)}X_u(s)ds\right\|_{\mathcal{C}^{2-\frac{3\kappa}{2}}} \\
&\overset{(9.26)}{\lesssim} \|w_b^{\mathcal{L}}(t)\|_{H^\theta}(\|w_b^{\mathcal{L}}(t)\|_{H^\theta} + N_t^\kappa)N_t^\kappa.
\end{aligned}
$$

We omit the estimate on $\Delta f \circledS_a g$ here and refer the interested readers to [359, Proposition 7] for details. We will elaborate later in Section 9.1.3 the difficulty that arises in the case of a fractional Laplacian due to the absence of Leibniz rule. □

As mentioned, Proposition 9.1.10, together with [359, Propositions 5–7] on $\mathbb{X}_2$, $\mathbb{X}_3$, and $\mathbb{X}_4$, which we explained partially before Proposition 9.1.10, allows us to conclude the following bound.

Proposition 9.1.11 ([359, Propositions 8–9]). *Define $i_0(u^{\text{in}}, b^{\text{in}})$ by (9.34). Fix λ_t from (9.36) with $\mathfrak{a} = 3$ and $\kappa_0 \in (0, \frac{1}{44}]$. Then there exist a constant $C_1 > 0$ and increasing continuous functions C_2 and C_3 from $\mathbb{R}_{\geq 0}$ to $\mathbb{R}_{\geq 0}$ such that for all $\kappa \in (0, \kappa_0]$, $i \in \mathbb{N}_0$ such that $i \geq i_0(u^{\text{in}}, b^{\text{in}})$, and $t \in [T_i, T_{i+1})$,*

$$
\begin{aligned}
\partial_t\|(w_u^{\mathcal{L}}, w_b^{\mathcal{L}})(t)\|_{L^2}^2 \leq -\frac{\upsilon}{2}\|(w_u^{\mathcal{L}}, w_b^{\mathcal{L}})(t)\|_{\dot{H}^1}^2 + (C_1\ln(\lambda_t) + C_2(N_t^\kappa)) \\
\times [\|(w_u^{\mathcal{L}}, w_b^{\mathcal{L}})(t)\|_{L^2}^2 + \|(w_u^{\mathcal{L}}, w_b^{\mathcal{L}})(T_i)\|_{L^2}^2] + C_3(N_t^\kappa),
\end{aligned}
$$

so that

$$
\begin{aligned}
&\sup_{t\in[T_i,T_{i+1})}\|(w_u^{\mathcal{L}}, w_b^{\mathcal{L}})(t)\|_{L^2}^2 + \frac{\upsilon}{2}\int_{T_i}^{T_{i+1}}\|(w_u^{\mathcal{L}}, w_b^{\mathcal{L}})(s)\|_{\dot{H}^1}^2 ds \\
&\leq e^{(T_{i+1}-T_i)(C_2(N_{T_{i+1}}^\kappa)+C_1\ln(\lambda_{T_i}))}(\|(w_u^{\mathcal{L}}, w_b^{\mathcal{L}})(T_i)\|_{L^2}^2 + C_3(N_{T_{i+1}}^\kappa)).
\end{aligned}
$$

Consequently, if $T_{i+1} < T^{\max} \wedge t$, then for all $\kappa \in (0, \kappa_0]$, there exist constants $C(N_t^\kappa)$ and $\tilde{C}(N_t^\kappa)$ large enough such that

$$
T_{i+1} - T_i \geq \frac{1}{\tilde{C}(N_t^\kappa)(1+\ln(1+i))}\ln\left(\frac{i^2+2i-C(N_t^\kappa)}{i^2+\tilde{C}(N_t^\kappa)}\right). \tag{9.59}
$$

Informally stated, given that our initial data in Theorem 9.1.2 lies in $\mathcal{C}^{-1+\kappa}(\mathbb{T}^2) \times \mathcal{C}^{-1+\kappa}(\mathbb{T}^2)$, considering the embeddings $H^\kappa(\mathbb{T}^2) \hookrightarrow B_{2,\infty}^\kappa(\mathbb{T}^2) \hookrightarrow \mathcal{C}^{-1+\kappa}(\mathbb{T}^2)$ (Lemmas 3.2.1 and 2.3.5), we need to perform higher-regularity estimate, which is the content of the following proposition.

Proposition 9.1.12 ([359, Proposition 10]). *Fix λ_t from* (9.36) *with $\mathfrak{a} = 3$ and $\kappa_0 \in (0, \frac{1}{44}]$. Then for any $\kappa \in (0, \kappa_0)$ and $\epsilon \in (0, \kappa)$, we have the following. Suppose that there exist $M > 1$ and $T > 0$ such that*

$$\|(w_u^{\mathcal{L}}, w_b^{\mathcal{L}})(0)\|_{\dot{H}^\epsilon}^2 + \sup_{s \in [0, T \wedge T^{\max}]} \|(w_u^{\mathcal{L}}, w_b^{\mathcal{L}})(s)\|_{L^2}^2 + \nu \int_0^{T \wedge T^{\max}} \|(w_u^{\mathcal{L}}, w_b^{\mathcal{L}})(s)\|_{\dot{H}^1}^2 ds \leq M.$$

Then there exists $C(T, M, N_T^\kappa) \in (0, \infty)$ such that

$$\sup_{t \in [0, T \wedge T^{\max}]} \|(w_u^{\mathcal{L}}, w_b^{\mathcal{L}})(t)\|_{\dot{H}^\epsilon}^2 \leq C(T, M, N_T^\kappa).$$

With this higher-regularity estimates and Proposition 9.1.4, we can deduce the following blowup criterion.

Proposition 9.1.13 ([359, Proposition 11]). *Suppose that $u^{\mathrm{in}}, b^{\mathrm{in}} \in L^2_\sigma \cap \mathcal{C}^{-1+\kappa}(\mathbb{T}^2)$ for some $\kappa > 0$. If $T^{\max} < \infty$, then the corresponding solution (w_u, w_b) satisfies $\limsup_{t \nearrow T^{\max}} \|(w_u, w_b)(t)\|_{L^2} = +\infty$.*

Finally, we are ready to conclude the proof of Theorem 9.1.2.

Proof of Theorem 9.1.2. Suppose that $T^{\max} < \infty$. By Proposition 9.1.13 this implies $\limsup_{t \nearrow T^{\max}} \|(w_u, w_b)(t)\|_{L^2} = +\infty$. As a consequence of (9.33), we have $T_i < T^{\max}$ for all $i \in \mathbb{N}$. Because $T^{\max} < +\infty$, (9.59) gives us

$$T_{i+1} - T_i \geq \frac{1}{\widetilde{C}(N_{T^{\max}}^\kappa)(1 + \ln(1+i))} \ln\left(\frac{i^2 + 2i - C(N_{T^{\max}}^\kappa)}{i^2 + \widetilde{C}(N_{T^{\max}}^\kappa)} \right).$$

Summing both sides over $i \in \mathbb{N}$, the left-hand side remains finite because

$$\sum_{i=1}^{\infty} T_{i+1} - T_i = \lim_{i \to \infty} T_i - T_1 \leq T_{\max} - T_1 < \infty. \tag{9.60}$$

Thus a contradiction follows if the sum over the right-hand side of (9.60) over $i \in \mathbb{N}$ blows up to $+\infty$. To this end, it is not difficult to see that it suffices to show the divergence of

$$\sum_{m \geq 1} \frac{1}{1 + \ln(i)} \ln\left(\frac{(i+1)^2}{i^2} \right) = 2 \sum_{m \geq 1} \frac{1}{1 + \ln(i)} \ln\left(\frac{i+1}{i} \right),$$

which follows from $\int_{10}^{\infty} \frac{dx}{x \ln(x) \ln(\ln(x))} = \infty$. □

9.1.3 Global solution theory for the 3D singular Navier–Stokes equations with energy critical diffusion

There are multiple novelties in the approach of [176], as well as challenging problems to extend further. First, in the deterministic case, energy estimates lead to the global regularity of the 2D Navier–Stokes equations, whereas the 2D Navier–Stokes equations forced by STWN lead to the emergence of logarithmically growing renormalization constant (recall (9.43)). Precisely for this reason, identically to Tao [314], who proved the global regularity of logarithmically supercritical Navier–Stokes equations (recall Section 5.2), Hairer and Rosati split the Fourier frequency by a time-dependent threshold (cf. (5.19) and Definition 9.1.5). Although the case of singular SPDEs has additional difficulties such as Anderson Hamiltonian, the similarity between the approaches of Tao [314] and Hairer–Rosati [176] is striking, even though they arose completely independently. Unfortunately, this observation also leads us to realizing the severe restriction for the Hairer–Rosati approach; just as the Tao's global regularity phenomena only apply in case of logarithmically supercritical diffusion (recall (5.16)–(5.17)), Hairer–Rosati's approach also seems to hinge heavily on the assumption that the renormalization constant only grows logarithmically. This rules out the possibility of directly applying [176] approach to, e. g., the 1D KPZ equation (recall Example 16).

Nevertheless, the approach of [176] offers flexibility unseen before, as already demonstrated by Theorem 9.1.2 on the 2D MHD system. To investigate further possibilities, let us return to (8.8):

$$\partial_t u + \operatorname{div}(u \otimes u) + \nabla \pi + \nu \Lambda^{m_1} u = \Lambda^{m_2} \mathcal{P}_L \mathbb{P}_{\neq 0} \xi, \quad \nabla \cdot u = 0.$$

Computing the corresponding renormalization constant by the procedure identical to the proof of Proposition 9.1.7 shows

$$r_\lambda^{i,l}(t) = \sum_{k \in \mathbb{Z}^d \setminus \{0\}} \mathfrak{l}\left(\frac{|k|}{\lambda}\right)^2 \left(\frac{1 - e^{-2\nu |k|^{m_1} t}}{2\nu |k|^{m_1}}\right)\left(\frac{\nu |k|^{m_1}}{2} + 1\right)^{-1} |k|^{2m_2} k_i k_l.$$

Then, analogously to (9.43), we can bound this by

$$\int_1^\lambda \frac{1}{\eta^{m_1}} \eta^{-m_1 + 2m_2 + 2} \eta^{d-1} d\eta \approx \ln(\lambda)$$

if and only if $m_2 = m_1 - \frac{d+2}{2}$. Keen readers realize that when $m_2 = 0$, this matches the threshold of Lions in (5.5). In other words, the strength of diffusion required for the generalized Navier–Stokes equations to become $L^2(\mathbb{T}^d)$-critical and the renormalization constant to grow logarithmically in the presence of STWN are the same.

With this observation in mind, the 1D Burgers equation forced by derivatives of STWN

$$\partial_t \theta + \frac{1}{2}\partial_x \theta^2 - \partial_x^2 \theta = \Lambda^{\frac{1}{2}} \xi, \quad x \in \mathbb{T}$$

and the 3D Navier–Stokes equations forced by STWN and $L^2(\mathbb{T}^3)$-critical diffusion

$$\partial_t u + \mathcal{P}_L \operatorname{div}(u \otimes u) + \upsilon \Lambda^{\frac{5}{2}} u = \mathcal{P}_L \mathbb{P}_{\neq 0} \xi, \quad x \in \mathbb{T}^3, \tag{9.61}$$

were studied in [366, 367], respectively. Here we focus on (9.61) for brevity. As we already described, (9.61) is $L^2(\mathbb{T}^3)$-critical (recall (5.5)), and the corresponding renormalization constant is expected to grow logarithmically. This suggests that the key step (a) in Section 9.1.1 concerning the renormalization constant growth can be achieved. This is interesting because the approach of Da Prato–Debussche [113] on (9.61) seems unfit because an analogue of (3.139) in the 3D case is false,

$$\int_{\mathbb{T}^3} (u \cdot \nabla) u \cdot \Delta u dx \neq 0; \tag{9.62}$$

yet, the approach of [176] raises hope. However, interestingly, major difficulties arise in both key steps (b) Anderson Hamiltonian and (c) Estimates in Section 9.1.1 due to the presence of the fractional Laplacian $\Lambda^{\frac{5}{2}}$ in (9.61), which we will describe next.

Heuristically, the difficulty can be understood immediately; the Leibniz rule that led to (9.57) no longer applies, so that there is no cancelation in

$$\Lambda^s(fg) - (\Lambda^s f)g - f(\Lambda^s g). \tag{9.63}$$

As a result, the corresponding commutator will be more complex than (9.56); in fact, it cannot be simplified any further than

$$\begin{aligned} \mathcal{C}^{\circledcirc_s}(w, Q^{\mathcal{H}}) = {} & (\partial_t w + \upsilon \Lambda^{\frac{5}{2}} w) \circledcirc_s Q^{\mathcal{H}} \\ & + \upsilon(\Lambda^{\frac{5}{2}}(w \circledcirc_s Q^{\mathcal{H}}) - (\Lambda^{\frac{5}{2}} w) \circledcirc_s Q^{\mathcal{H}} - w \circledcirc_s \Lambda^{\frac{5}{2}} Q^{\mathcal{H}}) \end{aligned}$$

(see [367, Equation (106a)]). As we described in detail through (9.27), (9.28), (9.29), (9.30), and (9.31), such a commutator arises due to the nature of the paracontrolled ansatz

$$w \triangleq -\mathcal{P}_L \operatorname{div}(w \circledcirc_s Q) + w^{\sharp}, \tag{9.64}$$

and such a complex structure seems inevitable in the case of fractional Laplacian. Consequently, merely relying on (8.7a) as we did in (9.58) will not be sufficient. This calls for an actual commutator estimate, and we realize that (9.63) is precisely of the form of commutator in Lemma 3.8.2. Yet, it turns out that Lemma 3.8.2 is not applicable to our case because the hypothesis of Lemma 3.8.2 requires $s \in [0, 1)$, whereas (9.61) requires $s = \frac{5}{2}$; for $s < \frac{5}{2}$, the global-in-time solution theory even in the deterministic case is unknown. As we mentioned immediately after Lemma 3.8.2, D'Ancona [112] extended to the case $s \in (0, 2)$, but even that falls far out of reach for us since $s = \frac{5}{2}$.

This issue was recently overcome in [367], where the author proved the global-in-time existence of a unique solution to (9.61) (see Theorem 9.1.15). Therein, the author faced (9.63) with $s = \frac{5}{2}$ but made the following key observation. Specifically, the author faced

$$\|\Lambda^{\frac{5}{2}}(w \circledS_s Q^{\mathcal{H}}) - (\Lambda^{\frac{5}{2}} w) \circledS_s Q^{\mathcal{H}} - w \circledS_s \Lambda^{\frac{5}{2}} Q^{\mathcal{H}}\|_{H^r} \tag{9.65}$$

for various $r \in \mathbb{R}$. As we pointed out, (9.63) with $s = \frac{5}{2}$ would not allow us to apply Lemma 3.8.2; the key is to realize that (9.65) has advantage over (9.63) due to $\circledS_s$ within. Traditionally, the purpose of (9.63) is to avoid the worst case that the highest-order derivative falls on f and g. It turns out that we can split (9.65) into two parts:

$$\|(\Lambda^{\frac{5}{2}} w) \circledS_s Q^{\mathcal{H}}\|_{H^r} + \|\Lambda^{\frac{5}{2}}(w \circledS_s Q^{\mathcal{H}}) - w \circledS_s \Lambda^{\frac{5}{2}} Q^{\mathcal{H}}\|_{H^r}.$$

Then we can apply (8.7a) to $\|(\Lambda^{\frac{5}{2}} w) \circledS_s Q^{\mathcal{H}}\|_{H^r}$ and tame the $\Lambda^{\frac{5}{2}}$ applied on w by $\|\Lambda^{\frac{5}{2}} w\|_{H^{\alpha}} \|Q^{\mathcal{H}}\|_{\mathcal{C}^{1-\alpha}}$ for $\alpha < 0$ sufficiently negative, whereas we can apply the Kato–Ponce commutator (3.95), which is valid for all $s > 0$, particularly $s = \frac{5}{2}$, to $\|\Lambda^{\frac{5}{2}}(w \circledS_s Q^{\mathcal{H}}) - w \circledS_s \Lambda^{\frac{5}{2}} Q^{\mathcal{H}}\|_{H^r}$.

Let us recap the main ideas. For singular SPDEs with fractional derivatives, the commutators of the type (9.63) would inevitably arise for which neither Lemma 3.8.2 nor [112] is applicable if $s > 2$. However, due to $\circledS_s$ within paracontrolled ansatz (9.64), a combination of Bony's estimates (8.7a) and commutator estimate (3.95) valid *for all* $s > 0$ can be applied. The commutator similar to (9.65) also appears in the key step in Section 9.1.1 (c) Anderson Hamiltonian, and we refer the interested readers to [367] for details. After [367] was completed, discussions with In-Jee Jeong led to Proposition 3.8.3, which may be used here too, but at the expense of additional terms seen in (3.98). We conclude this section with the formal statement of result on (9.61) from [367].

Proposition 9.1.14 ([367, Proposition 2.1]). *Consider* (9.61) *and*

$$\partial_t X + \upsilon \Lambda^{\frac{5}{2}} X = \mathcal{P}_L \mathbb{P}_{\neq 0} \xi, \quad X(0, \cdot) = 0,$$

so that $v \triangleq u - X$ *satisfies*

$$\partial_t v + \mathcal{P}_L \operatorname{div}(v + X)^{\otimes 2} + \upsilon \Lambda^{\frac{5}{2}} v = 0, \quad v(0, \cdot) = u^{\text{in}}(\cdot). \tag{9.66}$$

There exists a null set $\mathcal{N} \subset \Omega$ *such that for all* $\omega \in \Omega \setminus \mathcal{N}$ *and* $\kappa > 0$, *we have the following. For every* $u^{\text{in}} \in \mathcal{C}^{-\frac{5}{4}+\kappa}(\mathbb{T}^3)$ *that is divergence-free and mean-zero, there exist* $T^{\max}(\omega, u^{\text{in}}) \in (0, \infty]$ *and a unique maximal mild solution* $v(\omega)$ *to* (9.66) *on* $[0, T^{\max}(\omega, u^{\text{in}}))$ *such that* $v(\omega, 0, x) = u^{\text{in}}(x)$.

Theorem 9.1.15 ([367, Theorem 2.2]). *There exists a null set $\mathcal{N}' \subset \Omega$ such that we have the following. For any $\kappa > 0$ and every $u^{\text{in}} \in \mathcal{C}^{-1+\kappa}(\mathbb{T}^3)$ that is divergence-free and mean-zero, $T^{\max}(\omega, u^{\text{in}})$ from Proposition* 9.1.14 *satisfies $T^{\max}(\omega, u^{\text{in}}) = \infty$ for all $\omega \in \Omega \setminus \mathcal{N}'$.*

9.2 Nonuniqueness

In harmony with the topic of this chapter, we continue to discuss the global-in-time solution theory but now in the nonunique case, using convex integration technique. We briefly review in Sections 9.2.1, 9.2.2, and 9.2.3 the works [189, 190, 183] on the 3D Navier–Stokes equations forced by STWN, 2D SQG equations forced by white-in-time noise, and 2D SQG equations forced by STWN, respectively.

9.2.1 Example of the 3D Navier–Stokes equations forced by STWN

Let us first recall that Zhu and Zhu [374] provided the proofs of the local-in-time solution theory of the 3D Navier–Stokes equations forced by STWN via both theories of RS and paracontrolled distributions:

$$\partial_t u + \mathcal{P}_L \operatorname{div}(u \otimes u) = \nu \Delta u + \mathcal{P}_L \mathbb{P}_{\neq 0} \xi, \quad x \in \mathbb{T}^3. \tag{9.67}$$

Its global-in-time solution theory, unique or nonunique, was completely open; the approach of Da Prato and Debussche [113] is inapplicable due to (9.62), and the global-in-time solution theory in Theorem 9.1.15 hinges on hyperdissipation $\Lambda^{\frac{5}{2}}$.

Hofmanová, Zhu, and Zhu [189] instead constructed infinitely many global-in-time nonunique solutions to (9.67) from a prescribed initial data in $L^2 \cup \mathcal{C}^{-1+\kappa}(\mathbb{T}^3)$. We briefly sketch its main ideas here. We consider again, what is so familiar by now, the Da Prato–Debussche trick, specifically

$$z^{\mathbf{|}}, \quad u_1 \triangleq u - z^{\mathbf{|}}, \quad z^{\mathbf{Y}}$$

in (8.31), (8.32), and (8.33). Instead of u_2, for clarity, we define

$$h \triangleq u_1 - z^{\mathbf{Y}},$$

so that h solves the same equation as u_2 in (8.34):

$$\begin{aligned} &\partial_t h - \nu \Delta h \\ &= -\mathcal{P}_L \operatorname{div}\big((h + z^{\mathbf{Y}})^{\otimes 2} + (h + z^{\mathbf{Y}}) \otimes z^{\mathbf{|}} + z^{\mathbf{|}} \otimes (h + z^{\mathbf{Y}})\big). \end{aligned} \tag{9.68}$$

We know from (8.41) that h has the regularity of $\beta < \frac{1}{2}$, implying that construction on equations involving h is possible, although the nonlinear term such as $z^{\mathrm{I}} \otimes h$ is still ill-defined. The next step is to decompose

$$h = v^1 + v^2$$

so that (9.68) can also be split into

$$\partial_t v^1 - \upsilon \Delta v^1 + \operatorname{div} \underbrace{(z^{\mathrm{Y}} \otimes z^{\mathrm{Y}} + V^1)}_{\text{Ill-defined products}} + \nabla p^1 = 0,$$
$$\partial_t v^2 - \upsilon \Delta v^2 + \operatorname{div} \underbrace{((v^1 + v^2) \otimes (v^1 + v^2) + V^2)}_{\text{Well-defined products}} + \nabla p^2 = 0.$$

The next idea is analogous to Remark 7.3.1. We consider the following convex integration scheme:

$$\partial_t v_q^1 - \upsilon \Delta v_q^1 + \operatorname{div}(z^{\mathrm{Y}} \otimes z^{\mathrm{Y}} + V_q^1) + \nabla p_q^1 = 0, \tag{9.69a}$$
$$\partial_t v_q^2 - \upsilon \Delta v_q^2 + \operatorname{div}((v_q^1 + v_q^2) \otimes (v_q^1 + v_q^2) + V_q^2) + \nabla p_q^2 = \operatorname{div} \mathring{R}_q. \tag{9.69b}$$

Then we can

(a) assume the inductive hypothesis at level q,
(b) construct v_{q+1}^2 via convex integration building blocks,
(c) plug this v_{q+1}^2 into (9.69a) at level $q+1$ and derive a unique v_{q+1}^1 via the theory of paracontrolled distributions,
(d) plug this v_{q+1}^1 into (9.69b) at level $q+1$, so that we can solve for $\mathring{R}_{q+1}$ via the inverse divergence operator (cf. Lemma 6.2.2),
(e) complete the induction estimate at level $q+1$ and move onto $q+2$.

This allows us to solve for h that is a weak solution. Combined with z and z^{Y} that solves their own equations mildly, we get

$$u = z^{\mathrm{I}} + z^{\mathrm{Y}} + h.$$

Then the main idea is to fix any divergence-free initial data $u^{\text{in}} \in L^2(\mathbb{T}^3) \cup \mathcal{C}^{-1+\kappa}(\mathbb{T}^3)$ for $\kappa > 0$, obtain a local-in-time solution up to the explosion time τ (recall notations from Theorem 8.5.5), apply convex integration on h, and thus construct a nonunique solution. Extending such a convex integration solution to SPDE can be done via the stopping-time argument, which we described in Section 7.4. Finally, we mention that extension of [189] to the 2D case was achieved in [247, 248].

9.2.2 Example of the SQG equations with white-in-space noise

The convex integration technique led [189] to construct a global-in-time solution theory, although nonunique, to the 3D Navier–Stokes equations forced by STWN for the first time. We recall from our discussions in Section 8.5 that this remains in the locally subcritical case. Shortly after [189], Hofmanová, Zhu, and Zhu [190], for the first time, applied the convex integration technique to construct solutions to locally critical/supercritical singular SPDEs, namely the following singular SQG equations: for $x \in \mathbb{T}^2$, $\upsilon \geq 0$,

$$\partial_t \theta + \mathcal{R}^\perp \theta \cdot \nabla \theta + \upsilon \Lambda^\gamma \theta = \Lambda^\alpha \xi, \quad u = \mathcal{R}^\perp \theta, \alpha \in [0,1), \gamma \in \left[0, \frac{3}{2}\right), \tag{9.70}$$

where ξ is white-in-space. The main result [190, Theorems 1.2 and 3.2] is the construction of infinitely many analytically weak solutions

$$\theta \in L^2(0,\infty; \dot{H}^{-\frac{1}{2}}(\mathbb{T}^2)) \cap C([0,\infty); B_{\infty,1}^{-\frac{1}{2}-\delta}(\mathbb{T}^2)) \tag{9.71}$$

from a given initial data θ^{in}, which again leads to nonuniqueness in law, specifically

$$\begin{aligned}\langle \theta(t), \psi\rangle_{\dot{H}^{-\frac{1}{2}} \times \dot{H}^{\frac{1}{2}}} &+ \int_0^t \frac{1}{2} \langle \Lambda^{-\frac{1}{2}} \theta, \Lambda^{\frac{1}{2}} [\mathcal{R}^\perp \cdot \nabla, \nabla \psi] \theta \rangle_{\dot{H}^{-\frac{1}{2}} \times \dot{H}^{\frac{1}{2}}} ds \\ &+ \int_0^t \langle \upsilon \Lambda^\gamma \theta, \psi \rangle_{\dot{H}^{-\frac{1}{2}} \times \dot{H}^{\frac{1}{2}}} ds = \langle \theta^{\text{in}}, \psi \rangle_{\dot{H}^{-\frac{1}{2}} \times \dot{H}^{\frac{1}{2}}} + \int_0^t \langle \Lambda^\alpha \xi(s), \psi \rangle_{\dot{H}^{-\frac{1}{2}} \times \dot{H}^{\frac{1}{2}}} ds\end{aligned}$$

for all $\psi \in C^\infty(\mathbb{T}^2)$. The nonlinear term is finite due to the Calderón commutator estimate from Proposition 3.8.4 as long as $\theta \in L^2(0,\infty; \dot{H}^{-\frac{1}{2}}(\mathbb{T}^2))$, which is satisfied due to (9.71).

There are three aspects in this work, all of which are related but still interesting individually.

(i) The result covers the locally critical/supercritical cases. Indeed, because the noise is additive and white only in space, let us deduce the requirement for local subcriticality via rescaling. If $\widetilde{\xi}(x) = \lambda \xi(\lambda x)$, then the rescaled solution is $\widetilde{\theta}(t,x) = \lambda^{1+\alpha-\gamma} \theta(\lambda^\gamma t, \lambda x)$, so that they satisfy

$$\partial_t \widetilde{\theta} + \lambda^{2\gamma-2-\alpha} \mathcal{R}^\perp \widetilde{\theta} \cdot \nabla \widetilde{\theta} + \upsilon \Lambda^\gamma \widetilde{\theta} = \Lambda^\alpha \widetilde{\xi}, \tag{9.72}$$

clearly indicating that local subcriticality requires

$$2\gamma - \alpha - 2 > 0. \tag{9.73}$$

Because [190, Theorem 1.2] holds for all $\alpha \in [0,1)$ and $\gamma \in [0, \frac{3}{2})$, we observe that the threshold for local criticality can clearly be violated, implying that neither theory of RS nor the paracontrolled distributions alone seems applicable to construct even a local-in-time solution theory.

(ii) The proof consisted of no renormalization constants. Let us explain why. Hofmanová, Zhu, and Zhu worked directly on (9.70); to be precise, they worked on the equation of $f = \Lambda^{-1}\theta$ analogously to v that satisfies $-\nabla^{\perp} \cdot v = \theta$ and solved (1.6):

$$-\partial_t \mathcal{R} f + \Lambda f \nabla^{\perp} f \approx \Lambda^{\gamma-1} \nabla f - \mathcal{R}\Lambda^{-1+\alpha}\xi, \quad f(0) = \Lambda^{-1}\theta^{\text{in}}, \tag{9.74}$$

where temporarily for this section and Section 9.2.2,

$$v \approx w \quad \text{if } v, w : \mathbb{R}^2 \mapsto \mathbb{R}^2 \quad \text{and} \quad v = w + \nabla^{\perp} p. \tag{9.75}$$

The noise is white only in space and additive; thus no renormalization was needed, and only the spatial irregularity had to be dealt with. The crucial ingredient was the convex integration scheme, especially the building blocks, of [86] by Cheng, Kwon, and Li on the stationary solutions (time independent) to the 2D deterministic SQG equation. Using such building blocks, any ill-defined products could be avoided; specifically, for any product of two distributions with compact support that are sufficiently far apart, no resonant terms would arise, and hence the product would be well-defined.

(iii) As we pointed out in (9.71), the solution constructed in [190] possessed the regularity of

$$L^2(0, \infty; \dot{H}^{-\frac{1}{2}}(\mathbb{T}^2)) \cap C([0, \infty); B_{\infty,1}^{-\frac{1}{2}-\delta}(\mathbb{T}^2))$$

for all $\delta > 0$. This is another interesting aspect. Heuristic argument that is standard in singular SPDEs leads us to feel that the regularity of the solution can be at best

$$\theta \in \mathcal{C}^{\beta}(\mathbb{T}^2) \quad \text{for } \beta < -1 - \alpha + \gamma \quad \mathbb{P}\text{-a. s.}$$

Considering the particular case of $\gamma = 0$ and $\alpha = 1 - \delta$ for any $\delta > 0$, it shows that $\theta \in \mathcal{C}^{\beta}(\mathbb{T}^2)$ for $\beta < -2 + \delta$ $\mathbb{P}$-a. s. In sharp contrast, the solution constructed in [190] via convex integration lies in $B_{\infty,1}^{-\frac{1}{2}-\delta}(\mathbb{T}^2)$, which satisfies $B_{\infty,1}^{-\frac{1}{2}-\delta}(\mathbb{T}^2) \subset \mathcal{C}^{-\frac{1}{2}-\delta}(\mathbb{T}^2) \subset \mathcal{C}^{-2+\delta}(\mathbb{T}^2)$ for all $\delta \in (0, \frac{3}{4})$. The reason why this happens is simply because the authors use convex integration to construct analytically weak solution, not a mild solution constructed through the theory of RS of paracontrolled distributions. This is interesting because for many solutions to deterministic PDEs in fluid mechanics, solutions constructed via convex integration technique are considered rough and thus unphysical; in sharp contrast, when applied to singular SPDEs, the solutions constructed by convex integration can have better regularity, pointing another justification why the technique of convex integration is fit to be applied to singular SPDEs.

Finally, we go back to Remark 3.10.7. If the solution constructed by the convex integration technique satisfies the energy inequality, considering the worst case $\alpha = 1-\kappa$ for some $\kappa > 0$ and $\gamma = 0$, we may consider

$$\int_0^t \int_{\mathbb{T}^2} \Lambda^{\alpha}\xi(s)(-\Delta)^{-\frac{1}{2}}\theta dx ds \lesssim \int_0^t \|\xi(s)\|_{\dot{H}^{\frac{1}{2}-\kappa}} \|\theta\|_{\dot{H}^{-\frac{1}{2}}} ds$$

and feel that ξ must lie in $L^1(0,t;\dot{H}^{\frac{1}{2}-\kappa}(\mathbb{T}^2))$ $\mathbb{P}$-a. s. Yet, ξ has a far worse regularity, specifically $\xi \in \mathcal{C}^{\beta}(\mathbb{T}^2)$ for $\beta < -1$ $\mathbb{P}$-a. s. As already pointed out, the misconception is that the solution constructed by the convex integration technique does not necessarily satisfy the energy inequality.

9.2.3 Example of the SQG equations forced by STWN

In this section, we briefly describe the work of Hofmanová, Luo, Zhu, and Zhu [183] on (9.70), but this time with an STWN ξ, albeit under slightly more restrictive range of parameters:

$$\partial_t\theta + \mathcal{R}^{\perp}\theta \cdot \nabla\theta + \nu\Lambda^{\gamma}\theta = \Lambda^{\alpha}\xi, \quad u = \mathcal{R}^{\perp}\theta, \alpha \in \left[0, \frac{1}{4}\right), \gamma \in \left[0, \frac{3}{2}\right). \tag{9.76}$$

With the scaling of $\mathfrak{s} = (\gamma, 1, 1)$ from the equation, we know that $\Lambda^{\alpha}\xi \in \mathcal{C}_{\mathfrak{s}}^{\beta}$ for $\beta < -1 - \frac{\gamma}{2} - \alpha + \kappa$ $\mathbb{P}$-a. s. Local subcriticality would require

$$-3 + \gamma - 2\alpha + 2\kappa > -1 - \frac{\gamma}{2} - \alpha + \kappa$$

or, equivalently, $\frac{3\gamma}{2} + \kappa > \alpha + 2$, which is easily violated considering the flexibility of $\gamma \in [0, \frac{3}{2})$ and $\alpha \in [0, \frac{1}{4})$, implying that the theory of RS or paracontrolled distributions alone cannot construct even local-in-time solution theory for (9.76).

In contrast to [190], it is difficult to work directly on (9.76), because now ξ is a distribution even in time. In such a case the previous approach would be to consider a linear fractional heat equation forced by $\Lambda^{-1+\alpha}\xi$ and subtract such an equation from an analogue of (9.74) solved by $f = \Lambda^{-1}\theta$ to obtain a random PDE. However, this approach is doomed because we recall Section 8.5.2, where we described that the first Wick product already blows up in the locally critical/supercritical case. A novel idea in [183] is that they actually assign z to solve

$$\partial_t z + \nu\Lambda^{\gamma_1} z = \Lambda^{-1+\alpha}\xi,$$

where $\gamma \in (1 + 2\alpha, (2 - 2\alpha) \wedge (2\eta + 2\alpha))$ (see [183, eq. (3.5)]. Then, interestingly, $\xi \in \mathcal{C}_{\mathfrak{s}'}^{\beta}$ for $\beta < -\frac{2+\gamma_1}{2}$ $\mathbb{P}$-a. s., where $\mathfrak{s}' = (\gamma_1, 1, 1)$, and [183, Proposition 4.1] proved that $z \in$

$C([0,T]; \mathcal{C}^{\beta}(\mathbb{T}^2))$ for $\beta < \frac{\gamma_1}{2} - \alpha$, which is consistent with Schauder's estimate applied on $\Lambda^{\alpha-1}\xi \in \mathcal{C}^{\beta}_{\mathfrak{s}'}$ for $\beta < -\frac{\gamma_1}{2} - \alpha$ $\mathbb{P}$-a. s. With this in mind, in contrast to [190], the authors first define $v \triangleq \Lambda^{-1}\theta$ so that

$$-\partial_t \mathcal{R} v + \Lambda v \nabla^{\perp} v \approx \Lambda^{\gamma-1}\nabla v - \mathcal{R}\Lambda^{-1+\alpha}\xi, \quad v(0) = \Lambda^{-1}\theta^{\text{in}}, \tag{9.77}$$

which is just equation (9.74) with f replaced by v (recall (9.75)), and then define $f \triangleq v - z$ to apply convex integration technique on the equation of f (see [183, eq. (3.6)]). We refer the interested readers to [183] for further details.

9.3 Recent developments and open problems

Previous discussions in Section 9.2 strongly suggest that perhaps the convex integration technique has the possibility to be applied to other singular SPDEs such as the Φ^4 model (1.19) and stochastic Yang–Mills equation (1.18), for which solution theory in dimensions $d \geq 4$ is completely open, along with other singular SPDEs in Section 1.2. As pointed out in Remark 1.2.1, the nonlinearities in the Yang–Mills equation consist of Burgers' and cubic types. Considering Problem 5 and the difficulties of applying convex integration to higher-dimensional Burgers equation, attempting such an idea on the simpler Φ^4 model is a natural first step, which leads to a question of whether convex integration has any chance of being applied on the Φ^4 model.

Interestingly, the answer to this question is negative, as shown in [129] by Dong and the author. More specifically, unless a new type of convex integration technique that is very different from our previous discussions is introduced, convex integration technique seems not applicable to the Φ^4 model in any dimension $d \geq 2$. In fact, a much more general and stronger statement can be made, as we will describe the logic behind [129] in the following section.

9.3.1 Convex integration cannot be applied on the Φ^4 model

As we described in detail in Chapter 7, the presence of a random noise certainly does not help the convex integration scheme to prove the nonuniqueness; it creates the potential need for the Da Prato–Debussche trick, and the additional error must be ultimately estimated. Thus, if the convex integration technique is already deemed inapplicable to the following deterministic heat equation with power nonlinearity

$$\partial_t u + u^n = \Delta u, \quad u|_{t=0} = u^{\text{in}} \in L^n(\mathbb{T}^d), \tag{9.78}$$

where $n \in \mathbb{N}$ is any odd number ≥ 3, then no result in the literature at the time of writing this monograph even suggests any idea how the addition of noise can lead to success in the application of convex integration. Now the principle of convex integration technique is to consider weak solutions with minimum regularity needed for the weak formulation to hold (e. g., (3.121)). Then considering the nonlinearity u^n within (9.78), we realize that its minimum regularity needed just for its weak formulation to hold must be $u \in L^n(0,\infty;L^n(\mathbb{T}^d))$, analogously to $L^2(0,\infty;L^2(\mathbb{T}^d))$ being the minimum regularity of the solution to the Navier–Stokes equations to satisfy weak formulation, which was precisely the regularity level of convex integration solution constructed by Scheffer [296] and Shnirelman [300]; cf. (6.9) for the NLS with cubic nonlinearity. This motivates the following definition.

Definition 9.3.1. Given initial data $u^{\text{in}} \in L^n(\mathbb{T}^d)$, a scalar field $u \in L^n(0,\infty;L^n(\mathbb{T}^d))$ is a weak solution to (9.78) if

$$\int_{\mathbb{T}^d} u(t,x)\psi(t,x)dx - \int_{\mathbb{T}^d} u^{\text{in}}(x)\psi(0,x)dx - \int_0^t\int_{\mathbb{T}^d} u(s,x)\partial_s\psi(s,x)dxds$$
$$+ \int_0^t\int_{\mathbb{T}^d} u^n(s,x)\psi(s,x)dxds = \int_0^t\int_{\mathbb{T}^d} u(s,x)\Delta\psi(s,x)dxds$$

for all test functions $\psi \in C_c^\infty((-\infty,T)\times\mathbb{T}^d)$.

Because the solutions constructed by convex integration are always nonunique, if we prove the uniqueness of any weak solution defined by Definition 9.3.1, then any application of convex integration to (9.78) can be ruled out, and this is the content of the following result.

Theorem 9.3.2 ([129, Theorem 4.1]). *Let $n \in \mathbb{N}$ be any odd number ≥ 3. Suppose that $u \in L^n(0,\infty;L^n(\mathbb{T}^d))$ is a weak solution to* (9.78) *according to Definition* 9.3.1. *Then u is unique.*

We consider the same spatial mollifier described in Section 2.5 and periodization (recall Remark 2.5.2), so that $f_\epsilon(x) \triangleq (f *_x \rho_\epsilon)(x)$ identically to (2.25). We note that simply assuming the existence of two weak solutions from the same initial data and performing $L^2(\mathbb{T}^d)$-estimate would not prove Theorem 9.3.2 because upon passing to the limit as $\epsilon \searrow 0$ in the nonlinear term, the dominated convergence theorem requires a uniform bound, and the low regularity $L^n(0,\infty;L^n(\mathbb{T}^d))$ is not sufficient. Thus, instead, we perform an $L^1(\mathbb{T}^d)$-estimate with care, using the Steklov average from Section 2.6.

Proof of Theorem 9.3.2. We fix $T > 0$, assume that given $u^{\text{in}} \in L^n(\mathbb{T}^d)$, $u_j, j \in \{1,2\}$, are both weak solutions from the same initial data according to Definition 9.3.1, so that using $\psi(T) \equiv 0$,

$$- \int_{\mathbb{T}^d} u^{\text{in}}(x)\psi(0,x)dx - \int_0^T \int_{\mathbb{T}^d} u_j(s,x)\partial_s \psi(s,x)dxds$$

$$= - \int_0^T \int_{\mathbb{T}^d} u_j^n(s,x)\psi(s,x)dxds + \int_0^T \int_{\mathbb{T}^d} u_j(s,x)\Delta\psi(s,x)dxds.$$

We extend $u_j \equiv 0$ to $(-\infty, 0)$, denote $f_h(t,x) \triangleq \frac{1}{h}\int_{t-h}^t f(s,x)ds$, so that $u_{j,h}(0,x) = \frac{1}{h}\int_{-h}^0 u_j(s,x)ds = 0$, and obtain after integration by parts

$$- \int_{\mathbb{T}^d} u^{\text{in}}(x)v(0,x)dx + \int_0^t \int_{\mathbb{T}^d} \partial_s u_{j,h}(s,x)v(s,x)dxds$$

$$= - \int_0^t \int_{\mathbb{T}^d} u_{j,h}^n(s,x)v(s,x)dxds + \int_0^t \int_{\mathbb{T}^d} u_{j,h}(s,x)\Delta v(s,x)dxds$$

for all $v \in C_0^\infty(\mathbb{R} \times C^\infty(\mathbb{T}^d))$ and $t \in (0,T)$. We define $w \triangleq u_1 - u_2$ and notice that

$$[u_1^n - u_2^n]/[u_1 - u_2] = u_1^{n-1} + u_1^{n-2}u_2 + \cdots + u_2^{n-1}.$$

Because $(\frac{w_{h,\epsilon}}{\sqrt{\delta + w_{h,\epsilon}^2}})_\epsilon$ is C^∞ in space and $\partial_t w_h \in L_{T,x}^n$ for all $h > 0$ fixed by Lemma 2.6.2(a), by approximation we can choose $v = (\frac{w_{h,\epsilon}}{\sqrt{\delta + w_{h,\epsilon}^2}})_\epsilon$ and deduce

$$\int_0^t \int_{\mathbb{T}^d} \partial_s w_h(s,x)\left(\frac{w_{h,\epsilon}}{\sqrt{\delta + w_{h,\epsilon}^2}}\right)_\epsilon (s,x)dxds$$

$$= - \int_0^t \int_{\mathbb{T}^d} \left[w_h \sum_{l=0}^{n-1} u_{1,h}^{n-1-l} u_{2,h}^l \right] \left(\frac{w_{h,\epsilon}}{\sqrt{\delta + w_{h,\epsilon}^2}}\right)_\epsilon (s,x)dxds$$

$$+ \int_0^t \int_{\mathbb{T}^d} w_h(s,x)\Delta\left(\frac{w_{h,\epsilon}}{\sqrt{\delta + w_{h,\epsilon}^2}}\right)_\epsilon (s,x)dxds.$$

We then shift the mollification, integrate by parts on the time derivative, and use the fact that

$$w_h(0) = u_{1,h}(0) - u_{2,h}(0) = \int_{-h}^0 u_1(s)ds - \int_{-h}^0 u_2(s)ds = 0,$$

so that $w_{h,\epsilon}(0) = 0$, to deduce

$$\int_{\mathbb{T}^d} w_{h,\epsilon}(t,x)\Big(\frac{w_{h,\epsilon}}{\sqrt{\delta + w_{h,\epsilon}^2}}\Big)(t,x)dx - \int_0^t\int_{\mathbb{T}^d} w_{h,\epsilon}(s,x)\frac{\partial_s w_{h,\epsilon}\delta}{(\delta + w_{h,\epsilon}^2)^{\frac{3}{2}}}dxds$$
$$= -\int_0^t\int_{\mathbb{T}^d}\Big[w_h\sum_{l=0}^{n-1} u_{1,h}^{n-1-l}u_{2,h}^l\Big]_\epsilon\Big(\frac{w_{h,\epsilon}}{\sqrt{\delta + w_{h,\epsilon}^2}}\Big)(s,x)dxds$$
$$+ \int_0^t\int_{\mathbb{T}^d} w_{h,\epsilon}(s,x)\Delta\Big(\frac{w_{h,\epsilon}}{\sqrt{\delta + w_{h,\epsilon}^2}}\Big)(s,x)dxds. \tag{9.79}$$

Integration by parts shows

$$\int_0^t\int_{\mathbb{T}^d} w_{h,\epsilon}(s,x)\Delta\Big(\frac{w_{h,\epsilon}}{\sqrt{\delta + w_{h,\epsilon}^2}}\Big)(s,x)dxds \le 0. \tag{9.80}$$

For the nonlinear term, we can split it into

$$-\int_0^t\int_{\mathbb{T}^d}\Big[w_h\sum_{l=0}^{n-1} u_{1,h}^{n-1-l}u_{2,h}^l\Big]_\epsilon\Big(\frac{w_{h,\epsilon}}{\sqrt{\delta + w_\epsilon^2}}\Big)dxds = I_{1,h,\epsilon} + I_{2,h,\epsilon} + I_{3,h,\epsilon}, \tag{9.81}$$

where

$$I_{1,h,\epsilon} = -\int_0^t\int_{\mathbb{T}^d}\Big(\Big[w_h\sum_{l=0}^{n-1} u_{1,h}^{n-1-l}u_{2,h}^l\Big]_\epsilon - w_h\sum_{l=0}^{n-1} u_{1,h}^{n-1-l}u_{2,h}^l\Big)\frac{w_{h,\epsilon}}{\sqrt{\delta + w_{h,\epsilon}^2}}dxds, \tag{9.82a}$$

$$I_{2,h,\epsilon} = -\int_0^t\int_{\mathbb{T}^d}(w_h - w_{h,\epsilon})\sum_{l=0}^{n-1} u_{1,h}^{n-1-l}u_{2,h}^l\frac{w_{h,\epsilon}}{\sqrt{\delta + w_{h,\epsilon}^2}}dxds, \tag{9.82b}$$

$$I_{3,h,\epsilon} = -\int_0^t\int_{\mathbb{T}^d} w_{h,\epsilon}\sum_{l=0}^{n-1} u_{1,h}^{n-1-l}u_{2,h}^l\frac{w_{h,\epsilon}}{\sqrt{\delta + w_{h,\epsilon}^2}}dxds, \tag{9.82c}$$

and show that

$$\lim_{\epsilon\searrow 0} I_{1,h,\epsilon} \le 0, \quad \lim_{\epsilon\searrow 0} I_{2,h,\epsilon} \le 0, \quad \text{and} \quad I_{3,h,\epsilon} \le 0 \quad \forall\, \epsilon > 0$$

(see [129, eqs. (4.10), (4.12), and (4.15)]). Heuristically, this is possible only because the nonlinearities have a damping effect. This, along with (9.80) applied to (9.79), gives

$$\int_{\mathbb{T}^d} w_h(t,x)\Big(\frac{w_h}{\sqrt{\delta + w_h^2}}\Big)(t,x)dx - \int_0^t\int_{\mathbb{T}^d} w_h(s,x)\partial_s w_h \frac{\delta}{(\delta + w_h^2)^{\frac{3}{2}}}dxds \leq 0.$$

Taking the limit as $\delta \searrow 0$ gives us $\|w_h(t)\|_{L^1(\mathbb{T}^d)} = 0$, so that $\|w_h\|_{L^1((0,T)\times\mathbb{T}^d)} = 0$. Taking the limit as $h \searrow 0$ then gives us $\|w\|_{L^1((0,T)\times\mathbb{T}^d)} = 0$ according to Lemma 2.6.2(c). This implies that $\|w(t)\|_{L^1(\mathbb{T}^d)} = 0$ for almost every t, so the proof is complete. □

Although the Yang–Mills equation (1.18) consists of nonlinearity of a cubic type (besides the Burgers type), we believe that nothing definitive can be stated at the moment concerning its potential application of the convex integration technique due to its complex structure. The proof of Theorem 9.3.2 crucially depended on the sign of the nonlinearity, and the cubic terms in (1.18) are not necessarily damping.

We conclude this section with two open problems.

Problem 8. The local-in-time solution theory for the 2D stochastic Yang–Mills equation (1.18) was proven in [74]. Because it lacks explicit knowledge of invariant measure, nonlinearities are not damping in general, and no favorable transformation is known in contrast respectively to the 2D Navier–Stokes equations, Φ^4 model, and KPZ equation, a new approach seems necessary to prove its global-in-time solution theory. Considering Remark 1.2.1, the approach of [176] seems hopeful to this endeavor. However, at the time of writing this monograph, the global-in-time solution theory of the 2D stochastic Yang–Mills equations (1.18) is open.

Problem 9. As we described, Da Prato and Debussche [113] constructed global-in-time solution theory of the 2D Navier–Stokes equations forced by STWN by taking advantage of explicit knowledge of invariant measure ν starting from almost every initial data w. r. t. ν, whereas Hairer and Rosati [176] provided a new approach that relies on estimates and is independent of invariant measure. On the other hand, in the deterministic case, even the 2D Euler equations admit unique global-in-time solutions, although initial data must have bounded vorticity. Can we extend [113, 176] to the 2D Navier–Stokes equations forced by STWN with diffusion of the form Λ^m for $m < 2$ by potentially combining the advantages from both approaches of [113, 176]?

10 Conclusion with a list of problems

In this monograph, we covered a wide range of topics: preliminaries on measure theory, functional analysis, harmonic analysis, stochastic analysis, and studies on singular SPDEs, which include the revolutionary theories of RS and paracontrolled distributions in Chapters 2, 3, 4, and 8. Difficult problems that have remained open for more than a decade concerning deterministic 2D MHD system were presented in Chapter 5. The breakthrough technique of convex integration was discussed in Chapter 6, along with applications to SPDEs in Chapter 7. Recent advances in the global-in-time solution theory, unique and nonunique, were discussed in Chapter 9. This monograph is by no means intended to be comprehensive; in countless occasions, we were intentionally sketchy, only giving ideas and suggesting references.

Finally, we conclude with a list of open problems aforementioned for readers' convenience. The author acknowledges that there are many other important interesting open problems in the research area of PDEs and SPDEs in fluid mechanics and mathematical physics, and those listed below represent only a handful of most relevance to discussions in this monograph.

- **Problem 1**: Does the 2D deterministic MHD system that is inviscid but resistive with a full Laplacian admit a global-in-time unique smooth solution for all time starting from any sufficiently smooth initial data, or does there exist initial data from which its corresponding solution has a finite-time blowup? Analogous question can be asked for the 2D deterministic MHD system that is viscous with a full Laplacian but nonresistive.
- **Problem 2**: Does the 2D deterministic MHD system with viscous and magnetic diffusion respectively given by $\upsilon\Lambda^{m_1}u$ and $\eta\Lambda^{m_2}b$ with $\eta > 0$, $m_2 < 2$, and $m_1 + m_2 \in [2, 4)$ admit a global-in-time unique smooth solution for all time starting from any sufficiently smooth initial data, or does there exist initial data from which its corresponding solution has a finite-time blowup?
- **Problem 3**: In either $2\frac{1}{2}$D or 3D case, for any given arbitrary initial data in H_x^s for $s > 0$ sufficiently large, does there exist $T > 0$ such that a unique strong solution to the Hall-MHD system (1.9) with $\upsilon = \eta = 0$ or the electron MHD system (1.10) with $\eta = 0$ exists on $[0, T)$?
- **Problem 4**: With $\upsilon > 0$ and $\eta > 0$ in the $2\frac{1}{2}$D case, for any given initial data in H_x^s for $s > 0$ sufficiently large, does there exist a unique smooth solution for all time $t > 0$ to the Hall-MHD system (1.9) or the electron MHD system (1.10)?
- **Problem 5**: Can convex integration technique be systematically extended to PDEs with nonlinearity in nondivergence form in general?
- **Problem 6**: Does the 2D deterministic Kuramoto–Sivashinsky equation admit a global-in-time unique smooth solution starting from an arbitrary smooth initial data? Can we construct a global-in-time weak solution to the 2D KSE, unique or nonunique?

https://doi.org/10.1515/9783111349985-011

- **Problem 7**: Can we construct any solution theory, even local-in-time, for the Hall-MHD system forced by STWN, even in the $2\frac{1}{2}$D case? This is a special case of the greater problem of constructing any solution theory in locally critical and super-critical SPDEs.
- **Problem 8**: Can we prove the global-in-time solution theory for the 2D stochastic Yang–Mills equation (1.18)?
- **Problem 9**: Can we extend [113, 176] and prove the global-in-time solution theory for the 2D Navier–Stokes equations forced by STWN with diffusion Λ^m for $m < 2$?

A Appendix

A.1 Proof of Lemma 2.1.2

The proof is the same as usual. Let $S \triangleq \sum_{j=n}^{m} r^j, r \in \mathbb{C} \setminus \{1\}$, so that $rS = \sum_{j=n+1}^{m+1} r^j$. This implies $(r-1)S = r^{m+1} - r^n$, and therefore $S = \frac{r^{m+1}-r^n}{r-1}$.

A.2 Proof of Lemma 2.1.4

The case $p = 1$ is trivial. For $p > 1, f(t) \triangleq t^p$ is convex, so that $(\frac{a+b}{2})^p \leq \frac{1}{2}(a^p + b^p)$, which verifies the claim. For the case $p \in [0,1)$, we fix $a \geq 0$ and define

$$f(x) \triangleq a^p + x^p - (a+x)^p. \tag{A.1}$$

It follows that $f(0) = 0$ and $f'(x) = p(x^{p-1} - (a+x)^{p-1})$, which is nonnegative because $p - 1 < 0$. Consequently, $f(x) \geq 0$, which verifies the claim.

A.3 Proof of Lemma 3.4.8

Without loss of generality, we assume $s \in (\frac{d}{p}, 1 + \frac{d}{p}]$ and compute

$$\begin{aligned} \|f\|_{L^\infty} &\overset{(3.43)(3.24)}{\lesssim} \|\Delta_{-1}f\|_{L^\infty} + M\|f\|_{\dot{B}^{\frac{d}{2}}_{2,\infty}} + \sum_{j=M}^{\infty} 2^{j(\frac{d}{p}-s)+js}\|\Delta_j f\|_{L^p} \\ &\overset{(2.2)}{\lesssim} \|f\|_{L^2} + M\|f\|_{\dot{B}^{\frac{d}{2}}_{2,\infty}} + \|f\|_{B^s_{p,\infty}} 2^{M(\frac{d}{p}-s)}. \end{aligned} \tag{A.2}$$

We set $M \approx \frac{1}{s-\frac{d}{p}} \log_2(2 + \|f\|_{B^s_{p,\infty}})$, which, thanks to (3.52b), leads to the desired result as follows:

$$\|f\|_{L^\infty} \lesssim \|f\|_{L^2} + \|f\|_{\dot{B}^{\frac{d}{2}}_{2,\infty}} \log_2(2 + \|f\|_{B^s_{p,\infty}}) + 1.$$

A.4 Proof of Proposition 4.5.7

We can write

$$|v^m(t)|^2 = \sum_{j=1}^{m} (v^m(t), e_j)^2. \tag{A.3}$$

https://doi.org/10.1515/9783111349985-012

Applying Itô's formula to (4.34) and then summing over $j \in \{1, \dots, m\}$, we deduce

$$d|v^m(t)|^2 + 2\|v^m\|_V^2 dt = 2(G(v^m), v^m) d\overline{W} + \sum_{j=1}^{m} ((G(v^m), e_j) d\overline{W})^2. \tag{A.4}$$

Applying Itô's formula to (A.4), we deduce, for all $p \in [4, \infty)$,

$$\begin{aligned} &d|v^m|^p + p|v^m|^{p-2}\|v^m\|_V^2 dt \\ &= p\Big(\frac{p}{2} - 1\Big)|v^m|^{p-4}((G(v^m), v^m) d\overline{W})^2 \\ &\quad + \frac{p}{2}|v^m|^{p-2} \sum_{j=1}^{m} ((G(v^m), e_j) d\overline{W})^2 + p|v^m|^{p-2}(G(v^m), v^m) d\overline{W}. \end{aligned} \tag{A.5}$$

Using (4.33) and (A.3), we estimate

$$\left| \sum_{j=1}^{m} ((G(v^m), e_j) d\overline{W})^2 \right| \le \sum_{k=1}^{\infty} |c_k v^m(t)| dt, \tag{A.6a}$$

$$|(G(v^m), v^m) d\overline{W})^2| \le \sum_{i=1}^{\infty} |c_i v^m|^2 |v^m|^2 dt. \tag{A.6b}$$

Consequently, integrating (A.5) over $[0, \tau]$, applying (A.6), taking the supremum over $\tau \in [0, t \wedge \tau_N]$, where τ_N was defined in (4.37), and then squaring both sides of the resulting inequality and relying on Lemma 2.1.4, we arrive at (4.36). Taking the mathematical expectation w. r. t. $\overline{\mathbb{P}}$ on (4.36), we estimate

$$\overline{\mathbb{E}}\left[\left(\int_0^{t\wedge\tau_N} |v^m|^{p-2} \sum_{i=1}^{\infty} |c_i v^m|^2 d\tau\right)^2\right] \overset{(4.30)}{\lesssim} \overline{\mathbb{E}}\left[\int_0^{t\wedge\tau_N} |v^m|^{2p} d\tau\right],$$

$$\overline{\mathbb{E}}\left[\left(\sup_{\tau\in[0,t\wedge\tau_N]} \left|\int_0^{\tau} |v^m|^{p-2} G(v^m, v^m) d\overline{W}(\delta)\right|\right)^2\right]$$

$$\overset{(4.33)}{\lesssim_p} \sum_{i=1}^{\infty} \overline{\mathbb{E}}\left[\left\langle\!\!\left\langle \int_0^{t\wedge\tau_N} |v^m|^{p-2}(c_i v^m d\beta_i(\delta), v^m) \right\rangle\!\!\right\rangle\right] \lesssim \overline{\mathbb{E}}\left[\int_0^{t\wedge\tau_N} |v^m|^{2p} d\tau\right],$$

so that

$$\overline{\mathbb{E}}\Big[\sup_{\tau\in[0,t\wedge\tau_N]} |v^m(\tau)|^{2p}\Big] \lesssim |v^m(0)|^{2p} + \overline{\mathbb{E}}\left[\int_0^{t\wedge\tau_N} |v^m|^{2p} d\tau\right].$$

We can apply Gronwall's inequality (Lemma 2.1.6) and let $N \to \infty$ to obtain $v^m \in L^p(\overline{\Omega}; L^\infty(0, T; H))$, which is the first claim in (4.35).

Next, integrating (A.4) over $[0, \tau]$, we obtain, for some $c \geq 0$,

$$|v^m(t)|^2 + 2\int_0^\tau \|v^m\|_V^2 d\delta \leq |v^m(0)|^2 + 2\left|\int_0^\tau (G(v^m), v^m) d\overline{W}(\delta)\right| + c\int_0^\tau |v^m|^2 d\delta.$$

We drop the first term, take the supremum over $\tau \in [0, T]$ on the right-hand side and then on the left-hand side, raise to the pth power for $p \in [2, \infty)$, and take the mathematical expectation w. r. t. $\overline{\mathbb{P}}$ and estimate

$$\overline{\mathbb{E}}\left[\left(\sup_{\tau\in[0,T]}\left|\int_0^\tau (G(v^m), v^m) d\overline{W}(\delta)\right|\right)^p\right] \lesssim \sum_{i=1}^\infty \overline{\mathbb{E}}\left[\left(\int_0^T |c_i v^m(\delta)|^2 |v^m|^2 d\delta\right)^{\frac{p}{2}}\right]$$
$$\lesssim \overline{\mathbb{E}}\left[\sup_{\delta\in[0,T]} |v^m(\delta)|^{2p}\right] \lesssim_p 1,$$
$$\overline{\mathbb{E}}\left[\left(\int_0^T |v^m|^2 d\tau\right)^p\right] \lesssim \overline{\mathbb{E}}\left[\sup_{\tau\in[0,T]} |v^m(\tau)|^{2p}\right] \lesssim 1,$$

so that we deduce the second claim of (4.35) from

$$2^p\overline{\mathbb{E}}\left[\left(\int_0^T \|v^m\|_V^2 d\tau\right)^p\right] \lesssim_p |v^m(0)|^{2p} + 1 \lesssim_p |v^{\text{in}}|^{2p} + 1. \tag{A.7}$$

A.5 Proof of Proposition 5.1.1

Local-in-time solution theory and blowup criterion for the solution to (5.2) follow from Section 3.9, and thus we work on a priori estimates. First, we consider the difficult case $\upsilon > 0$, $m_1 = 2 + d$, $\eta = 0$, so that (5.3b) implies

$$u \in L_T^\infty L_x^2 \cap \dot{H}_x^{1+\frac{d}{2}} \quad \text{and} \quad b \in L_T^\infty L_x^2.$$

The $H^1(\mathbb{R}^d)$-estimate would not work in this case, but for clarity, we compute it, with the presence of magnetic diffusion for later use; the case $\upsilon > 0$, $m_1 = 2 + d$, and $\eta = 0$ below is the case we are currently considering. Taking the $L^2(\mathbb{R}^d)$-inner products with $(-\Delta u, -\Delta b)$ on (5.2) and integrating by parts, we get

$$\frac{1}{2}\frac{d}{dt}\|(\nabla u, \nabla b)\|_{L^2}^2 + \upsilon\|\nabla u\|_{\dot{H}^{\frac{m_1}{2}}}^2 + \eta\|\nabla b\|_{\dot{H}^{\frac{m_2}{2}}}^2 \lesssim \int_{\mathbb{R}^d} |\nabla u|(|\nabla u|^2 + |\nabla b|^2) dx, \tag{A.8}$$

where we used that

$$\sum_{i,j,k=1}^{d}\int_{\mathbb{R}^d} b_i\partial_i\partial_k b_i\partial_k u_j + b_i\partial_i\partial_k u_j\partial_k b_j dx$$
$$= \sum_{i,j,k=1}^{d}\int_{\mathbb{R}^d} b_i\partial_i(\partial_k b_i\partial_k u_j)dx = 0. \tag{A.9}$$

Due to the absence of magnetic diffusion, the only option is to bound by $\|\nabla u\|_{L^\infty}(\|\nabla u\|_{L^2}^2 + \|\nabla b\|_{L^2}^2)$ via Hölder's inequality; however, Sobolev embedding does not allow us to bound $\|\nabla u\|_{L^\infty} \lesssim \|u\|_{H^{1+\frac{d}{2}}}$, and hence we will not be able to close this $H^1(\mathbb{R}^d)$-estimate. The trick here is that we aim for a higher estimate immediately and use the Brezis–Wainger-type inequality (Lemmas 3.4.7–3.4.8). Before doing so, let us point out that the proof of Lemma 3.4.8 shows that

$$\|\nabla u\|_{L^\infty} \lesssim \|u\|_{L^2} + \|\nabla u\|_{\dot{H}^{\frac{d}{2}}} \log_2(2 + \|u\|_{H^s}) + 1 \tag{A.10}$$

for all $s > 1 + \frac{d}{2}$, because in (A.2), we could have computed $\|\Delta_{-1}\nabla f\|_{L^\infty} \lesssim \|f\|_{L^2}$ instead of $\|\Delta_{-1} f\|_{L^\infty} \lesssim \|f\|_{L^2}$. We take the $L^2(\mathbb{R}^d)$-inner products in (5.2) with $(\Lambda^{2s}u, \Lambda^{2s}b)$ for $s > 1+\frac{d}{2}$ and, similarly to (3.108), compute

$$\int_{\mathbb{R}^d} (b\cdot\nabla)\Lambda^s b\cdot\Lambda^s u + (b\cdot\nabla)\Lambda^s u\cdot\Lambda^s b dx = \int_{\mathbb{R}^d}(b\cdot\nabla)(\Lambda^s b\cdot\Lambda^s u)dx = 0 \tag{A.11}$$

to deduce by (3.95), (A.10), (3.63), and Young's inequality

$$\begin{aligned}
&\frac{1}{2}\frac{d}{dt}\|(u,b)\|_{\dot{H}^s}^2 + \nu\|\Lambda^s u\|_{\dot{H}^{1+\frac{d}{2}}}^2 \\
&= -\int_{\mathbb{R}^d} [\Lambda^s((u\cdot\nabla)u) - (u\cdot\nabla)\Lambda^s u]\cdot\Lambda^s u + [\Lambda^s((u\cdot\nabla)b) - (u\cdot\nabla)\Lambda^s b]\cdot\Lambda^s b \\
&\quad - [\Lambda^s((b\cdot\nabla)b) - (b\cdot\nabla)\Lambda^s b]\cdot\Lambda^s u - [\Lambda^s((b\cdot\nabla)u) - (b\cdot\nabla)\Lambda^s u]\cdot\Lambda^s b dx \\
&\lesssim (\|u\|_{L^2} + \|\nabla u\|_{\dot{H}^{\frac{d}{2}}}\log_2(2+\|u\|_{H^s}) + 1)\|(\Lambda^s u, \Lambda^s b)\|_{L^2}^2 \\
&\quad + \|b\|_{L^2}^{\frac{2s-2-d}{2s}}\|\Lambda^s b\|_{L^2}^{\frac{d+2}{2s}}\|\Lambda^{1+\frac{d}{2}}u\|_{L^2}^{\frac{1+\frac{d}{2}}{s}}\|\Lambda^{s+1+\frac{d}{2}}u\|_{L^2}^{\frac{s-1-\frac{d}{2}}{s}}\|\Lambda^s b\|_{L^2} \\
&\leq \frac{\nu}{2}\|\Lambda^{s+1+\frac{d}{2}}u\|_{L^2}^2 + C(1 + \|u\|_{\dot{H}^{1+\frac{d}{2}}}\log_2(2+\|u\|_{H^s}) + 1)\|(u,b)\|_{\dot{H}^s}^2 \\
&\quad + C\|\Lambda^s b\|_{L^2}^2\|\Lambda^{1+\frac{d}{2}}u\|_{L^2}^{2(\frac{1+\frac{d}{2}}{s+1+\frac{d}{2}})}.
\end{aligned} \tag{A.12}$$

Subtracting $\frac{\nu}{2}\|\Lambda^{s+1+\frac{d}{2}}u\|_{L^2}^2$ from both sides of (A.12) and applying Gronwall's inequality (Lemma 2.1.6), we obtain the desired $H^s(\mathbb{R}^d)$-bound.

The other case $\eta > 0$ is easier. If $\eta > 0$, $m_2 \in (0, 1+\frac{d}{2})$, $\nu > 0$, and $m_1 = 2+d-m_2$, then from (5.3b) we have $u \in L_T^\infty L_x^2 \cap L_T^2\dot{H}_x^{\frac{m_1}{2}}$ and $b \in L_T^\infty L_x^2 \cap L_T^2\dot{H}_x^{\frac{m_1}{2}}$, and we can continue from (A.8). The term without ∇b is easier because $m_1 \geq m_2$ by hypothesis; for the term with $|\nabla b|$, by the Sobolev embeddings $\dot{H}^{\frac{m_1}{2}-1}(\mathbb{R}^d) \hookrightarrow L^{\frac{2d}{m_2}}(\mathbb{R}^d)$ and $\dot{H}^{\frac{m_2}{2}}(\mathbb{R}^d) \hookrightarrow L^{\frac{2d}{d-m_2}}(\mathbb{R}^d)$ and Young's inequality we estimate

$$\begin{aligned}\int_{\mathbb{R}^d} |\nabla u||\nabla b|^2 dx &\leq \|\nabla u\|_{L^{\frac{2d}{m_2}}} \|\nabla b\|_{L^{\frac{2d}{d-m_2}}} \|\nabla b\|_{L^2} \lesssim \|u\|_{\dot{H}^{\frac{m_1}{2}}} \|\nabla b\|_{\dot{H}^{\frac{m_2}{2}}} \|\nabla b\|_{L^2} \\ &\leq \frac{\eta}{4}\|\nabla b\|^2_{\dot{H}^{\frac{m_2}{2}}} + C\|\nabla b\|^2_{L^2}\|u\|^2_{\dot{H}^{\frac{m_1}{2}}},\end{aligned}$$

so that we can close this estimate. We deduce higher estimates from $H^1(\mathbb{R}^d)$ similarly to (A.12).

A.6 Proof of Lemma 7.1.1

For $\zeta \in (0,2]$, $\sigma > 0$ from hypothesis, and $\alpha \in (0, \frac{3\sigma}{2\zeta})$, we define

$$Y(s) \triangleq \frac{\sin(\pi\alpha)}{\pi} \int_0^s e^{-\Lambda^\zeta (s-r)}(s-r)^{-\alpha} dW(r) \tag{A.13}$$

and using (4.71) and (7.7), compute

$$\int_0^t (t-s)^{\alpha-1} e^{-\Lambda^\zeta (t-s)} Y(s) ds = z(t). \tag{A.14}$$

Then for any $l \geq 1$, using the Gaussian hypercontractivity theorem (Lemma 4.6.4), (4.12), Lemma 2.4.11(2b), the hypothesis $\mathrm{Tr}(\Lambda^{d+2-\zeta+4\sigma}GG^*) < \infty$, and (2.23), we can estimate

$$\begin{aligned}\mathbb{E}[\|Y(s)\|^{2l}_{\dot{H}^{\frac{2+d+\sigma}{2}}}] &\overset{\text{(A.13)}}{\lesssim_l} \left(\mathbb{E}\left[\int_0^s \|\Lambda^{\zeta(\frac{2+d+\sigma}{2\zeta})} e^{-\Lambda^\zeta (s-r)}(s-r)^{-\alpha} G\|^2_{L_2(U_0,H)} dr\right]\right)^l \\ &\lesssim_l \left(\int_0^s (s-r)^{-2\alpha}(s-r)^{-2(\frac{1}{2}-\frac{3\sigma}{2\zeta})} dr\right)^l \lesssim_l 1.\end{aligned}$$

We now integrate over $[0,T]$ and deduce that for all $l \in \mathbb{N}$,

$$\mathbb{E}\left[\int_0^T \|Y(s)\|^{2l}_{\dot{H}^{\frac{2+d+\sigma}{2}}} ds\right] \lesssim_l 1. \tag{A.15}$$

This allows us to take $l > \frac{1}{2\alpha}$ and by (A.13), (A.14), and (A.15) deduce that

$$\mathbb{E}\Big[\sup_{t\in[0,T]} \|z(t)\|^{2l}_{\dot{H}^{\frac{2+d+\sigma}{2}}}\Big] \lesssim_l 1.$$

For the second claim, for $\alpha \in (0, \frac{1}{2})$, we see that $\alpha < \frac{1}{2} \leq \frac{1}{\zeta} + \frac{3\sigma}{2\zeta}$, so that the by the Gaussian hypercontractivity theorem (Lemma 4.6.4), (4.12), Lemma 2.4.11(2b), the hypothesis $\mathrm{Tr}(\Lambda^{d+2-\zeta+4\sigma}GG^*) < \infty$, and (2.23) we obtain

$$\mathbb{E}[\|Y(s)\|^{2l}_{H_x^{\frac{d+\sigma}{2}}}] \overset{(A.13)}{\lesssim} \left(\mathbb{E}\left[\left\|\Lambda^{2(\frac{d+\sigma}{4})}\int_0^s e^{-\Lambda^{\zeta}(s-r)}(s-r)^{-\alpha}dW(r)\right\|^2_{L^2_x}\right]\right)^l$$
$$\lesssim \left(\int_0^s (s-r)^{-2\alpha}(s-r)^{-2(\frac{1}{2}-\frac{1}{\zeta}-\frac{3\sigma}{2\zeta})}dr\right)^l \lesssim 1.$$

Integrating over $[0,T]$ and applying Fubini's theorem show that

$$\mathbb{E}\left[\int_0^T \|Y(s)\|^{2l}_{\dot{H}^{\frac{d+\sigma}{2}}} ds\right] < \infty. \tag{A.16}$$

Lemma 2.4.12(b) now shows that the operator $Y \mapsto z$ is bounded and linear from $L^{2l}(0,T;\dot{H}^{\frac{d+\sigma}{2}}(\mathbb{T}^d))$ to $C^{\delta}([0,T];\dot{H}^{\frac{d+\sigma}{2}}(\mathbb{T}^d))$ for any $\delta \in (0, \alpha - \frac{1}{2l})$, from which the claim follows.

A.7 Feynman diagrams and derivations of Examples 14–15

We follow explanations from [351], which relied on [201]. First, we fix a probability space $(\Omega, \mathcal{F}, \mathbb{P})$ and recall the following formula for mixed moments of normal variables, which is called the Wick theorem.

Theorem A.7.1 (Wick's theorem, [201, Theorem 1.28], [301, Proposition 1.2], [282, Theorem 2.8]). *Let $\xi_1, \ldots, \xi_n$ be centered jointly normal variables. Then*

$$\mathbb{E}[\xi_1 \ldots \xi_n] = \sum \prod_k \mathbb{E}[\xi_{i_k}\xi_{j_k}],$$

where the summation is over all the partitions of $\{1, \ldots, n\}$ into disjoint pairs $\{i_k, j_k\}$.

Let us now introduce the definition of a Feynman diagram.

Definition A.7.2 ([201, Definition 1.35]; also [282] and [152, Chapter 8]). A Feynman diagram of order $n \geq 0$ and rank $r \geq 0$ is a graph consisting of a set of n vertices and a set of r edges without common endpoints. Thus there are r disjoint pairs of vertices, each joined by an edge, and $n-2r$ unpaired vertices. The Feynman diagram is complete if $r=\frac{n}{2}$ so that all vertices are paired off and incomplete if $r<\frac{n}{2}$ so that some vertices are unpaired. A Feynman diagram labeled by n random variables $\xi_1, \dots, \xi_n$ is a Feynman diagram of order n with vertices $1, \dots, n$, where we think of ξ_i as attached to vertex i. The value of such a labeled Feynman diagram γ with edges (i_k, j_k), $k=1,\dots,r$, and unpaired vertices $\{i : i \in A\}$ is

$$\nu(\gamma) \triangleq \prod_{k=1}^{r} \mathbb{E}[\xi_{i_k}\xi_{j_k}] \prod_{i\in A} \xi_i.$$

Similarly, we denote the order and rank of a Feynman diagram γ by $n(\gamma)$ and $r(\gamma)$, respectively.

Theorem A.7.1 can be rewritten in terms of Feynman diagrams as follows.

Theorem A.7.3 ([201, Theorem 1.36] and [282, eq. (13)]). *Let $\{\xi_i\}_{i=1}^n$ be centered jointly normal random variables. Then*

$$\mathbb{E}[\xi_1 \dots \xi_n] = \sum_{\gamma} \nu(\gamma),$$

where the summation is over all complete Feynman diagrams γ labeled by $\{\xi_i\}_{i=1}^n$.

Now we give a formal definition of a Wick product.

Definition A.7.4 ([201, Definitions 2.1 and 3.1]). Let H be a Gaussian Hilbert space defined on a probability space $(\Omega, \mathcal{F}, \mathbb{P})$. For $n \geq 0$, we denote by $\overline{P}_n(H)$ the closure in $L^2(\Omega, \mathcal{F}, \mathbb{P})$ of the linear space

$$P_n(H) = \{p(\xi_1, \dots, \xi_m) : p \text{ is a polynomial of degree } \leq n,\ \xi_1, \dots, \xi_m \in H, m \in \mathbb{N}\}$$

and define

$$H^{:n:} \triangleq \overline{P}_n(H) \cap \overline{P}_{n-1}(H)^{\perp}.$$

By convention, $H^{:0:} = \overline{P}_0(H)$, the space of constants; moreover, $H^{:-1:} = \overline{P}_{-1}(H) = \{0\}$. We also denote by π_n the orthogonal projection of $L^2(\Omega, \mathcal{F}, \mathbb{P})$ onto $H^{:n:}$.

Additionally if $\xi_1, \dots, \xi_n$ is a finite sequence of elements of H, then we define their Wick product as

$$:\xi_1 \dots \xi_n: \triangleq \pi_n(\xi_1, \dots, \xi_n),$$

and we extend it to the case $n=0$ via convention of $:: \triangleq 1 \in H^{:0:}$.

The following useful formula can be used to derive Example 14.

Lemma A.7.5 ([201, Theorem 3.4]). *The Wick product is given by*

$$:\xi_1 \dots \xi_n := \sum_{\gamma} (-1)^{r(\gamma)} v(\gamma),$$

where the summation is over all Feynman diagrams γ labeled by $\{\xi_i\}_{i=1}^n$.

Finally, the following useful formula allows us to derive Example 15.

Lemma A.7.6 ([201, Theorem 3.12]). *Let $Y_i =: \xi_{i1} \dots \xi_{il_i}:$, where $\{\xi_{ij}\}_{1\le i\le k, 1\le j\le l_i}$ are centered jointly normal variables with $k \ge 0$ and $l_1, \dots, l_k \ge 0$. Then*

$$\mathbb{E}[Y_1 \dots Y_k] = \sum_{\gamma} v(\gamma),$$

where the summation is over all complete Feynman diagrams γ labeled by $\{\xi_{ij}\}_{i,j}$ such that no edge joins two variables with $\xi_{i_1 j_1}$ and $\xi_{i_2 j_2}$ where $i_1 = i_2$.

A.8 Derivation of $\mathcal{F}_0$ in (8.79)

We complete details of the elements of $\mathcal{F}_0$ here for convenience.

$$\mathcal{F}_0 = \{G_j\}_{j=1}^{12},$$

where

$$
\begin{aligned}
G_1 &\triangleq \{1\}, && \text{(A.17a)}\\
G_2 &\triangleq \{\mathcal{I}^{ii_1}(\Theta_{u,i_1}), \mathcal{I}^{ii_1}(\Theta_{b,i_1})\}, && \text{(A.17b)}\\
G_3 &\triangleq \{\mathcal{I}^{ii_1}(\Theta_{u,i_1})\mathcal{I}^{jj_1}(\Theta_{u,j_1}), \mathcal{I}^{ii_1}(\Theta_{b,i_1})\mathcal{I}^{jj_1}(\Theta_{b,j_1}),\\
&\qquad \mathcal{I}^{ii_1}(\Theta_{u,i_1})\mathcal{I}^{jj_1}(\Theta_{b,j_1})\}, && \text{(A.17c)}\\
G_4 &\triangleq \{\mathcal{I}_j^{ii_1}(\mathcal{I}^{i_1i_2}(\Theta_{u,i_2})), \mathcal{I}_j^{ii_1}(\mathcal{I}^{i_1i_2}(\Theta_{b,i_2}))\}, && \text{(A.17d)}\\
G_5 &\triangleq \{\mathcal{I}_j^{ii_1}(\mathcal{I}^{i_1i_2}(\Theta_{u,i_2})\mathcal{I}^{jj_1}(\Theta_{u,j_1})), \mathcal{I}_j^{ii_1}(\mathcal{I}^{i_1i_2}(\Theta_{b,i_2})\mathcal{I}^{jj_1}(\Theta_{b,j_1})),\\
&\qquad \mathcal{I}_j^{ii_1}(\mathcal{I}^{i_1i_2}(\Theta_{u,i_2})\mathcal{I}^{jj_1}(\Theta_{b,j_1})), \mathcal{I}_j^{ii_1}(\mathcal{I}^{i_1i_2}(\Theta_{b,i_2})\mathcal{I}^{jj_1}(\Theta_{u,j_1}))\}, && \text{(A.17e)}\\
G_6 &\triangleq \{\mathcal{I}_j^{ii_1}(\mathcal{I}^{jj_1}(\Theta_{u,j_1})), \mathcal{I}_j^{ii_1}(\mathcal{I}^{jj_1}(\Theta_{b,j_1})), \}, && \text{(A.17f)}\\
G_7 &\triangleq \{\mathcal{I}_k^{ii_1}(\mathcal{I}^{i_1i_2}(\Theta_{u,i_2}))\mathcal{I}^{jj_1}(\Theta_{u,j_1}), \mathcal{I}_k^{ii_1}(\mathcal{I}^{i_1i_2}(\Theta_{b,i_2}))\mathcal{I}^{jj_1}(\Theta_{u,j_1}),\\
&\qquad \mathcal{I}_k^{ii_1}(\mathcal{I}^{i_1i_2}(\Theta_{u,i_2}))\mathcal{I}^{jj_1}(\Theta_{b,j_1}), \mathcal{I}_k^{ii_1}(\mathcal{I}^{i_1i_2}(\Theta_{b,i_2}))\mathcal{I}^{jj_1}(\Theta_{b,j_1})\}, && \text{(A.17g)}\\
G_8 &\triangleq \{\mathcal{I}_k^{ii_1}(\mathcal{I}^{kk_1}(\Theta_{u,k_1}))\mathcal{I}^{jj_1}(\Theta_{u,j_1}), \mathcal{I}_k^{ii_1}(\mathcal{I}^{kk_1}(\Theta_{b,k_1}))\mathcal{I}^{jj_1}(\Theta_{u,j_1}),\\
&\qquad \mathcal{I}_k^{ii_1}(\mathcal{I}^{kk_1}(\Theta_{u,k_1}))\mathcal{I}^{jj_1}(\Theta_{b,j_1}), \mathcal{I}_k^{ii_1}(\mathcal{I}^{kk_1}(\Theta_{b,k_1}))\mathcal{I}^{jj_1}(\Theta_{b,j_1})\}, && \text{(A.17h)}\\
G_9 &\triangleq \{\mathcal{I}_k^{ii_1}(\mathcal{I}^{i_1i_2}(\Theta_{u,i_2})\mathcal{I}^{kk_1}(\Theta_{u,k_1}))\mathcal{I}^{jj_1}(\Theta_{u,j_1}),
\end{aligned}
$$

$$\begin{aligned}
&\mathcal{I}_k^{ii_1}(\mathcal{I}^{i_1i_2}(\Theta_{b,i_2})\mathcal{I}^{kk_1}(\Theta_{b,k_1}))\mathcal{I}^{jj_1}(\Theta_{u,j_1}),\\
&\mathcal{I}_k^{ii_1}(\mathcal{I}^{i_1i_2}(\Theta_{b,i_2})\mathcal{I}^{kk_1}(\Theta_{u,k_1}))\mathcal{I}^{jj_1}(\Theta_{b,j_1}),\\
&\mathcal{I}_k^{ii_1}(\mathcal{I}^{i_1i_2}(\Theta_{u,i_2})\mathcal{I}^{kk_1}(\Theta_{b,k_1}))\mathcal{I}^{jj_1}(\Theta_{b,j_1}),\\
&\mathcal{I}_k^{ii_1}(\mathcal{I}^{i_1i_2}(\Theta_{u,i_2})\mathcal{I}^{kk_1}(\Theta_{u,k_1}))\mathcal{I}^{jj_1}(\Theta_{b,j_1}),\\
&\mathcal{I}_k^{ii_1}(\mathcal{I}^{i_1i_2}(\Theta_{b,i_2})\mathcal{I}^{kk_1}(\Theta_{b,k_1}))\mathcal{I}^{jj_1}(\Theta_{b,j_1}),\\
&\mathcal{I}_k^{ii_1}(\mathcal{I}^{i_1i_2}(\Theta_{b,i_2})\mathcal{I}^{kk_1}(\Theta_{u,k_1}))\mathcal{I}^{jj_1}(\Theta_{u,j_1}),\\
&\mathcal{I}_k^{ii_1}(\mathcal{I}^{i_1i_2}(\Theta_{u,i_2})\mathcal{I}^{kk_1}(\Theta_{b,k_1}))\mathcal{I}^{jj_1}(\Theta_{u,j_1})\},
\end{aligned} \tag{A.17i}$$

$$\begin{aligned}
G_{10} \triangleq \{&\mathcal{I}_k^{ii_1}(\mathcal{I}^{i_1i_2}(\Theta_{u,i_2})\mathcal{I}^{kk_1}(\Theta_{u,k_1}))\mathcal{I}_l^{jj_1}(\mathcal{I}^{j_1j_2}(\Theta_{u,j_2})\mathcal{I}^{ll_1}(\Theta_{u,l_1})),\\
&\mathcal{I}_k^{ii_1}(\mathcal{I}^{i_1i_2}(\Theta_{u,i_2})\mathcal{I}^{kk_1}(\Theta_{u,k_1}))\mathcal{I}_l^{jj_1}(\mathcal{I}^{j_1j_2}(\Theta_{b,j_2})\mathcal{I}^{ll_1}(\Theta_{b,l_1})),\\
&\mathcal{I}_k^{ii_1}(\mathcal{I}^{i_1i_2}(\Theta_{b,i_2})\mathcal{I}^{kk_1}(\Theta_{b,k_1}))\mathcal{I}_l^{jj_1}(\mathcal{I}^{j_1j_2}(\Theta_{b,j_2})\mathcal{I}^{ll_1}(\Theta_{b,l_1})),\\
&\mathcal{I}_k^{ii_1}(\mathcal{I}^{i_1i_2}(\Theta_{u,i_2})\mathcal{I}^{kk_1}(\Theta_{b,k_1}))\mathcal{I}_l^{jj_1}(\mathcal{I}^{j_1j_2}(\Theta_{u,j_2})\mathcal{I}^{ll_1}(\Theta_{b,l_1})),\\
&\mathcal{I}_k^{ii_1}(\mathcal{I}^{i_1i_2}(\Theta_{u,i_2})\mathcal{I}^{kk_1}(\Theta_{b,k_1}))\mathcal{I}_l^{jj_1}(\mathcal{I}^{j_1j_2}(\Theta_{b,j_2})\mathcal{I}^{ll_1}(\Theta_{u,l_1})),\\
&\mathcal{I}_k^{ii_1}(\mathcal{I}^{i_1i_2}(\Theta_{b,i_2})\mathcal{I}^{kk_1}(\Theta_{u,k_1}))\mathcal{I}_l^{jj_1}(\mathcal{I}^{j_1j_2}(\Theta_{b,j_2})\mathcal{I}^{ll_1}(\Theta_{u,l_1})),\\
&\mathcal{I}_k^{ii_1}(\mathcal{I}^{i_1i_2}(\Theta_{u,i_2})\mathcal{I}^{kk_1}(\Theta_{u,k_1}))\mathcal{I}_l^{jj_1}(\mathcal{I}^{j_1j_2}(\Theta_{u,j_2})\mathcal{I}^{ll_1}(\Theta_{b,l_1})),\\
&\mathcal{I}_k^{ii_1}(\mathcal{I}^{i_1i_2}(\Theta_{u,i_2})\mathcal{I}^{kk_1}(\Theta_{u,k_1}))\mathcal{I}_l^{jj_1}(\mathcal{I}^{j_1j_2}(\Theta_{b,j_2})\mathcal{I}^{ll_1}(\Theta_{u,l_1})),\\
&\mathcal{I}_k^{ii_1}(\mathcal{I}^{i_1i_2}(\Theta_{b,i_2})\mathcal{I}^{kk_1}(\Theta_{b,k_1}))\mathcal{I}_l^{jj_1}(\mathcal{I}^{j_1j_2}(\Theta_{u,j_2})\mathcal{I}^{ll_1}(\Theta_{b,l_1})),\\
&\mathcal{I}_k^{ii_1}(\mathcal{I}^{i_1i_2}(\Theta_{b,i_2})\mathcal{I}^{kk_1}(\Theta_{b,k_1}))\mathcal{I}_l^{jj_1}(\mathcal{I}^{j_1j_2}(\Theta_{b,j_2})\mathcal{I}^{ll_1}(\Theta_{u,l_1}))\},
\end{aligned} \tag{A.17j}$$

$$G_{11} \triangleq \{\mathcal{I}_l^{ii_1}(\mathcal{I}_k^{i_1i_2}(\mathcal{I}^{i_2i_3}(\Theta_{u,i_3})\mathcal{I}^{kk_1}(\Theta_{u,k_1}))\mathcal{I}^{ll_1}(\Theta_{u,l_1}))\mathcal{I}^{jj_1}(\Theta_{u,j_1}), \tag{A.17k}$$

$$\begin{aligned}
&\mathcal{I}_l^{ii_1}(\mathcal{I}_k^{i_1i_2}(\mathcal{I}^{i_2i_3}(\Theta_{b,i_3})\mathcal{I}^{kk_1}(\Theta_{b,k_1}))\mathcal{I}^{ll_1}(\Theta_{u,l_1}))\mathcal{I}^{jj_1}(\Theta_{u,j_1}),\\
&\mathcal{I}_l^{ii_1}(\mathcal{I}_k^{i_1i_2}(\mathcal{I}^{i_2i_3}(\Theta_{u,i_3})\mathcal{I}^{kk_1}(\Theta_{b,k_1}))\mathcal{I}^{ll_1}(\Theta_{b,l_1}))\mathcal{I}^{jj_1}(\Theta_{u,j_1}),\\
&\mathcal{I}_l^{ii_1}(\mathcal{I}_k^{i_1i_2}(\mathcal{I}^{i_2i_3}(\Theta_{b,i_3})\mathcal{I}^{kk_1}(\Theta_{u,k_1}))\mathcal{I}^{ll_1}(\Theta_{b,l_1}))\mathcal{I}^{jj_1}(\Theta_{u,j_1}),\\
&\mathcal{I}_l^{ii_1}(\mathcal{I}_k^{i_1i_2}(\mathcal{I}^{i_2i_3}(\Theta_{u,i_3})\mathcal{I}^{kk_1}(\Theta_{u,k_1}))\mathcal{I}^{ll_1}(\Theta_{b,l_1}))\mathcal{I}^{jj_1}(\Theta_{b,j_1}),\\
&\mathcal{I}_l^{ii_1}(\mathcal{I}_k^{i_1i_2}(\mathcal{I}^{i_2i_3}(\Theta_{b,i_3})\mathcal{I}^{kk_1}(\Theta_{b,k_1}))\mathcal{I}^{ll_1}(\Theta_{b,l_1}))\mathcal{I}^{jj_1}(\Theta_{b,j_1}),\\
&\mathcal{I}_l^{ii_1}(\mathcal{I}_k^{i_1i_2}(\mathcal{I}^{i_2i_3}(\Theta_{u,i_3})\mathcal{I}^{kk_1}(\Theta_{b,k_1}))\mathcal{I}^{ll_1}(\Theta_{u,l_1}))\mathcal{I}^{jj_1}(\Theta_{b,j_1}),\\
&\mathcal{I}_l^{ii_1}(\mathcal{I}_k^{i_1i_2}(\mathcal{I}^{i_2i_3}(\Theta_{b,i_3})\mathcal{I}^{kk_1}(\Theta_{u,k_1}))\mathcal{I}^{ll_1}(\Theta_{u,l_1}))\mathcal{I}^{jj_1}(\Theta_{b,j_1}),\\
&\mathcal{I}_l^{ii_1}(\mathcal{I}_k^{i_1i_2}(\mathcal{I}^{i_2i_3}(\Theta_{u,i_3})\mathcal{I}^{kk_1}(\Theta_{u,k_1}))\mathcal{I}^{ll_1}(\Theta_{u,l_1}))\mathcal{I}^{jj_1}(\Theta_{b,j_1}),\\
&\mathcal{I}_l^{ii_1}(\mathcal{I}_k^{i_1i_2}(\mathcal{I}^{i_2i_3}(\Theta_{b,i_3})\mathcal{I}^{kk_1}(\Theta_{b,k_1}))\mathcal{I}^{ll_1}(\Theta_{u,l_1}))\mathcal{I}^{jj_1}(\Theta_{b,j_1}),\\
&\mathcal{I}_l^{ii_1}(\mathcal{I}_k^{i_1i_2}(\mathcal{I}^{i_2i_3}(\Theta_{u,i_3})\mathcal{I}^{kk_1}(\Theta_{b,k_1}))\mathcal{I}^{ll_1}(\Theta_{b,l_1}))\mathcal{I}^{jj_1}(\Theta_{b,j_1}),\\
&\mathcal{I}_l^{ii_1}(\mathcal{I}_k^{i_1i_2}(\mathcal{I}^{i_2i_3}(\Theta_{b,i_3})\mathcal{I}^{kk_1}(\Theta_{u,k_1}))\mathcal{I}^{ll_1}(\Theta_{b,l_1}))\mathcal{I}^{jj_1}(\Theta_{b,j_1}),\\
&\mathcal{I}_l^{ii_1}(\mathcal{I}_k^{i_1i_2}(\mathcal{I}^{i_2i_3}(\Theta_{u,i_3})\mathcal{I}^{kk_1}(\Theta_{u,k_1}))\mathcal{I}^{ll_1}(\Theta_{b,l_1}))\mathcal{I}^{jj_1}(\Theta_{u,j_1}),
\end{aligned}$$

$$
\begin{aligned}
&\mathcal{I}_l^{ii_1}(\mathcal{I}_k^{i_1i_2}(\mathcal{I}^{i_2i_3}(\Theta_{b,i_3})\mathcal{I}^{kk_1}(\Theta_{b,k_1}))\mathcal{I}^{ll_1}(\Theta_{b,l_1}))\mathcal{I}^{jj_1}(\Theta_{u,j_1}),\\
&\mathcal{I}_l^{ii_1}(\mathcal{I}_k^{i_1i_2}(\mathcal{I}^{i_2i_3}(\Theta_{u,i_3})\mathcal{I}^{kk_1}(\Theta_{b,k_1}))\mathcal{I}^{ll_1}(\Theta_{u,l_1}))\mathcal{I}^{jj_1}(\Theta_{u,j_1}),\\
&\mathcal{I}_l^{ii_1}(\mathcal{I}_k^{i_1i_2}(\mathcal{I}^{i_2i_3}(\Theta_{b,i_3})\mathcal{I}^{kk_1}(\Theta_{u,k_1}))\mathcal{I}^{ll_1}(\Theta_{u,l_1}))\mathcal{I}^{jj_1}(\Theta_{u,j_1})\},
\end{aligned}
\tag{A.17l}
$$

$$
\begin{aligned}
G_{12} \triangleq \{&\mathcal{I}_l^{ii_1}(\mathcal{I}_k^{ll_1}(\mathcal{I}^{l_1l_2}(\Theta_{u,l_2})\mathcal{I}^{kk_1}(\Theta_{u,k_1}))\mathcal{I}^{i_1i_2}(\Theta_{u,i_2}))\mathcal{I}^{jj_1}(\Theta_{u,j_1}),\\
&\mathcal{I}_l^{ii_1}(\mathcal{I}_k^{ll_1}(\mathcal{I}^{l_1l_2}(\Theta_{b,l_2})\mathcal{I}^{kk_1}(\Theta_{b,k_1}))\mathcal{I}^{i_1i_2}(\Theta_{u,i_2}))\mathcal{I}^{jj_1}(\Theta_{u,j_1}),\\
&\mathcal{I}_l^{ii_1}(\mathcal{I}_k^{ll_1}(\mathcal{I}^{l_1l_2}(\Theta_{u,l_2})\mathcal{I}^{kk_1}(\Theta_{b,k_1}))\mathcal{I}^{i_1i_2}(\Theta_{b,i_2}))\mathcal{I}^{jj_1}(\Theta_{u,j_1}),\\
&\mathcal{I}_l^{ii_1}(\mathcal{I}_k^{ll_1}(\mathcal{I}^{l_1l_2}(\Theta_{b,l_2})\mathcal{I}^{kk_1}(\Theta_{u,k_1}))\mathcal{I}^{i_1i_2}(\Theta_{b,i_2}))\mathcal{I}^{jj_1}(\Theta_{u,j_1}),\\
&\mathcal{I}_l^{ii_1}(\mathcal{I}_k^{ll_1}(\mathcal{I}^{l_1l_2}(\Theta_{u,l_2})\mathcal{I}^{kk_1}(\Theta_{u,k_1}))\mathcal{I}^{i_1i_2}(\Theta_{b,i_2}))\mathcal{I}^{jj_1}(\Theta_{b,j_1}),\\
&\mathcal{I}_l^{ii_1}(\mathcal{I}_k^{ll_1}(\mathcal{I}^{l_1l_2}(\Theta_{b,l_2})\mathcal{I}^{kk_1}(\Theta_{b,k_1}))\mathcal{I}^{i_1i_2}(\Theta_{b,i_2}))\mathcal{I}^{jj_1}(\Theta_{b,j_1}),\\
&\mathcal{I}_l^{ii_1}(\mathcal{I}_k^{ll_1}(\mathcal{I}^{l_1l_2}(\Theta_{u,l_2})\mathcal{I}^{kk_1}(\Theta_{b,k_1}))\mathcal{I}^{i_1i_2}(\Theta_{u,i_2}))\mathcal{I}^{jj_1}(\Theta_{b,j_1}),\\
&\mathcal{I}_l^{ii_1}(\mathcal{I}_k^{ll_1}(\mathcal{I}^{l_1l_2}(\Theta_{b,l_2})\mathcal{I}^{kk_1}(\Theta_{u,k_1}))\mathcal{I}^{i_1i_2}(\Theta_{u,i_2}))\mathcal{I}^{jj_1}(\Theta_{b,j_1}),\\
&\mathcal{I}_l^{ii_1}(\mathcal{I}_k^{ll_1}(\mathcal{I}^{l_1l_2}(\Theta_{u,l_2})\mathcal{I}^{kk_1}(\Theta_{u,k_1}))\mathcal{I}^{i_1i_2}(\Theta_{u,i_2}))\mathcal{I}^{jj_1}(\Theta_{b,j_1}),\\
&\mathcal{I}_l^{ii_1}(\mathcal{I}_k^{ll_1}(\mathcal{I}^{l_1l_2}(\Theta_{b,l_2})\mathcal{I}^{kk_1}(\Theta_{b,k_1}))\mathcal{I}^{i_1i_2}(\Theta_{u,i_2}))\mathcal{I}^{jj_1}(\Theta_{b,j_1}),\\
&\mathcal{I}_l^{ii_1}(\mathcal{I}_k^{ll_1}(\mathcal{I}^{l_1l_2}(\Theta_{u,l_2})\mathcal{I}^{kk_1}(\Theta_{b,k_1}))\mathcal{I}^{i_1i_2}(\Theta_{b,i_2}))\mathcal{I}^{jj_1}(\Theta_{b,j_1}),\\
&\mathcal{I}_l^{ii_1}(\mathcal{I}_k^{ll_1}(\mathcal{I}^{l_1l_2}(\Theta_{b,l_2})\mathcal{I}^{kk_1}(\Theta_{u,k_1}))\mathcal{I}^{i_1i_2}(\Theta_{b,i_2}))\mathcal{I}^{jj_1}(\Theta_{b,j_1}),\\
&\mathcal{I}_l^{ii_1}(\mathcal{I}_k^{ll_1}(\mathcal{I}^{l_1l_2}(\Theta_{u,l_2})\mathcal{I}^{kk_1}(\Theta_{u,k_1}))\mathcal{I}^{i_1i_2}(\Theta_{b,i_2}))\mathcal{I}^{jj_1}(\Theta_{u,j_1}),\\
&\mathcal{I}_l^{ii_1}(\mathcal{I}_k^{ll_1}(\mathcal{I}^{l_1l_2}(\Theta_{b,l_2})\mathcal{I}^{kk_1}(\Theta_{b,k_1}))\mathcal{I}^{i_1i_2}(\Theta_{b,i_2}))\mathcal{I}^{jj_1}(\Theta_{u,j_1}),\\
&\mathcal{I}_l^{ii_1}(\mathcal{I}_k^{ll_1}(\mathcal{I}^{l_1l_2}(\Theta_{u,l_2})\mathcal{I}^{kk_1}(\Theta_{b,k_1}))\mathcal{I}^{i_1i_2}(\Theta_{u,i_2}))\mathcal{I}^{jj_1}(\Theta_{u,j_1}),\\
&\mathcal{I}_l^{ii_1}(\mathcal{I}_k^{ll_1}(\mathcal{I}^{l_1l_2}(\Theta_{b,l_2})\mathcal{I}^{kk_1}(\Theta_{u,k_1}))\mathcal{I}^{i_1i_2}(\Theta_{u,i_2}))\mathcal{I}^{jj_1}(\Theta_{u,j_1})\},
\end{aligned}
\tag{A.17m}
$$

$$
\begin{aligned}
G_{13} \triangleq \{&\mathcal{I}_j(\mathcal{I}^{ii_1}(\Theta_{u,i_1})\mathcal{I}^{jj_1}(\Theta_{u,j_1})), \mathcal{I}_j(\mathcal{I}^{ii_1}(\Theta_{b,i_1})\mathcal{I}^{jj_1}(\Theta_{b,j_1})),\\
&\mathcal{I}_j(\mathcal{I}^{ii_1}(\Theta_{u,i_1})\mathcal{I}^{jj_1}(\Theta_{b,j_1})), \mathcal{I}_j(\mathcal{I}^{ii_1}(\Theta_{b,i_1})\mathcal{I}^{jj_1}(\Theta_{u,j_1}))\}.
\end{aligned}
\tag{A.17n}
$$

Here, in every set, $i, j, k, l, i_1, i_2, i_3, j_1, j_2, k_1, l_1, l_2$ run through all elements in $\{1, 2, 3\}\}$, but we omitted them for simplicity.

A.9 Analysis of G_{11} in (A.17l)

Here we include details on how to handle terms in G_{11} in (A.17l), which was omitted in [352]. Without loss of generality, we work on

$$
\tau = \mathcal{I}_l^{ii_1}(\mathcal{I}_k^{i_1i_2}(\mathcal{I}^{i_2i_3}(\Theta_{u,i_3})\mathcal{I}^{kk_1}(\Theta_{u,k_1}))\mathcal{I}^{ll_1}(\Theta_{b,l_1}))\mathcal{I}^{jj_1}(\Theta_{b,j_1})
\tag{A.18}
$$

from (A.17l). Similarly to (8.92), using (8.91) with (8.80d) instead of (8.80f), we deduce

$$\langle \widehat{\Pi}_0^{(\epsilon,t)} \tau, \phi \rangle = \int \Pi_0^{(\epsilon,t)} [\mathcal{I}_l^{ii_1} \left(\mathcal{I}_k^{i_1 i_2} (\mathcal{I}^{i_2 i_3} (\Theta_{u,i_3}) \mathcal{I}^{kk_1} (\Theta_{u,k_1})) \mathcal{I}^{ll_1} (\Theta_{b,l_1}) \right) \times \mathcal{I}^{jj_1} (\Theta_{b,j_1})] \phi(y) dy - \int C^3_{ii_1 i_2 i_3 ll_1 kk_1 jj_1} \phi(y) dy. \tag{A.19}$$

Similarly to (8.93), because

$$|\mathcal{I}_k^{i_1 i_2} (\mathcal{I}^{i_2 i_3} (\Theta_{u,i_3}) \mathcal{I}^{kk_1} (\Theta_{u,k_1})) \mathcal{I}^{ll_1} (\Theta_{b,l_1})|_{\mathfrak{s}} = 7 + 3\alpha,$$

(8.73)–(8.74) and (8.77)–(8.78) allow us to continue from (A.19) by

$$\begin{aligned} &\langle \widehat{\Pi}_0^{(\epsilon,t)} \tau, \phi \rangle \\ &= \int \Big[\int D_l K^{ii_1} (t - s_1, y - y_1) \\ &\times \Pi_0^{(\epsilon,s_1)} \Sigma_0^{s_1 t} (\mathcal{I}_k^{i_1 i_2} (\mathcal{I}^{i_2 i_3} (\Theta_{u,i_3}) \mathcal{I}^{kk_1} (\Theta_{u,k_1})) \mathcal{I}^{ll_1} (\Theta_{b,l_1}))(s_1, y_1) dy_1 ds_1 \\ &- \int D_l K^{ii_1} (t - s_1, -y_1) \\ &\times \Pi_0^{(\epsilon,s_1)} \Sigma_0^{s_1 t} \left(\mathcal{I}_k^{i_1 i_2} (\mathcal{I}^{i_2 i_3} (\Theta_{u,i_3}) \mathcal{I}^{kk_1} (\Theta_{u,k_1})) \mathcal{I}^{ll_1} (\Theta_{b,l_1}) \right) (s_1, y_1) dy_1 ds_1 \Big] \\ &\times [\int K^{jj_1} (t - s_3, y - y_3) \xi^{j_1}_{b,\epsilon} (s_3, y_3) dy_3 ds_3] \phi(y) dy \\ &- \int C^3_{ii_1 i_2 i_3 ll_1 kk_1 jj_1} \phi(y) dy. \end{aligned} \tag{A.20}$$

We continue to write using $z = (t, y)$, $z_1 = (s_1, y_1)$, and $z_3 = (s_3, y_3)$:

$$\begin{aligned} &\langle \widehat{\Pi}_0^{(\epsilon,t)} \tau, \phi \rangle \\ &= \int\!\!\int\!\!\int\!\!\int\!\!\int\!\!\int\!\!\int I_1 (S_{z_2})^{l_1+3} I_1 (S_{z_3})^{j_1+3} I_1^{i_3} (S_{z_4}) I_1 (S_{z_5})^{k_1} \phi(y) \\ &\times \Big[D_l K^{ii_1} (z - z_1) D_k K^{i_1 i_2} (z_1 - z_6) K_\epsilon^{i_2 i_3} (z_6 - z_4) \\ &\times K_\epsilon^{kk_1} (z_6 - z_5) K_\epsilon^{ll_1} (z_1 - z_2) K_\epsilon^{jj_1} (z - z_3) \\ &- D_l K^{ii_1} ((t, 0) - z_1) D_k K^{i_1 i_2} (z_1 - z_6) K_\epsilon^{i_2 i_3} (z_6 - z_4) \\ &\times K_\epsilon^{kk_1} (z_6 - z_5) K_\epsilon^{ll_1} (z_1 - z_2) K_\epsilon^{jj_1} (z - z_3) \Big] \prod_{j=1}^{6} dz_j dy \\ &- \int C^3_{ii_1 i_2 i_3 ll_1 kk_1 jj_1} \phi(y) dy. \end{aligned} \tag{A.21}$$

We can apply (8.99) to deduce the fourth, second, and zeroth Wiener chaos as follows. The fourth Wiener chaos takes the form

$(\widehat{\mathcal{W}}^{(\epsilon;4)}\tau)(z) =$ 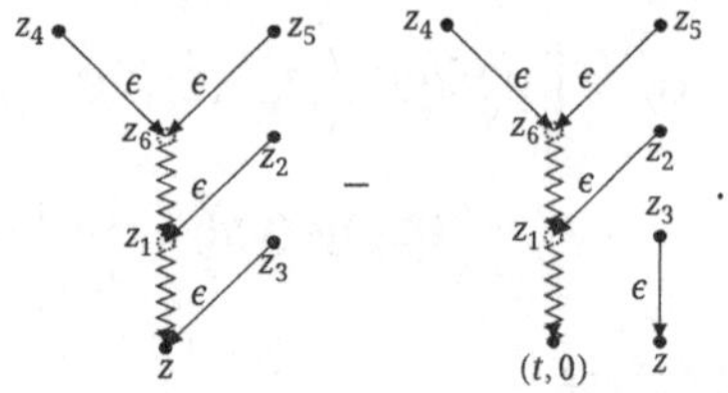

Corresponding to I_2, we have

$$(\widehat{\mathcal{W}}^{(\epsilon;2)}\tau)(z) = \sum_{j=1}^{5}(\widehat{\mathcal{W}}_j^{(\epsilon;2)}\tau)(z),$$

where

$(\widehat{\mathcal{W}}_1^{(\epsilon;2)}\tau)(z) =$ 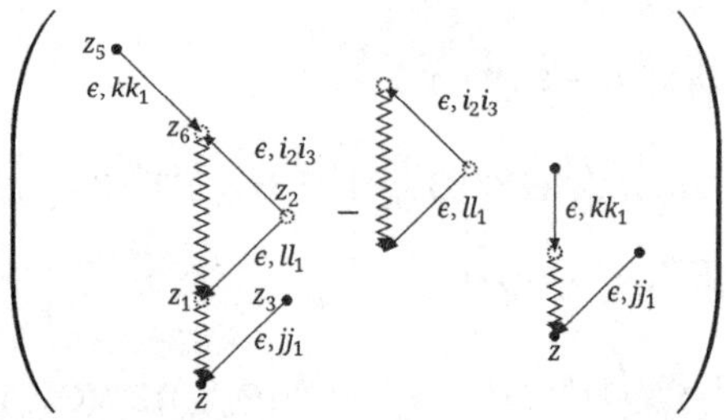

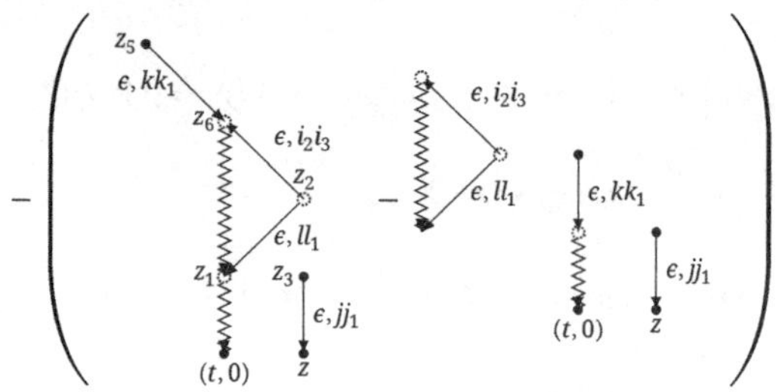

$(\widehat{\mathcal{W}}_2^{(\epsilon;2)}\tau)(z) =$ 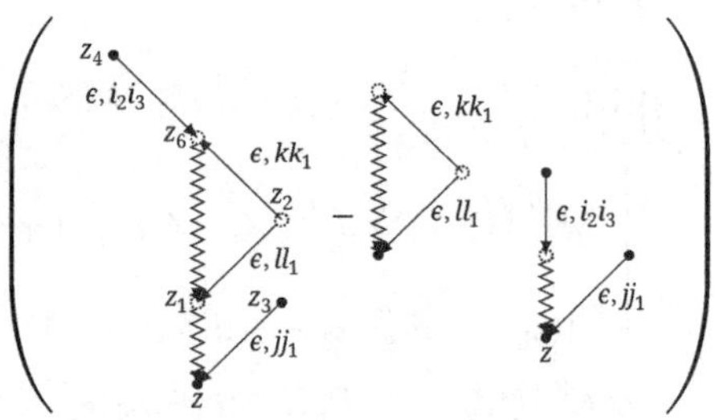

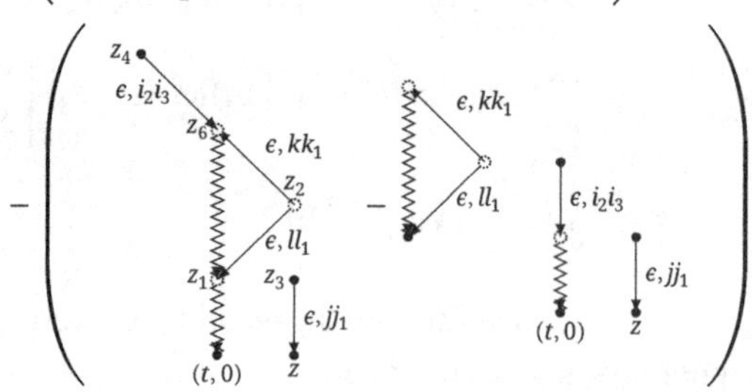

$(\widehat{\mathcal{W}}_3^{(\epsilon;2)}\tau)(z) =$

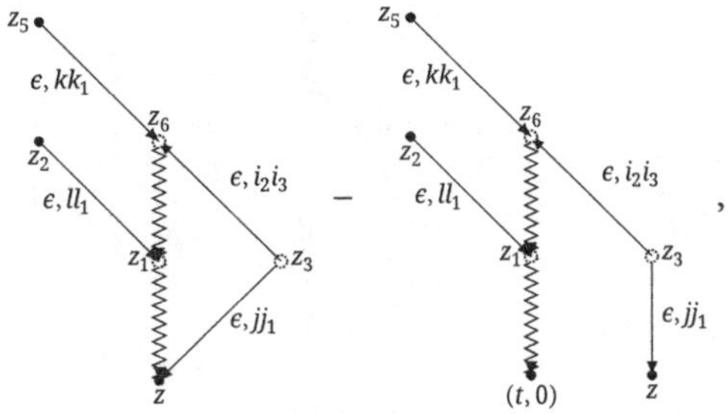

$(\widehat{\mathcal{W}}_4^{(\epsilon;2)}\tau)(z) =$

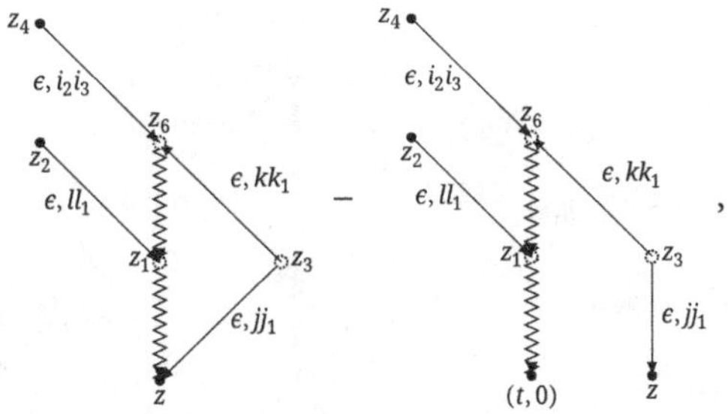

$(\widehat{\mathcal{W}}_5^{(\epsilon;2)}\tau)(z) =$

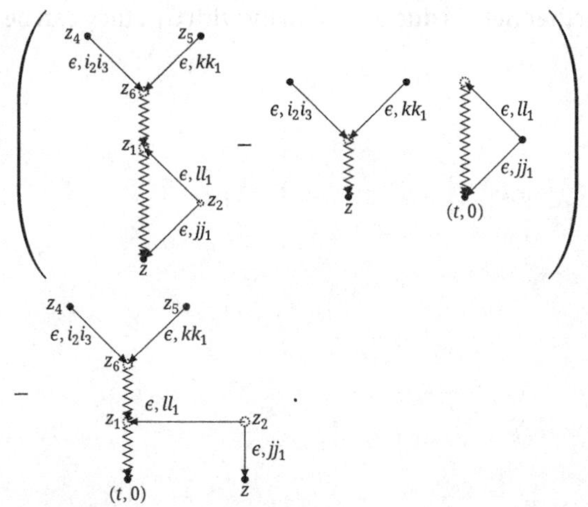

Finally, corresponding to I_0, by applying (8.99) to (A.21) we have

$(\widehat{\mathcal{W}}^{(\epsilon;0)}\tau)(z) = -$

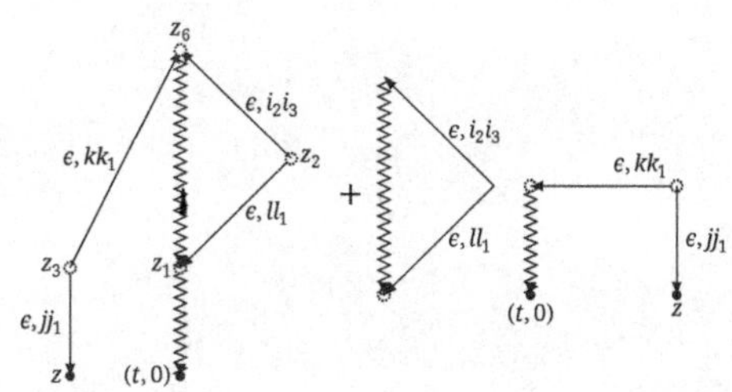

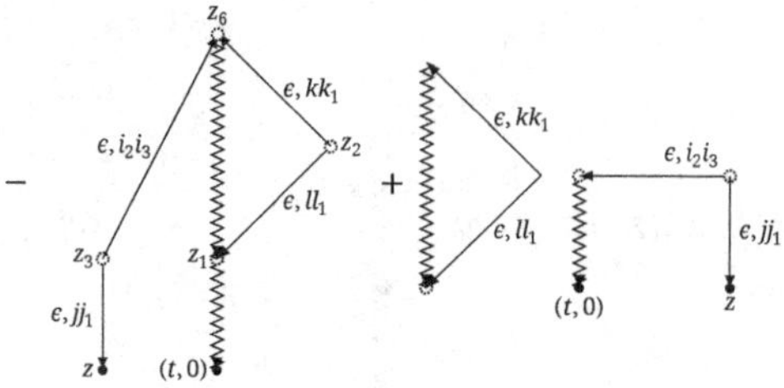

if we choose

$$C^3_{ii_1 i_2 i_3 ll_1 kk_1 jj_1} \triangleq$$

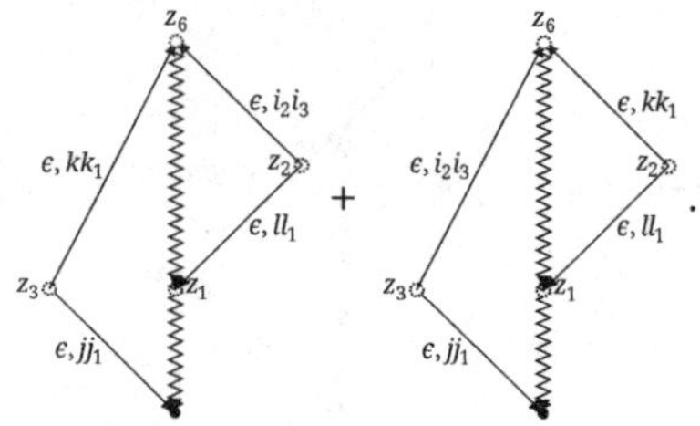

We omit further details due to similarity with G_{12}; they can be found in [352, Section B.9].

B Appendix

B.1 Preliminaries used in Example 22

Here we state the lemmas that were used in Example 22.

Lemma B.1.1 ([375, Lemma 3.2]). *Define* Δ_ϵ *by* (8.109) *and* P_t^ϵ *by* (8.112b). *Let* $f \in \mathcal{C}^\alpha(\mathbb{T}^d)$ *for* $\alpha \in \mathbb{R}$. *Then for all* $\delta \geq 0$, $\kappa > 0$, *and* $t > 0$,

$$\sup_{\epsilon\in(0,1)} \|P_t^\epsilon f\|_{\mathcal{C}^{\alpha+\delta-\kappa}} \lesssim t^{-\frac{\delta}{2}} \|f\|_{\mathcal{C}^\alpha}.$$

Lemma B.1.2 ([375, Lemma 3.3]). *Define* P_t^ϵ *by* (8.112b) *and* P_t *by* (8.9). *Let* $f \in \mathcal{C}^{\alpha+\eta}(\mathbb{T}^d)$ *for* $\alpha \in \mathbb{R}$ *and* $\eta \in (0,1)$. *Then for all* $\delta \geq 0$, $\kappa > 0$, *and* $\epsilon \in (0,1)$,

$$\|(P_t^\epsilon - P_t)f\|_{\mathcal{C}^{\alpha+\delta-\kappa}} \lesssim \epsilon^\eta t^{-\frac{\delta}{2}} \|f\|_{\mathcal{C}^{\alpha+\eta}}.$$

Lemma B.1.3 ([375, Lemma 3.7]). *Define* D^ϵ *by* (8.109). *Let* $f \in \mathcal{C}^{\alpha+1}(\mathbb{T}^d)$ *for some* $\alpha \in \mathbb{R}$. *Then for every* $\kappa > 0$,

$$\sup_{\epsilon\in(0,1)} \|D^\epsilon f\|_{\mathcal{C}^{\alpha-\kappa}} \lesssim \|f\|_{\mathcal{C}^{\alpha+1}}.$$

Lemma B.1.4 ([375, Lemma 3.8]). *Define* D^ϵ *by* (8.109). *Let* $f \in \mathcal{C}^{\alpha+1+\eta}(\mathbb{T}^d)$ *for some* $\alpha \in \mathbb{R}$ *and* $\eta > 0$. *Then for every* $\kappa > 0$,

$$\|(D^\epsilon - D)f\|_{\mathcal{C}^{\alpha-\kappa}} \lesssim \epsilon^\eta \|f\|_{\mathcal{C}^{\alpha+1+\eta}}.$$

https://doi.org/10.1515/9783111349985-013

Index

https://doi.org/10.1515/9783111349985-014

Bibliography

[1] M. Acheritogaray, P. Degond, A. Frouvelle, and J-G. Liu, *Kinetic formulation and global existence for the Hall-Magneto-hydrodynamics system*, Kinet. Relat. Models, **4** (2011), pp. 901–918.

[2] G. Ahlers, M. C. Cross, P. C. Hohenberg, and S. Safran, *The amplitude equation near the convective threshold: application to time-dependent heating experiments*, J. Fluid Mech., **110** (1981), pp. 297–334.

[3] D. Albritton, E. Brué, and M. Colombo, *Non-uniqueness of Leray solutions of the forced Navier–Stokes equations*, Ann. of Math., **196** (2022), pp. 415–455.

[4] A. Alexanderian, *On non-existence of Lebesgue-like measure in infinite dimensions*, August 2, 2025, https://aalexan3.math.ncsu.edu/articles/infdim_meas.pdf.

[5] H. Alfvén, *On the existence of electromagnetic-hydrodynamic waves*, Ark. Mat., Astron. Fys., **29 B** (1942), pp. 1–7.

[6] N. Alibaud, J. Droniou, and J. Vovelle, *Occurrence and non-appearance of shocks in fractal Burgers equations*, J. Hyperbolic Differ. Equ., **04** (2007), pp. 479–499.

[7] R. Allez and K. Chouk, *The continuous Anderson hamiltonian in dimension two*, arXiv:1511.02718 [math.PR], 2015.

[8] D. M. Ambrose and A. L. Mazzucato, *Global solutions of the two-dimensional Kuramoto-Sivashinsky equation with a linearly growing mode in each direction*, J. Nonlinear Sci., **31** (2021), https://doi.org/10.1007/s00332-021-09748-8.

[9] L. Ambrosio, *Transport equation and Cauchy problem for BV vector fields*, Invent. Math., **158** (2004), pp. 227–260.

[10] L. Ambrosio, *Well posedness of ODE's and continuity equations with nonsmooth vector fields, and applications*, Rev. Mat. Complut., **30** (2017), pp. 427–450.

[11] D. Applebaum, *Lévy Processes and Stochastic Calculus*, 2nd ed., Cambridge University Press, New York, 2009.

[12] H. Bahouri, J.-Y. Chemin, and R. Danchin, *Fourier Analysis and Nonlinear Partial Differential Equations*, Springer-Verlag, Berlin Heidelberg, 2011.

[13] V. Barbu and G. Da Prato, *Existence and ergodicity for the two-dimensional stochastic magneto-hydrodynamics equations*, Appl. Math. Optim., **56** (2007), pp. 145–168.

[14] D. Barbato, F. Morandin, and M. Romito, *Global regularity for a logarithmically supercritical hyperdissipative dyadic equation*, Dyn. Partial Differ. Equ., **11** (2014), pp. 39–52.

[15] D. Barbato, F. Morandin, and M. Romito, *Global regularity for a slightly supercritical hyperdissipative Navier–Stokes system*, Anl. PDE, **7** (2014), pp. 2009–2027.

[16] H. Bateman, *Some recent researches on the motion of fluids*, Monthly Weather Review, **43** (1915), pp. 163–170.

[17] J. T. Beale, T. Kato, and A. Majda, *Remarks on the breakdown of smooth solutions for the 3-D Euler equations*, Comm. Math. Phys., **94** (1984), pp. 61–66.

[18] F. Bechtold, T. Lange, and J. Wichmann, *On convex integration solutions to the surface quasi-geostrophic equation driven by generic additive noise*, Electron. J. Probab., **29** (2024), pp. 1–38.

[19] R. Beekie, T. Buckmaster, and V. Vicol, *Weak solutions of ideal MHD which do not conserve magnetic helicity*, Annals of PDE, **6** (2020), pp. 1–40.

[20] B. Bekmaganbetov and H. Dong, *Singular-degenerate parabolic systems with the conormal boundary condition on the upper half space*, arXiv:2509.18418 [math.AP], 2025.

[21] A. Bensoussan and R. Temam, *Equations stochastiques du type Navier–Stokes*, J. Funct. Anal., **13** (1973), pp. 195–222.

[22] Á. Bényi and T. Oh, *The Sobolev inequality on the torus revisited*, Publ. Math. Debrecen, **83** (2013), pp. 359–374.

[23] J. Bergh and J. Löfström, *Interpolation Spaces, an Introduction*, Springer-Verlag, Berlin, Heidelberg, New York, 1976.

https://doi.org/10.1515/9783111349985-015

[24] N. Berglund and C. Kuehn, *Model spaces of regularity structures for space-fractional SPDEs*, J. Stat. Phys., **168** (2017), pp. 331–368.

[25] S. E. Berkemeier, *On the 3D Navier–Stokes equations with a linear multiplicative noise and prescribed energy*, J. Evol. Equ., **23** (2023), https://doi.org/10.1007/s00028-023-00884-0.

[26] S. E. Berkemeier, *Existence and non-uniqueness of ergodic Leray-Hopf solutions to the stochastic power-law flows*, arXiv:2412.08622 [math.AP], 2024.

[27] L. Bertini, N. Cancrini, and G. Jona-Lasinio, *The stochastic Burgers equation*, Comm. Math. Phys., **165** (1994), pp. 211–232.

[28] L. Bertini and G. Giacomin, *Stochastic Burgers and KPZ equations from particle systems*, Comm. Math. Phys., **183** (1997), pp. 571–607.

[29] H. Bessaih, *Martingale solutions for stochastic Euler equations*, Stoch. Anal. Appl., **17** (1999), pp. 713–725.

[30] F. Biagini, Y. Hu, B. Øksendal, and T. Zhang, *Stochastic Calculus For Fractional Brownian Motion and Applications*, Springer-Verlag, London Limited, 2008.

[31] S. Bianchini and P. Bonicatto, *A uniqueness result for the decomposition of vector fields in* $\mathbb{R}^d$, Invent. Math., **220** (2020), pp. 255–393.

[32] P. Billingsley, *Probability and Measure*, 3rd ed., Wiley series in probability and mathematical statistics, 1995.

[33] D. Breit, E. Feireisl, and M. Hofmanová, *Stochastically Forced Compressible Fluid Flows*, De Gruyter Series in Applied and Numerical Mathematics, 2018.

[34] D. Breit, E. Feireisl, and M. Hofmanová, *On solvability and ill-posedness of the compressible Euler system subject to stochastic forces*, Anal. PDE, **13** (2020), pp. 371–402.

[35] Y. Brenier, R. Natalini, and M. Puel, *On a relaxation approximation of the incompressible Navier–Stokes equations*, Proc. Amer. Math. Soc., **132** (2003), pp. 1021–1028.

[36] H. Brezis, *Functional Analysis, Sobolev Spaces and Partial Differential Equations*, Springer Science+Business Media, LLC, 2011.

[37] H. Brezis and S. Wainger, *A note on limiting cases of Sobolev embeddings and convolution inequalities*, Comm. Partial Differential Equations, **5**, 7 (1980), pp. 773–789.

[38] A. M. Bruckner, J. B. Bruckner, and B. S. Thomson, *Real Analysis*, Prentice-Hall, Inc., 1997.

[39] E. Brué, R. Jin, Y. Li, and D. Zhang, *Non-uniqueness in law of Leray solutions to 3D forced stochastic Navier–Stokes equations*, arXiv:2309.09753v2 [math.PR], 2024.

[40] Y. Bruned, M. Hairer, and L. Zambotti, *Algebraic renormalisation of regularity structures*, Invent. Math., **215** (2019), pp. 1039–1156.

[41] Y. Bruned, A. Chandra, I. Chevyrev, and M. Hairer, *Renormalising SPDEs in regularity structures*, J. Eur. Math. Soc., **23** (2021), pp. 869–947.

[42] Z. Brzeźniak and E. Motyl, *The existence of martingale solutions to the stochastic Boussinesq equations*, Global and Stochastic Analysis, **1** (2014), pp. 175–216.

[43] T. Buckmaster, M. Colombo and V. Vicol, *Wild solutions of the Navier–Stokes equations whose singular sets in time have Hausdorff dimension strictly less than 1*, J. Eur. Math. Soc., **24** (2022), pp. 3333–3378.

[44] T. Buckmaster, C. De Lellis, P. Isett, and L. Székelyhidi Jr., *Anomalous dissipation for* 1/5*-Hölder Euler flows*, Ann. of Math., **182** (2015), pp. 127–172.

[45] T. Buckmaster, A. Nahmod, G. Staffilani, and K. Widmayer, *The surface quasi-geostrophic equation with random diffusion*, Int. Math. Res. Notices, **2020** (2020), pp. 9370–9385.

[46] T. Buckmaster, S. Shkoller, and V. Vicol, *Nonuniqueness of weak solutions to the SQG equations*, Comm. Pure Appl. Math., **LXXII** (2019), pp. 1809–1874.

[47] T. Buckmaster and V. Vicol, *Nonuniqueness of weak solutions to the Navier–Stokes equation*, Ann. of Math., **189** (2019), pp. 101–144.

[48] T. Buckmaster and V. Vicol, *Convex integration and phenomenologies in turbulence*, EMS Surveys in Mathematical Sciences, **6** (2019), pp. 173–263.

[49] T. Buckmaster and V. Vicol, *Convex integration constructions in hydrodynamics*, Bull. Amer. Math. Soc., **58** (2020), pp. 1–14.
[50] J. Burczak, S. Modena, and L. Székelyhidi Jr., *Non uniqueness of power-law flows*, Comm. Math. Phys., **388** (2021), pp. 199–243.
[51] J. M. Burgers, *A mathematical model illustrating the theory of turbulence*, Advances in Applied Mechanics, **1** (1948), pp. 171–199. https://doi.org/10.1016/S0065-2156(08)70100-5.
[52] L. Caffarelli and A. Vasseur, *Drift diffusion equations with fractional diffusion and the quasi-geostrophic equation*, Ann. of Math. **171** (2010), pp. 1903–1930.
[53] J. W. Cahn and J. E. Hilliard, *Free energy of a nonuniform system I. Interfacial free energy*, J. Chem. Phys., **28** (1958), pp. 258–267.
[54] S. J. Camargo and H. Tasso, *Renormalization group in magnetohydrodynamic turbulence*, Phys. Fluids B, **4** (1992), pp. 1199–1212.
[55] L. M. B. C. Campos, *On hydromagnetic waves in atmospheres with application to the sun*, Theoret. Comput. Fluid Dynamics, **10** (1998), pp. 37–70.
[56] W. Cao, Y. Li, and D. Zhang, *Existence and non-uniqueness of probabilistically strong solutions to 3D stochastic magnetohydrodynamic equations*, arXiv:2408.05450 [math.AP], 2024.
[57] C. Cao, J. Wu, and B. Yuan, *The 2D incompressible magnetohydrodynamics equations with only magnetic diffusion*, SIAM J. Math. Anal., **46** (2014), pp. 588–602.
[58] W. Cao, Z. Zeng, and D. Zhang, *Non-Leray-Hopf solutions to 3D stochastic hyper-viscous Navier–Stokes equations: beyond the lions exponent*, Stochastic Process. Appl., **193** (2026), 104836.
[59] J. Carlson, A. Jaffe, and A. Wiles, *The Millennium Prize Problems*, Providence, American Mathematical Society and Clay Mathematics Institute, Rhode Island, 2006.
[60] M. Carrassi and A. Morro, *A modified Navier–Stokes equation and its consequence on sound dispersion*, Il Nuovo Cimento, **9 b** (1972), pp. 321–342.
[61] R. Catellier and K. Chouk, *Paracontrolled distributions and the 3 dimensional stochastic quantization equation*, Ann. Probab., **46** (2018), pp. 2621–2679.
[62] C. Cattaneo, *Sulla conduzione del calore*, Atti Sem. Mat. Fis. Univ. Modena, **3** (1949), pp. 83–101.
[63] C. Cattaneo, *Sur une forme de l'équation de la chaleur éliminant le paradoxe d'une propagation instantanée*, C. R. Acad. Sci. Paris, **247** (1958), pp. 431–433.
[64] D. Chae, P. Constantin, D. Córdoba, F. Gancedo, and J. Wu, *Generalized surface quasi-geostrophic equations with singular velocities*, Comm. Pure Appl. Math., **LXV** (2012), pp. 1037–1066.
[65] D. Chae, P. Constantin, and J. Wu, *Inviscid models generalizing the two-dimensional Euler and the surface quasi-geostrophic equations*, Arch. Ration. Mech. Anal., **202** (2011) pp. 35–62.
[66] D. Chae, P. Degond, and J.-G. Liu, *Wel-posedness for Hall-magnetohydrodynamics*, Ann. Inst. H. Poincaré Anal. Non Linéaire, **31** (2014), pp. 555–565.
[67] D. Chae and J. Lee, *On the blow-up criterion and small data global existence for the Hall-magnetohydrodynamics*, J. Differential Equations **256** (2014), pp. 3835–3858.
[68] D. Chae and M. Schonbek, *On the temporal decay for the Hall-magnetohydrodynamic equations*, J. Differential Equations, **255** (2013), pp. 3971–3982.
[69] D. Chae, R. Wan, and J. Wu, *Local well-posedness for the Hall-MHD equations with fractional magnetic diffusion*, J. Math. Fluid Mech., **17** (2015), pp. 627–638.
[70] D. Chae and S. Weng, *Singularity formation for the incompressible Hall-MHD equations without resistivity*, Ann. Inst. H. Poincaré Anal. Non Linéaire, **33** (2016), pp. 1009–1022.
[71] D. Chae and J. Wolf, *On partial regularity for the steady Hall magnetohydrodynamics system*, Comm. Math. Phys., **339** (2015), pp. 1147–1166.
[72] D. Chae and J. Wolf, *On partial regularity for the 3D nonstationary Hall magnetohydrodynamics equations on the plane*, SIAM J. Math. Anal., **48** (2016), pp. 443–469.
[73] J. Q. Chagas, N. M. Diehl, and P. L. Guidolin, *Some properties for the Steklov averages*, arXiv:1707.06368 [math.AP], 2017.

[74] A. Chandra, I. Chevyrev, M. Hairer, and H. Shen, *Langevin dynamic for the 2D Yang-Mills measure*, Publications mathématiques de l'IHÉS **136** (2022), pp. 1–147.

[75] A. Chandra, I. Chevyrev, M. Hairer, and H. Shen, *Stochastic quantisation of Yang-Mills-Higgs in 3D*, Invent. Math., **237** (2024), pp. 541–696.

[76] A. Chandra and M. Hairer, *An analytic BPHZ theorem for regularity structures*, arXiv:1612.08138 [math.PR], 2018.

[77] J. Charney, *On the scale of atmospheric motions*, Geofys. Pub. Oslo, **17** (1948), pp. 251–265.

[78] S. Chatterjee, *Yang-Mills for probabilists*. In *Probability and Analysis in Interacting Physical Systems*, **283** Springer, Proc. Math. Stat., Springer (2019), pp. 1–16.

[79] J.-Y. Chemin, *Perfect Incompressible Fluids*, Clarendon Press, Oxford, 1998.

[80] J.-Y. Chemin and P. Zhang, *On the critical one component regularity for 3-D Navier–Stokes system*, Annales Scientifiques de L'ens, **49** (2016), pp. 131–167.

[81] W. Chen, Z. Dong, and Y. Zheng, *Non-uniqueness of Leray-Hopf solutions to forced stochastic hyperdissipative Navier–Stokes equations up to Lions index*, arXiv:2503.18041 [math.PR], 2025.

[82] W. Chen, Z.Dong, and X. Zhu, *Sharp non-uniqueness of solutions to stochastic Navier–Stokes equations*, SIAM J. Math. Anal., **56** (2024), pp. 2248–2285.

[83] K. Chen and J. Liu, *Weak solutions of the three-dimensional hypoviscous elastodynamics with finite kinetic energy*, J. Differential Equations, **339** (2022), pp. 637–665.

[84] Q. Chen, C. Miao, and Z. Zhang, *The Beale-Kato-Majda criterion for the 3D magneto-hydrodynamics equations*, Comm. Math. Phys., **275** (2007), pp. 861–872.

[85] Q. Chen, C. Miao, and Z. Zhang, *A new Bernstein's inequality and the 2D dissipative quasi-geostrophic equation*, Comm. Math. Phys., **271** (2007), pp. 821–838.

[86] X. Cheng, H. Kwon, and D. Li, *Non-uniqueness of steady-state weak solutions to the surface quasi-geostrophic equations*, Comm. Math. Phys., **388** (2021), pp. 1281–1295.

[87] A. S. Cherny, *On the uniqueness in law and the pathwise uniqueness for stochastic differential equations*, Theory Probab. Appl. **46** (2003), pp. 406–419.

[88] A. Cheskidov and X. Luo, *L^2-critical nonuniqueness for the 2D Navier–Stokes equations*, Ann. PDE, **9** (2023), https://doi.org/10.1007/s40818-023-00154-9.

[89] A. Cheskidov, Z. Zeng, and D. Zhang, *Existence and non-uniqueness of weak solutions with continuous energy to the 3D deterministic and stochastic Navier–Stokes equations*, arXiv:2407.17463 [math.AP], 2024.

[90] E. Chiodaroli, E. Feireisl, and F. Flandoli, *Ill posedness for the full Euler system driven by multiplicative white noise*, Indiana Univ. Math. J., **70** (2021), pp. 1267–1282. https://doi.org/10.1512/iumj.2021.70.8591.

[91] M. Christ, *Nonuniqueness of weak solutions of the nonlinear Schrödinger equation*, arXiv:math/0503366 [math.AP], 2005.

[92] M. Christ, *Nonuniqueness of generalized solutions of the Navier–Stokes equation*, July 8, 2025, available at https://math.berkeley.edu/~mchrist/Papers/stokes.pdf.

[93] I. Chueshov and A. Millet, *Stochastic 2D hydrodynamical type systems: well posedness and large deviations*, Appl. Math. Optim., **61** (2010), pp. 379–420.

[94] R. R. Coifman, P. L. Lions, Y. Meyer, and S. Semmes, *Compensated compactness and Hardy spaces*, J. Math. Pures et Appl., **72** (1992), pp. 247–286.

[95] J. F. Colombeau, *Multiplication of Distributions, A Tool in Mathematics, Numerical Engineering and Theoretical Physics*, Springer-Verlag, Berlin Heidelberg New York, 1992.

[96] M. Colombo, C. De Lellis, and L. De Rosa, *Ill-posedness of Leray solutions for the hypodissipative Navier–Stokes equations*, Comm. Math. Phys., **362** (2018), pp. 659–688.

[97] P. Constantin and C. Foias, *Navier–Stokes Equations*, The University of Chicago Press, 1989.

[98] P. Constantin, G. Iyer, and J. Wu, *Global regularity for a modified critical dissipative quasi-geostrophic equation*, Indiana U. Math. J., **57** (2008), pp. 2681–2692.

[99] P. Constantin, W. E, and E. S. Titi, *Onsager's conjecture on the energy conservation for solutions of Euler's equation*, Comm. Math. Phys., **165** (1994), pp. 207–209.

[100] P. Constantin, A. J. Majda, and E. Tabak, *Formation of strong fronts in the 2D quasi-geostrophic thermal active scalar*, Nonlinearity, **7** (1994), pp. 1495–1533.

[101] P. Constantin and V. Vicol, *Nonlinear maximum principles for dissipative linear nonlocal operators and applications*, Geom. Funct. Anal., **22** (2012), pp. 1289–1321.

[102] P. Constantin and J. Wu, *Behavior of solutions of 2D quasi-geostrophic equations*, SIAM J. Math. Anal., **30** (1999), pp. 937–948.

[103] P. Constantin and J. Wu, *Regularity of Hölder continuous solutions of the supercritical quasi-geostrophic equation*, Ann. Inst. H. Poincaré Anal. Non Linéaire, **25** (2008), pp. 1103–1110.

[104] P. Constantin and J. Wu, *Hölder continuity of solutions of supercritical dissipative hydrodynamic transport equations*, Ann. Inst. H. Poincaré Anal. Non Linéaire, **26** (2009), pp. 159–180.

[105] S. Conti, C. De Lellis, and L. Székelyhidi Jr., *h-principle and rigidity for $C^{1,\alpha}$ isometric embeddings*. In H. Holden and K. H. Karlsen(Eds.) *Nonlinear Partial Differential Equations, Abel Symposia 7*, Springer-Verlag Berlin Heidelberg, (2012), pp. 83–116.

[106] A. Córdoba and D. Córdoba, *Maximum principle applied to quasi-geostrophic equations*, Comm. Math. Phys., **249** (2004), pp. 511–528. https://doi.org/10.1007/s00220-004-1055-1.

[107] D. Córdoba, D. Faraco, and F. Gancedo, *Lack of uniqueness for weak solutions of the incompressible porous media equation*, Arch. Ration. Mech. Anal., **200** (2011), pp. 725–746.

[108] D. D. Córdoba, L. Martinez-Zoroa, and F. Zheng, *Finite time singularities to the 3D incompressible Euler equations for solutions in $C^\infty(\mathbb{R}^3 \setminus \{0\}) \cap C^{1,\alpha} \cap L^2$*, Ann. PDE, **11** (2025), https://doi.org/10.1007/s40818-025-00214-2.

[109] M. C. Zelati, M. Dolce, Y. Feng, and A. L. Mazzucato, *Global existence for the two-dimensional Kuramoto-Sivashinsky equation with a shear flow*, J. Evol. Equ., **21** (2021), pp. 5079–5099.

[110] O. Couland, I. Hachicha, and G. Raugel, *Hyperbolic quasilinear Navier–Stokes equations in $\mathbb{R}^2$, J. Dynam*, Differential Equations, **34** (2022), pp. 2749–2758.

[111] V. Dallas and A. Alexakis, *The signature of initial conditions on magnetohydrodynamic turbulence*, The Astrophysical Journal Letters, **788** (2014), pp. 1–4.

[112] P. D'Ancona, *A short proof of commutator estimates*, J. Fourier Anal. Appl., **25** (2019), pp. 1134–1146.

[113] G. Da Prato and A. Debussche, *Two-dimensional Navier–Stokes equations driven by a space-time white noise*, J. Funct. Anal., **196** (2002), pp. 180–210.

[114] G. Da Prato and A. Debussche, *Strong solutions to the stochastic quantization equations*, Ann. Probab., **31** (2003), pp. 1900–1916.

[115] G. Da Prato and A. Debussche, *Ergodicity for the 3D stochastic Navier–Stokes equations*, J. Math. Pures Appl., **82** (2003), pp. 877–947.

[116] G. Da Prato, A. Debussche, and R. Temam, *Stochastic Burgers' equation*, NoDEA Nonlinear Differential Equations Appl., **1** (1994), pp. 389–402.

[117] G. Da Prato, J. Zabczyk, *Ergodicity for Infinite Dimensional Systems*, Cambridge University Press, Cambridge, 1996.

[118] G. Da Prato and J. Zabczyk, *Stochastic Equations in Infinite Dimensions*, Cambridge University Press, United Kingdom, 2014.

[119] M. Dai, V. Giri, and R.-O. Radu, *An Onsager-type theorem for SQG*, arXiv:2407.02582 [math.AP], 2024.

[120] M. Dai and C. Wu, *Dissipation wavenumber and regularity for electron magnetohydrodynamics*, J. Differential Equations, **376** (2023), pp. 655–681.

[121] C. De Lellis and L. Székelyhidi Jr., *The Euler equations as a differential inclusion*, Ann. of Math., **170** (2009), pp. 1417–1436.

[122] C. De Lellis and L. Székelyhidi Jr., *On admissibility criteria for weak solutions of the Euler equations*, Arch. Ration. Mech. Anal., **195** (2010), pp. 225–260.

[123] C. De Lellis and L. Székelyhidi Jr., *Dissipative continuous Euler flows*, Invent. Math., **193** (2013), pp. 377–407.

[124] L. De Rosa, *Infinitely many Leray-Hopf solutions for the fractional Navier–Stokes equations*, Comm. Partial Differential Equations, **44** (2019), pp. 335–365.

[125] A. Debussche, *Ergodicity results for the stochastic Navier–Stokes equations: an introduction*. In P. Constantin, A. Debussche, G. P. Galdi, M. Růžička and G. Seregin (Eds.) *Topics in Mathematical Fluid Mechanics*, Lecture Notes in Mathematics, **2073**, Springer-Verlag, Berlin, Heidelberg (2013), pp. 23–108.

[126] R. J. DiPerna and P. L. Lions, *Ordinary differential equations, transport theory and Sobolev spaces*, Invent. Math., **98** (1989), pp. 511–547.

[127] S. Donato, S. Servidio, P. Dmitruk, V. Carbone, M. A. Shay, P. A. Cassak, and W. H. Matthaeus, *Reconnection events in two-dimensional Hall magnetohydrodynamic turbulence*, Phys. Plasmas, **19** 092307 (2012).

[128] H. Dong, D. Du, and D. Li, *Finite time singularities and global well-posedness for fractal Burgers equations*, Indiana Univ. Math. J., **58** (2009), pp. 807–821.

[129] H. Dong and K. Yamazaki, *Remarks on the convex integration technique applied to singular stochastic partial differential equations*, Contemp. Math., to appear. Also available at arXiv:2601.09990 [math.PR], 2026.

[130] T. M. Elgindi, *Finite-time singularity formation for $C^{1,\alpha}$ solutions to the incompressible Euler equations on* $\mathbb{R}^3$, Ann. Math., **194** (2021), pp. 647–727.

[131] L. Escauriaza, G. Seregin, and V. Šverak, *$L_{3,\infty}$-solutions of Navier–Stokes equations and backward uniqueness* (In Russian), Usp. Mat. Nauk, **58** 350 (2003), 3–44: translation in Russ. Math. Surv., **58** (2003), pp. 211–250.

[132] L. C. Evans, *Partial Differential Equations*, American Mathematical Society, Volume 19, 1998.

[133] G. L. Eyink, *Energy dissipation without viscosity in ideal hydrodynamics, I. Fourier analysis and local energy transfer*, Phys. D, **78** (1994), pp. 222–240.

[134] J. Fan, H. Malaikah, S. Monaquel, G. Nakamura, and Y. Zhou, *Global Cauchy problem of 2D generalized MHD equations*, Monatsch. Math., **175** (2014), pp. 127–131.

[135] D. Faraco and S. Lindberg, *Proof of Taylor's conjecture on magnetic helicity conservation*, Comm. Math. Phys., **373** (2020), pp. 707–738.

[136] D. Faraco, S. Lindberg, and L. Székelyhidi Jr., *Bounded solutions of ideal MHD with compact support in space-time*, Arch. Ration. Mech. Anal., **239** (2021), pp. 51–93.

[137] D. Faraco, S. Lindberg, and L. Székelyhidi Jr., *Magnetic helicity, weak solutions and relaxation of ideal MHD*, Comm. Pure Appl. Math., **77** (2024), pp. 2387–2412.

[138] C. L. Fefferman, D. S. McCormick, J. C. Robinson and J. L. Rodrigo, *Higher order commutator estimates and local existence for the non-resistive MHD equations and related models*, J. Funct. Anal., **267** (2014), pp. 1035–1056.

[139] F. Flandoli, *Dissipativity and invariant measures for stochastic Navier–Stokes equations*, NoDEA Nonlinear Differential Equations Appl., **1** (1994), pp. 403–423.

[140] F. Flandoli, *An Introduction to 3D Stochastic Fluid Dynamics*: In: G. Da Prato, M. Rückner (Eds.) *SPDE in Hydrodynamic: Recent Progress and Prospects*. Lecture Notes in Mathematics, **1942**, Springer, Berlin Heidelberg, (2008), pp. 51–150.

[141] F. Flandoli, D. Gatarek, *Martingale and stationary solutions for stochastic Navier–Stokes equations*, Probab. Theory Relat. Fields, **102** (1995), pp. 367–391.

[142] F. Flandoli, M. Gubinelli, and E. Priola, *Well-posedness of the transport equation by stochastic perturbation*, Invent. Math., **180** (2010), pp. 1–53.

[143] F. Flandoli and M. Rehmeier, *Remarks on regularization by noise, convex integration and spontaneous stochasticity*, Milan J. Math., **92** (2024), pp. 349–370.

[144] F. Flandoli and M. Romito, *Markov selections for the 3D stochastic Navier–Stokes equations*, Probab. Theory Related Fields, **140** (2008), pp. 407–458.

[145] F. Flandoli and B. Schmalfuß, *Weak solutions and attractors for three-dimensional Navier–Stokes equations with nonregular force*, J. Dyn. Differential Equations, **11** (1999), pp. 355–398.

[146] C. Foias and R. Temam, *Homogeneous statistical solutions of Navier–Stokes equations*, Indiana Univ. Math. J., **29** (1980), pp. 913–957.
[147] G. B. Folland, *Real Analysis: Modern Techniques and Their Applications*, John Wiley & Sons, Inc., Canada, 1999.
[148] D. Forster, D. R. Nelson, and M. J. Stephen, *Large-distance and long-time properties of a randomly stirred fluid*, Phys. Rev. A, **16** (1977), pp. 732–749.
[149] P. Friz and N. Victoir, *Differential equations driven by Gaussian signals*, Ann. Inst. Henri Poincaré Probab. Stat., **46** (2010), pp. 369–413.
[150] P. Friz and N. Victoir, *Multidimensional Stochastic Processes as Rough Paths*, Cambridge University Press, United Kingdom, 2010.
[151] V. Giri and R.-Oc. Radu, *The 2D Onsager conjecture: a Newton-Nash iteration*, Invent. Math., **238** (2024), pp. 691–768.
[152] J. Glimm and A. Jaffe, *Quantum Physics, A Functional Integral Point of View*, 2nd ed., Springer-Verlag, 1987.
[153] B. Goldys, M. Röckner, and X. Zhang, *Martingale solutions and Markov selections for stochastic partial differential equations*, Stochastic Process. Appl., **119** (2009), pp. 1725–1764.
[154] L. Grafakos, *Classical Fourier Analysis*, 3rd ed., Springer, 2014.
[155] R. Graham and H. Pleiner, *Mode-mode coupling theory of the heat convection threshold*, The Physics of Fluids, **18** (1975), pp. 130–140.
[156] M. Gromov, *Partial Differential Relations*, Springer-Verlag, Berlin, Heidelberg, 1986.
[157] M. Gromov, *Local and global in geometry* International Balzan Prize Foundation, 9 Sep. 2020, https://www.balzan.org/en/prizewinners/mikhael-gromov.
[158] M. Gubinelli, *Controlling rough paths*, J. Funct. Anal., **216** (2004), pp. 86–140.
[159] M. Gubinelli, P. Imkeller and N. Perkowski, *Paracontrolled distributions and singular PDEs*, Forum Math., Pi, **3** (2015), pp. 1–75.
[160] M. Gubinelli and N. Perkowski, *KPZ reloaded*, Comm. Math. Phys., **349** (2017), pp. 165–269.
[161] M. Gubinelli and S. Tindel, *Rough evolution equations*, Ann. Probab., **38** (2010), pp. 1–75.
[162] M. Gubinelli, B. Ugurcan, and I. Zachhuber, *Semilinear evolution equations for the Anderson Hamiltonian in two and three dimensions*, Stoch. PDE: Anal. Comp., **8** (2020), pp. 82–149.
[163] J. Guillod and V. Šverák, *Numerical investigations of non-uniqueness for the Navier–Stokes initial value problem in borderline spaces*, J. Math. Fluid Mech., **25** (2023), https://doi.org/10.1007/s00021-023-00789-5.
[164] M. Hairer, *Rough stochastic PDEs*, Comm. Pure Appl. Math., **LXIV** (2011), pp. 1547–1585.
[165] M. Hairer, *Singular perturbations to semilinear stochastic heat equations*, Probab. Theory Relat. Fields, **152** (2012), pp. 265–297.
[166] M. Hairer, *Solving the KPZ equation*, Ann. of Math., **178** (2013), pp. 559–664.
[167] M. Hairer, *A theory of regularity structures*, Invent. Math., **198** (2014), pp. 269–504.
[168] M. Hairer, *Singular stochastic PDEs*. In S. Y. Jang, Y. R. Kim, D.-W. Lee, and I. Yie, (Eds.), *Proceedings of the International Congress of Mathematicians*, SEOUL ICM 2014 Organizing Committee.
[169] M. Hairer, *Introduction to regularity structures*, Braz. J. Probab. Stat., **29** (2015), pp. 175–210.
[170] M. Hairer and C. Labbé *The reconstruction theorem in Besov spaces*, J. Funct. Anal., **273** (2017), pp. 2578–2618.
[171] M. Hairer and J. Maas, *A spatial version of the Ito-Stratonovich correction*, Ann. Probab., **40** (2012), pp. 1675–1714.
[172] M. Hairer, J. Maas, and H. Weber, *Approximating rough stochastic PDEs*, Comm. Pure Appl. Math., **67** (2014), pp. 776–870.
[173] M. Hairer and K. Matetski, *Discretisations of rough stochastic PDEs*, Ann. Probab., **46** (2018), pp. 1651–1709.
[174] M. Hairer and J. C. Mattingly, *Ergodicity of the 2D Navier–Stokes equations with degenerate stochastic forcing*, Ann. Math., **164** (2006), pp. 993–1032.

[175] M. Hairer and J. C. Mattingly, *The strong Feller property for singular stochastic PDEs*, Ann. Inst. H. Poincaré Probab. Stat., **54** (2018), pp. 1314–1340.

[176] M. Hairer and T. Rosati, *Global existence for perturbations of the 2D stochastic Navier–Stokes equations with space-time white noise*, Annals of PDE, **10** (2024), pp. 1–46.

[177] M. Hairer and J. Voss, *Approximations to the stochastic Burgers equation*, J. Nonlinear Sci., **21** (2011), pp. 897–920.

[178] M. Hairer and H. Weber, *Rough Burgers-like equations with multiplicative noise*, Probab. Theory Relat. Fields, **155** (2013), pp. 71–126.

[179] A. Hasegawa, *Self-organization processes in continuous media*, Adv. Phys., **34** (1985), pp. 1–42.

[180] T. Hmidi, *On a maximum principle and its application to logarithmically critical Boussinesq system*, Anal. PDE, **4** (2011) pp. 247–284.

[181] P. C. Hohenberg and J. B. Swift, *Effects of additive noise at the onset of Rayleigh-Bénard convection*, Physical Review A, **46** (1992), pp. 4773–4785.

[182] M. Hofmanová, T. Lange, and U. Pappalettera, *Global existence and non-uniqueness of 3D Euler equations perturbed by transport noise*, Probab. Theory Related Fields, **188** (2024), pp. 1183–1255.

[183] M. Hofmanová, X. Luo, R. Zhu, and X. Zhu, *Surface quasi-geostrophic equation perturbed by derivatives of space-time white noise*, Math. Ann. (2024), https://doi.org/10.1007/s00208-024-02881-1.

[184] M. Hofmanová, U. Pappalettera, R. Zhu, and X. Zhu, *Kolmogorov 4/5 law for the forced 3D Navier–Stokes equations*, arXiv:2304.14470 [math.AP], 2023.

[185] M. Hofmanová, U. Pappalettera, R. Zhu, and X. Zhu, *Anomalous and total dissipation due to advection by solutions of randomly forced Navier–Stokes equations*, arXiv:2305.08090 [math.AP], 2023.

[186] M. Hofmanová, R. Zhu, and X. Zhu, *Non-uniqueness in law of stochastic 3D Navier–Stokes equations*, J. Eur. Math. Soc. (2023), https://doi.org/10.4171/JEMS/1360.

[187] M. Hofmanová, R. Zhu, and X. Zhu, *On ill- and well-posedness of dissipative martingale solutions to stochastic 3D Euler equations*, Comm. Pure Appl. Math., **75** (2022), pp. 2446–2510.

[188] M. Hofmanová, R. Zhu, and X. Zhu, *Global-in-time probabilistically strong and Markov solutions to stochastic 3D Navier–Stokes equations: existence and non-uniqueness*, Ann. Probab., **51** (2023), pp. 524–579.

[189] M. Hofmanová, R. Zhu, and X. Zhu, *Global existence and non-uniqueness for 3D Navier–Stokes equations with space-time white noise*, Arch. Ration. Mech. Anal., **46** (2023), https://doi.org/10.1007/s00205-023-01872-x.

[190] M. Hofmanová, R. Zhu, and X. Zhu, *A class of supercritical/critical singular stochastic PDEs: existence, non-uniqueness, non-Gaussianity, non-unique ergodicity*, J. Funct. Anal., **285** (2023), pp. 1–43.

[191] M. Hofmanová, R. Zhu, and X. Zhu, *Non-unique ergodicity for deterministic and stochastic 3D Navier–Stokes and Euler equations*, arXiv:2208.08290 [math.PR], 2022.

[192] M. Hofmanová, R. Zhu, and X. Zhu, *Non-uniqueness of Leray-Hopf solutions for stochastic forced Navier–Stokes equations*, Electron. J. Probab., **29** (2024), pp. 1–27.

[193] E. Hopf, *Über die Anfangswertaufgabe für die hydrodynamischen Grundgleichungen*, Math. Nachr., **4** (1951), pp. 213–231.

[194] T. Hou, Y. Wang, and C. Yang, *Nonuniqueness of Leray-Hopf solutions to the unforced incompressible 3D Navier–Stokes equation*, arXiv:2509.25116 [math.AP], 2025.

[195] P. D. Iftimie, *The 3D Navier–Stokes equations seen as a perturbation of the 2D Navier–Stokes equations*, Bull. Soc. Math. France, **127** (1999), pp. 473–517.

[196] P. D. Iftimie, *The resolution of the Navier–Stokes equations in anisotropic spaces*, Rev. Mat. Iberoamericana, **15** (1999), pp. 1–36.

[197] P. Isett, *A proof of Onsager's conjecture*, Ann. of Math., **188** (2018), pp. 871–963.

[198] P. Isett and S.-Z. Looi, *A proof of Onsager's conjecture for the SQG equation*, arXiv:2407.02578 [math.AP], 2024.

[199] P. Isett and A. Ma, *A direct approach to nonuniqueness and failure of compactness for the SQG equation*, Nonlinearity, **34**, 3122 (2021).

[200] P. Isett and V. Vicol, *Hölder continuous solutions of active scalar equations*, Ann. PDE, **1** (2015), pp. 1–77.
[201] S. Janson, *Gaussian Hilbert Spaces*, Cambridge University Press, United Kingdom, 1997.
[202] I.-J. Jeong and S.-J. Oh, *On illposedness of the Hall and electron magnetohydrodynamic equations without resistivity on the whole space*, arXiv:2404.13790 [math.AP], 2024.
[203] R. Ji, J. Li, L. Tian, and J. Wu, *Global solutions of a general hyperbolic Navier–Stokes equations*, Frontiers of Mathematics (2025), https://doi.org/10.1007/s11464-024-0007-7.
[204] H. Jia and V. Śverák, *Local-in-space estimates near initial time for weak solutions of the Navier–Stokes equations and forward self-similar solutions*, Invent. Math., **196** (2014), pp. 233–265.
[205] H. Jia and V. Śverák, *Are the incompressible 3d Navier–Stokes equations locally ill-posed in the natural energy space?* J. Funct. Anal., **268** (2015), pp. 3734–3766.
[206] Q. Jiu and J. Zhao, *A remark on global regularity of 2D generalized magnetohydrodynamic equations*, J. Math. Anal. Appl., **412** (2014), pp. 478–484.
[207] Q. Jiu, J. Zhao, *Global regularity of 2D generalized MHD equations with magnetic diffusion*, Z. Angew. Math. Phys., **66** (2015), pp. 677–687.
[208] Y. Kaneda, T. Ishihara, M. Yokokawa, K. Itakura, and A. Uno, *Energy dissipation rate and energy spectrum in high resolution direct numerical simulations of turbulence in a periodic box*, Physics of Fluids, **15** (2003), pp. 21–24.
[209] I. Karatzas and S. E. Shreve, *Brownian Motion and Stochastic Calculus*, 2nd ed., Springer-Verlag, New York, Inc., 1991.
[210] M. Kardar, G. Parisi, and Y.-C. Zhang, *Dynamic scaling of growing interfaces*, Phys. Rev. Lett., **56** (1986), pp. 889–892.
[211] T. Kato, *Liapunov functions and monotonicity in the Navier–Stokes equation, Functional-analytic methods for partial differential equations*, lecture notes in mathematics, **1450** (1990), pp. 53–63.
[212] T. Kato and G. Ponce, *Commutator estimates and Euler and Navier–Stokes equations*, Comm. Pure Appl. Math., **41** (1988), pp. 891–907.
[213] K. Kawasaki, *Diffusion constants near the critical point for time-dependent Ising models. I*, Phys. Rev., **145** (1966), pp. 224–230.
[214] C. Kenig, G. Ponce, and L. Vega, *Well-posedness and scattering for the generalized Korteweg-de Vries equation via the contraction principle*, Comm. Pure Appl. Math., **46** (1993), pp. 527–620.
[215] J. U. Kim, *On the stochastic quasi-linear symmetric hyperbolic system*, J. Differential Equations, **250** (2011), pp. 1650–1684.
[216] K. Kinra and U. Koley, *Non-uniqueness of Hölder continuous solutions for stochastic Euler and hypodissipative Navier–Stokes equations*, arXiv:2407.20270 [math.AP], 2024.
[217] A. Kiselev, *Regularity and blow-up for active scalars*, Math. Model. Nat. Phenom., **5** (2010), pp. 225–255.
[218] A. Kiselev, *Nonlocal maximum principles for active scalars*, Adv. Math., **277** (2011), pp. 1806–1826.
[219] A. Kiselev and F. Nazarov, *Variation on a theme of Caffarelli and Vasseur*, J. Math. Sci., **166** (2010), pp. 31–39.
[220] A. Kiselev, F. Nazarov, and R. Shterenberg, *Blow up and regularity for fractal Burgers equation*, Dyn. Partial Differ. Equ., **5** (2008), pp. 211–240.
[221] A. Kiselev, F. Nazarov, and A. Volberg, *Global well-posedness for the critical 2D dissipative quasi-geostrophic equation*, Invent. Math., **167** (2007), pp. 445–453.
[222] U. Koley and K. Yamazaki, *Non-uniqueness in law of transport-diffusion equation forced by random noise*, J. Differential Equations, **416** (2025), pp. 82–142.
[223] A. N. Kolmogorov, *Dissipation of energy in the locally isotropic turbulence*, C. R. (Doklady) Acad. Sci. URSS (N.S.), **32** (1941), pp. 16–18.
[224] A. N. Kolmogorov, *The local structure of turbulence in incompressible viscous fluid for very large Reynolds numbers*, C. R. (Doklady) Acad. Sci. URSS (N.S.), **30** (1941), pp. 301–305.
[225] E. Kreyszig, *Introductory Functional Analysis with Applications*, John Wiley & Sons, 1978.
[226] N. H. Kuiper, *On C^1-isometric embeddings. I*, Nederl. Akad. Wetensch. Proc. A, **58** (1955), pp. 545–556.
[227] H. Kunita, *Stochastic Flows and Stochastic Differential Equations*, Cambridge University Press, 1990.

[228] Y. Kuramoto and T. Tsuzuki, *On the formation of dissipative structures in reaction-diffusion*, Prog. Theore. Phys., **54** (1975), pp. 687–699.

[229] Y. Kuramoto and T. Tsuzuki, *Persistent propagation of concentration waves in dissipative media far from equilibrium*, Prog. Theore. Phys., **55** (1976), pp. 365–369.

[230] C. Labbé, *The continuous Anderson hamiltonian in $d \leq 3$*, J. Funct. Anal., **277** (2019), pp. 3187–3235.

[231] O. Ladyzhenskaya, *On the uniqueness and on the smoothness of weak solutions of the Navier–Stokes equation*, Zapiski Nauchnykh Seminarov POMI, **5** (1967), pp. 169–185.

[232] T. Lange, M. Rehmeier, and A. Schenke, *Non-uniqueness of Leray-Hopf solutions for the 3D fractional Navier–Stokes equations perturbed by transport noise*, arXiv:2412.16532 [math.AP], 2024.

[233] L. D. Landau and E. M. Lifshitz, *Hydrodynamic fluctuations*, J. Exptl. Theoret. Phys. (U.S.S.R.) **32** (1957), pp. 618–619.

[234] A. Larios, M. M. Rahman, and K. Yamazaki, *Regularity criteria for the Kuramoto-Sivashinsky equation in dimensions two and three*, J. Nonlinear Sci., **32** (2022), https://doi.org/10.1007/s00332-022-09828-3.

[235] A. Larios and K. Yamazaki, *On the well-posednesss of an anisotropically-reduced two-dimensional Kuramoto-Sivashinsky equation*, Phys. D, **411** (2020), 132560.

[236] P. G. Lemarié-Rieusset, *Recent Developments in the Navier–Stokes Problem*, Chapman & Hall/CRC, 2002.

[237] J. Leray, *Essai sur le mouvement d'un fluide visqueux emplissant l'espace*, Acta Math., **63** (1934), pp. 193–248.

[238] D. Li, *On Kato-Ponce and fractional Leibniz*, Rev. Mat. Iberoam., **35** (2019), pp. 23–100.

[239] Y. Li, Z. Zeng, and D. Zhang, *Non-uniqueness of weak solutions to 3D magnetohydrodynamic equations*, J. Math. Pures Appl., **165** (2022), pp. 232–285.

[240] M. J. Lighthill, F. R. S., Studies on magneto-hydrodynamic waves and other anisotropic wave motions, Philos. Trans. R. Soc. Lond. Ser. A, **252** (1960), pp. 397–430.

[241] J.-L. Lions, *Quelques résultats d'existence dans des équations aux dérivées partielles non linéaires*, Bull. Soc. Math. France, **87** (1959), pp. 245–273.

[242] J.-L. Lions, *Quelquels Méthodes de Résolution des Problémes aux Limites Non Linéaires*, Dunod, Gauthiers-Villars, Paris, 1969.

[243] H. Lü, L. Lü, and R. Zhu, *A proof of Onsager's conjecture for the stochastic 3D Euler equations*, arXiv:2505.06915 [math.PR], 2025.

[244] H. Lü, M. Röckner, and X. Zhu, *Non-uniqueness of (Stochastic) Lagrangian trajectories for Euler equations*, arXiv:2504.16687v2 [math.AP], 2025.

[245] H. Lü and M. Röckner, *Sharp non-uniqueness in law for stochastic differential equations on the whole line*, arXiv:2510.08248 [math.PR], 2025.

[246] L. Lü and R. Zhu, *Stationary solutions to stochastic 3D Euler equations in Hölder space*, Stoch. Process. Appl., **177** (2024), https://doi.org/10.1016/j.spa.2024.104465.

[247] H. Lü and X. Zhu, *Sharp non-uniqueness of solutions to 2D Navier–Stokes equations with space-time white noise*, arXiv:2304.06526v2 [math.PR], 2023.

[248] H. Lü and X. Zhu, *Non-unique ergodicity for the 2D stochastic Navier–Stokes equations with derivative of space-time white noise*, arXiv:2402.13743 [math.PR], 2024.

[249] H. Lü and Y. Zhu, *Non-uniqueness in law of stochastic 3D Navier–Stokes equations with general multiplicative noise*, arXiv:2505.07181 [math.PR], 2025.

[250] L. Lü, *Hölder continuous solutions to stochastic 3D Euler equations via stochastic convex integration*, arXiv:2407.19671v2 [math.PR], 2025.

[251] T. Luo and P. Qu, *Non-uniqueness of weak solutions to 2D hypoviscous Navier–Stokes equations*, J. Differential Equations, **269** (2020), pp. 2896–2919.

[252] T. Luo, T. Tao, and L. Zhang, *Finite energy weak solutions of 2D Boussinesq equations with diffusive temperature*, Discrete Contin. Dyn. Syst., **40** (2020), pp. 3737–3765.

[253] T. Luo and E. S. Titi, *Non-uniqueness of weak solutions to hyperviscous Navier–Stokes equations - on sharpness of J.-L. Lions exponent*, Calc. Var. Partial Differential Equations, **59** (2020), https://doi.org/10.1007/s00526-020-01742-4.

[254] T. Lyons, *Differential equations driven by rough signals (I): an extension of an inequality of L. C. Young*, Math. Res. Lett., **1** (1994), pp. 451–464.
[255] T. Lyons, *Differential equations driven by rough signals*, Rev. Mat. Iberoam., **14** (1998), pp. 215–310.
[256] T. Lyons and Z. Qian, *System Control and Rough Paths*, Clarendon Press, Oxford, 2002.
[257] S.-k.Ma and G. F. Mazenko, *Critical dynamics of ferromagnets in* $6-\epsilon$ *dimensions: General discussion and detailed calculation*, Phys. Rev. B, **11** (1975), pp. 4077–4100.
[258] A. Majda, *Introduction to PDEs and Waves for the Atmosphere and Ocean*, American Mathematical Society, Rhode Island, U.S.A., 2003.
[259] A. J. Majda and A. L. Bertozzi, *Vorticity and Incompressible Flow*, Cambridge University Press, 2002.
[260] S. J. A. Malham, *Regularity assumptions and length scales for the Navier–Stokes equations*, Ph.D. thesis, University of London, 1997.
[261] V. Mandrekar and B. Rüdiger, *Existence and uniqueness of path wise solutions for stochastic integral equations driven by Lévy noise on separable Banach spaces*, Stochastics, **78** (2006), pp. 189–212.
[262] F. Marchand, *Existence and regularity of weak solutions to the quasi-geostrophic equations in the spaces* L^p *or* $\dot{H}^{-\frac{1}{2}}$, Comm. Math. Phys., **277** (2008), pp. 45–67.
[263] P. A. Meyer, *Probability and Potentials*, Blaisdell publishing company, U.S.A., 1966.
[264] P. D. Mininni and A. Pouquet, *Finite dissipation and intermittency in magnetohydrodynamics*, Phys. Rev. E, **80** 025401 (R), (2009), pp. 1–4.
[265] A. Miranville, *The Cahn-Hilliard Equation: Recent Advances and Applications*, CBMS-NSF Regional Conference Series in Applied Mathematics, Society for Industrial and Applied Mathematics, 2019.
[266] S. Modena and A. Schenke, *Local nonuniqueness for stochastic transport equations with deterministic drift*, arXiv:2306.08758 [math.PR], 2023.
[267] H. K. Moffatt, *Some remarks on topological fluid mechanics.* In R. L. Ricca (ed.) *An Introduction to the Geometry and Topology of Fluid Flows*, Kluwer Academic Publishers, Netherlands (2001), pp. 3–10.
[268] J.-C. Mourrat and H. Weber, *Global well-posedness of the dynamic* Φ^4 *model in the plane*, Ann. Probab., **45** (2017), pp. 2398–2476.
[269] J.-C. Mourrat and H. Weber, *The dynamic* Φ^4_3 *model comes down from infinity*, Comm. Math. Phys., **356** (2017), pp. 673–753.
[270] C. Mueller, *On the support of solutions to the heat equation with noise*, Stochastics and Stochastic Reports, **37** (1991), pp. 225–245.
[271] S. Müller and V. Šverák, *Unexpected solutions of first and second order partial differential equations*, Proc. of ICM-1998, Vol. II, p. 691–702, Berlin, 1998.
[272] *S. Müller and V. Śverák,* Convex integration for Lipschitz mappings and counterexamples to regularity, Ann. of Math., **157** (2003), pp. 715–742.
[273] J. Nash, C^1 *isometric imbeddings*, Ann. of Math., **60** (1954), pp. 383–395.
[274] D. Nualart, *The Malliavin Calculus and Related Topics*, Springer-Verlag, New York, Berlin, Heidelberg, 1995.
[275] B. Øksendal, *Stochastic Differential Equations*, 6th ed., Springer-Verlag, Berlin, Heidelberg, New York, 2003.
[276] L. Onsager, *Statistical hydrodynamics*, Nuovo Cim, **6** (1949), pp. 279–287.
[277] K. Osterwalder and R. Schrader, *Axioms for Euclidean Green's functions II*, Comm. Math. Phys., **42** (1975), pp. 281–305.
[278] U. Pappalettera, *Global existence and non-uniqueness for the Cauchy problem associated to 3D Navier–Stokes equations perturbed by transport noise*, Stoch. PDE: Anal. Comp. (2023), https://doi.org/10.1007/s40072-023-00318-5.
[279] G. Parisi and Y. Wu, *Perturbation theory without gauge fixing*, Scientia Sinica, **XXIV** (1981), pp. 483–496.
[280] A. Pazy, *Semigroups of Linear Operators and Applications to Partial Differential Equations*, Springer-Verlag New York, Inc., 1983.
[281] H. Politano, A. Pouquet, and P.-L. Sulem, *Current and vorticity dynamics in three-dimensional magnetohydrodynamic turbulence*, Phys. Plasmas, **2** (1995), pp. 2931–2939.

[282] M. Polyak, *Feynman diagrams for pedestrians and mathematicians*, in *Graphs and Patterns in Mathematics and Theoretical Physics*, Proc. Sympos. Pure Math. **73**, Amer. Math. Soc., Providence, RI, 2005, pp. 15–42.

[283] B. C. Pooley and J. C. Robinson, *Well-posedness for the diffusive 3D Burgers equation with initial data in $H^{1/2}$*, in: *Recent Progress in the Theory of the Euler and Navier–Stokes equations*, in: London Math. Soc. Lecture Note Ser., **430**, Cambridge Univ. Press, Cambridge, 2016, pp. 137–153.

[284] G. Prodi, *Un teorema di unicitá per le equazioni di Navier–Stokes*, Ann. Mat. Pura Appl., **48** (1959), pp. 173–182.

[285] M. M. Rahman and K. Yamazaki, *Remarks on the global regularity issue of the two and a half dimensional Hall-magnetohydrodynamics system*, Z. Angew. Math. Phys., **73** (2022), pp. 1–29.

[286] M. M. Rahman and K. Yamazaki, *Another remark on the global regularity issue of the Hall-magnetohydrodynamics system*, J. Evol. Equ., **24** (2024), https://doi.org/10.1007/s00028-024-01000-6.

[287] M. Reed and B. Simon, *Methods of Modern Mathematical Physics, I: Functional Analysis*, Academic Press, California, 1980.

[288] M. Rehmeier and A. Schenke, *Nonuniqueness in law for stochastic hypodissipative Navier–Stokes equations*, Nonlinear Anal., **227** (2023), https://doi.org/10.1016/j.na.2022.113179.

[289] S. Resnick, *Dynamical problems in nonlinear advective partial differential equations*, Ph.D. thesis, University of Chicago, 1995.

[290] D. Revuz and M. Yor, *Continuous Martingales and Brownian Motion*, Springer-Verlag, Berlin, Heidelberg, 1999.

[291] M. Röckner, R. Zhu, and X. Zhu, *Local existence and non-explosion of solutions for stochastic fractional partial differential equations driven by multiplicative noise*, Stochastic Process. Appl., **124** (2014), pp. 1974–2002.

[292] T. Roy, *Global existence of smooth solutions of a 3D log-log energy-supercritical wave equation*, Anal. PDE, **2** (2009), pp. 261–280.

[293] B. Rüdiger and G. Ziglio, *Itô formula for stochastic integrals w. r. t. compensated Poisson random measures on separable Banach spaces*, Stochastics, **78** (2006), pp. 377–410.

[294] W. Rudin, *Real and Complex Analysis*, 3rd ed., McGraw-Hill Book Company, Singapore, 1966.

[295] G. Rudolph and M. Schmidt, *Differential Geometry and Mathematical Physics, Part II. Fibre Bundles, Topology and Gauge Fields*, Springer Science+Business Media Dordrecht, 2017.

[296] V. Scheffer, *An inviscid flow with compact support in space-time*, J. Geom. Anal., **3** (1993), pp. 343–401.

[297] A. Schenke, *The stochastic tamed MHD equations: existence, uniqueness and invariant measures*, Stoch. PDE: Anal. Comp. (2021), https://doi.org/10.1007/s40072-021-00205-x.

[298] J. Serrin, *On the interior regularity of weak solutions of the Navier–Stokes equations*, Arch. Ration. Mech. Anal., **9** (1962), pp. 187–195.

[299] J. Serrin, *The initial value problem for the Navier–Stokes equations*, in *Nonlinear Problems* (Proc. Sympos., Madison, Wis., 1962), Univ. of Wisconsin Press, Madison, Wis., 1963, pp. 69–98.

[300] A. Shnirelman, *On the nonuniqueness of weak solution of the Euler equation*, Comm. Pure Appl. Math., **50** (1997), pp. 1261–1286.

[301] B. Simon, *The $P(\phi)_2$ Euclidean (Quantum) Field Theory*, Princeton University Press, Princeton, New Jersey, 1974.

[302] J. Simon, *Compact sets in the space $L^p(0,T;B)$*, Ann. Mat. Pure Appl. (4), **146** (1986), pp. 65–96.

[303] J. Simon, *Nonhomogeneous viscous incompressible fluids: existence of velocity, density, and pressure*, SIAM J. Math. Anal., **21** (1990), pp. 1093–1117.

[304] G. I. Sivashinsky, *Nonlinear analysis of hydrodynamic instability in laminar flames. Derivation of basic equations*, Acta Astronaut., **4** (1977), pp. 1177–1206.

[305] A. V. Skorokhod, *Studies in the Theory of Random Processes*, Dover Publications, Inc., New York, 1965.

[306] S. S. Sritharan, P. Sundar, *The stochastic magneto-hydrodynamic system*, Infin. Dimens. Anal. Quantum Probab. Relat. Top., **2** (1999), pp. 241–265.

[307] E. M. Stein, *Singular Integrals and Differentiability Properties of Functions*, Princeton University Press, New Jersey, 1970.

[308] E. M. Stein, *Harmonic analysis: real-variable methods, orthogonality, and oscillatory integrals*, Princeton University Press, Princeton, New Jersey, 1993.

[309] E. M. Stein and R. Shakarchi, *Fourier Analysis: An Introduction*, Princeton University Press, 2003.

[310] D. W. Stroock and S. R. S. Varadhan, *Multidimensional Diffusion Processes*, Springer-Verlag, Berlin, Heidelberg, New York, 1997.

[311] J. Swift and P. C. Hohenberg, *Hydrodynamic fluctuations at the convective stability*, Phys. Rev. A, **15** (1977), pp. 319–328.

[312] E. Tadmor, *The well-posedness of the Kuramoto-Sivashinsky equation*, SIAM J. Math. Anal., **17** (1986), pp. 884–893.

[313] T. Tao, *Global regularity for a logarithmically supercritical defocusing nonlinear wave equation for spherically symmetric data*, J. Hyperbolic Differ. Equ., **4** (2007), pp. 259–266.

[314] T. Tao, *Global regularity for a logarithmically supercritical hyperdissipative Navier–Stokes equation*, Anal. PDE, **2** (2009), pp. 361–366.

[315] T. Tao, *255B, Notes 2: Onsager's conjecture.* What's New, 8 Jan. 2019, https://terrytao.wordpress.com/tag/onsagers-conjecture/.

[316] J. B. Taylor, *Relaxation of toroidal plasma and generation of reverse magnetic fields*, Phys. Rev. Lett., **33** (1974), pp. 1139–1141.

[317] J. B. Taylor, *Relaxation and magnetic reconnection in plasmas*, Rev. Modern Phys., **58** (1986), pp. 741–763.

[318] R. Temam, *Navier–Stokes equations: theory and numerical analysis*, North-Holland Publishing Company, 1977.

[319] R. Temam, *Infinite-Dimensional Dynamical Systems in Mechanics and Physics*, 2nd ed., Springer-Verlag New York, Inc., 1997.

[320] C. V. Tran, X. Yu and Z. Zhai, *On global regularity of 2D generalized magnetohydrodynamics equations*, J. Differential Equations, **254** (2013), pp. 4194–4216.

[321] C. V. Tran, X. Yu, and Z. Zhai, *Note on solution regularity of the generalized magnetohydrodynamic equations with partial dissipation*, Nonliner Anal., **85** (2013), pp. 43–51.

[322] H. Triebel, *Interpolation Theory, Function Spaces, Differential Operators*, North-Holland Publishing Company, Amsterdam, New York, Oxford, 1978.

[323] P. Vernotte, *Les paradoxes de la théorie continue de l'équation de la chaleur*, Comptes Rendus Acad. Sci., **246** (1958), pp. 3154–3155.

[324] M. Vishik, *Instability and non-uniqueness in the Cauchy problem for the Euler equations of an ideal incompressible fluid. Part I*, arXiv:1805.09426 [math.AP], 2018.

[325] M. Vishik, *Instability and non-uniqueness in the Cauchy problem for the Euler equations of an ideal incompressible fluid. Part II*, arXiv:1805.09440 [math.AP], 2018.

[326] H.-H. Vo and S. Kim, *Convex integration for scalar conservation laws in one space dimension*, SIAM J. Math. Anal., **50** (2018), pp. 3122–3146.

[327] E. Walker and K. Yamazaki, *Surface quasi-geostrophic equations forced by random noise: prescribed energy and non-unique Markov selections*, arXiv:2407.00920 [math.AP], 2024.

[328] M. Wardle, *Star formation and the Hall effect*, Astrophys. Space Sci., **292** (2004), pp. 317–323.

[329] C. J. Wareing and R. Hollerback, *Forward and inverse cascades in decaying two-dimensional electron magnetohydrodynamic turbulence*, Physics of Plasmas, **16** (2009), 042307.

[330] S. Watanabe and T. Yamada, *On the uniqueness of solutions of stochastic differential equations II*, J. Math. Kyoto Univ., **11** (1971), pp. 553–563.

[331] W. Walter, *Ordinary Differential Equations*, Springer-Verlag New York, Inc., 1998.

[332] L. Woltjer, *A theorem on force-free magnetic fields*, Proc. Natl. Aca. Sci. USA, **44** (1958), pp. 489–491.

[333] J. Wu, *The generalized MHD equations*, J. Differential Equations, **195** (2003), pp. 284–312.

[334] J. Wu, *Lower bounds for an integral involving fractional Laplacians and the generalized Navier–Stokes equations in Besov spaces*, Comm. Math. Phys., **263** (2005), pp. 803–831.

[335] J. Wu, *Global regularity for a class of generalized magnetohydrodynamic equations*, J. Math. Fluid Mech., **13** (2011), pp. 295–305.

[336] J. Wu and K. Yamazaki, *Convex integration solution of two-dimensional hyperbolic Navier–Stokes equations*, Nonlinearity, **37** (2024), 115014, pp. 1–48.

[337] V. Yakhot and S. A. Orszag, *Renormalization group analysis of turbulence. I. basic theory*, J. Sci. Comput., **1** (1986), pp. 3–51.

[338] T. Yamada and S. Watanabe, *On the uniqueness of solutions of stochastic differential equations*, J. Math. Kyoto Univ., **11** (1971), pp. 155–167.

[339] K. Yamazaki, *Remarks on the global regularity of two-dimensional magnetohydrodynamics system with zero dissipation*, Nonlinear Anal., **94** (2014), pp. 194–205.

[340] K. Yamazaki, *Global regularity of logarithmically supercritical MHD system with zero diffusivity*, Appl. Math. Lett., **29** (2014), pp. 46–51.

[341] K. Yamazaki, *On the global regularity of two-dimensional generalized magnetohydrodynamics system*, J. Math. Anal. Appl., **416** (2014), pp. 99–111.

[342] K. Yamazaki, *On the global regularity of N-dimensional generalized Boussinesq system*, Appl. Math., **60** (2015), pp. 109–133.

[343] K. Yamazaki, *On the three-dimensional magnetohydrodynamics system in scaling-invariant spaces*, Bull. Sci. Math., **140** (2016), pp. 575–614.

[344] K. Yamazaki, *Stochastic Hall-magneto-hydrodynamics system in three and two and a half dimension*, J. Stat. Phys., **166** (2017), pp. 368–397.

[345] K. Yamazaki, *Global regularity of logarithmically supercritical MHD system with improved logarithmic powers*, Dyn. Partial Differ. Equ., **15** (2018), pp. 147–173.

[346] K. Yamazaki, *Second proof of the global regularity of the two-dimensional MHD system with full diffusion and arbitrary weak dissipation*, Methods Appl. Anal., International Press of Boston, **25** (2018), pp. 73–96.

[347] K. Yamazaki, *On the Navier–Stokes equations in scaling-invariant spaces in any dimension*, Rev. Mat. Iberoam., **34** (2018), pp. 1515–1540.

[348] K. Yamazaki, *On the global regularity issue of the two-dimensional magnetohydrodynamics system with magnetic diffusion weaker than a Laplacian*, in *Nonlinear Dispersive Waves and Fluids*, Contemp. Math., **725**, Amer. Math. Soc., Providence, RI, 2019, pp. 251–264.

[349] K. Yamazaki, *Markov selections for the magnetohydrodynamics system and Hall-magnetohydrodynamics system*, J. Nonlinear Sci., **29** (2019), pp. 1761–1812.

[350] K. Yamazaki, *Irreducibility of the three, and two and a half dimensional Hall-magnetohydrodynamics system*, Phys. D, **401** (2020), 13299.

[351] K. Yamazaki, *A note on the applications of Wick products and Feynman diagrams in the study of singular partial differential equations*, J. Comput. Appl. Math., **388** (2021), 113338.

[352] K. Yamazaki, *Strong Feller property of the magnetohydrodynamics system forced by space-time white noise*, Nonlinearity, **34** (2021), pp. 1–91.

[353] K. Yamazaki, *Remarks on the non-uniqueness in law of the Navier–Stokes equations up to the J.-L. Lions' exponent*, Stochastic Process. Appl., **147** (2022), pp. 226–269.

[354] K. Yamazaki, *Non-uniqueness in law for two-dimensional Navier–Stokes equations with diffusion weaker than a full Laplacian*, SIAM J. Math. Anal., **54** (2022), pp. 3997–4042.

[355] K. Yamazaki, *Non-uniqueness in law for the Boussinesq system forced by random noise*, Calc. Var. Partial Differential Equations, **61** (2022), pp. 1–65.

[356] K. Yamazaki, K. Yamazaki, *Ergodicity of Galerkin approximations of surface quasi-geostrophic equations and Hall-magnetohydrodynamics system forced by degenerate noise*, Nonlinear Differential Equations and Applications NoDEA, **29** (2022), https://doi.org/10.1007/s00030-022-00753-8.

[357] K. Yamazaki, *Three-dimensional magnetohydrodynamics system forced by space-time white noise*, Electron. J. Probab., **28** (2023), pp. 1–66.
[358] K. Yamazaki, *Non-uniqueness in law of three-dimensional Navier–Stokes equations diffused via a fractional Laplacian with power less than one half*, Stoch. PDE: Anal. Comp., **12** (2024), pp. 794–855.
[359] K. Yamazaki, *Remarks on the two-dimensional magnetohydrodynamics system forced by space-time white noise*, Stochastic Process. Appl., (2026), https://doi.org/10.1016/j.spa.2026.104893.
[360] K. Yamazaki, *Non-uniqueness in law of three-dimensional magnetohydrodynamics system forced by random noise*, Potential Anal., **61** (2024), pp. 775–847.
[361] K. Yamazaki, *Approximating three-dimensional magnetohydrodynamics system forced by space-time white noise*, J. Differential Equations, **402** (2024), pp. 35–179.
[362] K. Yamazaki, *Non-uniqueness in law of the two-dimensional surface quasi-geostrophic equations forced by random noise*, Ann. Inst. Henri Poincare Probab. Stat., **65** (2025), pp. 457–509.
[363] K. Yamazaki, *Non-uniqueness in law of the surface quasi-geostrophic equations: the case of linear multiplicative noise*, J. Anal. Math. (2026), https://doi.org/10.1007/s11854-026-0439-y.
[364] K. Yamazaki, *Stochastic magnetohydrodynamics system: cross and magnetic helicity in ideal case; non-uniqueness up to Lions' exponents from prescribed initial data*, Math. Models Methods Appl. Sci., **35** (2025), pp. 3011–3082.
[365] K. Yamazaki, *Recent developments on convex integration technique applied to stochastic partial differential equations*, (2025). In: M. Breit-Goodwin, M. S. Im, K. Jabbusch, and K.-N. Lin (Eds.) *Advances in Mathematical Sciences*. Association for Women in Mathematics Series, 38. Springer, Cham. https://doi.org/10.1007/978-3-031-93669-2_7.
[366] K. Yamazaki, *Global unique solution to the perturbation of the Burgers' equation forced by derivatives of space-time white noise*, arXiv:2503.00343 [math.AP], 2025.
[367] K. Yamazaki, *Remarks on the three-dimensional Navier–Stokes equations with Lions' exponent forced by space-time white noise*, arXiv:2508.16906 [math.AP], 2025.
[368] K. Yamazaki and M. Mohan, *Well-posedness of Hall-magnetohydrodynamics system forced by Levy noise*, Stoch. PDE: Anal. Comp., **7** (2019), pp. 331–378.
[369] L. C. Young, *An inequality of the Hölder type, connected with Stieltjes integration*, Acta. Math., **67** (1936), pp. 251–282.
[370] B. Yuan and L. Bai, *Remarks on global regularity of 2D generalized MHD equations*, J. Math. Anal. Appl., **413** (2014), pp. 633–640.
[371] V. Yudovich, *Non stationary flows of an ideal incompressible fluid*, Zhurnal Vych Matematika, **3** (1963), pp. 1032–1066.
[372] V. M. Zaítsev and M. I. Shliomis, *Hydrodynamic fluctuations near the convection threshold*, Soviet Physics JETP, **32** (1971), pp. 866–870.
[373] R. Zhu, *SDE and BSDE on Hilbert spaces: applications to quasi-linear evolution equations and the asymptotic properties of the stochastic quasi-geostrophic equation*, Ph.D. thesis, Bielefeld University, 2012.
[374] R. Zhu and X. Zhu, *Three-dimensional Navier–Stokes equations driven by space-time white noise*, J. Differential Equations, **259** (2015), pp. 4443–4508.
[375] R. Zhu and X. Zhu, *Approximating 3D Navier–Stokes equations driven by space-time white noise*, Infin. Dimens. Anal. Quantum Probab. Relat. Top., **20** (2017), 1750020, pp. 1–77.
[376] R. Zhu and X. Zhu, *Strong-Feller property for Navier–Stokes equations driven by space-time white noise*, arXiv:1709.09306v2 [math.PR], 2017.
[377] X. Zhu, *SDE in infinite dimensions: the reflection problem and the stochastic quasi-geostrophic equation*, Ph.D. thesis, Bielefeld University, 2012.

About the author

Kazuo Yamazaki

Kazuo Yamazaki received a PhD in Applied Mathematics from the Department of Mathematics, Oklahoma State University, in 2014. He was a postdoctoral research fellow at the Department of Mathematics and Statistics, Washington State University, during 2014–2016. He was then a Visiting Assistant Professor at the Department of Mathematics, University of Rochester, during 2016–2019. Afterward, he was an Assistant Professor at the Department of Mathematics and Statistics, Texas Tech University, during 2019–2023. Currently, he is an Assistant Professor in the Department of Mathematics, University of Nebraska-Lincoln, U.S.A.

https://doi.org/10.1515/9783111349985-016

De Gruyter Series in Nonlinear Analysis and Applications

Volume 47
Victor L. Mironov, Sergey V. Mironov
Space-Time Algebra of Sedeons. Electrodynamics, Hydrodynamics, Field Theory, and Quantum Mechanics, 2025
ISBN 978-3-11-162517-1, e-ISBN (PDF) 978-3-11-162519-5, e-ISBN (EPUB) 978-3-11-162524-9

Volume 46
Svetlin G. Georgiev
Partial Dynamic Equations. Wave, Parabolic and Elliptic Equations on Time Scales, 2025
ISBN 978-3-11-163551-4, e-ISBN (PDF) 978-3-11-163614-6, e-ISBN (EPUB) 978-3-11-163615-3

Volume 45
Khaled Zennir, Svetlin G. Georgiev
Distributional Nonlinear Wave Equations. Well-Posedness and Stabilizability, 2025
ISBN 978-3-11-163368-8, e-ISBN (PDF) 978-3-11-163378-7, e-ISBN (EPUB) 978-3-11-163403-6

Volume 44
Yong Zhou, Jia Wei He
Nonautonomous Fractional Evolution Equations, 2024
ISBN 978-3-11-139057-4, e-ISBN (PDF) 978-3-11-139124-3, e-ISBN (EPUB) 978-3-11-139218-9

Volume 43
Ranis Ibragimov
Lie Group Analysis of Differential Equations. Invariant Solutions of Nonlinear Physical Phenomena, 2024
ISBN 978-3-11-138699-7, e-ISBN (PDF) 978-3-11-138749-9, e-ISBN (EPUB) 978-3-11-138780-2

Volume 42
Catherine Bandle, Alfred Wagner
Shape Optimization. Variations of Domains and Applications, 2023
ISBN 978-3-11-102526-1, e-ISBN (PDF) 978-3-11-102543-8, e-ISBN (EPUB) 978-3-11-102581-0

Volume 41
Jesús Garcia-Falset, Khalid Latrach
Nonlinear Functional Analysis and Applications, 2023
ISBN 978-3-11-103096-8, e-ISBN (PDF) 978-3-11-103181-1, e-ISBN (EPUB) 978-3-11-103208-5